9
Plant Cell Monographs

Series Editor: David G. Robinson

Available online at
SpringerLink.com

Plant Cell Monographs

Recently Published Titles

Cell Division Control in Plants
Volume Editors: Verma, D. P. S., Hong, Z.
Vol. 9, 2008

Endosperm
Volume Editor: Olsen, O.-A.
Vol. 8, 2007

Viral Transport in Plants
Volume Editors: Waigmann, E., Heinlein, M.
Vol. 7, 2007

Nitric Oxide in Plant Growth,
Development and Stress Physiology
Volume Editors: Lamattina, L., Polacco, J.
Vol. 6, 2007

The Expanding Cell
Volume Editors: Verbelen, J.-P., Vissenberg, K.
Vol. 5, 2007

The Plant Endoplasmic Reticulum
Volume Editor: Robinson, D. G.
Vol. 4, 2006

The Pollen Tube
A Cellular and Molecular Perspective
Volume Editor: Malhó, R.
Vol. 3, 2006

Somatic Embryogenesis
Volume Editors: Mujib, A., Šamaj, J.
Vol. 2, 2006

Plant Endocytosis
Volume Editors:
Šamaj, J., Baluška, F., Menzel, D.
Vol. 1, 2005

Cell Division Control in Plants

Volume Editors: Desh Pal S. Verma, Zonglie Hong

With 58 Figures and 5 Tables

 Springer

Volume Editors:

Dr. Desh Pal S. Verma
Ohio State University
Dept. Molecular Genetics &
Plant Biotechnology Center
Carmack Rd. 1060
Columbus, OH 43210
USA
verma.1@osu.edu

Dr. Zonglie Hong
University of Idaho
Dept. Microbiology,
Molecular Biology & Biochemistry
Moscow, ID 83844
USA
zhong@uidaho.edu

Series Editor:

Professor Dr. David G. Robinson
Ruprecht-Karls-University of Heidelberg
Heidelberger Institute for Plant Sciences (HIP)
Department Cell Biology
Im Neuenheimer Feld 230
D-69120 Heidelberg
Germany

Library of Congress Control Number: 2007938900

ISSN 1861-1370
ISBN 978-3-540-73486-4 Springer Berlin Heidelberg New York
DOI 10.1007/978-3-540-73487-1

This work is subject to copyright. All rights are reserved, whether the whole or part of the material is concerned, specifically the rights of translation, reprinting, reuse of illustrations, recitation, broadcasting, reproduction on microfilm or in any other way, and storage in data banks. Duplication of this publication or parts thereof is permitted only under the provisions of the German Copyright Law of September 9, 1965, in its current version, and permission for use must always be obtained from Springer. Violations are liable for prosecution under the German Copyright Law.

Springer is a part of Springer Science+Business Media

springer.com

© Springer-Verlag Berlin Heidelberg 2008

The use of registered names, trademarks, etc. in this publication does not imply, even in the absence of a specific statement, that such names are exempt from the relevant protective laws and regulations and therefore free for general use.

Editor: Dr. Christina Eckey, Heidelberg
Desk Editor: Anette Lindqvist, Heidelberg
Cover design: WMXDesign GmbH, Heidelberg
Typesetting and Production: LE-TeX Jelonek, Schmidt & Vöckler GbR, Leipzig

Printed on acid-free paper   149/3180 YL – 5 4 3 2 1 0

Editor

Desh Pal S. Verma is a full professor at the Ohio State University, USA. He obtained his B.Sc. degree in biology and chemistry, and M.Sc. degree in botany from Agra University, India, and his Ph.D. degree in plant physiology and biochemistry from the University of Western Ontario, Canada. He is a Fellow of the Royal Society of Canada and a Fellow of the Third World Academy of Sciences, Italy. His pioneering research work includes the identification and characterization of nudulins and phragmoplastin, and genes responsible for proline and callose biosynthesis in plants. He has served on the editorial boards for several international journals, edited eleven scholarly books, and published over 160 original research papers.

Zonglie Hong is an assistant professor at the University of Idaho, USA. He obtained his B.Sc. degree in agronomy from Fujian Agricultural University, China, and his M.Sc. and Ph.D. degrees in biochemistry from the University of Novi Sad, Serbia. As a postdoctoral fellow and senior research associate in the Department of Molecular Genetics and Plant Biotechnology Center of the Ohio State University he was engaged in molecular and cell biology studies of nodule and cell plate formation, phosphoinositide signaling, and proline and callose biosynthesis pathways. His laboratory focuses on the biochemical and genetic analyses of protein sumoylation at the cell plate, and callose synthesis during flower development.

Preface

The molecular mechanisms controlling cell cycle progression are conserved in essentially the same form in all eukaryotes studied. Among the best-characterized examples of the regulators involved here are the cyclin-dependent kinases, CDKs, which were originally identified as cell cycle-related protein kinases. These enzymes are activated at appropriate times in the cell cycle by association with a phase-specific cyclin to form a cyclin-dependent kinase complex that promotes progression through successive G1, S, and M phases.

Although plants employ the same cell-cycle regulatory proteins as other eukaryotes, they have also evolved unique molecular mechanisms that allow integration of environmental, physiological, and developmental signals into networks to control proper cell division and expansion. Our understanding in these fields has come, and could only have come, from experimental observations on plants, although studies on other model eukaryotes have provided very useful knowledge in the control of cell cycle in general.

The circadian rhythmicity of cell division in unicellular algae has been known for a long time and research into the circadian clock and cell cycle control in higher plants is just beginning to emerge. Whereas the biological clocks of plants and animals share common features in their organization and functionality, most clock genes are different between these kingdoms. Furthermore, since the contribution made by each plant cell to the shape of the surrounding tissue is determined solely by the processes of cytokinesis and differential cell expansion, cell-to-cell communication plays a critical role in forcing division in a given plane, either periclinally or anticlinally. Whereas unicellular organisms constantly face the divide-or-die challenge, higher plants have mainly relied on the cell division in meristems to grow. They have evolved regulatory mechanisms that allow the integration of processes at the cellular level with those related to organogenesis. Such a coordination between cell division and differentiation is required in order to develop a functional plant. While organogenesis in animals occurs during embryogenesis, organ initiation and growth in plants is a post-embryonic and continuous process that occurs over the entire lifespan of the organism.

Cytokinesis in plants, involving the establishment of the cell plate, is morphologically distinct from cytokinesis in yeast and animals. Significant progress in plant cytokinesis research has been achieved primarily due to the discovery of protein markers and to the implementation of new technologies such as confocal microscopy and tomographic electron microscopy. Among the several hundred proteins that are believed to be involved in building the cell plate, only a few dozen have been identified. Some of these proteins, such as phragmoplastin and knolle, are specific to this transient subcellular organelle, while many others may also have functions in other membranes and organelle building processes.

The scope of this book is focused on the molecular aspects of cell division control in higher plants. Publications in this field have increased exponentially in the past decade and this trend will certainly continue in the post genomics era. Further research on cell division control in plants involving signaling systems as well as details of the machinery of cell wall architecture holds tremendous promise for advancing our understanding of this fundamental process essential for building the plant body.

We are greatly indebted to all our contributors for their individual chapters and for their patience during the unanticipated delay of the edition of this volume. We also thank the National Science Foundation for their support of our research.

May 2007

Desh Pal S. Verma
Zonglie Hong

Contents

Commitment for Cell Division and Cell Cycle Control

Circadian Regulation of Cell Division
F. ç.-Y. Bouget · M. Moulager · F. Corellou 3

Transcriptional Control of the Plant Cell Cycle
P. Doerner 13

Division Plane Orientation in Plant Cells
A. J. Wright · L. G. Smith 33

G1/S Transition and the Rb-E2F Pathway
W.-H. Shen 59

The Endoreduplication Cell Cycle: Regulation and Function
P. A. Sabelli · B. A. Larkins 75

Dynamics of Chromosomes, Cytoskeleton and Organelles During Cell Division

Chromosome Dynamics in Meiosis
A. Ronceret · M. J. Sheehan · W. P. Pawlowski 103

Cytoskeletal and Vacuolar Dynamics During Plant Cell Division: Approaches Using Structure-Visualized Cells
T. Sano · N. Kutsuna · T. Higaki · Y. Oda · A. Yoneda
F. Kumagai-Sano · S. Hasezawa 125

Mitotic Spindle Assembly and Function
J. C. Ambrose · R. Cyr 141

Cytoskeletal Motor Proteins in Plant Cell Division
Y.-R. J. Lee · B. Liu 169

Organelle Dynamics During Cell Division
A. Nebenführ 195

Open Mitosis: Nuclear Envelope Dynamics
A. Rose 207

Cell Plate Formation and Cytokinesis

MAP Kinase Signaling During M Phase Progression
M. Sasabe · Y. Machida 233

**Plant Cytokinesis – Insights Gained
from Electron Tomography Studies**
J. M. Seguí-Simarro · M. S. Otegui · J. R. Austin II · L. A. Staehelin 251

Vesicle Traffic at Cytokinesis
A. Sanderfoot 289

Molecular Analysis of the Cell Plate Forming Machinery
Z. Hong · D. P. S. Verma 303

Cell Division and Differentiation

Asymmetric Cell Divisions: Zygotes of Fucoid Algae as a Model System
S. R. Bisgrove · D. L. Kropf 323

Stomatal Patterning and Guard Cell Differentiation
K. U. Torii 343

Genetic Control of Anther Cell Division and Differentiation
C. L. H. Hord · H. Ma 361

Coordination of Cell Division and Differentiation
C. Gutierrez 377

Subject Index 395

Part A
Commitment for Cell Division and Cell Cycle Control

Circadian Regulation of Cell Division

François-Yves Bouget (✉) · Mickael Moulager · Florence Corellou

UMR 7628 CNRS, Université Paris VI, Laboratoire Arago,
Modèles en Biologie Cellulaire et Evolutive, BP44, 66651 Banyuls sur Mer, France
fy.bouget@obs-banyuls.fr

Abstract Plants are photosynthetic organisms, which use light, as main source of energy, for growth and development. Because sun-light is also mutagenic, many organisms, from most kingdoms, have evolved an internal time-tracking system, so-called the circadian, which restricts cell division to specific times of the day in when DNA is less exposed to UV damage ("escape from mutagenic light" theory). While the circadian regulation of cell division has been extensively characterized in animals and unicellular green algae, little is known about the photoperiodic regulation of cell division in land plants. Recent findings about the possible links between cell division, the circadian clock and the DNA damage checkpoint are discussed.

1
Introduction

Biological clocks are pacemakers, which play an essential role in regulating the physiology and behavior of living organisms. Such clocks are the cell division cycle (CDC) and the circadian clock, which coexist in many cells from organisms including plants, animals, fungi and cyanobacteria (Bell-Pedersen et al. 2005; McClung 2006; Naef 2005). Each day, living organisms are exposed to changes in light (photoperiod) and temperature due to the rotation of the earth. Furthermore the relative day to night lengths vary along the year. The circadian clock is an autonomous system, which gives the time and can be entrained by light or temperature cycles. This clock allows the organism to adapt to the predictable daily environmental changes by anticipating them. The first evidence of a circadian clock was first demonstrated in 1729 by de Mairan, who showed that a heliotrop plant still exhibits robust rhythms of leaf movement when placed in constant darkness. Since then, the circadian clock was shown to regulate a wide array of plant physiology processes including photosynthesis, metabolism, and at the molecular level, a significant proportion of gene expression (Gardner et al. 2006). The circadian clock is likely to confer an adaptive advantage to multicellular organisms since it is found in all kingdoms including animals, plants and fungi. Among unicellular free-living organisms circadian rhythms have been described almost exclusively in photosynthetic ones.

Cells often divide with a period of 24 hours, leading sometimes to a confusion between the circadian clock and the CDC. The rhythmicity of cell division has been known for a long time in unicellular algae such as the flagellate Euglena (Edmunds et al. 1982), the green alga *Chlamydomonas* (Bruce 1972; Goto and Johnson 1995) or the marine dinoflagellate Gonyaulax (Sweeney 1958). These cells usually divide once per day or with a multiple of 24 hours. However, when they are allowed to divide more than once per day, the rhythmicity of cell division is lost in *Euglena* or in *Gonyaulax*. It was concluded, according to the "circadian/infradian" rule that circadian rhythms of division exist only in cells dividing in a circadian or infradian way, that is, once per day or less. According to this rule, both the circadian clock and the CDC are tightly interconnected and cell division might even be controlled by an autonomous clock that is completely independent of a circadian clock. Experiments in the prokaryotic cyanobacteria *Synechococcus* contradicted this rule, since cells dividing more than once per day still exhibit robust circadian rhythms of cell division and gene expression (Kondo et al. 1997).

Plants are photosynthetic organisms, which heavily rely on light as a source of energy and it is sometime difficult to discriminate between the effect of light as a signal through activating specific signalization pathways and as a source of energy that is a required for a cell to grow and divide when it reaches a critical size. The molecular basis of both the cell cycle core machinery (Inze 2005) and the circadian clock are well understood in plants (McClung 2006). However, neither a direct regulation of cell division by light nor a circadian regulation of the CDC have been demonstrated in higher plants. Recent results from our group, using the firefly luciferase reporter protein fused to *Arabidopsis* specific promoters indicate that the expression of cell cycle regulated genes such as *histone H4* and of central CDC genes such as *cyclinB* occurs at the end of the day and is under circadian control in the *Arabidopsis* shoot apical meristem (unpublished data). However, direct studies of the CDC remain difficult to perform *in planta* because only a limited number of cells divide in growing plants and cell division is restricted to meristems. Cell division studies are much easier to perform in unicellular organisms, such as unicellular algae, which can be naturally synchronized by the light/dark cycle. Circadian regulation of cell division has been known for over 30 years in green algae such as *Chlorophyta chlamydomonas*. *Chlorophyta* belong to the green lineage and are a sister group to *Streptophyta* which encompasses higher plants.

Recently, several important findings have started to unravel the molecular basis of circadian regulation of cell division in animals (Fu et al. 2002; Matsuo et al. 2003). In this review we will first focus on the circadian regulation of cell division in unicellular photosynthetic eukaryotes, especially those evolutionarily related to higher plants, that is green algae. We will also summarize the actual knowledge of molecular basis of circadian regulation of the CDC based on recent work in animals. Finally, we will speculate on what we can expect from plants.

2
Circadian Gating of Cell Division

In *Chlamydomonas*, the question whether or not the CDC is under circadian control, has been a never-ending story. When *Chlamydomonas* cells entrained by light/dark cycles are placed under constant light, cell division persists with a period of 24 hours, suggesting that cell division is under control of a circadian clock (Bruce 1972). Subsequent studies challenged the former studies suggesting that the CDC is directly regulated by light, being dependent on energy available though photosynthesis (Spudich and Sager 1980). The proposed model is that cell division is dependent on a timer and a sizer (Donnan and John 1983). Cells divide only when they have reached a critical size at a specific phase relative to the synchronizing light/dark cycle but the timer is not a circadian oscillator. Twenty five years later Goto and Johnson clearly demonstrated that cell division is under circadian control, fulfilling the three criteria of a circadian regulation: (1) entrainment by different photoperiods, (2) persistence under constant conditions, (3) temperature compensation (Goto and Johnson 1995). The current view is that the circadian clock gates cell division. Every day a circadian-regulated gate is opened only during a limited time, allowing cell division. *Chlamydomonas* divides by multiple fission, i.e. a cell is able to divide several times in a row, when it has reached a critical size. As a result, when the gate is opened in the middle of the night cells can divide several times in a row, if they were large enough. Similar gating of cell division has been described in *Euglena*, *Synechococcus*, or in animal cells suggesting that gating of cell division, which restricts cell division to specific phases of the day, is widely distributed in living organisms (Cardone and Sassone-Corsi 2003). What is the meaning of the circadian gating of cell division? A seducing hypothesis is that the circadian clock prevents cells from dividing when DNA is the most exposed to mutations. In agreement with this hypothesis, *Chlamydomonas* cells exposed to UV exhibit a circadian-regulated rhythm of survival, the most sensitive phases corresponding to the time of nuclear division (Nikaido and Johnson 2000). The importance of circadian gating of cell division in tumor suppression has been established in mouse where the central clock gene *Period2* is required to arrest the CDC progression in animals exposed to γ-radiation (Fu et al. 2002). Finally, a tight connection between the cell cycle control and the circadian clock was recently reported in the fungus *Neurospora* (Pregueiro et al. 2006). In this organism, the clock protein Period4 is a checkpoint kinase2 homolog, which is involved in both the DNA damage checkpoint and cell cycle progression, suggesting that the cell cycle can feedback to the clock though activation of a common checkpoint.

3
Evidence for Different Types of Circadian Checkpoints

The CDC core cell cycle machinery is well conserved in eukaryotes, relying mainly on Cyclin-dependent-kinases (CDKs) which are required for the main stages of CDC progression including DNA replication during the S phase and segregation of chromosomes at mitosis (Inze 2005). CDKs are positively regulated by association to cyclins and negatively regulated by small inhibitors (CKI). CDKs are also strongly regulated by phosphorylation and dephosphorylation. For example, in G2 the WEE1 kinase down-regulates mitotic CDKs by phosphorylation whereas the antagonistic CDC25 phosphatase activates CDK at the G2/M transition. CDKs are the main targets of the various checkpoints, which ensure that CDC progression arrests if something wrong happens such as incomplete DNA replication, DNA damage or spindle defect (Millband et al. 2002; Weinert 1998). Another critical checkpoint is the cell-size checkpoint also called "sizer" (Kellogg 2003). During development, cell growth and division must be tightly coordinated to maintain a specific size. This size control is well known in yeasts. In budding yeast, a critical restriction point START or (R) has been defined in G1, when cells become irreversibly engaged in cell division when they reach a critical size. In contrast fission yeast grow mainly in G2 and the control of CDC progression is exerted at the G2/M transition (Kellogg 2003). Photosynthetic organisms rely on light as the source of energy, to reach a critical size for cell division. The regulation of cell division by light was dissected in *Chlamydomonas*. Two points were defined: in early G1, a restriction point called primary arrest (A) when cells become light-dependent; in late G1, a "transition point" (T) when cell division becomes independent of light (Spudich and Sager 1980). Though it remains to be demonstrated clearly that gating of cell division occurs before entry into the S phase in *Chlamydomonas*, it is likely to be so, since arrests outside of the G1 phase are never observed when cells are moved from light to darkness (Fig. 1). In contrast, light-dependent restriction mechanisms were shown to exist both in G1, S and G2 phases of the cell cycle in *Euglena* cells transferred from light to darkness (Hagiwara et al. 2002) and a circadian gating of CDC progression has been observed from G2 to mitosis but also at the S/G2 and G1/S transitions in this organism (Bolige et al. 2005). The targets of the circadian checkpoints remain elusive in *Euglena* or *Chlamydomonas*. The only demonstration of a direct regulation of cell division by the circadian clock come from animals. In hepatocytes re-entering cell cycle upon hepatectomy, the circadian clock gates cell cycle progression at the G2/M transition, through a direct transcriptional regulation of the WEE1 kinase (Matsuo 2003). A negative regulation of cell cycle by the circadian clock at the G1/S transition was demonstrated in osteoblasts where the clock inhibits the G1 cyclin D1 (Fu et al. 2002). In summary, the circadian clock ap-

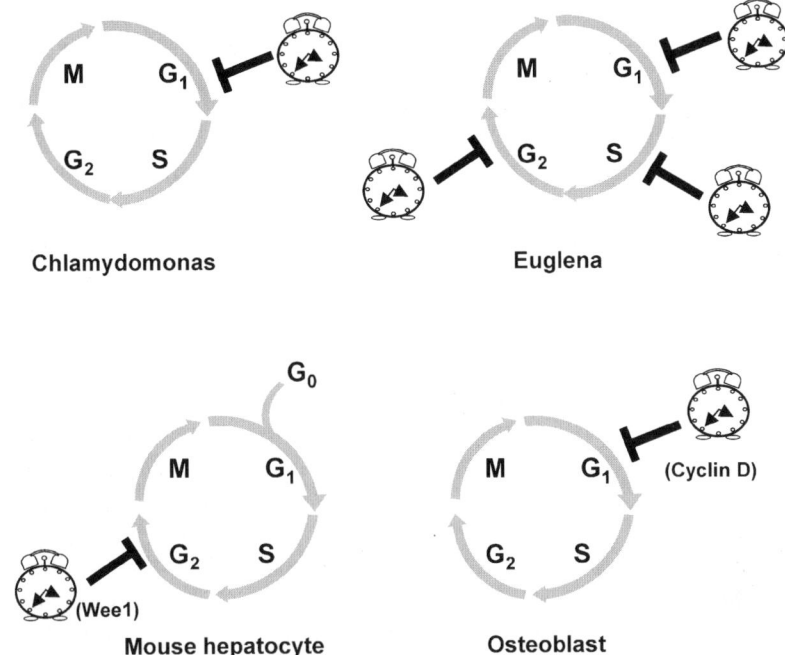

Fig. 1 Circadian gating of cell division in different cell types. In *Chlamydomonas* or osteoblasts the clock gates cell division at the G1/S transition. In contrast, in hepatocytes, a circadian regulation of cell division is observed at the G2/M transition through the regulation of the CDK inhibitor *WEE1*. In *Euglena*, the clock can arrest the CDC at all major transitions

pears to gate the CDC at various stages of cell cycle progression, including G1/S, S/G2 and G2/M (Fig. 1).

Finally, it should be mentioned that while the molecular link between the circadian clock and the CDC has been characterized in cells re-entering or exiting the CDC and upon checkpoint activation, the molecular basis of circadian gating of cell division in cycling cells, remains to be established.

4
What about Plants?

Direct evidences for a circadian regulation of cell division in higher plants are still lacking. However, based on the ubiquitous nature of circadian regulation of cell division in multicellular organisms, the essential role of the circadian clock in regulating plant physiology, it is likely that such a control will also exist in plants. On the other hand "green" organisms such as *Chlamydomonas*, which belong to the green lineage, are potentially valuable systems to study circadian regulation of cell cycle in plants, however, the molecular mechanisms of the circadian clock are not known in *Chlamydomonas*.

Except for cryptochromes and casein kinase II, the molecular actors of the circadian clock are not conserved between kingdoms suggesting that clocks have emerged independently during evolution of plant, animals and fungi (Gardner et al. 2006). Nevertheless, the principles of the clock organization is highly conserved between organisms. Coupled oscillators, which rely on interconnected feedback loops produce robust oscillations. These oscillators are entrained by environmental cycles such as light cycles but also gate input to the clock by regulating the expression of photoreceptors (Fig. 2). Most of our knowledge about the plant circadian oscillator comes from *Arabidopsis*. This pacemaker consists of two interlocked oscillators. The first transcriptional feedback loop consists of the pseudo-response regulator TOC1 (Time of CAB expression1) and the two MYB transcription factors of the REVEILLE family, CCA1 (Circadian Clock Associated 1)/LHY (Late Elongated Hypocotyl) (Alabadi et al. 2001). The transcription of CCA1/LHY is induced by light in the morning. CCA1/LHY bind to an evening element (EE) found in the promoter of TOC1 and other genes expressed in the evening (Harmer et al. 2000), re-

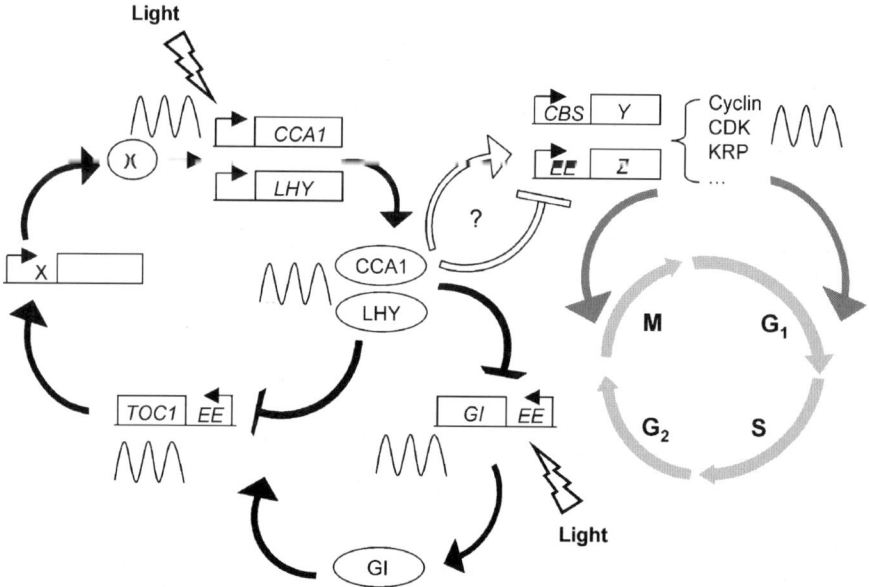

Fig. 2 A speculative model of CDC control by the circadian clock in plants. The clock consists of two interconnected loops. *CCA1/LHY* and *GI* are transcribed in response to light in the morning. CCA1/LHY and GI have opposite inhibiting and activating roles in the regulation of *TOC1* expression. CCA1/LHY may regulate the transcription of *CDC* genes such as cyclins, CDK or CDK inhibitors (*KRP*) through a binding to specific circadian elements in their promoters

pressing the expression of TOC1. On the other hand, TOC1 is responsible for the activation of CCA1/LHY, though probably indirectly, since there is a delay between the peak of TOC1 expression and the transcription of CCA1/LHY. This picture of the plant circadian clock is far from being complete, since this simple loop does not account for several experimental observations (Locke et al. 2005). In particular TOC1 overexpression led to an unpredicted decrease in CCA1 (Hayama and Coupland 2003). Furthermore, *cca1/lhy* double mutants still exhibit rhythmicity (Mizoguchi et al. 2002). A second loop was recently identified by a mathematical modelization approach. A predicted candidate Gigantea (GI) is also induced by light leading to the activation of *TOC1* and is transcriptionally repressed by CCA1 (Locke et al. 2005). Though the molecular roles of the respective actors remain to be precisely assessed, this two-loop model accounts much better for experimental data, than the one-loop model does.

The question then is, how does the circadian clock regulate cell division in plants? In hepatocytes, the central CLOCK-BMAL1 complex binds the E box response element (CACGTG) resulting in a direct activation of *WEE1* transcription (Matsuo 2003). In *Arabidopsis*, *cis*-acting elements mediating gene expression have been identified in the promoters of circadian-regulated genes (ref). The EE (AAAATATCT), mediates the transcriptional repression of *TOC1* by CCA1 at the heart of the clock. A single mutation in the EE to the AAAAAATCT CBS sequence (CCA1 binding site element) switches the phase of catalase expression from evening to morning (Michael and McClung 2002). We have performed a genome-wide analysis of the main core cell cycle gene promoters in *Arabidopsis* (Table 1). Both EE and CBS sequences were iden-

Table 1 *In silico* identification of potential circadian elements (EE and CBS) in *Arabidopsis* core cell cycle genes

Gene	Accession number	Motif	Location	Function
CDKA,1	At3g48750	AAATATCT (EE)	− 1130	G1/S G2/M
CDKB1,2	At2g38620	AAATATCT (EE)	− 200	G2/M
		AAAAATCT (CBS)	− 300	
CDKD,2	At1g66750	AAAATATCT (EE)	− 570	G2/M
cyclin D3,1	At4g34160	AATATCT (EE)	− 280	G1/S
cyclinA3,1	At5g43080	AAAATATCT (EE)	− 1000	S/M?
		AAAAATCT (CBS)	− 1800	S
DEL2	At5g14960	AAAATATCT (EE)	− 20	G1/S
Retinoblastoma	At1g22803	AAATATCT (EE)	− 500	
		AAAATCT (CBS)	− 500	
KRP2	At3g50630	AATATCT (EE)	− 1600	G2/M
KRP3	At5g48820	AAATATCT (EE)	− 2200	G1/S G2/M?

tified in cell cycle genes involved at various stages of cell cycle progression (Table 1), suggesting that the transcription of several cell cycle genes may be regulated by circadian clock components such as CCA1/LHY. A highly speculative view of how such a regulation of the clock may regulate cell division is represented in Fig. 2.

5
Conclusion: A Future for Unicellular Algae in Circadian Studies of Cell Cycle?

The circadian regulation of cell division has been known for a long time in unicellular algae of the green lineage. Recently, the genomes of two green algae, the Prasinophyta *Ostreococcus tauri* and the Chlorophyta *Chlamydomonas*, have been sequenced (Derelle et al. 2006). Genetic tools are available in both species and luciferase reporter allows automated monitoring of in vivo expression of circadian clock and CDC genes in *Ostreococcus* (our unpublished results). Both organisms display a minimum set of cell cycle genes, most cyclins, CDK and regulating kinases and phosphatases being present as a single copy (Bisova et al. 2005; Robbens et al. 2005). These genes are regulated by light/dark cycles in *Chlamydomonas*. Interestingly, both *Chlamydomonas* and *Ostreococcus* contain plant-specific CDC proteins such as a B-type CDK, which plays a major role in the control of mitosis in *Ostreococcus* (Corellou et al. 2005). *In silico* analysis, has revealed the presence of CCA1/LHY and TOC1-like protein homologues in *Chlamydomonas* and *Ostreococcus* (Breton and Kay 2006, our unpublished data).

These unique features of green algae may promote them as models of choice to study circadian regulation of cell division.

References

Alabadi D, Oyama T, Yanovsky MJ, Harmon FG, Mas P, Kay SA (2001) Reciprocal regulation between TOC1 and LHY/CCA1 within the Arabidopsis circadian clock. Science 293:880–883

Bell-Pedersen D, Cassone VM, Earnest DJ, Golden SS, Hardin PE, Thomas TL, Zoran MJ (2005) Circadian rhythms from multiple oscillators: lessons from diverse organisms. Nat Rev Genet 6:544–556

Bisova K, Krylov DM, Umen JG (2005) Genome-wide annotation and expression profiling of cell cycle regulatory genes in *Chlamydomonas reinhardtii*. Plant Physiol 137:475–491

Bolige A, Hagiwara SY, Zhang Y, Goto K (2005) Circadian G2 arrest as related to circadian gating of cell population growth in Euglena. Plant Cell Physiol 46:931–936

Breton G, Kay SA (2006) Circadian rhythms lit up in Chlamydomonas. Genome Biol 7:215

Bruce VG (1972) Mutants of the biological clock in *Chlamydomonas reinhardi*. Genetics 70:537–548

Cardone L, Sassone-Corsi P (2003) Timing the cell cycle. Nat Cell Biol 5:859–861
Corellou F, Camasses A, Ligat L, Peaucellier G, Bouget FY (2005) Atypical regulation of a green lineage-specific B-type cyclin-dependent kinase. Plant Physiol 138:1627–1636
Derelle E, Ferraz C, Rombauts S, Rouze P, Worden AZ, Robbens S, Partensky F, Degroeve S, Echeynie S, Cooke R, Saeys Y, Wuyts J, Jabbari K, Bowler C, Panaud O, Piegu B, Ball SG, Ral JP, Bouget FY, Piganeau G, De Baets B, Picard A, Delseny M, Demaille J, Van de Peer Y, Moreau H (2006) Genome analysis of the smallest free-living eukaryote Ostreococcus tauri unveils many unique features. Proc Natl Acad Sci USA 103:11647–11652
Donnan L, John PC (1983) Cell cycle control by timer and sizer in *Chlamydomonas*. Nature 304:630–633
Edmunds LN, Tay DE, Laval-Martin DL (1982) Cell division cycles and circadian clocks: Phase-response curves for light perturbations in synchronous cultures of *Euglena*. Plant Physiol 70:297–302
Fu L, Pelicano H, Liu J, Huang P, Lee C (2002) The circadian gene Period2 plays an important role in tumour suppression and DNA damage response in vivo. Cell 111:41–50
Gardner MJ, Hubbard KE, Hotta CT, Dodd AN, Webb AA (2006) How plants tell the time. Biochem J 397:15–24
Goto K, Johnson CH (1995) Is the cell division cycle gated by a circadian clock? The case of Chlamydomonas reinhardtii. J Cell Biol 129:1061–1069
Hagiwara SY, Bolige A, Zhang Y, Takahashi M, Yamagishi A, Goto K (2002) Circadian gating of photoinduction of commitment to cell-cycle transitions in relation to photoperiodic control of cell reproduction in Euglena. Photochem Photobiol 76:105–115
Harmer SL, Hogenesch JB, Straume M, Chang HS, Han B, Zhu T, Wang X, Kreps JA, Kay SA (2000) Orchestrated transcription of key pathways in *Arabidopsis* by the circadian clock. Science 290:2110–2113
Hayama R, Coupland G (2003) Shedding light on the circadian clock and the photoperiodic control of flowering. Curr Opin Plant Biol 6:13–19
Inze D (2005) Green light for the cell cycle. EMBO J 24:657–662
Kellogg DR (2003) Wee1-dependent mechanisms required for coordination of cell growth and cell division. J Cell Sci 116:4883–4890
Kondo T, Mori T, Lebedeva NV, Aoki S, Ishiura M, Golden SS (1997) Circadian rhythms in rapidly dividing cyanobacteria. Science 275:224–227
Locke JC, Southern MM, Kozma-Bognar L, Hibberd V, Brown PE, Turner MS, Millar AJ (2005) Extension of a genetic network model by iterative experimentation and mathematical analysis. Mol Syst Biol 1:2005.0013
Matsuo T (2003) Control mechanism of the circadian clock for timing of cell division in vivo. Science 302:234–235
Matsuo T, Yamaguchi S, Mitsui S, Emi A, Shimoda F, Okamura H (2003) Control mechanism of the circadian clock for timing of cell division in vivo. Science 302:255–259
McClung CR (2006) Plant circadian rhythms. Plant Cell 18:792–803
Michael TP, McClung CR (2002) Phase-specific circadian clock regulatory elements in *Arabidopsis*. Plant Physiol 130:627–638
Millband DN, Campbell L, Hardwick KG (2002) The awesome power of multiple model systems: interpreting the complex nature of spindle checkpoint signalling. Trends Cell Biol 12:205–209
Mizoguchi T, Wheatley K, Hanzawa Y, Wright L, Mizoguchi M, Song HR, Carre IA, Coupland G (2002) *LHY* and *CCA1* are partially redundant genes required to maintain circadian rhythms in *Arabidopsis*. Dev Cell 2:629–641

Moulager M, Monnier A, Jesson B, Bouvet R, Mosser J, Schwartz C, Garnier L, Corellou F, Bouget F-Y (2007) Light-dependent Regulation of Cell Division in Ostreococcus: Evidence for a Major Transcriptional Input. Plant Physiol, July issue

Naef F (2005) Circadian clocks go in vitro: purely post-translational oscillators in cyanobacteria. Mol Syst Biol 1:2005.0019

Nikaido SS, Johnson CH (2000) Daily and circadian variation in survival from ultraviolet radiation in Chlamydomonas reinhardtii. Photochem Photobiol 71:758–765

Pregueiro AM, Liu Q, Baker CL Dunlap JC, Loros JJ (2006) The *Neurospora* checkpoint kinase 2: a regulatory link between the circadian and cell cycles. Science 313:644–649

Robbens S, Khadaroo B, Camasses A, Derelle E, Ferraz C, Inze D, Van de Peer Y, Moreau H (2005) Genome-wide analysis of core cell cycle genes in the unicellular green alga *Ostreococcus tauri*. Mol Biol Evol 22:589–597

Spudich JL, Sager R (1980) Regulation of the *Chlamydomonas* cell cycle by light and dark. J Cell Biol 85:136–145

Sweeney BM, Hastings JW (1958) Rhythmic cell division in populations of *Goyaunolax polyhedrea*. J Protozool 5:217–224

Weinert T (1998) DNA damage checkpoints update: getting molecular. Curr Opin Genet Dev 8:185–193

Transcriptional Control of the Plant Cell Cycle

Peter Doerner

Institute of Molecular Plant Science, School of Biological Sciences,
Daniel Rutherford Building, King's Buildings, University of Edinburgh,
Edinburgh EH9 3JH, UK
Peter.Doerner@ed.ac.uk

Abstract Progression through the eukaryotic cell cycle is driven and controlled by interlocked oscillating mechanisms, including reversible protein phosphorylation, protein degradation and temporally regulated gene expression. Advances in cell synchronization techniques have enabled the genome-wide characterization of temporally regulated gene expression patterns in Arabidopsis, but the cognate transcription factors remain largely unknown. In other eukaryotic model organisms, the network components involved in orchestrating cell cycle phase-specific gene expression are being identified and their regulatory logic delineated. This chapter reviews progress in our understanding of cell cycle regulated gene expression in plants.

1
Introduction and Background

Progression through the eukaryotic cell cycle is controlled by several mechanisms whose activity varies periodically with cell cycle phase. These oscillating regulatory mechanisms are involved in controlling all aspects of cell cycle progression, including: commitment, directionality, arrest in the event of DNA damage, incomplete replication or unattached spindles and responsivity to the cellular environment. The best studied of these mechanisms are reversible protein phosphorylation, protein degradation and temporally regulated gene expression, but in plants there are further oscillatory processes, such as cytokinin hormone metabolism, that likely contribute to cell cycle control.

In this chapter, I will focus on genome-wide studies of cell cycle regulated gene expression and the mechanisms that underpin periodic, cell cycle phase-dependent changes in gene expression. I will emphasize the transcriptional regulation of *regulatory* genes that participate in the control of cell cycle progression, e.g. cyclins, over the phase-specific expression of *structural* genes, for example histones, which are required for the appropriate execution of cell cycle-dependent events at specific times during the cycle. This distinction does not preclude the parsimonious use of some factors in both types of processes, where the same factor may interact with different partners on distinct promoters.

Much progress has recently been made in the identification of transcripts that vary periodically according to cell cycle phase in plants and to identify the entire set of genes regulated in this fashion. However, much less is known about the transcription factors that mediate cell cycle regulated gene expression in plants than about the key components involved in phosphorylation – the cyclin-dependent kinase (CDK) complexes. This has two reasons. First, the transcription factors responsible for cell cycle phase-specific expression of cell cycle regulatory genes are poorly conserved. Second, this is also due to a now dated view of cell cycle regulation as a hierarchy of processes orchestrated by a succession of cyclin–CDK complexes that "drive" the cell cycle forward by enslaving dependent processes such as protein degradation and temporally regulated gene expression.

Recent studies of the budding and fission yeast cell cycle (Bähler 2005) and of factors that contribute to robustness of the cell cycle (Li et al. 2004; Jensen et al. 2006; Braunewell and Bornholdt 2007), lead me to propose a different view of the relation between the distinct regulatory mechanisms that in concert mediate cell cycle control. Such studies reveal that these mechanisms are coupled by regulatory interactions, for example: cell cycle phase-specific expression of individual cyclins conditions high-level CDK activity at specific times, which in turn controls the activity of the proteasome that regulates the stability of cell cycle transcription factors. In other words, each of the major oscillating processes (protein phosphorylation, degradation and temporally regulated gene expression) comprises a complete and self-regulating cycle that is interlocked and intersects with other oscillating mechanisms for enhanced stability. Such coupled oscillators are more robust, and inherently less prone to interference by inevitable stochastic noise caused by fluctuations in abundance of individual proteins and their activity (Raser and O'Shea 2004; Braunewell and Bornholdt 2007); interconnected oscillators are also more easily regulated, in a coordinate manner, as they continually entrain each other.

In contrast to the highly conserved cyclin–CDK modules, and the core components of the cellular proteolysis machinery, the key transcriptional regulators that mediate cell cycle phase-specific gene expression programs are poorly conserved between different eukaryotic model systems. This has made it substantially more difficult to identify the components of the transcriptional networks involved in plant cell cycle control as reverse genetic approaches are not likely to succeed. A recent study of the evolution of cell cycle regulatory mechanisms has revealed that the regulated subunits (i.e. those that change their abundance or activity) of different regulatory complexes vary between organisms (Jensen et al. 2006). However, the same study found that these regulated subunits were likely to be regulated both transcriptionally and post-transcriptionally. This striking convergence of different regulatory mechanisms involved in cell cycle control onto shared targets is expected to increase the overall robustness of cell cycle regulation (Braunewell and Bornholdt 2007), and provides further support to the notion that the different mechanisms involved in cell cycle control contribute equally to its overall regulation.

1.1
Oscillating Processes Involved in Cell Cycle Control

Three canonical mechanisms have been identified in cell cycle control of all eukaryotic systems so far analyzed in sufficient detail: co-regulated and sequential expression of specific gene sets; cyclin–CDK mediated protein phosphorylation and proteasome-mediated proteolysis. Each of these processes constitute a genuine, closed cycle such that periodic changes in activity states promote, directly or indirectly, conditions that favor the following state, with the "last" state promoting the first state of the series again, to form a closed regulatory cycle. As each of these regulatory paradigms has recently been the subject of excellent reviews, I shall restrict the discussion to an outline of the general principles of the more universal mechanisms.

1.1.1
Transcriptional Networks in the Eukaryotic Cell Cycle

To review the logic of transcriptional control of cell cycle regulated gene expression, our current understanding of such control in yeast is primarily discussed here, as our current knowledge in plants is too incomplete. The transcriptional networks in yeast that mediate periodic, cell cycle associated gene expression are made up of at least nine serially regulated transcription factors, most of which interact in various heteromeric complexes (Simon et al. 2001; Bähler 2005).

Periodic gene expression at the G1/S transition is mediated by two transcription factor complexes, MBF and SBF, which accumulate and bind to their target promoters from early G1 onwards. They remain inactive until the Cln3p-Cdk1p cyclin–CDK complex that mediates entry into a new cycle at START activates them in a two-step process. SBF and MBF bind to the promoter of Swi4p, which is part of the SBF complex, thereby leading to a decisive, positive reinforcement of the decision to traverse START. During S-phase, these complexes mediate transcription of S-phase-specific genes including B-type cyclins required for progression into G2/M. In late S- and G2-phase, SBF and MBF activity is down-regulated by CDK-mediated phosphorylation that depends on cyclins transcribed earlier in the cycle by SBF/MBF complexes.

The periodic expression of G2/M genes is governed by a ternary transcription factor complex comprising Mcm1p; one of two partially redundant factors, Fkh1p or Fkh2p, and Ndd1p. Ndd1p is the regulated component of the complex, as the former two subunits stay bound to their targets throughout the cell cycle. Ndd1p is regulated transcriptionally by SBF/MBF but also post-transcriptionally by CDK-dependent phosphorylation and proteolysis (Loy et al. 1999; Reynolds et al. 2003), and Fkh2p is also phosphorylated in a CDK-dependent manner. The expression targets of this ternary complex include the mitotic cyclins required for chromosome segregation and cytokinesis.

At M/G1, another wave of gene expression occurs to mediate exit from mitosis, and reset the cellular state such that the cell becomes responsive again to the accumulation of mass by committing to a new round of cell division. Expression of this suite of genes depends on the activity of two types of transcription factors: Mcm1p; and the related and partially redundant Swi5p and Ace2p proteins (Dohrmann et al. 1992). The latter two are themselves transcriptionally regulated by the G2/M ternary complex factor, and post-translationally regulated by CDK-dependent phosphorylation. In contrast, two negatively acting homeodomain co-regulators that bind target promoters in the vicinity of the Mcm1p binding sites act to restrict the autonomous, M/G1-specific function of Mcm1p to this cell cycle transition. Their periodic expression is mediated by SBF and later in the cycle presumably, by the G2/M ternary complex factor.

The general principles of the budding yeast network of transcriptional regulators which periodically activate waves of cell cycle phase-specific gene expression are therefore: First, *serial regulation* such that transcription factors functioning early in the cell cycle regulate subsequently acting transcription factors, to form a fully connected regulatory cycle. Second, *indirect coupling* between the factors responsible for sequential waves of transcription, which is usually mediated by components of other oscillating mechanisms, e.g. CDK-dependent phosphorylation or proteasome-mediated proteolysis, that themselves are subject to prior transcriptional control. Indirect feedback and feed-forward regulation are mechanisms to delay the initiation of the next periodic wave of transcription. This is because activation of heteromeric protein kinases such as cyclin–CDKs depends on the attainment of concentration thresholds of the individual components and is furthermore subjected to ultrasensitive control by series of activating kinases and counter-acting protein phosphatases. A corollary is that the majority of the transcription factors comprising the cell cycle transcription network themselves need not be transcriptionally regulated in a cell cycle phase-specific manner. Third, the individual transcription factors that comprise the core regulatory network *themselves control series of dependent transcription factors* that mediate the expression of co-regulated sets of genes necessary for the successful completion of individual cell cycle phases (Horak et al. 2002).

Due to the paucity of data, it is currently not possible to gauge the relative importance of the cell cycle transcriptional network vis à vis CDK-dependent protein phosphorylation and ubiquitin-dependent proteolysis in controlling cell cycle progression in different eukaryotic model systems. It is conceivable that the relative contribution of transcriptional control compared to the other canonical processes varies in different organisms, but due to the inherent logic of control of CDK-dependent and ubiquitin-dependent mechanisms, which requires phase specific transcription and accumulation of at least some key regulators, it is very unlikely that it plays only a minor role.

1.1.2
Protein Phosphorylation in the Eukaryotic Cell Cycle

Heteromeric protein kinase complexes comprising various catalytic subunits (CDK proteins) in obligate association with one of many distinct activating cyclin subunits that accumulate and decay periodically in the course of the cell cycle mediate major cell cycle transitions: at START, the commitment to a cell cycle conditional on the determination of adequate cell mass; at G1/S, the commitment to a round of nuclear DNA replication; at G2/M, the dissolution of the nuclear compartment; and at the metaphase to anaphase transition, the irreversible separation of sister chromatids for segregation into two new cells. The cyclin–CDK cycle has recently been reviewed, therefore, I shall focus on the basic principles of CDK-dependent cell cycle regulation here and not invest them with details (see Inzé and Veylder (2006) for an excellent recent review of cell cycle control in plants, and Morgan (2007) for an outstanding and comprehensive review of the principles of cell cycle regulation).

The key function of cyclin–CDK complexes in the cell cycle is to orchestrate and decisively mediate progression through crucial control points. This role is reflected in their regulation, which occurs in a series of primed steps and culminates in a runaway, exponential increase of cyclin–CDK activity at the onset of the relevant phase transition.

To exemplify this, I will discuss the processes in budding yeast during G1 phase that lead up to the key cell cycle transition at G1/S in detail; our understanding of the plant cell cycle, and of its biochemistry in particular, is still too fragmentary to provide a comprehensive picture at the present time. For those biological systems in which cells grow and accrue mass in G1, which includes budding yeast and animal cells, but not fission yeast (where growth occurs during G2), it is useful to consider G1 phase as two distinct parts: In the first part, which commences as soon as cells have completed their exit from mitosis, control of cell cycle progression is ceded to the mechanisms that assess cell mass. The mechanism by which cell growth is measured in plants, and whether there is a single such mechanism in all plants, is not clear yet. In budding yeast, these mechanisms are thought to result in the gradual accumulation of cyclin3 protein (Cln3p), proportional to cell growth; culminating in the accumulation of sufficient Cln3p-CDK activity to propel it into the next part of G1. This occurs by Cln3p-CDK phosphorylation of an inhibitor of the SBF transcription factor complex, which leads to the inactivation of the inhibitor. As explained in the previous section, SBF activity is central to the transcriptional network that launches S-phase. In this second part of G1, different cyclin–CDK complexes take the reins to control the crucial decision to enter a round of DNA replication. The now-active SBF stimulates the transcription of *two* distinct types of cyclins (late G1-type and S-phase type) sequentially required for full-blown commitment and entry into S-phase. This proceeds by a build-up of the *potential* for S-phase CDK activity by progressive accumulation of cyclin subunits required for S-phase CDK activity and their association with appropriate CDK proteins. Importantly, these complexes initially lack protein kinase activity, as the nascent complexes are held inactive by association with CDK inhibitory proteins (CKI). The late G1 cyclin–CDK complexes that gradually accumulate in an SBF-dependent manner mediate the activation of these latent S-phase cyclin–CDK complexes. When sufficient amounts of late G1-type cyclins have accumulated, the cyclin–CDK complexes they are part of surpass an activation threshold and phosphorylate CKI. CKI phosphorylation leads to its recognition by an SCF-dependent ubiquitin ligase (see next section), resulting in its destruction by the proteasome. The now liberated S-phase cyclin–CDK complexes also recognize CKI as a substrate, resulting in a runaway activation of S-phase cyclin–CDK complexes that constitutes the initiation of S-phase.

Thus, the principal contribution of cyclin–CDK complexes to cell cycle control is to mediate key cell cycle transitions *decisively*. CDK activity is controlled in a two-stage activation process. This is analogous to the cocking, and subsequent firing of a gun by the activation of a trigger mechanism. The trigger is provided by the accumulation of sufficient late G1-type (but not Cln3p) cyclin–CDK activity to phosphorylate the CKI proteins holding S-phase cyclin–CDK complexes in check. This multistage activation process allows the final activation of cyclin–CDK complexes at a phase transition to be subject to many controls. A similar process of stepwise activation occurs prior to mitosis, although here, it is generally the phosphorylation state of the CDK subunit that regulates its activity; additional protein kinases, which are controlled by cyclin–CDK complexes, function as well.

1.1.3
Proteolysis in the Eukaryotic Cell Cycle

Regulated stability of regulatory proteins is a crucial aspect of eukaryotic cell cycle control to impart irreversibility on cell cycle transitions. Proteolysis occurs by ubiquitin-dependent mechanisms that are distinguished by their mode of regulation. In G1 and S-phase, the SCF-mediated mechanism dominates; while in M-phase and for exit from M-phase, the APC-dependent mechanism prevails. The SCF complex has ubiquitin ligase activity and recognizes its substrates when these are phosphorylated; in contrast, subunits of the APC complex are activated by phosphorylation, which leads to substrate recognition and subsequent degradation.

SCF-dependent protein degradation is coupled to CDK-dependent cell cycle control at G1/S through shared substrates. At this transition, SCF-mediated proteolysis complements and reinforces the CDK-dependent switch mechanism: the degradation of CDK inhibitors prevents the cell cycle from dithering or going backwards. Similarly, at the onset of M-phase with the breakdown of the nuclear membrane and at the metaphase–anaphase transition, the APC complex is activated, directly or indirectly, by CDK-dependent phosphorylation. APC fulfills an essential role in destroying the proteins that hold the sister chromatids joined until the start of anaphase, and in destroying mitotic cyclins at M/G1, without which cells could not exit from mitosis. In these functions, the exquisite balance between ubiquitination and de-ubiquitination activities is critical for the precise timing of cell cycle progress (Stegmeier et al. 2007), while CDK-dependent phosphorylation of a regulatory subunit of APC is required for its inactivation late in G1 to allow subsequent accumulation of mitotic cyclins. Taken together, proteolysis acts to complement, reinforce and re-set cyclin–CDK regulation. Its activity is closely coupled to CDK regulation.

1.1.4
Other Oscillating Parameters in the Plant Cell Cycle

Proliferation of plant cells requires growth regulators such as auxins and cytokinins. However, the requirement for cytokinin supplements is not absolute, as some cell cultures, including the tobacco BY2 cell line, are "habituated", which means that endogenous cytokinin production is adequate to sustain proliferation.

Interestingly, recent work indicates that cytokinin synthesis, degradation and perception are exquisitely regulated in the course of cell cycle progression: Peak transcript accumulation of one of the three cytokinin receptors in Arabidopsis, *CRE1* (also known as *AHK4*, or *WOL1*) occurs in G1 (Menges et al. 2003). In cells treated with low concentrations of lovastatin, which specifically suppresses synthesis of cytokinin isoprenoid side-chains, only zeatin supplements can rescue cells from the ensuing cell cycle arrest (Laureys et al. 1998). Cytokinin levels, specifically of *trans*-zeatin, increase in a very intriguing pattern in the course of cell cycle progression: a broad peak is observed in S-phase and a very transient, and high peak is observed at the G2/M transition (Dobrev et al. 2002). Other reports find additional peaks of cytokinin accumulation associated with all cell cycle phase transitions in the tobacco BY2 cell cycle (Hartig and Beck 2005). Furthermore, these striking dynamics of cytokinin accumulation are complemented by fluctuations of cytokinin-degrading cytokinin oxidase activity (Dobrev et al. 2002; Hartig and Beck 2005), suggesting that the observed sharp transients are the net result of waves of cytokinin synthesis and degradation. These observations raise the intriguing possibility that fluctuations in cytokinin abundance and perception are an additional oscillating mechanism contributing to overall robustness in plant cell cycle control. Although cytokinins have been shown to promote de-phosphorylation of the phospho-tyrosine inhibitory for CDK activity in animal systems, and in *Nicotiana plumbaginifolia* (Zhang et al. 1996, 2005), there is no evidence that this plays an important role in "normal" cell cycle progression in Arabidopsis (De Schutter et al. 2007). However, cytokinins play an important role in plant source–sink relationships, transducing environmental cues (Miyawaki et al. 2004), balancing the activities of shoot and root meristems (Werner et al. 2001, 2003), and determining meristem size and growth rate (Dello Ioio et al. 2007); all functions related to growth control of the organism. It will be interesting to more precisely identify the mechanisms of cytokinin action during cell cycle progression.

1.2
Conclusions

Taken together, the currently available data indicates that the canonical cell cycle mechanisms are not hierarchically organized but rather are equally important, specialized and complementary functions required for robust control of the eukaryotic cell cycle. While each of these mechanisms is a closed cycle inasmuch as their activities oscillate in a phase-specific manner, and early steps in each cycle promote the activation of later steps, they are not independent of each other. The canonical mechanisms are coupled through regulatory interactions that allow them to continually entrain each other. However, the coupling mechanisms can vary between model species. In a scenario of equitable but specialized functions, the transcriptional regulatory network has the role of propelling cell cycle progression; while the CDK-dependent phosphorylation cascade provides the switches to move from one phase to the next and the complementary ubiquitin-dependent proteolysis machinery insures the directionality of the cycle.

2
Cell Cycle Regulated Gene Expression in Plants

Comprehensive studies on cell cycle regulated gene expression have been conducted on two systems: the readily synchronized tobacco Bright Yellow 2 (BY2) cell culture system (Nagata et al. 1992), and a recently developed synchronized Arabidopsis cell culture system (Menges and Murray 2002). The degree of synchrony achievable with the Arabidopsis system is not quite as high and persistent as with the BY2 system. However, it has the important advantage of representing a completely sequenced genome and many array tools for gene expression analysis are available in Arabidopsis. These tools greatly facilitate the unequivocal identification of genes with dynamic expression patterns in the course of the cell cycle.

When interpreting experimental data from cell cycle time courses, it is worthwhile considering potential pitfalls inherent with existing synchronization methods: Inhibitor-based synchronization methods are susceptible to non-specific or unexpected side effects. For example, the commonly used fungal toxin Aphidicolin, which inhibits replicative DNA polymerases, will elicit some of the cellular responses controlled by the mechanisms that monitor the completion of DNA replication. A good example is the observation of S-phase induction of cyclin B1;1 (but not of other cyclins of the cyclin B1 clade) expression in Aphidicolin-treated Arabidopsis cells, but not in cells synchronized by withdrawal and provision of sucrose (Menges et al. 2005). Cyclin B1;1 expression has been shown to be specifically induced by DNA damage and incomplete replication check points (Culligan et al. 2006). Other inhibitors, including the mitotic spindle inhibitors (propyzamide, colchicine, amiprophos methyl, oryzalin), are likely to elicit similar artifacts, but have not yet been analyzed in detail in this respect. Other methods for synchronization, involving the withdrawal and re-provision of nutrients (Sano et al. 1999; Menges et al. 2003) or hormones (Planchais et al. 1997) might cause fewer such artifacts. Therefore, data genuinely reflecting cell cycle regulated gene expression is likely to emerge only from the consensus of experiments performed with different synchronization protocols.

2.1
Analysis of Cell Cycle Regulated Gene Expression in Tobacco BY2 Cells

The tobacco BY2 cell system was the first to offer high levels of synchrony (Nagata et al. 1992). Synchronization in these cells can be achieved by several treatments: Aphidicolin treatments, which result in an early S-phase block; treatments with microtubule antagonists, leading to an arrest at the metaphase to anaphase transition; and withdrawal and re-provision of nutrients (Sano et al. 1999; Menges et al. 2003) or hormones (Planchais et al. 1997).

In a landmark experiment, Breyne and colleagues arrested BY2 cells with Aphidicolin and followed changes in cell cycle gene expression at hourly intervals (Breyne et al. 2002). The analysis was performed by cDNA AFLP and the abundance of ~1300 transcripts was found to significantly vary over time in the synchronized cells. To unequivocally identify transcripts observed to fluctuate during the cell cycle, the transcript tags were re-amplified and sequenced. Although only about half of the identified genes could be placed into functional categories, numerous genes whose function is associated with the cell cycle (e.g. functional categories such as "cell cycle regulation", "cell wall", "cytoskeleton", "protein synthesis", "replication"), and gene expression (e.g. "transcription factors") were identified. Their abundance distribution across the canonical cell cycle phases was analyzed and striking phase-specific patterns emerged (Breyne et al. 2002). Significant oscillation of transcript abundance in the course of the cell cycle was observed for core cell cycle regulators such as A- and B-type cyclins, B- and C-type CDK genes, as well of many genes associated with DNA replication (e.g. ribonucleotide reductase, histones). This experiment showed for the first time that transcriptional regulation is a widespread, important and integrated aspect of overall cell cycle control in plants, just as has been observed in other eukaryotic model systems. However, this experiment suffered from the unavailability of a tobacco genome sequence, ambiguities in gene identification due to the amphidiploid nature of tobacco and, despite the tedium of individual handling of each transcript, many remaining unidentified transcripts.

2.2
Analysis of Cell Cycle Regulated Gene Expression in Arabidopsis Cells

Taking advantage of the full suite of genomics tools available for Arabidopsis, Menges and colleagues (2002) developed and validated two cell culture systems derived from a long established Arabidopsis cell culture line. These were used in experiments in which cells were synchronized by Aphidicolin treatments, by withdrawal and provision of sucrose to either freshly subcultured cells or cells in the mid-exponential phase of the culture cycle (Menges et al. 2003, 2005).

Using rigorous statistics to validate their results, Menges and colleagues confirm that the transcripts of the vast majority of canonical cell cycle regulators are expressed in the cell culture system and fluctuate in the course of the cell division cycle. The transcripts whose levels periodically changed with sinusoidal kinetics were defined as "cell cycle regulated". By contrast, those genes whose abundance changed greater than threefold, and that were expressed over background levels for at least three time points over the entire time course but not with sinusoidal kinetics were classified as "cell cycle associated". In the absence of detailed understanding of the cognate transcription factors involved in cell cycle regulated gene expression in Arabidopsis, the significance of this distinction is currently not readily apparent.

With the ATH1 gene chip, ∼ 1100 genes were found to be cell cycle regulated or cell cycle associated, and were regulated in cells synchronized by either Aphidicolin or by sucrose withdrawal and supply, corresponding to ∼ 7% of all genes detected in the cell culture sample. A substantially higher proportion of all genes expressed in cell cultures showed a similar magnitude of change in cells synchronized by sucrose starvation (31%). In the latter time course, a large number of genes are presumably responding to changes in metabolism and not to cell cycle phase per se.

Overall, two major waves of periodic transcript accumulation could be identified: the first was observed at late G1 and into S-phase, the second at late G2 and into M-phase (Menges et al. 2002, 2003, 2005). In contrast to budding yeast, where a clear peak is also associated with exit from the cell cycle at the M/G1 transition, this is not unequivocally clear in the synchronized Arabidopsis cells. However, in yeast, high levels of cell cycle synchrony persist for more than two cycles, whereas in plant cell cultures, including Arabidopsis, tight synchrony does not persist for an entire cycle.

There were $\sim$ 60 transcription factors amongst the cohort of $\sim$ 1100 genes that were cell cycle regulated or cell cycle associated, but only 14 of these were clearly cell cycle regulated. With two likely exceptions (3RMYB genes, see below), it is presently unclear whether any of these transcription factors mediates the phase-specific expression of core cell cycle regulators. In this scenario, they would be central to the transcription control cycle that is interlocked with the CDK and proteolysis cycles. Alternatively, these factors could be required to orchestrate the expression of effector genes associated with phase-specific physiological processes such as DNA or cell wall synthesis. Interestingly, eight of the cell cycle regulated factors showed only very modest amplitudes ($\sim$ two- to threefold) of transcript oscillation. This property would be expected for at least some transcription factors that target core cell cycle regulators, as their activity, subcellular localization or stability is likely to be regulated post-transcriptionally by CDK-dependent phosphorylation as a mechanism to interlock these regulatory cycles.

3
Cell Cycle Associated Transcription Factors in Plants

Three regulatory circuits involved in cell cycle phase-specific gene expression are currently known in plants. These are at present only defined by the *cis*-regulatory elements identified in promoters of some periodically expressed genes and not yet conclusively by the cognate proteins that bind these elements. These are: (i) expression of histone genes in S-phase depends on octamer (OCT) and hexamer (HEX) motifs (Chaubet et al. 1996; Taoka et al. 1999; Minami et al. 2000); (ii) the promoters of many genes that are up-regulated at G1/S or in S-phase contain motifs known to bind E2F/DP factors (Ramirez-Parra et al. 2003; Vandepoele et al. 2005); and (iii) expression of some genes at G2/M depends on MSA elements, to which a class of transcription factors with three repeats of the canonical Myb-type DNA binding motif bind (3RMybs) (Ito et al. 1998, 2001). However, with the exception of the latter class of Myb-type transcription factors (Ito et al. 2001; Haga et al. 2007), far too little is known about the transcription factors that comprise the regulatory network responsible for cell cycle phase-specific gene expression in plants.

3.1
Histone Expression in S-Phase

Histone gene expression is strongly stimulated in S-phase. Detailed promoter analysis of histone genes in maize, Arabidopsis and wheat has revealed the presence of clusters of *cis*-elements termed the OCTAMER (OCT) element and the HEXAMER (HEX) element, respectively. Both of these were shown to be necessary for S-phase periodic expression (Ohtsubo et al. 1997), and as these binding sites are usually in close proximity, it is possible they interact in a complex. The factors binding to HEX elements have been identified and belong to the basic–leucine zipper (bZIP) class of transcription factors (Tabata et al. 1989, 1991). However, bZIP factors define a large gene family, and the specific members of this family involved in periodic, S-phase specific gene expression of histones and other genes in vivo have yet to be identified. It is possible that some HEX-binding factors have a more auxiliary role in supporting, but not mediating, periodic gene expression: the distribution of HEX and closely-related motifs is much more widespread, while OCT elements are present in fewer genes, albeit not restricted to histones.

Gel shift analysis suggests that nuclear proteins from wheat binding to the OCT element and HEX binding protein 1a (HBP-1a) might mediate periodic S-phase RNA accumulation: the formation of S-phase specific complexes was observed for one of three OCT-specific complexes and for one of three HBP-1a-type complexes detected, respectively (Minami et al. 2000). However, the genes encoding these factors have not yet been identified, and therefore, it is still not clear whether these factors have a broader role in mediating S-phase periodic transcription or whether it is restricted to one aspect of the S-phase gene expression program.

3.2
E2F/DP/Rb

E2F transcription factors were first discovered in animal cell extracts as positive regulators of S-phase gene expression that interact with the tumor suppressor gene product of the Retinoblastoma locus (pRb) (Mudryj et al. 1990; Chellappan et al. 1991). The interaction with the Rb protein inhibits E2F transcriptional activity. To bind to their target promoters, most E2F gene products interact with a co-activator DP that shares a homologous DNA binding domain, although recently E2F-type transcription factors were identified in plants and animals that have two DNA binding domains and can bind their targets autonomously. E2F proteins mediate the expression of S-phase genes, and ectopic expression of activating E2F proteins is sufficient to force S-phase specific gene expression; but E2F proteins are also involved in regulating the transition to cell differentiation and keeping stem cells in an undifferentiated state. E2F activity is regulated through the relief of pRb-imposed inhibition, which is mediated by CDK-dependent hyperphosphorylation. The specific functions of E2F proteins in controlling cell cycle progression are still contentious (Attwooll et al. 2004), but they are undoubtedly effectors of cell cycle gene expression networks. For a more detailed discussion of plant E2F genes and their function, see the chapter by Prof. Gutierrez in this volume.

3.3
3RMyb

Functional analysis of mitotic cyclin expression revealed that their periodic accumulation at G2/M was regulated transcriptionally (Shaul et al. 1996; Ito et al. 1997). Subsequent characterization of a *Catharanthus roseus* mitotic cyclin gene promoter led to the identification of a nine base pair sequence that was sufficient to confer G2/M phase-specific stimulation of transcript accumulation (Ito et al. 1998), which was designated the M-specific activator (MSA) sequence. This sequence (YCYAACGGYYA, where Y is either T or C) has similarities to the sequences that animal and plant Myb-type transcription factors bind to. Using this sequence as a probe in a one-hybrid screen, three Myb-type transcription factors were identified (Ito et al. 2001). In contrast to the vast majority of plant transcription factors, which have only two of the canonical Myb-repeats in their DNA binding domains, these factors have three such repeats, just like the Myb genes in the animal kingdom (Ito et al. 2001).

The expression of one of these, NtmybB, was constitutive during the cell cycle, whereas the transcripts of the other two, NtmybA1 and NtmybA2, accumulated periodically, commencing in late S-phase and peaking in G2 (Ito et al. 2001). Using transfected tobacco BY2 cells to assay the ability of these genes to affect expression of genes that accumulate maximally at G2/M, it was shown that NtmybB represses, while NtmybA1 and NtmybA2 activate, expression of such genes (Ito et al. 2001). Moreover, co-transfection assays showed that these two classes of 3RMyb factors could compete for occupancy of *cis*-elements. Thus, the peak of MSA activity at the G2/M transition was likely mediated by the transcriptional regulation of NtmybA1 and NtmybA2 factors. This suggests that the pronounced but transient activation of gene expression conferred by the MSA element is the result of accumulation of NtmybA-type activity against a constant background of NtmybB-mediated repression, leading to a sharp maximum once this threshold is overcome (Ito et al. 2001). Thus, the control of MSA-mediated gene expression at G2/M by 3RMyb factors carries all the hallmarks of cell cycle-regulated gene expression. Interestingly, evidence for regulation of NtmybA2 by cyclin–CDK complexes that effect a positive feedback loop was recently found: the non-conserved activation domain of NtmybA2 was found to contain a repressor domain, whose activity was relieved by CDK-dependent phosphorylation (Araki et al. 2004). Only cyclins that are expressed at G2/M were able to activate the cognate CDK complexes, but not cyclins that accumulate more broadly during the cell cycle, suggesting that this feedback coupling is critical for full-blown commitment to M-phase.

Recently, the putative Arabidopsis orthologs of NtmybA1 and NtmybA2, MYB3R1 and MYB3R4, respectively, were characterized (Haga et al. 2007): loss-of-function mutants in these genes are impaired in cytokinesis, often resulting in multinucleate cells. Molecular analysis revealed that KNOLLE

(KN), a cytokinesis-specific syntaxin necessary for cell plate formation that has MSA elements in its promoter, is under transcriptional control of 3RMyb genes; this indicates that 3RMyb genes are crucial factors in a periodic gene expression network involved in not only entry into G2/M but also key mechanisms of M-phase (Haga et al. 2007).

3.4
TCP

Expression of the proliferating cell nuclear antigen (PCNA) gene is cell cycle regulated with a broad but shallow peak in S-phase (Menges et al. 2002), consistent with its function as a processivity factor for DNA polymerase. Its transcriptional regulation was first studied in rice which led to the discovery of the proliferating cell nuclear antigen factors 1 and 2 (PCF1/2), the founding members of a novel class of plant-specific group of transcription factors (Kosugi and Ohashi 1997). Further in silico and experimental evidence showed that the binding site for PCF1/2-type transcription factors (GCCCR, where R is either G or A) was present in many additional genes associated with growth and cell division (Tremousaygue et al. 2003; Li et al. 2005). Detailed functional analysis of the Arabidopsis cyclin B1;1 promoter showed that the GCCCR element was required only for the normal magnitude of expression, but not for periodic, cell cycle phase-specific expression, which was in this case conferred by the MSA element (Li et al. 2005). Moreover, the majority of the genes with the cognate *cis* element in their promoter are not periodically expressed, and those that are, are expressed in different phases. Together, this indicates that while TCP genes contribute to the expression of cell cycle regulated genes, they do not mediate periodic expression maxima.

3.5
Others

Many other transcription factor genes have been implicated in the expression of cell cycle regulators. For example, *AINTEGUMENTA* (*ANT*), a member of the plant-specific AP2 class of transcription factors, has been shown to stimulate growth of leaf organs in Arabidopsis and, when over-expressed, leads to increased expression of cyclin D3 (Mizukami and Fischer 2000). Likewise, the related *PLETHORA* (*PLT*) transcription factors are implicated in the maintenance of cell cycle gene expression (Aida et al. 2004). Similar functions are ascribed to *JAGGED* (*JAG*) zinc-finger, KNAT-type homeodomain transcription factors, and the novel *ULTRAPETALA* (*ULT*) genes (Lincoln et al. 1994; Ohno et al. 2004; Carles et al. 2005). However, a direct role in cell cycle gene expression has not been demonstrated for any of these and presently, there is no strong evidence that they are involved in periodically active cell cycle transcription networks.

4
Approaches to Identify Transcription Networks

Advances in genome-wide technologies have recently and will continue to provide novel opportunities for dissecting the mechanisms and identifying the components of the cell cycle gene expression networks. The following sections will highlight current progress and identify key areas where advances are required to address the challenging tasks ahead.

4.1
Expression Analysis

To better understand the regulation of plant genes expressed periodically during the cell cycle, advances in several areas are needed: improved synchronization methods, the analysis of gene expression in additional species, as well as novel approaches to analyze cell cycle phase-specific expression.

Current synchronization methods are not optimal, because synchronization methods based on inhibitor-mediated cell cycle arrest and release do not constrain cell cycle progression from in-built cell cycle control points, and are therefore likely to activate additional unrelated responses. This has been clearly shown to be the case for Aphidicolin-based cell cycle synchronization, which elicits replication stress responses. It is possible that using nutrient withdrawal/supply will cause fewer such artifacts. Ultimately, it will be necessary to compare results obtained with different synchronization methods.

It would also be extremely useful to develop and characterize cell cycle gene expression from additional plant systems with sequenced genomes. Initial developments have been made for rice and other cereal synchronization systems, but these clearly still require improvements (Kaeppler et al. 1997; Minami et al. 2000; Lendvai et al. 2002). The tobacco genome is currently being sequenced, facilitating progress with this excellent system for cell cycle studies (Nagata et al. 1992). Recently, the genome sequence of the unicellular green alga Ostreococcus was determined, revealing that this very simple eukaryote at the base of the green plant lineage has a simple complement of core cell cycle regulators (Derelle et al. 2006). This system carries a lot of potential, as its cell cycle is entrained by the circadian clock (Farinas et al. 2006), which will provide unique opportunities.

Comprehensive analysis of periodically expressed genes has only been performed in suspension cultured cells, which lack many of the developmental pathways that control and interface with cell cycle regulation in intact plants. New technologies, such as laser microdissection microscopy, are increasingly being used to sample the transcript profile, proteome and even metabolites from small numbers of well-defined cells within a tissue context (Nelson et al. 2006). When used with appropriate markers, this approach could be used to characterize cell cycle phase-specific expression in free-running cells, which is ultimately the most valid approach.

4.2
Identification of Motifs

Much effort and hope has been invested into the in silico discovery of shared regulatory motifs in co-regulated genes with the aim of using these to identify the cognate transcriptional regulators. The early methods just took the promoter sequences of gene arrays into account, but more recently developed programs provide the option to include expression data (i.e. to restrict the analysis to co-regulated genes) and to exclude motifs present in promoters of genes expressed at other cell cycle phases (see, for example, WORDSPY: Wang et al. 2005; Wang and Zhang 2006). The diverse approaches have recently been comparatively evaluated (not including more recent programs: Tompa et al. 2005); the bottom-line is that when using real data sets, these computational approaches to dissecting gene regulatory networks are still far from being predictive and convincing starting points for experimental validation.

4.3
Novel Experiments: Accessing the Entire Regulatory Space

Since the various regulatory networks controlling cell cycle progression intersect and are coupled, it is not surprising that transcriptional control of key transcriptional regulators involved in periodic gene expression is only one avenue for their regulation. Numerous examples from plant, yeast and animal systems show that post-translational control mechanisms play important roles in controlling their activity (Bähler 2005; Inze and Veylder 2006; Morgan 2007). Therefore, systems-wide approaches to concomitantly determine not only changes to transcript abundance, but also to phosphorylation, regulated subcellular localization and regulated protein stability (to name just a few) will be crucial for experimental strategies to identify the components and functional logic of gene regulatory networks that governs periodic gene expression during the plant cell cycle.

5
Conclusions and Perspectives

Recent conceptual advances and experimental data suggest that the conventional, hierarchical view of cell cycle control has reached the limit of its utility. This view, in which the oscillations of CDK/cyclin activity are considered to be at the nexus of cell cycle control, with periodic bursts of proteolytic activity and transcription factor activity acting as enforcers of cell cycle phase transitions, or as executors of phase-specific waves of gene expression, does not sufficiently account for the regulatory interactions observed in cells or the interdependence of these processes. In contrast, our current knowledge is much more consistent with the view that these processes comprise closed regulatory cycles that are interlocked and coupled with each other through numerous regulatory interactions. With advances at the technological and conceptual level, there are now exciting opportunities to characterize these regulatory circuits in unprecedented detail. This will lead to a better mechanistic understanding of plant cell cycle control, and enable new approaches to modify plant growth and increase agricultural productivity in the service of human welfare.

Acknowledgements The support of the Biotechnology and Biological Sciences Research Council (BBSRC), the Royal Society, the Darwin Trust, and the Samuel Roberts Noble Foundation for work in the Doerner Lab is gratefully acknowledged. I gratefully acknowledge all stimulating discussions with present and former members of the laboratory. I apologize to colleagues whose work was not mentioned due to space restrictions.

References

Aida M, Beis D, Heidstra R, Willemsen V, Blilou I, Galinha C, Nussaume L, Noh YS, Amasino R, Scheres B (2004) The *PLETHORA* genes mediate patterning of the Arabidopsis root stem cell niche. Cell 119:109–120

Araki S, Ito M, Soyano T, Nishihama R, Machida Y (2004) Mitotic cyclins stimulate the activity of c-Myb-like factors for transactivation of G2/M phase-specific genes in tobacco. J Biol Chem 279:32979–32988

Attwooll C, Lazzerini Denchi E, Helin K (2004) The E2F family: specific functions and overlapping interests. EMBO J 23:4709–4716

Bähler J (2005) Cell-cycle control of gene expression in budding and fission yeast. Annu Rev Genet 39:69–94

Braunewell S, Bornholdt S (2007) Superstability of the yeast cell-cycle dynamics: Ensuring causality in the presence of biochemical stochasticity. J Theor Biol 245:638–643

Breyne P, Dreesen R, Vandepoele K, De Veylder L, Van Breusegem F, Callewaert L, Rombauts S, Raes J, Cannoot B, Engler G, Inze D, Zabeau M (2002) Transcriptome analysis during cell division in plants. Proc Natl Acad Sci USA 99:14825–14830

Carles CC, Choffnes-Inada D, Reville K, Lertpiriyapong K, Fletcher JC (2005) ULTRAPETALA1 encodes a SAND domain putative transcriptional regulator that controls shoot and floral meristem activity in Arabidopsis. Development 132:897–911

Chaubet N, Flenet M, Clement B, Brignon P, Gigot C (1996) Identification of cis-elements regulating the expression of an Arabidopsis histone H4 gene. Plant J 10:425–435

Chellappan SP, Hiebert S, Mudryj M, Horowitz JM, Nevins JR (1991) The E2F transcription factor is a cellular target for the RB protein. Cell 65:1053–1061

Culligan KM, Robertson CE, Foreman J, Doerner P, Britt AB (2006) ATR and ATM play both distinct and additive roles in response to ionizing radiation. Plant J 48:947–961

De Schutter K, Joubes J, Cools T, Verkest A, Corellou F, Babiychuk E, Van Der Schueren E, Beeckman T, Kushnir S, Inze D, De Veylder L (2007) Arabidopsis WEE1 kinase controls cell cycle arrest in response to activation of the DNA integrity checkpoint. Plant Cell 19:211–225

Dello Ioio R, Linhares FS, Scacchi E, Casamitjana-Martinez E, Heidstra R, Costantino P, Sabatini S (2007) Cytokinins determine Arabidopsis root-meristem size by controlling cell differentiation. Curr Biol 17:678–682

Derelle E, Ferraz C, Rombauts S, Rouze P, Worden AZ, Robbens S, Partensky F, Degroeve S, Echeynie S, Cooke R, Saeys Y, Wuyts J, Jabbari K, Bowler C, Panaud O, Piegu B, Ball SG, Ral JP, Bouget FY, Piganeau G, De Baets B, Picard A, Delseny M, Demaille J, Van de Peer Y, Moreau H (2006) Genome analysis of the smallest free-living eukaryote Ostreococcus tauri unveils many unique features. Proc Natl Acad Sci USA 103:11647–11652

Dobrev P, Motyka V, Gaudinova A, Malbeck J, Travnickova A, Kaminek M, Vankova R (2002) Transient accumulation of cis- and trans-zeatin type cytokinins and its relation to cytokinin oxidase activity during cell cycle of synchronized tobacco BY-2 cells. Plant Physiol Biochem 40:333–337

Dohrmann PR, Butler G, Tamai K, Dorland S, Greene JR, Thiele DJ, Stillman DJ (1992) Parallel pathways of gene regulation: homologous regulators SWI5 and ACE2 differentially control transcription of HO and chitinase. Genes Dev 6:93–104

Farinas B, Mary C, de OMCL, Bhaud Y, Peaucellier G, Moreau H (2006) Natural synchronisation for the study of cell division in the green unicellular alga Ostreococcus tauri. Plant Mol Biol 60:277–292

Haga N, Kato K, Murase M, Araki S, Kubo M, Demura T, Suzuki K, Muller I, Voss U, Jurgens G, Ito M (2007) R1R2R3-Myb proteins positively regulate cytokinesis through activation of KNOLLE transcription in Arabidopsis thaliana. Development 134:1101–1110

Hartig K, Beck E (2005) Endogenous cytokinin oscillations control cell cycle progression of tobacco BY-2 cells. Plant Biol (Stuttg) 7:33–40

Horak CE, Luscombe NM, Qian J, Bertone P, Piccirrillo S, Gerstein M, Snyder M (2002) Complex transcriptional circuitry at the G1/S transition in Saccharomyces cerevisiae. Genes Dev 16:3017–3033

Inze D, Veylder LD (2006) Cell cycle regulation in plant development. Annu Rev Genet 40:77–105

Ito M, Marie-Claire C, Sakabe M, Ohno T, Hata S, Kouchi H, Hashimoto J, Fukuda H, Komamine A and Watanabe A (1997) Cell-cycle-regulated transcription of A- and B-type plant cyclin genes in synchronous cultures. Plant J 11:983–992

Ito M, Iwase M, Kodama H, Lavisse P, Komamine A, Nishihama R, Machida Y, Watanabe A (1998) A novel cis-acting element in promoters of plant B-type cyclin genes activates M phase-specific transcription. Plant Cell 10:331–341

Ito M, Araki S, Matsunaga S, Itoh T, Nishihama R, Machida Y, Doonan JH, Watanabe A (2001) G2/M-phase-specific transcription during the plant cell cycle is mediated by c-Myb-like transcription factors. Plant Cell 13:1891–1905

Jensen LJ, Jensen TS, de Lichtenberg U, Brunak S, Bork P (2006) Co-evolution of transcriptional and post-translational cell-cycle regulation. Nature 443:594–597

Kaeppler HF, Kaeppler SM, Lee JH, Arumuganathan K (1997) Synchronization of cell division in root tips of seven major cereal species for high yields of metaphase chromosomes for flow-cytometric analysis and sorting. Plant Mol Biol Rep 15:141–147

Kosugi S, Ohashi Y (1997) PCF1 and PCF2 specifically bind to cis elements in the rice proliferating cell nuclear antigen gene. Plant Cell 9:1607–1619

Laureys F, Dewitte W, Witters E, Van Montagu M, Inze D, Van Onckelen H (1998) Zeatin is indispensable for the G2-M transition in tobacco BY-2 cells. FEBS Lett 426:29–32

Lendvai A, Nikovics K, Bako' L, Dudits D, Gyorgyey J (2002) Synchronization of Oryza sativa L. cv Taipei-309 cell suspension culture. Acta Biol Szeged 46:39–41

Li C, Potuschak T, Colon-Carmona A, Gutierrez RA, Doerner P (2005) Arabidopsis TCP20 links regulation of growth and cell division control pathways. Proc Natl Acad Sci USA 102:12978–12983

Li F, Long T, Lu Y, Ouyang Q, Tang C (2004) The yeast cell-cycle network is robustly designed. Proc Natl Acad Sci USA 101:4781–4786

Lincoln C, Long J, Yamaguchi J, Serikawa K, Hake S (1994) A knotted1-like homeobox gene in Arabidopsis is expressed in the vegetative meristem and dramatically alters leaf morphology when overexpressed in transgenic plants. Plant Cell 6:1859–1876

Loy CJ, Lydall D, Surana U (1999) NDD1, a high-dosage suppressor of cdc28-1N, is essential for expression of a subset of late-S-phase-specific genes in Saccharomyces cerevisiae. Mol Cell Biol 19:3312–3327

Menges M, Hennig L, Gruissem W, Murray JA (2002) Cell cycle-regulated gene expression in *Arabidopsis*. J Biol Chem 277:41987–42002

Menges M, Murray JA (2002) Synchronous Arabidopsis suspension cultures for analysis of cell-cycle gene activity. Plant J 30:203–212

Menges M, Hennig L, Gruissem W, Murray JA (2003) Genome-wide gene expression in an *Arabidopsis* cell suspension. Plant Mol Biol 53:423–442

Menges M, de Jager SM, Gruissem W, Murray JA (2005) Global analysis of the core cell cycle regulators of Arabidopsis identifies novel genes, reveals multiple and highly specific profiles of expression and provides a coherent model for plant cell cycle control. Plant J 41:546–566

Minami M, Meshi T, Iwabuchi M (2000) S phase-specific DNA-binding proteins interacting with the Hex and Oct motifs in type I element of the wheat histone H3 promoter. Gene 241:333–339

Miyawaki K, Matsumoto-Kitano M, Kakimoto T (2004) Expression of cytokinin biosynthetic isopentenyltransferase genes in Arabidopsis: tissue specificity and regulation by auxin, cytokinin, and nitrate. Plant J 37:128–138

Mizukami Y, Fischer RL (2000) Plant organ size control: AINTEGUMENTA regulates growth and cell numbers during organogenesis. Proc Natl Acad Sci USA 97:942–947

Morgan D (2007) The cell cycle. Oxford University Press, Corby

Mudryj M, Hiebert SW, Nevins JR (1990) A role for the adenovirus inducible E2F transcription factor in a proliferation dependent signal transduction pathway. EMBO J 9:2179–2184

Nagata T, Nemoto Y, Hasezawa S (1992) Tobacco BY-2 cell line as the "HeLa" cell in the cell biology of higher plants. Int Rev Cytol 132:1–29

Nelson T, Tausta SL, Gandotra N, Liu T (2006) Laser microdissection of plant tissue: what you see is what you get. Annu Rev Plant Biol 57:181–201

Ohno CK, Reddy GV, Heisler MG, Meyerowitz EM (2004) The Arabidopsis JAGGED gene encodes a zinc finger protein that promotes leaf tissue development. Development 131:1111–1122

Ohtsubo N, Nakayama T, Kaya H, Terada R, Shimamoto K, Meshi T, Iwabuchi M (1997) Cooperation of two distinct cis-acting elements is necessary for the S phase-specific activation of the wheat histone H3 promoter. Plant J 11:1219–1225

Planchais S, Glab N, Trehin C, Perennes C, Bureau JM, Meijer L, Bergounioux C (1997) Roscovitine, a novel cyclin-dependent kinase inhibitor, characterizes restriction point and G2/M transition in tobacco BY-2 cell suspension. Plant J 12:191–202

Ramirez-Parra E, Frundt C, Gutierrez C (2003) A genome-wide identification of E2F-regulated genes in Arabidopsis. Plant J 33:801–811

Raser JM, O'Shea EK (2004) Control of stochasticity in eukaryotic gene expression. Science 304:1811–1814

Reynolds D, Shi BJ, McLean C, Katsis F, Kemp B, Dalton S (2003) Recruitment of Thr 319-phosphorylated Ndd1p to the FHA domain of Fkh2p requires Clb kinase activity: a mechanism for CLB cluster gene activation. Genes Dev 17:1789–1802

Sano T, Kuraya Y, Amino S, Nagata T (1999) Phosphate as a limiting factor for the cell division of tobacco BY-2 cells. Plant Cell Physiol 40:1–8

Shaul O, Mironov V, Burssens S, Van Montagu M, Inze D (1996) Two Arabidopsis cyclin promoters mediate distinctive transcriptional oscillation in synchronized tobacco BY-2 cells. Proc Natl Acad Sci USA 93:4868–4872

Simon I, Barnett J, Hannett N, Harbison CT, Rinaldi NJ, Volkert TL, Wyrick JJ, Zeitlinger J, Gifford DK, Jaakkola TS, Young RA (2001) Serial regulation of transcriptional regulators in the yeast cell cycle. Cell 106:697–708

Stegmeier F, Rape M, Draviam VM, Nalepa G, Sowa ME, Ang XL, McDonald ER, Li MZ, Hannon GJ, Sorger PK, Kirschner MW, Harper JW, Elledge SJ (2007) Anaphase initiation is regulated by antagonistic ubiquitination and deubiquitination activities. Nature 446:876–881

Tabata T, Takase H, Takayama S, Mikami K, Nakatsuka A, Kawata T, Nakayama T, Iwabuchi M (1989) A protein that binds to a cis-acting element of wheat histone genes has a leucine zipper motif. Science 245:965–967

Tabata T, Nakayama T, Mikami K, Iwabuchi M (1991) HBP-1a and HBP-1b: leucine zipper-type transcription factors of wheat. EMBO J 10:1459–1467

Taoka K, Kaya H, Nakayama T, Araki T, Meshi T, Iwabuchi M (1999) Identification of three kinds of mutually related composite elements conferring S phase-specific transcriptional activation. Plant J 18:611–623

Tompa M, Li N, Bailey TL, Church GM, De Moor B, Eskin E, Favorov AV, Frith MC, Fu Y, Kent WJ, Makeev VJ, Mironov AA, Noble WS, Pavesi G, Pesole G, Regnier M, Simonis N, Sinha S, Thijs G, van Helden J, Vandenbogaert M, Weng Z, Workman C, Ye C, Zhu Z (2005) Assessing computational tools for the discovery of transcription factor binding sites. Nat Biotechnol 23:137–144

Tremousaygue D, Garnier L, Bardet C, Dabos P, Herve C, Lescure B (2003) Internal telomeric repeats and "TCP domain" protein-binding sites co-operate to regulate gene expression in Arabidopsis thaliana cycling cells. Plant J 33:957–966

Vandepoele K, Vlieghe K, Florquin K, Hennig L, Beemster GTS, Gruissem W, Van De Peer Y, Inze D, De Veylder L (2005) Genome-wide identification of potential plant E2F target genes. Plant Physiol 139:316–328

Wang G, Yu T, Zhang W (2005) WordSpy: identifying transcription factor binding motifs by building a dictionary and learning a grammar. Nucleic Acids Res 33:W412–416

Wang G, Zhang W (2006) A steganalysis-based approach to comprehensive identification and characterization of functional regulatory elements. Genome Biol 7:R49

Werner T, Motyka V, Strnad M, Schmulling T (2001) Regulation of plant growth by cytokinin. Proc Natl Acad Sci USA 98:10487–10492

Werner T, Motyka V, Laucou V, Smets R, Van Onckelen H, Schmulling T (2003) Cytokinin-deficient transgenic *Arabidopsis* plants show multiple developmental alterations indicating opposite functions of cytokinins in the regulation of shoot and root meristem activity. Plant Cell 15:2532–2550

Zhang K, Letham DS, John PC (1996) Cytokinin controls the cell cycle at mitosis by stimulating the tyrosine dephosphorylation and activation of p34cdc2-like H1 histone kinase. Planta 200:2–12

Zhang K, Diederich L, John PC (2005) The cytokinin requirement for cell division in cultured Nicotiana plumbaginifolia cells can be satisfied by yeast Cdc25 protein tyrosine phosphatase: implications for mechanisms of cytokinin response and plant development. Plant Physiol 137:308–316

Division Plane Orientation in Plant Cells

Amanda J. Wright · Laurie G. Smith (✉)

Section of Cell and Developmental Biology, University of California, San Diego, 9500 Gilman Dr., La Jolla, CA 92093, USA
lgsmith@ucsd.edu

Abstract This review discusses current knowledge on division plane determination in plant cells and how division within this plane is executed during cytokinesis. Plants cells are unusual among eukaryotes in that their planes of division are established prior to the onset of mitosis. Factors that contribute to the initial selection of the division plane include extra-cellular signals, cell geometry and polarity, and nuclear position. During the G2 phase of the cell cycle, the formation of the preprophase band (PPB), a cortical assembly of microtubules and microfilaments, signals the future location of the division plane. Factors important for PPB formation and maturation include the actin cytoskeleton, changes in microtubule dynamics, and protein dephosphorylation. Prior to its disassembly at prometaphase, the PPB functions in more than one way to determine subsequent placement of the new cell wall. First, the PPB influences the initial orientation of the spindle, which facilitates subsequent cell wall formation within the division plane. Second, the PPB is thought to direct the formation of a cortical division site that persists after PPB breakdown and interacts during cytokinesis with the expanding phragmoplast, a cytoskeletal assembly that directs the deposition of the partitioning cell wall. Both negative and positive cortical markers are implicated in maintaining the memory of the former PPB site throughout mitosis and cytokinesis.

1
Introduction

Plants are sessile organisms composed of non-motile cells locked into position by rigid cell walls. Plants grow by a combination of cell elongation and cell division with no cell migration, making a cell's initial position relative to that of its neighbors difficult to adjust. Consequently, proper orientation of new cell walls during cell division is key to ensuring robust plant form and function. In contrast to animal cells, where cytokinesis is achieved via contraction of the plasma membrane between daughter nuclei (cleavage), plant cells divide by building a new cell wall between the daughter cells. Thus, it is perhaps not surprising that the mechanisms used by plant cells to orient their division planes also appear to be different from those of animal cells, where the division plane is determined by spindle position. In somatic plant cells, the division plane is established in the cell cortex prior to mitosis, and the new cell wall is inserted at this site upon completion of cytokinesis.

Initial establishment of the division plane is marked by a cytoskeletal structure unique to plant cells called the preprophase band (PPB), which appears during G2 and disappears before metaphase. The PPB is a cortical band of parallel microtubules (MTs) and microfilaments (MFs) that encircles the cell at the future division plane. The PPB breaks down upon formation of the mitotic spindle, which segregates chromosomes to daughter nuclei. Normally the spindle forms so that its axis is perpendicular to the plane delineated by the PPB. Upon completion of chromosome segregation, a plant-specific, cytokinetic apparatus called a phragmoplast forms between the daughter nuclei. A cytoskeleton-based structure containing both MTs and MFs, the phragmoplast, acts as scaffolding for the building of a new cell wall (cell plate) during cytokinesis. Following its initiation between the daughter nuclei, the phragmoplast expands laterally, taking on the shape of a donut or torus as MTs and MFs are disassembled from its interior, where the cell plate has already been deposited, and assembled at its exterior. Cytokinesis is completed when the phragmoplast has expanded to the cell periphery where the cell plate attaches to the mother cell wall at the former PPB site.

Understanding the spatial control of cytokinesis requires complete knowledge of how the PPB, spindle, and phragmoplast are formed and the forces that influence their positioning. The last major reviews related to this subject were published prior to 2002 and contain many valuable references (Mineyuki 1999; Kumagai and Hasezawa 2001; Brown and Lemmon 2001; Smith 2001). In this review, we emphasize discussion of work published in the past five years. During this period, imaging of GFP fusion proteins in living cells has confirmed and extended previous discoveries and has permitted new and informative observations regarding the spatial control of plant cell division. Further characterizations of new and old mutants and the corresponding gene products have also provided new insights into the mechanisms by which plant cells orient their division planes.

2
Selection and Establishment of the Division Plane

The PPB was originally described as a cortical band of MTs that encircles the mother cell perimeter at the future site of cell plate insertion (Pickett-Heaps and Northcote 1966a,b; Fig. 1c). Formed during an arbitrarily defined portion of the cell cycle called preprophase that corresponds to G2 or early prophase, the PPB persists throughout prophase (Wick and Duniec 1984; Venverloo and Libbenga 1987; Mineyuki et al. 1988). During preprophase and prophase, cortical MTs are found only within the PPB (Mineyuki et al. 1991; Granger and Cyr 2000). More recently, MFs have been identified as a component of the PPB. Prior to the onset of mitosis, MFs become co-aligned with PPB MTs while the density of MFs elsewhere in the cortex is reduced (Palevitz 1987; Traas et al. 1987; McCurdy and Gunning 1990; Sano et al. 2005). In large, vacuolated cells, formation of the PPB coincides with a global reorganization of the cytoplasm to form the phragmosome, a plate-like arrangement of transvacuolar cytoplasmic strands connecting the cortex/plasma membrane to the nucleus (Sinnott and Bloch 1940; Venverloo and Libbenga 1987). A variety of other cellular components and activities are enriched in the PPB/phragmoplast zone including Golgi (Nebenführ et al. 2000; Dixit and Cyr 2002a), endocytosis (Dhonukshe et al. 2005b), and endoplasmic reticulum in gymnosperms only (Zachariadis et al. 2001, 2003). Thus, while the PPB is often thought of as a MT structure, it is really a complex assembly associated with many local changes in cellular organization.

In the wide variety of cell types where PPBs have been observed, they faithfully predict the future division plane (for review, Mineyuki 1999). Moreover, pharmacological (Hoshino et al. 2003; Vanstraelen et al. 2006) or genetic (Traas et al. 1995) disruption of PPBs causes cells to divide in aberrant orientations, supporting the conclusion that the PPB plays a key role in determining division planes. The PPB has long been thought to function during prophase to establish a cortical "division site" that somehow guides the expanding phragmoplast (Pickett-Heaps and Northcote 1966a,b; Gunning 1982), but it is still a mystery how its position is determined, how it forms, and how it marks the cell cortex so that the expanding phragmoplast can be guided to its former location during cytokinesis.

2.1
Selection of the Division Plane

While selection of the division plane is not well understood, the preprophase nucleus, cell geometry, cell polarity, and extrinsic signals all appear to play a role.

2.1.1
A Role for the Nucleus

In addition to PPB/phragmosome formation, another important, early event in division plane establishment is migration of the nucleus into the division plane, if it is not already located there. Nuclear migration in symmetrically dividing cells is dependent on intact MTs, while most studies show little or no effect of actin depolymerizing drugs (Venverloo and Libbenga 1987; Mineyuki and Furuya 1986; Katsuta et al. 1990; but also see contrasting report of Miyake et al. 1997). MTs grow out from the nuclear surface to the cortex during preprophase/prophase, initially in all directions but becoming gradually restricted to the future division plane as the nucleus is centered and the PPB forms (Fig. 1c). This arrangement of cytoplasmic MTs is observed in both large, vacuolated cells with phragmosomes, where the MTs are present within phragmosomal strands, and also in small cells with no recognizable phragmosome (Fig. 1d; Wick and Duniec 1983; Katsuta et al. 1990; Kutsuna

Fig. 1 A transverse division in a "typical" plant cell is diagrammed. **a–c, e–h,** and ▶
k represent two-dimensional projections of the outer half of the cell, while **d, i,** and **j** represent midplane views. Genes important for the transition to each stage are indicated under the *arrows*. **a** Cells in interphase/G1 have peripheral nuclei, ordered cortical MT and MF arrays, cortical MFs arranged in a meshwork, and MFs that tether the actin-coated nucleus (as indicated by the *gray circle* around the nucleus) to the cortex. **b** G2 occurs after nuclear migration to the center of the cell in preparation for mitosis. MTs are nucleated from the nuclear surface (as indicated by the *black circle* around the nucleus) and connect the nucleus to the cortex in alignment with MFs. **c** In prophase, breakdown of the ordered cortical MT and MF arrays occurs in concert with PPB formation. The PPB consists of a cortical band of MTs and MFs. MTs and MFs extend from the nucleus to the PPB and to the poles of the cell. **d** This top down view highlights the cortical nature of the PPB and the MTs and MFs that connect the nucleus to the cortex. In vacuolated cells, these cytoskeleton components are contained within the transvacuolar strands that comprise the phragmosome. **e** During prophase, the MT PPB narrows while the MF PPB remains the same width. By late prophase, MTs nucleated at the nuclear surface have begun to organize themselves into a bipolar spindle. MTs extend from the new spindle poles to the PPB. **f** Before metaphase, the PPB disappears leaving behind an actin-depleted zone at the cortex. The MT spindle is surrounded by MFs that extend to the cortex at the former PPB site and to the poles to stabilize spindle position. **g** During telophase, the phragmoplast forms from the remnants of the spindle and is composed of two anti-parallel arrays of MTs and MFs. The new cell plate is deposited where the two arrays meet. MTs nucleated at the former spindle poles extend to the cortex, including the former site of the PPB. MFs connect the phragmoplast to the cortex, including the former site of the PPB. **h** During late cytokinesis, the phragmoplast makes adjustments so that it aligns with the former PPB site. MTs nucleated from the nucleus contribute to the expansion of the phragmoplast and continue to probe the cortex, but begin to focus on the former PPB site. MFs continue to connect the phragmoplast to PPB site. **i–j** Top down view of cytokinesis. Only one nucleus is visible since the other one is hidden beneath the expanding cell plate. **i** During symmetric cytokinesis, the phragmoplast is initiated in the center of the cell and expands to create a donut shape while depositing new cell plate. **j** Polarized cytokinesis begins off center so the new cell wall first becomes anchored to one side of the mother cell. The phragmoplast continues to expand along the other sides. **k** After formation of the new cell plate, the cytoskeletal arrangements in the daughter cells resemble that seen in G1 cells. This figure is based predominantly on results reported in Wick and Duniec 1984; Lloyd and Traas 1988; Katsuta et al. 1990; Mineyuki et al. 1991; Cleary 1995; Nogami et al. 1996; Cutler and Ehrhardt 2002; Dhonukshe et al. 2005b; Chan et al. 2005; Sano et al. 2005

and Hasezawa 2002; Dhonukshe et al. 2005b). As MTs are relatively stiff polymers, they may center the nucleus simply by pushing it away from the cell periphery as they extend from the nuclear surface and "hit" the plasma membrane. Maintaining the new nuclear position initially requires both MTs and MFs, but following the breakdown of the connecting cytoplasmic MTs during mitosis, MFs alone are sufficient to maintain the position of the spindle (Venverloo and Libbenga 1987; Lloyd and Traas 1988; Katsuta et al. 1990). Interestingly, in contrast to symmetrically dividing cells, nuclear migration to the division plane in asymmetrically dividing cells requires actin, but not MTs (Mineyuki and Palevitz 1990; Kennard and Cleary 1997).

Division Plane Orientation in Plant Cells

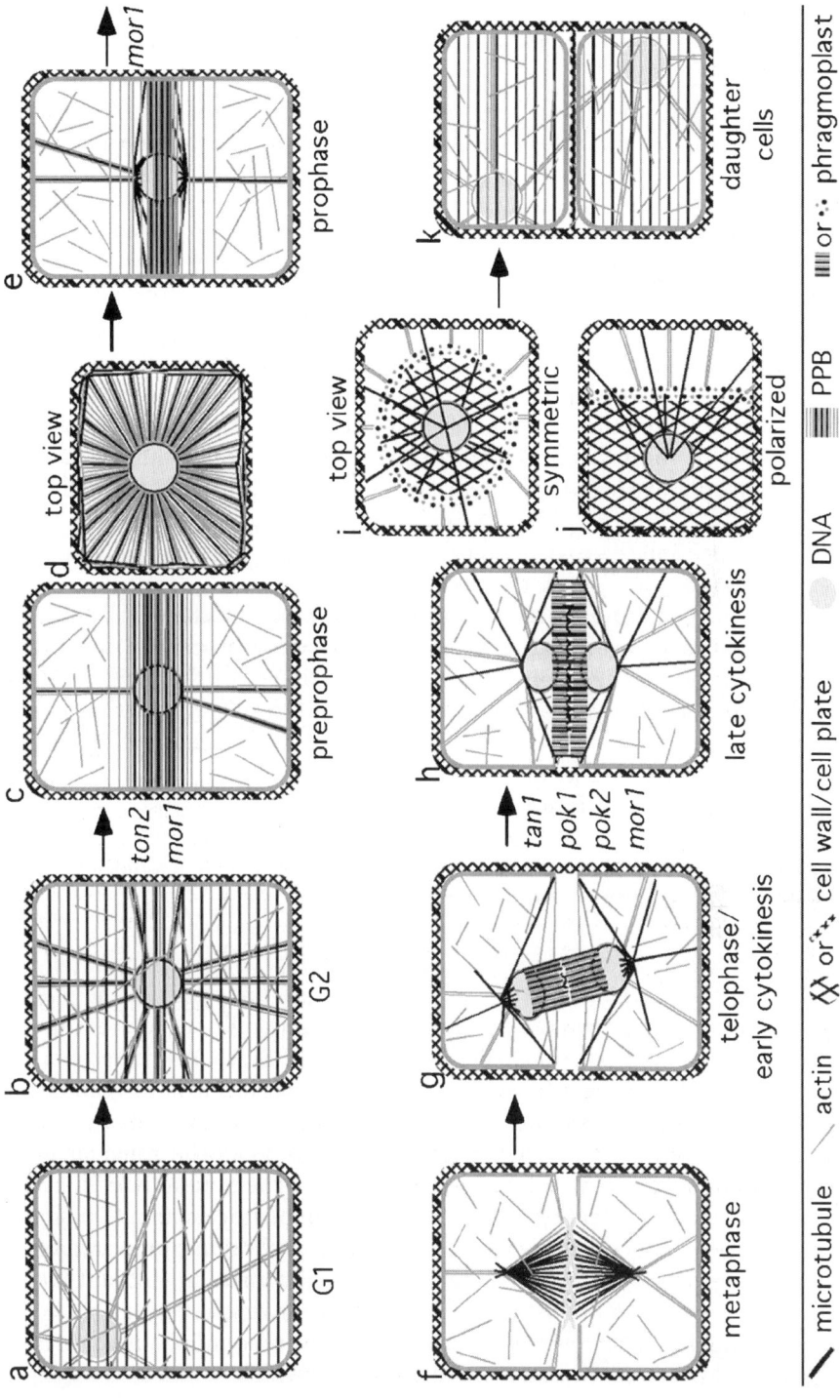

Clearly the position of the PPB and premitotic nucleus are interrelated, but what is the cause and effect relationship between them? When the premitotic nucleus of *Adiantum* protonema cells was displaced by centrifugation from its normal, apical position, the majority of cells formed a PPB around the displaced nucleus instead of at the apical location (Murata and Wada 1991). Similar results were previously reported for centrifuged wheat root cells (Burgess and Northcote 1968). These experiments clearly indicate an important role for the premitotic nucleus in dictating the site of PPB formation in some cells (Murata and Wada 1991). In contrast, when the premitotic nucleus was displaced in asymmetrically dividing cells, the PPB still formed in the usual location (Pickett-Heaps 1969; Galatis et al. 1984). Moreover, in asymmetrically dividing cells, the premitotic nucleus can occasionally be observed outside of the future division plane delineated by the PPB without experimental manipulation (e.g. Panteris et al. 2006). Thus, whether the nucleus leads or follows the PPB may depend on cell type, with the nucleus leading in symmetrically dividing cells and following in asymmetrically dividing cells. In this regard, it is interesting that nuclear migration to the division plane appears to depend primarily on MTs in symmetrically dividing cells and on MFs in asymmetrically dividing cells.

While the centering of the nucleus by a MT-based mechanism appears to be an important part of the division plane selection process in some cells, nuclear position alone cannot be sufficient to determine the division plane. For example, an elongated cell may divide symmetrically in either transverse or longitudinal planes—in either case, the premitotic nucleus will be located centrally. As discussed in the next sections, cell geometry and cell polarity may also be important factors in division plane selection.

2.1.2
A Role for Cell Geometry

Most plant cell divisions appear to be constrained by cell geometry in two key ways (reviewed by Lloyd 1991). First, cell plates do not attach to the mother cell wall at the same point as a mature, neighboring cell wall, preventing the formation of 4-way junctions (Fig. 2a; Sinnot and Bloch 1941). Second, the plane of cell division is often aligned with the shortest axis of the cell, although many exceptions to this rule exist (Hofmeister 1863). Experimental treatments in which round cells were forced to adopt an elongated shape by externally applied pressure have further reinforced the notion that cell geometry can be an important determinant in division plane selection (Lynch and Lintilhac 1997). Since it is known that the PPB and phragmosome predict the future plane of division, it is the position of their formation, and not that of the cell plate itself, that must be influenced by these rules.

How might cells "read" their geometries in order to follow these division plane rules? Cytoplasmic strands, and presumably the MTs and MFs they con-

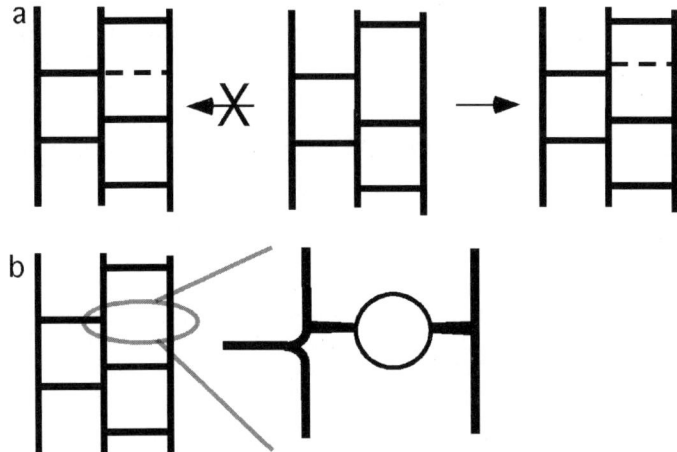

Fig. 2 Geometry and plant cell division. **a** A new cell wall (*dashed line*) avoids connecting to the mother cell wall at a pre-existing cell wall junction and forming a four-way junction (*on the left*). Instead the connection between the new cell wall and the mother cell will often form a three-way junction (*on the right*). **b** The avoidance of four-way junctions is hypothesized to be due to the cytoskeleton elements extending from the nucleus to the cortex seeking the shortest distance due to tension during PPB/phragmosome formation. Previous junctions have an out-pocketing which places the cortex in that area farther way from the nucleus

tain, are known to be under tension in preprophase (Hahne and Hoffman 1984; Goodbody et al. 1991). Since elements under tension with mobile attachment points will tend to adopt the shortest path across the cell, tension could provide a simple explanation for alignment of the division plane with the cell's short axis. Tension could also explain avoidance of four-way junctions. As a new cell wall ages and strengthens, it creates an inward-protruding vertex in its own cell file and a corresponding outward-protruding vertex in the neighboring files (Fig. 2b). Thus, cytoskeletal elements may simply avoid these vertexes because attaching at a vertex would mean spanning a longer distance (Flanders et al. 1990; Lloyd 1991). Since PPB position is correlated with the MTs extending from the nucleus to the cortex (often contained within phragmosomal strands), the connecting MTs may interpret the geometry via tension and then influence the position of the forming PPB (Flanders et al. 1990).

2.1.3
A Role for Cell Polarity

While it seems that geometrical rules can explain division plane selection in many cells, they may be descriptive rather than instructive. In Sect. 2.1.4, below, we discuss asymmetric divisions where cell polarity clearly influences division plane selection. However, it is interesting to consider the possibility that symmetric division planes are governed by polarizing cues as well. Hofmeister's rule states that new cell walls are typically formed perpendicular to the mother cell's axis of elongation (Hofmeister 1863). The connection between the plane of division and the growth polarity of a cell may be mediated in part by the hormone auxin (Dhonukshe et al. 2005a). This connection is suggested by the observation that mutations in genes encoding proteins involved in polar auxin transport often have disrupted planes of division (Mayer et al. 1993; Shevell et al. 1994; Willemsen et al. 2003). Additionally, application of auxin efflux inhibitors causes abnormal division planes, which are predicted by abnormally oriented PPBs, in elongated cells that would normally divide symmetrically and transversely (Petrásek et al. 2002; Dhonukshe et al. 2005b). Polarized auxin flow may link the plane of cell division with the polarity of cell growth, allowing for changes in growth and division patterns in response to developmental and environmental cues. Moreover, differing patterns of polarized auxin transport might explain why some symmetrically dividing, elongated cells divide transversely while others divide longitudinally.

2.1.4
A Role for Extracellular Signals

Some cells do not abide by the geometrical rules that predict the division planes of most plant cells. Divisions that occur in response to wounding are a clear example. Upon injury, cells close to the wound are induced to divide and the plane of division parallels the wound even if that plane is oblique relative to the axis of the dividing cell or the growth axis of the plant (Sinnott and Bloch 1941; Venverloo and Libbenga 1987; Goodbody and Lloyd 1990). It is unknown what signals orient wound-induced division planes. An intriguing recent study suggests that during wound-induced divisions in *Coleus*, both nuclear migration and PPB formation follow other, earlier determinants of the division plane that respond more directly to wound-induced extracellular cues. The cortical cytoplasmic ring, or CCR, is a ring of cytoplasm containing MFs and ER that encircles the future division plane following wounding of *Coleus* pith tissue (Panteris et al. 2004). Nuclear migration into the division plane occurs after CCR formation coincident with a proliferation of transvacuolar strands, some of which connect the nucleus to the CCR, forming a phragmosome. The MT component of the PPB forms in alignment with the CCR and phragmosome. It would be very interesting to know whether factors directing CCR formation have any role in more conventional modes of plant cell division.

Asymmetric divisions are another clear example where dividing cells do not follow standard geometrical rules in selecting their division planes (see Bisgrove and Kropf, this volume). In some cases, these divisions appear to be oriented in relation to extracellular cues. A good example is the asymmetric division of the subsidiary mother cells (SMCs) during stomatal development in grasses. Prior to dividing, SMCs become polarized towards the adjacent guard mother cell (GMC) as evidenced by the formation of a localized F-actin patch at a site on the plasma membrane adjacent to the GMC, migration of the SMC nucleus to this actin patch, and formation of an asymmetric PPB predicting the asymmetric division plane that will later form a small subsidiary cell adjacent to the GMC and a much larger sister cell (for review, Smith 2001). The orientation of the SMC division clearly violates the geometry rules because the division wall is curved, does not form perpendicular junctions with the mother cell, and is not orientated perpendicular to the axis of cell growth. However, in no case has a cell division-orienting signal been clearly identified, nor are there examples where the response to such a signal is understood as to how it positions the nucleus, PPB, or phragmosome.

2.2
PPB Formation

Studies of PPB formation have focused primarily on understanding the assembly of the MT component of this array. Prior to the decision to undergo mitosis and the formation of the PPB, interphase cells contain a cortical MT array (Fig. 1a,b). In elongated cells, this array consists of ordered, parallel MTs that are mostly orientated perpendicular to the axis of elongation. In isodiametric cells, the cortical MT array can be ordered or random. One of the mysteries of PPB formation is how PPB MTs are confined to a band at the cortex when previously the entire cortex was capable of supporting cortical MT growth. It is known that the MT PPB is created from newly polymerized MTs, not a rearrangement of pre-existing MTs (Cleary et al. 1992; Panteris et al. 1995). The lack of lateral MT movement also suggests that MT rearrangement via the action of motor proteins is not required for MT PPB formation (Shaw et al. 2003; Vos et al. 2004).

In highly vacuolated cells the PPB forms at the same time as the phragmosome (Venverloo and Libbenga 1987). Additionally, incomplete or complete PPBs were associated with incomplete or complete phragmosomes, respectively, suggesting interdependence in their formation (Venverloo and Libbenga 1987; Flanders et al. 1990). Unlike the PPB, the phragmosome is maintained throughout mitosis, suggesting that it may aid in phragmoplast guidance later during cytokinesis.

2.2.1
Insights from Studies of MT Dynamics in Living Cells

One theory to account for the spatially restricted appearance of the PPB is that its formation results from local changes in MT dynamics. To quantify differences in MT dynamics between interphase and preprophase cells, fluorescently labeled MTs were observed in live BY-2 cells. In one study, PPB MTs labeled at their plus ends with mammalian CLIP170:GFP had a faster growth rate and an increased catastrophe frequency when compared to similarly labeled interphase cortical MTs (Dhonukshe and Gadella 2003). Another study using a different fluorescently tagged MT binding protein (MAP4:GFP) also observed an increase in cortical MT growth rate and catastrophe frequency during preprophase as well as an increase in MT rescue frequency and a decrease in the shrinkage rate (Vos et al. 2004). These results of both studies suggest that alterations to MT dynamics during G2 could be a key force behind PPB formation. In one proposed model, the observed increase in catastrophe frequency selectively depolymerizes cortical MTs outside of the PPB zone, increasing the amount of tubulin available for the polymerization of PPB MTs and promoting the observed increase in growth rate (Dhonukshe and Gadella 2003). An alternative model proposes that changes in MT dynamics during preprophase actually increase average MT lengths throughout the cortex, and that MT bundling selectively stabilizes MTs within the PPB zone (Vos et al. 2004). This model is similar to the proposal for how MTs form aligned cortical arrays during interphase (Dixit et al. 2006).

Although the factors responsible for spatially restricted alterations to MT dynamics/stability needed to bring about PPB formation by either mechanism are largely unknown, members of the MAP65 family of MT bundling proteins, which associate with PPBs in preprophase/prophase cells (Smertenko et al. 2004) may be important for stabilization of PPB MTs.

2.2.2
Insights from Genetic and Pharmacological Perturbations

One advance in our understanding of how MT lengths are regulated during PPB formation has come from studies of *Arabidopsis* MOR1. Cells of *mor1* mutants have disorganized interphase MTs, short, disorganized spindles, short phragmoplasts that tend to fragment, and often lack PPBs (Kawamura et al. 2006; Whittington et al. 2001). The *MOR1* gene encodes a member of the well-characterized MAP215 family of MT binding proteins (Whittington et al. 2001). Alleles that give the previously described phenotypes are temperature-sensitive point mutations, while lesions expected to result in a complete loss of function cause sterility due to defects in cytokinesis following pollen mitosis I (Twell et al. 2002). Two different localization patterns for MOR1 in *Arabidopsis* have been reported. One anti-MOR1 antibody only recognizes the plus ends of MTs (Twell et al. 2002) while another labels the entire length of MTs (Kawamura et al. 2006). Kawamura et al. suggest that this discrepancy can be explained if the epitope recognized by the first antibody is masked everywhere except at the MT plus end. Together, the phenotypic and protein localization data suggest that MOR1, like its animal homologues, is required for the formation of long MTs (Whittington et al. 2001; Kawamura et al. 2006). The observation that only half of *mor1* cells form PPBs prior to mitosis suggests that MOR1-dependent MT lengthening is critical for PPB formation.

Analysis of another *Arabidopsis* mutant, *ton2*, suggests that regulation of MT stability by phosphorylation/dephosphorylation is essential for PPB formation. *ton2* mutants are sterile, dwarfed plants made up of misshapen cells with irregular planes of division (Traas et al. 1995). Interphase cortical MTs in these mutants are randomly oriented instead of aligned, and PPBs fail to form, although spindles and phragmoplasts appear relatively normal (Traas et al. 1995; McClinton and Sung 1997). The C terminal half of the TON2 protein is similar to a B″ regulatory subunit of the serine/threonine phosphatase, PP2A (Camilleri et al. 2002). In general, B″ subunits provide specificity to the PP2A holoenzymes they associate with by targeting the enzyme to specific substrates or subcellular localizations (for review, Janssens and Goris 2001). TON2 binds a member of the *Arabidopsis* PP2A holoenzyme family in a yeast two-hybrid assay, suggesting that TON2 acts as a PP2A regulator (Camilleri et al. 2002).

Consistent with this hypothesis, studies with kinase and phosphatase inhibitors indicate that the phosphorylation status of (unknown) proteins influencing MT stability plays a key role in the formation of PPBs and other MT arrays. Treatment with the PP2A inhibitor endothall results in disorganized cortical MTs, poorly assembled PPBs, premature spindle organization, multipolar spindles, and disrupted phragmoplasts in cultured alfalfa cells (Ayaydin et al. 2000). Similarly, after treatment with cantharidin, another phosphatase inhibitor, exposed cortical MTs in sectioned maize root cells depolymerized, while treatment with kinase inhibitors promoted MT stabilization (Tian et al. 2004). The implication of both studies is that dephosphorylation stabilizes MTs while phosphorylation de-stabilizes MTs. Thus, these studies are consistent with the conclusion that phosphatase activity promoted by TON2 could be critical for MT stabilization needed for the formation of PPBs and ordered MT arrays. In a survey of GFP fusions to proteins involved in cell division, TON2:GFP was reported to be in the nucleus and cytoplasm during interphase and localized to the phragmoplast during cytokinesis, but localization during preprophase/prophase was not described (Van Damme et al. 2004). Thus, it remains unclear where TON2 acts in the cell to influence PPB formation. An important advance for understanding TON2 function would be the identification of substrate(s) of the TON2-associated phosphatase.

2.3
PPB Maturation and Influence on Spindle Assembly/Orientation

After the initial formation of a broad PPB in G2, the MT PPB progressively narrows throughout prophase (Fig. 1c,e; Wick and Duniec 1983; Marcus et al.

2005). Like PPB formation, PPB narrowing requires de novo tubulin polymerization (Panteris et al. 1995). The MF component of the PPB does not undergo conspicuous narrowing, remaining much wider than the final MT PPB throughout prophase (McCurdy and Gunning 1990; Mineyuki and Palevitz 1990). While MTs are required for formation of the MF PPB, MFs are not required for the formation of the MT PPB (Palevitz 1987; McCurdy and Gunning 1990). However, application of actin depolymerizing drugs does inhibit MT PPB narrowing (in fact the MT PPB widens in this situation) suggesting MFs are needed both to initiate and to maintain narrow MT PPBs (Mineyuki and Palevitz 1990). Actin depolymerizing drugs also result in misaligned division planes, suggesting a possible role for PPB narrowing in division plane alignment (Venverloo and Libbenga 1987; Mineyuki and Palevitz 1990). To test this hypothesis, a MT depolymerizing drug was applied to BY-2 cells expressing a fluorescent MT marker after the initial formation of the PPB but before narrowing (Marcus et al. 2005). The drug was then washed out to allow for spindle re-formation, mitosis, and cytokinesis. While the re-forming spindle was not always perpendicular to the plane of the former PPB, the phragmoplast was able to successfully track back to the former PPB site, resulting in normally oriented cell division (Marcus et al. 2005). These results suggest that the key function(s) of the MT PPB in "marking" the division site for subsequent phragmoplast guidance are accomplished early, prior to PPB narrowing, but that the presence and/or narrowing of the PPB later in prophase is important for consistent spindle orientation (Marcus et al. 2005).

Other recent studies further support the notion that the PPB influences spindle assembly and orientation. The spindle normally forms so that its axis is perpendicular to the plane of division predicted by the PPB (Gunning et al. 1978a). Since spindle poles begin to form on the surface of the prophase nucleus while the PPB is still present and MTs connect the spindle poles to the PPB, it has long been thought that the PPB may influence the initial position of the spindle poles (Fig. 1e; Wick and Duniec 1984; Mineyuki et al. 1991; Nogami et al. 1996). To investigate this possibility, spindle behavior was examined in a unique population of *Arabidopsis* suspension cells expressing a MT marker, EB1:GFP. In this population, a significant percentage of cells fail to produce a PPB (Chan et al. 2005). In cells with a PPB, spindle poles marked by an accumulation of EB1-GFP formed prior to PPB and nuclear envelope disassembly, and the spindles that subsequently formed were always perpendicular to the plane of the PPB. In contrast, in cells that did not form a PPB, spindle formation was delayed until after nuclear envelope breakdown and spindle orientations were unpredictable (Chan et al. 2005).

Investigating the same issue by a different approach, Yoneda et al. (2005) examined spindle formation in BY-2 cells expressing a fluorescent tubulin marker that sometimes formed two PPBs after synchronization with aphidicolin. While cells with a single PPB formed the expected bipolar spindle, cells with two PPBs formed multipolar spindles. Extra spindle poles were most likely to be formed in the region between the double PPBs instead of directly underneath them. On the basis of this and other observations, it has been suggested that the bipolar nature of MT accumulations on the surface of the nuclear envelope in cells with a PPB could be due to inhibition by the PPB of MT polymerization or bundling in adjacent areas of the nuclear surface (Granger and Cyr 2001; Yoneda et al. 2005; Chan et al. 2005). Bipolar spindle formation also requires the kinesin ATK1 since *Arabidopsis* cells lacking this protein have multi-polar spindles (Marcus et al. 2003). Thus, through its influence on the initiation of spindle poles, the PPB appears to determine the initial orientation of the spindle.

Although the spindle does not determine the division plane in plant cells as it does in animal cells, its position can impact the final position of the cell plate. Since the phragmoplast arises from the remnants of the spindle, initiation and maintenance of the spindle in an orientation perpendicular to the former PPB ensures that the newly initiated phragmoplast is already aligned with the division plane, thereby facilitating attachment of the cell plate at the former PPB site (see further discussion, Sect. 4).

2.4
PPB Breakdown

Although it faithfully predicts the future division plane, the PPB disappears during mitosis well before the initiation of cytokinesis. The signal that initiates the breakdown of the PPB is unknown, but one hypothesis is that some factor released from the nucleus upon nuclear envelope breakdown is responsible (Mineyuki et al. 1991; Murata and Wada 1991). This hypothesis was suggested by experiments with centrifuged *Adiantum* protonemata. After PPB formation, centrifugation of these elongated cells to displace the nucleus from the site of the PPB caused PPB breakdown to be delayed until metaphase or after mitosis was complete (Murata and Wada 1991). Similar results were seen in centrifuged onion cells (Mineyuki et al. 1991). Following up on these results, Dixit and Cyr (2002b) determined precisely the temporal relationship between nuclear envelope breakdown and PPB disappearance in BY-2 cells. By observing cells expressing fluorescently tagged MTs and nuclear envelopes, they found that nuclear envelope breakdown precedes PPB breakdown by approximately 2.3 minutes. Interestingly, if nuclei were asymmetrically positioned within the plane of the PPB, the region of the PPB closest to the nucleus broke down first suggesting a signaling connection between the nucleus and the PPB that influences MT stability (Dixit and Cyr 2002b). Given the evidence discussed earlier in Sect. 2.2.2 that phosphorylation destabilizes PPB MTs, an interesting possibility to explain these findings is that release of a kinase from the nucleus upon nuclear envelope breakdown plays an important role in PPB disassembly.

3
How do Cells Remember the Division Plane Following PPB Disassembly?

In a wide variety of plant cell types, the phragmoplast attaches to the cell plate at the former location of the PPB despite the fact that the PPB was disassembled upon entry into mitosis. As discussed earlier, it has been suggested that the PPB leaves a mark on the plasma membrane during its brief existence that is later recognized by the phragmoplast. The nature of this mark remains to be fully elucidated, but several components have been identified.

3.1
Vesicles/Membrane Trafficking

In the vicinity of the PPB, vesicles and vesicle budding/fusion at the plasma membrane were observed using electron microscopy suggesting some form of vesicle transport might be required for establishment of the cortical mark recognized during cytokinesis (Gunning et al. 1978b; Galatis and Mitrakos 1979; Galatis et al. 1982; Eleftheriou 1996; Dhonukshe et al. 2005b). Testing a potential role for Golgi-dependent secretion, Dixit and Cyr (2002a) examined BY-2 cells expressing fluorescently labeled tubulin and N-acetylglucosaminyltransferase I, a resident Golgi enzyme. As previously observed by Nebenführ et al. (2000), they saw a small increase in the number of Golgi stacks in the PPB zone. However, treatment during PPB formation with brefeldin A, an inhibitor of Golgi-dependent secretion, followed by washing out to permit cytokinesis to occur, did not disrupt cell plate orientation. This result suggests that Golgi activity is not necessary for the formation of the PPB mark (Dixit and Cyr 2002a).

To probe a role for endocytosis, Dhonukshe et al. (2005b) applied FM-64, a dye that marks endocytotic vesicles, to BY-2 cells and observed a belt of labeled vesicles that co-localized with the PPB. One possible interpretation of this result is that endocytosis may be involved in establishing the cortical mark, perhaps by altering the characteristics of the plasma membrane in the vicinity of the PPB.

3.2
Actin Depleted Zone (ADZ)

While the MF component of the PPB breaks down along with the MT component, cortical actin is retained elsewhere, creating a zone of local F-actin depletion corresponding to the former PPB site which persists and "negatively" marks the division site throughout mitosis and cytokinesis. This so-called actin-depleted zone (ADZ) was first observed in living *Tradescantia* stamen hair cells injected with fluorescent phalloidin (Cleary et al. 1992) and in fixed root cells (Liu and Palevitz 1992). More recently, a similar

F-actin distribution was observed in BY-2 cells expressing an actin-binding domain from fimbrin fused to GFP (Sano et al. 2005). However, in these cells local accumulations of MF fluorescence were observed at both edges of the ADZ creating MF "twin peaks" (Sano et al. 2005). Although this feature was not pointed out, other published images of ADZs in living cells also show local enrichments of F-actin at the edges of the ADZ (e.g. Valster and Hepler 1997; Cleary 1995; Vanstraelen et al. 2006). Thus, MF twin peaks may be a general property of ADZs in living cells, but difficult to preserve by chemical fixation. In any case, a key question is what is the significance of the ADZ? To address this question, semi-synchronized cultures of BY-2 cells were treated with actin depolymerizing drugs then the drugs were washed out at various times during cell division (Hoshino et al. 2003; Sano et al. 2005). The drugs had a maximum disruption to cell plate orientation when applied for a period spanning the peak of ADZ formation, while drug treatment during cytokinesis had little effect. These experiments suggest that the presence of the ADZ during cytokinesis is not essential for phragmoplast guidance, but that either the ADZ or PPB F-actin plays an important role in establishment or maintenance of the division site prior to cytokinesis.

3.3
KCA1 Depleted Zone (KDZ)

A kinesin, KCA1, has recently been shown to serve as a second negative marker of the division site (Vanstraelen et al. 2006). KCA1 shows a strong increase in plasma membrane localization in dividing cells and also localizes to the newly forming cell plate during cytokinesis (Vanstraelen et al. 2004, 2006). Interestingly, from the time the PPB forms until the completion of cytokinesis, KCA1 localization is reduced at the future division site relative to the rest of the plasma membrane. The KCA1 depleted zone (KDZ) forms earlier than the ADZ but then co-localizes with it following PPB disassembly (Vanstraelen et al. 2006). Drug studies showed that while actin and MTs are not required for KDZ maintenance during mitosis and cytokinesis, MTs are required for KDZ formation (Vanstraelen et al. 2006). In cells where PPBs were destroyed by MT inhibitors and failed to reform upon drug wash out, no KDZ was present and cell plates failed to insert at the position of the former PPB (Vanstraelen et al. 2006). This experiment is consistent with the conclusion that KDZs play an important role in phragmoplast guidance but does not prove this point since MT PPBs were also destroyed by drug treatment. A definitive demonstration of the role of KCA1 in the spatial control of cytokinesis awaits further studies.

3.3.1
Tangled, POK1, and POK2

Recently, the first positive marker of the division site has been discovered. The *tangled1* (*tan1*) gene was originally studied in maize, where mutations in this gene cause a high frequency of misoriented cell divisions (Smith et al. 1996). This defect was attributed mainly to the failure of most phragmoplasts to be guided to former PPB sites, suggesting a role for TAN1 in division plane establishment and/or phragmoplast guidance (Cleary and Smith 1998). *Tan1* was cloned and found to encode a highly basic protein weakly similar to MT binding domains of vertebrate adenomatous polyposis coli (APC) proteins, and in vitro assays showed that TAN1 is capable of binding MTs (Smith et al. 2001). *tan1* is expressed in tissues where cells are actively dividing, but a specific protein localization could not be determined since anti-TAN1 antibodies apparently cross-reacted with other, TAN1-related proteins (Smith et al. 2001).

To overcome this difficulty, the localization pattern of the *Arabidopsis* TAN1 homolog (AtTAN) was determined by fusing it to YFP. Startlingly, AtTAN localizes to a ring coincident with the PPB in preprophase/prophase cells, which remains in position throughout mitosis and cytokinesis suggesting that AtTAN serves as a positive memory of the division site after the PPB breaks down (Walker K, Ehrhardt D, Smith LG, unpublished). Confirming an important role for AtTAN in spatial control of cytokinesis, root tips of plants with mutations in this gene have misoriented divisions attributable to misguided phragmoplasts. While maintenance of the AtTAN ring was found to be MT independent, *ton2* mutants, which lack PPBs, also lack AtTAN rings (Walker K, Ehrhardt D, Smith LG, unpublished).

PHRAGMOPLAST ORIENTING KINESIN 1 and 2 (POK1 and POK2) were originally identified in maize as interactors with TAN1 via a yeast two-hybrid screen (Müller et al. 2006). *Arabidopsis* mutants for both genes were isolated, and while neither single mutant has a phenotype, the *pok1;2* double mutant is dwarfed with misoriented planes of cell division suggesting a role for these kinesins in the orientation of the plane of division (Müller et al. 2006). As in maize *tan1* mutants, phragmoplasts often fail to be guided to former PPB sites in *pok1;2* double mutants. Additionally, *POK1/2* are needed for effective recruitment of AtTAN to the site of the PPB suggesting that the functionally redundant motor activities of POK1 and POK2 are used to localize AtTAN during preprophase (Muller S, Ehrhardt D, Smith LG, unpublished). Since TON2, the putative phosphatase regulatory subunit, is also required for AtTAN localization this suggests that the phosphorylation status of AtTAN, POK1, or POK2 may be relevant to the formation of the AtTAN ring. Alternatively, the dependence of AtTAN ring formation on TON2 maybe related to its role in PPB formation via hypothesized dephosphorylation of proteins needed for MT stability as previously discussed (Sect. 2.2.2).

4
Phragmoplast Guidance to the Division Site

After spindle breakdown, the phragmoplast arises from the remnants of the spindle and begins to expand, ultimately attaching the cell plate at the division site predicted by the former PPB. As discussed earlier, the spindle is initially suspended within the division plane with its long axis perpendicular to this plane and is usually maintained in this position throughout mitosis, apparently via cytoplasmic actin cables (Traas et al. 1987; Lloyd and Traas 1988; Sano et al. 2005). Thus, in most cells, the phragmoplast is already aligned with the division plane upon inception and relatively little guidance of its expansion may be needed for cell plate attachment at the former PPB site. This may be especially true if the phragmoplast expands within a phragmosome. However, if the spindle or early phragmoplast becomes misaligned with the division plane, the expanding phragmoplast can be guided to the former PPB site (Galatis et al. 1984; Granger and Cyr 2001). For example, in some cell types, spindles normally rotate to an oblique orientation during mitosis. When this occurs, the phragmoplast is initially oriented obliquely, but rotates as cytokinesis proceeds so that the cell plate attaches at the former PPB site (Palevitz and Hepler 1974a; Palevitz 1986; Cleary and Smith 1998). Furthermore, phragmoplasts displaced from the division plane via centrifugation can often migrate back to the former PPB site as they expand (Ôta 1961; Gunning and Wick 1985). However, under experimental conditions that result in the spindle forming far away from the former PPB site, the subsequent phragmoplast fails to be guided to this site. Thus, it appears that phragmoplast guidance can only operate within a relatively short distance of the cortical division site (Galatis et al. 1984; Granger and Cyr 2001). Very little is known about how the expanding phragmoplast "finds" the division site, but most investigation has focused on defining roles for MFs and MTs. Before discussing these studies, we first discuss two alternate modes of cytokinesis with different implications for mechanisms of phragmoplast guidance.

4.1
Symmetric Versus Polarized Cytokinesis

According to the "textbook" view of cytokinesis just discussed, the phragmoplast and associated nuclei are initially positioned in the middle of the cell and the phragmoplast expands extensively before making contact with the cell cortex (Fig. 1i). In this situation, initial guidance of the expanding phragmoplast would have to involve a long-range interaction with the cortex (e.g., as discussed later, perhaps mediated by MFs or MTs connecting the phragmoplast to the cell cortex at this stage). Subsequently, when the expanded phragmoplast makes contact with the cortex, phragmoplast position could be "fine tuned" via direct interactions between the phragmoplast edges and the cortical mark.

In contrast, a novel form of cytokinesis, termed asymmetric or polarized cytokinesis, is common in large, vacuolated cells as well as smaller, meristemic cells (Venverloo and Libbenga 1987; Cutler and Ehrhardt 2002; Panteris et al. 2004; Chan et al. 2005). In polarized cytokinesis, the prophase nucleus is found at the edge of the cell resulting in a lateral location of the subsequent spindle and phragmoplast (Cutler and Ehrhardt 2002). After the initial anchoring of the cell plate to the mother cell cortex, the phragmoplast continues to expand across the mother cell creating a plane that bisects it (Fig. 1j; Cutler and Ehrhardt 2002). In this mode of cytokinesis, the edges of the expanding phragmoplast may follow a track suggested by the cortical mark at the former location of the PPB without any need for long-range interactions between the phragmoplast and cortex. What causes the initial asymmetric positioning of the prophase nucleus that initiates polarized cytokinesis is of great interest (Cutler and Ehrhardt 2002). While the studies to be discussed in Sects. 4.2 and 4.3 have mostly not considered the question of whether the cells being examined were undergoing symmetric or polarized cytokinesis, future studies should consider this issue carefully because of possible mechanistic differences in phragmoplast guidance in these two types of divisions.

4.2
Role of Actin in Phragmoplast Guidance

MFs are a major component of the phragmoplast and they co-align with MTs. Nevertheless, phragmoplast MFs appear not to play as critical a role in cell plate formation as MTs because actin depolymerizing drugs cause distortion or misalignment of cell plates rather than disrupting cell plate formation per se (Palevitz and Helper 1974b; Gallagher and Smith 1999; Granger and Cyr 2001; Baluska et al. 2001). Actin bundles that could potentially be involved in phragmoplast guidance have been observed to connect the edge of the phragmoplast to the cortex in fixed cells, cells injected with florescent phalloidin, and cells expressing fluorescent actin binding proteins (Traas et al. 1987; Lloyd and Traas 1988; Valster and Hepler 1997; Sano et al. 2005). However, there are other MF populations that may be important for division plane control (e.g. the MF PPB or ADZ), and experiments in which drugs are applied continuously do not reveal which MF population(s) are critical for cell plate orientation.

Experiments employing timed treatments with actin depolymerizing drugs have been more informative as to the potential role of MFs in division plane determination. Palevitz and Hepler (1974b) applied cytochalasin to dividing cells after prophase was complete and found that the nuclear/phragmoplast complex was positioned correctly in a majority of *Allium* guard mother cells examined. Actin inhibitor experiments previously discussed in Sect. 3.2 also suggest that the important role played by MFs in division plane control is early in the cell cycle and that MFs do not play a large role in phragmoplast guidance during cytokinesis (Hoshino et al. 2003; Sano et al. 2005).

In addition to studies focusing on actin itself, other investigations have examined the potential role of myosin, an actin-based motor protein. Two different myosin inhibitors delayed or inhibited the lateral expansion of the phragmoplast beyond the width of the nuclei in *Tradescantia* stamen hair cells (Molchan et al. 2002). Application of BDM, an inhibitor suspected to affect a greater array of myosins, also resulted in tilting of the nuclear/phragmoplast complex (Molchan et al. 2002). These results suggest a role for myosins in lateral expansion of the phragmoplast and perhaps also in phragmoplast guidance, although drug application throughout the cell cycle limits the interpretation of these results since it is possible that myosins are needed for division plane establishment and/or maintenance. It was also observed that application of one of the inhibitors disrupted the structure of the phragmoplast, suggesting that the effects of myosin inhibitors on phragmoplast expansion/guidance may be indirect (Molchan et al. 2002). Thus, further work will be needed to clarify when and how myosins may be involved in cytokinesis or its spatial regulation. Genetic studies investigating the roles of myosins would be very helpful in this regard but are complicated by the large number of myosin genes, whose products may act redundantly.

4.3
Role of MTs in Phragmoplast Guidance

Phragmoplast MTs play an essential role in cell plate formation. Vesicles travel along phragmoplast MTs, fuse with the developing cell plate, and provide membrane and cell wall materials needed for its continuous expansion. Since MTs are essential for cell plate formation, destruction of phragmoplast MTs with MT depolymerizing drugs does not offer any clues about the roles of MTs in phragmoplast guidance. However, recent observations of MTs radiating out from the phragmoplast/nuclear complex suggest that MTs may indeed play a role. Chan et al. (2005) reported that in *Arabidopsis* cultured cells, EB1:GFP-labeled MTs link phragmoplast-associated nuclei to the cell poles, while Dhonukshe et al. (2005b) reported that in tobacco BY-2 cells, MTs labeled with either EB1:GFP or MAP4:GFP connect phragmoplast-associated nuclei to the former PPB site and other areas of the cortex (Fig. 1g,h). We have also observed MTs connecting the phragmoplast/nuclear complex to the cortex in fixed *Arabidopsis* root tips cells labeled with an anti-tubulin antibody (L.G. Smith, unpublished), alleviating concern that the GFP fusion proteins used in live cell studies might be stabilizing MTs to create a MT population that does not normally exist. Most likely, MT-based connections between the phragmoplast and the cell cortex have not been previously reported because they are easily overlooked in the presence of the dense, brightly labeled phragmoplast MT array.

MTs connecting the phragmoplast/nuclear complex to the cell cortex (referred to as "astral MTs" or "endoplasmic MTs") have been proposed to play a role in orienting the expanding phragmoplast, although evidence of the functional importance of this MT population has not yet been reported (Chan et al. 2005; Dhonukshe et al. 2005b). Certainly, this is an attractive idea, particularly given the lack of evidence that MFs can perform this function, and the well-established role of astral MTs in orienting the mitotic spindle via interactions with the cell cortex in animal cells, where spindle orientation determines the plane of division (cleavage). Interestingly, one mechanism for spindle orientation in animal cells involves a direct interaction of EB1 at MT plus ends with local, cortical accumulations of adenomatous polyposis coli (APC) protein (Lu et al. 2001; McCartney et al. 2001; Yamashita et al. 2003). The TANGLED protein, which is localized at the former PPB site in the cell cortex throughout mitosis and cytokinesis (Walker K, Ehrhardt D, Smith LG, unpublished), is distantly related to the basic domain of APC, though it lacks EB1-binding and other domains (Smith et al. 2001). Nonetheless, an intriguing model for future investigation is that EB1 at the plus ends of MTs radiating from the phragmoplast/nuclear complex interact (perhaps via an unknown linker protein) with TAN at the division site to help orient phragmoplast expansion, thereby aiding attachment of the cell plate at the former PPB site.

5
Conclusions

While the roles of MTs and MFs in the process of cell division generally and the orientation of the plane of division specifically are becoming ever more clear, there is still relatively little known about the identity and function of other proteins that participate in specifying the orientation of cell division. The challenge for the future will be to discover the proteins that regulate the formation and positioning of the PPB/phragmosome, spindle, and phragmoplast and determine how these proteins interpret polarity, external cues, and cell geometry to specify the correct positioning of cytoskeletal structures governing cell plate orientation.

Acknowledgements Work on cytokinesis in the laboratory of LGS was supported by NIH R01 GM53137 and AJW was supported by NIH GM68524 (UCSD/SDSU IRACDA).

References

Ayaydin F, Vissi E, Meszaros T, Miskolczi P, Kovacs I, Feher A, Dombradi V, Erdodi F, Gergely P, Dudits D (2000) Inhibition of serine/threonine-specific protein phosphatases causes premature activation of cdc2MsF kinase at G2/M transition and early mitotic microtubule organisation in alfalfa. Plant J 23:85–96

Baluska F, Jasik J, Edelmann HG, Salajová T, Volkmann D (2001) Latrunculin B-induced plant dwarfism: Plant cell elongation is F-actin-dependent. Dev Biol 231:113–124

Brown RC, Lemmon BE (2001) The cytoskeleton and spatial control of cytokinesis in the plant life cycle. Protoplasma 215:35–49

Burgess J, Northcote DH (1968) The relationship between the endoplasmic reticulum and microtubular aggregation and disaggregation. Planta 80:1–14

Camilleri C, Azimzadeh J, Pastuglia M, Bellini C, Grandjean O, Bouchez D (2002) The Arabidopsis *TONNEAU2* gene encodes a putative novel protein phosphatase 2A regulatory subunit essential for the control of the cortical cytoskeleton. Plant Cell 14:833–845

Chan J, Calder G, Fox S, Lloyd C (2005) Localization of the microtubule end binding protein EB1 reveals alternative pathways of spindle development in Arabidopsis suspension cells. Plant Cell 17:1737–1748

Cleary AL (1995) F-actin redistributions at the division site in living *Tradescantia* stomatal complexes as revealed by microinjection of rhodamine-phalloidin. Protoplasma 185:152–165

Cleary AL, Gunning BES, Wasteneys GO, Hepler PK (1992) Microtubule and F-actin dynamics at the division site in living *Tradescantia* stamen hair cells. J Cell Sci 103:977–988

Cleary AL, Smith LG (1998) The *Tangled1* gene is required for spatial control of cytoskeletal arrays associated with cell division during maize leaf development. Plant Cell 10:1875–1888

Cutler SR, Ehrhardt DW (2002) Polarized cytokinesis in vacuolate cells of *Arabidopsis*. Proc Natl Acad Sci USA 99:2812–2817

Dhonukshe P, Gadella TWJ (2003) Alteration of microtubule dynamic instability during preprophase band formation revealed by yellow fluorescent protein-CLIP170 microtubule plus-end labeling. Plant Cell 15:597–611

Dhonukshe P, Kleine-Vehn J, Friml J (2005a) Cell polarity, auxin transport, and cytoskeleton-mediated division planes: who comes first? Protoplasma 226:67–73

Dhonukshe P, Mathur J, Hulskamp M, Gadella TWJ (2005b) Microtubule plus-ends reveal essential links between intracellular polarization and localized modulation of endocytosis during division-plane establishment in plant cells. BMC Biol 3:11

Dixit R, Chang E, Cyr R (2006) Establishment of polarity during organization of the acentrosomal plant cortical microtubule array. Mol Biol Cell 17:1298–1305

Dixit R, Cyr R (2002a) Golgi secretion is not required for marking the preprophase band site in cultured tobacco cells. Plant J 29:99–108

Dixit R, Cyr RJ (2002b) Spatio-temporal relationship between nuclear-envelope breakdown and preprophase band disappearance in cultured tobacco cells. Protoplasma 219:116–121

Eleftheriou EP (1996) Developmental features of protophloem sieve elements in roots of wheat (*Triticum aestivum* L.). Protoplasma 193:204–212

Flanders DJ, Rawlins DJ, Shaw PJ, Lloyd CW (1990) Nucleus-associated microtubules help determine the division plane of plant epidermal cells: avoidance of four-way junctions and the role of cell geometry. J Cell Biol 110:1111–1122

Galatis B, Apostolakos P, Katsaros C (1984) Experimental studies on the function of the cortical cytoplasmic zone of the preprophase microtubule band. Protoplasma 122:11–26

Galatis B, Mitrakos K (1979) On the differential divisions and preprophase microtubule bands involved in the development of stomata of *Vigna sinensis* L. J Cell Sci 37:11–37

Galatis P, Apostolakos P, Katsaros C, Loukari H (1982) Pre-prophase microtubule band and local wall thickening in guard cell mother cells of some Leguiminosae. Ann Bot 50:779–791

Gallagher K, Smith LG (1999) *discordia* mutations specifically misorient asymmetric cell divisions during development of the maize leaf epidermis. Development 126:4623–4633

Goodbody KC, Lloyd CW (1990) Actin filaments line up across *Tradescantia* epidermal cells, anticipating wound-induced division planes. Protoplasma 157:92–101

Goodbody KC, Venverloo CJ, Lloyd CW (1991) Laser microsurgery demonstrates that cytoplasmic strands anchoring the nucleus across the vacuole of premitotic plant cells are under tension. Implications for division plane alignment. Development 113:931–939

Granger CL, Cyr RJ (2000) Microtubule reorganization in tobacco BY-2 cells stably expressing GFP-MBD. Planta 210:502–509

Granger CL, Cyr RJ (2001) Use of abnormal preprophase bands to decipher division plane determination. J Cell Sci 114:599–607

Gunning BES, Hardham AR, Hughes JE (1978a) Pre-prophase bands of microtubules in all categories of formative and proliferative cell division in *Azolla* roots. Planta 143:145–160

Gunning BES, Hardham AR, Hughes JE (1978b) Evidence for initiation of microtubules in discrete regions of the cell cortex in *Azolla* root tip cells, and a hypothesis on the development of cortical arrays of microtubules. Planta 134:161–179

Gunning BES, Wick SM (1985) Preprophase bands, phragmoplasts, and spatial control of cytokinesis. J Cell Sci Suppl 2:157–179

Gunning BES (1982) The cytokinetic apparatus: Its development and apatial regulation. In: Lloyd CW (ed) The Cytoskeleton in Plant Growth and Development. Academic Press, London, pp 229–292

Hahne G, Hoffman F (1984) The effect of laser microsurgery on cytoplamic strands and cytoplasmic streaming in isolated plant protoplasts. Eur J Cell Biol 33:175–179

Hofmeister W (1863) Zusätze und Berichtigungen zu den 1851 veröffentlichten Untersuchungen der Entwicklung höherer Kryptogamen. Jahrb Wiss Bot 3:259–293

Hoshino H, Yoneda A, Kumagai F, Hasezawa S (2003) Roles of actin-depleted zone and preprophase band in determining the division site of higher-plant cells, a tobacco BY-2 cell line expressing GFP-tubulin. Protoplasma 222:157–165

Janssens V, Goris J (2001) Protein phosphatase 2A: a highly regulated family of serine/threonine phosphatases implicated in cell growth and signalling. Biochem J 353:417–439

Katsuta J, Hashiguchi Y, Shibaoka H (1990) The role of the cytoskeleton in positioning of the nucleus in premitotic tobacco BY-2 cells. J Cell Sci 95:413–422

Kawamura E, Himmelspach R, Rashbrooke MC, Whittington AT, Gale KR, Collings DA, Wasteneys GO (2006) MICROTUBULE ORGANIZATION 1 regulates structure and function of microtubule arrays during mitosis and cytokinesis in the Arabidopsis root. Plant Physiol 140:102–114

Kennard JL, Cleary AL (1997) Pre-mitotic nuclear migration in subsidiary mother cells of *Tradescantia* occurs in G1 of the cell cycle and requires F-actin. Cell Motil Cytoskeleton 36:55–67

Kumagai F, Hasezawa S (2001) Dynamic organization of microtubules and microfilaments during cell cycle progression in higher plant cells. Plant Biol 3:4–16

Kutsuna N, Hasezawa S (2002) Dynamic organization of vacuolar and microtubule structures during cell cycle progression in synchronized tobacco BY-2 cells. Plant Cell Physiol 43:965–973

Liu B, Palevitz BA (1992) Organization of cortical microfilaments in dividing root cells. Cell Motil Cytoskeleton 23:252–264

Lloyd CW (1991) How does the cytoskeleton read the laws of geometry in aligning the division plane of plant cells? Dev Suppl 1:55–65

Lloyd CW, Traas JA (1988) The role of F-actin in determining the division plane of carrot suspension cells. Drug studies. Development 102:211–221

Lu B, Roegiers F, Jan LY, Jan YN (2001) Adherens junctions inhibit asymmetric division in the *Drosophila* epithelium. Nature 409:522–525

Lynch TM, Lintilhac PM (1997) Mechanical signals in plant development: a new method for single cell studies. Dev Biol 181:246–256

Marcus AI, Dixit R, Cyr RJ (2005) Narrowing of the preprophase microtubule band is not required for cell division plane determination in cultured plant cells. Protoplasma 226:169–174

Marcus AI, Li W, Ma H, Cyr RJ (2003) A kinesin mutant with an atypical bipolar spindle undergoes normal mitosis. Mol Biol Cell 14:1717–1726

Mayer U, Büttner G, Jürgens G (1993) Apical-basal pattern formation in the *Arabidopsis* embryo: studies on the role of the *gnom* gene. Development 117:149–162

McCartney BM, McEwen DG, Grevengoed E, Maddox P, Bejsovec A, Peifer M (2001) *Drosophila* APC2 and Armadillo participate in tethering mitotic spindles to cortical actin. Nat Cell Biol 3:933–938

McClinton RS, Sung ZR (1997) Organization of cortical microtubules at the plasma membrane in *Arabidopsis*. Planta 201:252–260

McCurdy DW, Gunning BES (1990) Reorganization of cortical actin microfilaments and microtubules at preprophase and mitosis in wheat root-tip cells: A double label immunofluorescence study. Cell Motil Cytoskeleton 15:76–87

Mineyuki Y (1999) The preprophase band of microtubules: Its function as a cytokinetic apparatus in higher plants. Int Rev Cytol 187:1–49

Mineyuki Y, Furuya M (1986) Involvement of colchicine-sensitive cytoplasmic element in premitotic nuclear positioning of *Adiantum* protonemata. Protoplasma 130:83–90

Mineyuki Y, Marc J, Palevitz BA (1991) Relationship between the preprophase band, nucleus, and spindle in dividing *Allium* cotyledon cells. J Plant Physiol 138:640–649

Mineyuki Y, Palevitz BA (1990) Relationship between preprophase band organization, F-actin, and the division site in *Allium*. J Cell Sci 97:283–295

Mineyuki Y, Wick SM, Gunning BES (1988) Preprophase bands of microtubules and the cell cycle: Kinetics and experimental uncoupling of their formation from the nuclear cycle in onion root-tip cells. Planta 174:518–526

Miyake T, Hasezawa S, Nagata T (1997) Role of cytoskeletal components in the migration of nuclei during the cell cycle transistion from G1 phase to S phase of tobacco BY-2 cells. J Plant Physiol 150:528–536

Molchan TM, Valster AH, Hepler PK (2002) Actomyosin promotes cell plate alignment and late lateral expansion in *Tradescantia* stamen hair cells. Planta 214:683–693

Müller S, Han S, Smith LG (2006) Two kinesins are involved in the spatial control of cytokinesis in *Arabidopsis thaliana*. Curr Biol 16:888–894

Murata T, Wada M (1991) Effects of centrifugation on preprophase-band formation in *Adiantum* protonemata. Planta 183:391–398

Nebenführ A, Frohlick JA, Staehelin LA (2000) Redistribution of Golgi stacks and other organelles during mitosis and cytokinesis in plant cells. Plant Physiol 124:135–151

Nogami A, Suzaki T, Shigenaka Y, Nagahama Y, Mineyuki Y (1996) Effects of cycloheximide on preprophase bands and prophase spindles in onion (*Allium cepa* L.) root tip cells. Protoplasma 192:109–121

Ôta T (1961) The role of cytoplasm in cytokinesis of plant cells. Cytologia 26:428–447

Palevitz BA (1986) Division plane determination in guard mother cells of *Alluim*: Video time-lapse analysis of nuclear movements and phragmoplast rotation in the cortex. Dev Biol 177:644-654

Palevitz BA (1987) Actin in the preprophase band of *Allium cepa*. J Cell Biol 104:1515-1519

Palevitz BA, Hepler PK (1974a) The control of the plane of division during stomatal differentiation in *Allium*. I. Spindle Reorientation. Chromosoma 46:297-326

Palevitz BA, Hepler PK (1974b) The control of the plane of division during stomatal differentiation in *Allium* II. Drug Studies. Chromosoma 46:327-341

Panteris E, Apostolakos P, Galatis B (1995) The effect of taxol on *Triticum* preprophase root cells: preprophase microtubule band organization seems to depend on new microtubule assembly. Protoplasma 186:72-78

Panteris E, Apostolakos P, Galatis B (2006) Cytoskeletal asymmetry in *Zea mays* subsidiary cell mother cells: a monopolar prophase microtubule half-spindle anchors the nucleus to its polar position. Cell Motil Cytoskeleton 63:696-709

Panteris E, Apostolakos P, Quader H, Galatis B (2004) A cortical cytoplasmic ring predicts the division plane in vacuolated cells of *Coleus*: the role of actomyosin and microtubules in the establishment and function of the division site. New Phytol 163:271-286

Petrásek J, Elckner M, Morris DA, Zazímalová E (2002) Auxin efflux carrier activity and auxin accumulation regulate cell division and polarity in tobacco cells. Planta 216:302-308

Pickett-Heaps JD (1969) Preprophase microtubules and stomatal differentiation; Some effects of centrifugation on symmetrical and asymmetrical cell division. J Ultrastruct Res 27:24-44

Pickett-Heaps JD, Northcote DH (1966a) Organization of microtubules and endoplasmic reticulum during mitosis and cytokinesis in wheat meristems. J Cell Sci 1:109-120

Pickett-Heaps JD, Northcote DH (1966b) Cell division in the formation of the stomatal complex of the young leaves of wheat. J Cell Sci 1:121-128

Sano T, Higaki T, Oda Y, Hayashi T, Hasezawa S (2005) Appearance of actin microfilament twin peaks in mitosis and their function in cell plate formation, as visualized in tobacco BY-2 cells expressing GFP-fimbrin. Plant J 44:595-605

Shaw SL, Kamyar R, Ehrhardt DW (2003) Sustained microtubule treadmilling in *Arabidopsis* cortical arrays. Science 300:1715-1718

Shevell DE, Leu WM, Gillmor CS, Xia G, Feldmann KA, Chua NH (1994) *EMB30* is essential for normal cell division, cell expansion, and cell adhesion in Arabidopsis and encodes a protein that has similarity to Sec7. Cell 77:1051-1062

Sinnott EW, Bloch R (1940) Cytoplasmic behavior during division of vacuolate plant cells. Proc Natl Acad Sci USA 26:223-227

Sinnott EW, Bloch R (1941) The relative position of cell walls in developing plant tissues. Am J Bot 28:607-617

Smertenko AP, Chang H-Y, Wagner V, Kaloriti D, Fenyk S, Sonobe S, Lloyd C, Hauser M-T, Hussey PJ (2004) The Arabidopsis microtubule associated protein AtMAP65-1: Molecular analysis of its microtubule bundling activity. Plant Cell 16:2035-2047

Smith LG (2001) Plant cell division: building walls in the right places. Nat Rev Mol Cell Biol 2:33-39

Smith LG, Gerttula SM, Han S, Levy J (2001) TANGLED1: A microtubule binding protein required for the spatial control of cytokinesis in maize. J Cell Biol 152:231-236

Smith LG, Hake S, Sylvester AW (1996) The *tangled-1* mutation alters cell division orientations throughout maize leaf development without altering leaf shape. Development 122:481-489

Tian GW, Smith D, Gluck S, Baskin TI (2004) Higher plant cortical microtubule array analyzed in vitro in the presence of the cell wall. Cell Motil Cytoskeleton 57:26–36

Traas J, Bellini C, Nacry P, Kronenberger J, Bouchez D, Caboche M (1995) Normal differentiation patterns in plants lacking microtubular preprophase bands. Nature 375:676–677

Traas JA, Doonan JH, Rawlins DJ, Shaw PJ, Watts J, Lloyd CW (1987) An actin network is present in the cytoplasm throughout the cell cycle of carrot cells and associates with the dividing nucleus. J Cell Biol 105:387–395

Twell D, Park SK, Hawkins TJ, Schubert D, Schmidt R, Smertenko A, Hussey PJ (2002) MOR1/GEM1 has an essential role in the plant-specific cytokinetic phragmoplast. Nat Cell Biol 4:711–714

Valster AH, Hepler PK (1997) Caffeine inhibition of cytokinesis: effect on the phragmoplast cytoskeleton in living *Tradescantia* stamen hair cells. Protoplasma 196:155–166

Van Damme D, Bouget FY, Van Poucke K, Inzé D, Geelen D (2004) Molecular dissection of plant cytokinesis and phragmoplast structure: a survey of GFP-tagged proteins. Plant J 40:386–398

Vanstraelen M, Torres Acosta JA, De Veylder L, Inzé D, Geelen D (2004) A plant-specific subclass of C-terminal kinesins contains a conserved A-type cyclin-dependent kinase site implicated in folding and dimerization. Plant Physiol 135:1417–1429

Vanstraelen M, Van Damme D, De Rycke R, Mylle E, Inze D, Geelen D (2006) Cell cycle-dependent targeting of a kinesin at the plasma membrane demarcates the division site in plant cells. Curr Biol 16:308–314

Venverloo CJ, Libbenga KR (1987) Regulation of the plane of cell division in vacuolated cells I. The function of nuclear positioning and phragmosome formation. J Plant Physiol 131:267–284

Vos JW, Dogterom M, Emons AMC (2004) Microtubules become more dynamic but not shorter during preprophase band formation: a possible search-and-capture mechanism for microtubule translocation. Cell Motil Cytoskeleton 57:246–258

Whittington AT, Vugrek O, Wei KJ, Hasenbein NG, Sugimoto K, Rashbrooke MC, Wasteneys GO (2001) MOR1 is essential for organizing cortical microtubules in plants. Nature 411:610–613

Wick SM, Duniec J (1983) Immunofluorescence microscopy of tubulin and microtubule arrays in plant cells. I. Preprophase band development and concomitant appearance of nuclear envelope-associated tubulin. J Cell Biol 97:235–243

Wick SM, Duniec J (1984) Immunofluorescence microscopy of tubulin and microtubule arrays in plant cells. II. Transition between the pre-prophase band and the mitotic spindle. Protoplasma 122:45–55

Willemsen V, Friml J, Grebe M, van den Toorn A, Palme K, Scheres B (2003) Cell polarity and PIN protein positioning in Arabidopsis require *STEROL METHYLTRANSFERASE1* function. Plant Cell 15:612–625

Yamashita YM, Jones DL, Fuller MT (2003) Orientation of asymmetric stem cell division by the APC tumor suppressor and centrosome. Science 301:1547–1550

Yoneda A, Akatsuka M, Hoshino H, Kumagai F, Hasezawa S (2005) Decision of spindle poles and division plane by double preprophase bands in a BY-2 cell line expressing GFP-tubulin. Plant Cell Physiol 46:531–538

Zachariadis M, Quader H, Galatis B, Apostolakos P (2001) Endoplasmic reticulum preprophase band in dividing root-tip cells of *Pinus brutia*. Planta 213:824–827

Zachariadis M, Quader H, Galatis B, Apostolakos P (2003) Organization of the endoplasmic reticulum in dividing cells of the gymnosperms *Pinus brutia* and *Pinus nigra*, and of the pterophyte *Asplenium nidus*. Cell Biol Int 27:31–40

G1/S Transition and the Rb-E2F Pathway

Wen-Hui Shen

Institut de Biologie Moléculaire des Plantes (IBMP),
Centre National de la Recherche Scientifique (CNRS),
Université Louis Pasteur de Strasbourg (ULP),
12 rue du Général Zimmer, 67084 Strasbourg, France
wen-hui.shen@ibmp-ulp.u-strasbg.fr

Abstract The G1/S transition appears central to the commitment to further cell division or differentiation in eukaryotic cells. The highly regulated G1/S transition requires the concerted action of specific cyclin-CDK kinases on specific target proteins. In plants as in animals, the Rb-E2F pathway represents the major target, its activation triggers transcription of a battery of genes involved in cell cycle control, DNA replication and cell metabolism. Recent molecular and genetic studies uncovered critical roles of the G1/S cell-cycle machinery and the Rb-E2F pathway in cell division and differentiation, bringing the cell cycle control in the context of plant growth and development. Correct and complete execution of DNA replication and packaging into chromatin not only ensures integral transmission of genetic information during cell division but also affects genome transcription.

1
Introduction

The cell division cycle proceeds in a highly ordered manner, which is driven by control systems at regulatory transitions named checkpoints. The checkpoint at the G1/S transition is called "Start" in yeast and "Restriction point" in animals. Once cells have passed this checkpoint, they become insensitive to external signals and undergo a complete cell cycle until they reach the next G1 phase. In unicellular algae, cell size and light-dependent restriction points at G1 have been documented (Hagiwara et al. 2001; Umen et al. 2001). In higher plants, application of chemical inhibitors reveals a crucial point of control in the G1 phase which commits cell division or differentiation (Planchais et al. 1997; Mourelatou et al. 2004). Because of its critical importance, the G1/S transition attracts particular attention in plant cell cycle research (Shen 2001; Gutierrez et al. 2002; Oakenfull et al. 2002; Rossi and Varotto 2002; Shen 2002). In this chapter, I will discuss recent research progress on the cell cycle machinery operating at the G1/S transition, the downstream Rb-E2F pathway, and also DNA and chromatin replication in higher plants.

2
The G1/S Cell Cycle Machinery

The universal drivers of cell-cycle transitions are cyclin-dependent kinases (CDKs), which are conserved from yeast to humans. The kinase activity of CDKs is accomplished by their association with cyclins and the cyclin-CDK complex is negatively regulated by binding with proteins named CKIs (CDK Kinase Inhibitors) in animals and ICKs (Inhibitor of CDK Kinases) or KRPs (KIP-Related Proteins) in plants. Specific classes of CDKs, cyclins and ICK/KRPs appear to be involved in the regulation of the G1/S transition in plants (Fig. 1).

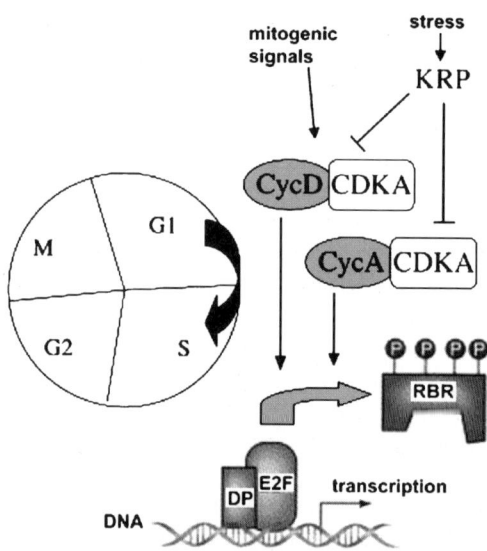

Fig. 1 Schematic view of G1/S transition in the plant cell cycle. Upon stimulation by environmental and developmental signals (mitogenic signals), D-type cyclins (CYCDs) and then A-type cyclins (CYCAs) are produced and associate with the A-type CDK (CDKA). The CYCD-CDKA and CYCA-CDKA are negatively regulated by the binding of a CDK inhibitor (KRP) whose production is induced by stress. Degradation of KRP by ubiquitin-dependent proteolysis allows re-activation of the CYCD-CDKA and CYCACDKA kinases. The CYCD-CDKA and CYCA-CDKA kinases activate the downstream Rb- E2F pathway by phosphorylation on Rb, which dissociates Rb from the Rb-E2F-DP complex and then the released E2F-DP activates transcription of E2F-target genes

2.1
CDKs

On the basis of sequence homology, plant CDKs have been classified into six types, namely, CDKA to CDKF (Vandepoele et al. 2002). CDKA is the prototype CDK in plants. It is constitutively expressed at all phases of the cell cycle and regulates both G1/S and G2/M transitions (Hemerly et al. 1995). Loss-of-function mutations of CDKA in *Arabidopsis* resulted in abnormal pollen formation (Iwakawa et al. 2006; Nowack et al. 2006). Homozygous mutant plants could not be obtained and analysis performed with heterozygous plants revealed that the mutant pollen contains only one sperm cell, instead of two, and consequently can fertilize only the egg cell. In the absence of fertilization of the central cell, which is required for normal endosperm development, the fertilized egg cell could proceed with only a limited number of cell divisions and embryogenesis stopped early during mutant seed development. These beautiful reverse genetic studies provide an example illustrating the crucial function of the cell cycle machinery in plant development. Wang and Chen (2004) previously reported that *HUA ENHANCER3 (HEN3)*, which is allelic to *CDKE*, is required for the specification of stamen and carpel identities and for the proper termination of stem cells in the floral meristem in *Arabidopsis*. Similar to that described for its homolog CDK8 in animals, the *Arabidopsis* CDKE has different substrate specificity than CDKA and phosphorylates the carboxylterminal domain (CTD) of the largest subunit of RNA polymerase II. However, distinct from CDK8 that associates with cyclin C, the *Arabidopsis* CDKE binds D-type cyclins (Wang and Chen 2004). Moreover, phenotypic analysis of the *hen3* mutant leaves revealed that CDKE is required for cell expansion rather than cell division. Thus, the precise role of CDKE in cell cycle progression needs future studies.

While CDKB is expressed from late S to M phases and controls G2/M transition (Boudolf et al. 2004), the role of CDKC, CDKD and CDKF in cell-cycle regulation is less well characterized. The *Arabidopsis* CDKC binds a cyclin T homolog and was proposed to be involved in transcription elongation (Barrôco et al. 2003). CDKD and CDKF belong to the CAK (CDK-activating Kinase) family that appear to be involved in regulation of transcription and cell division through phosphorylation of CDT and CDKs (Umeda et al. 2005).

Therefore, CDKA is the unique canonical CDK involved in the G1/S transition in plants, at least in *Arabidopsis*. This contrasts the situation in mammalian cells where the specific CDK4 and CDK6 bind cyclin D, and CDK2 binds cyclin E and cyclin A, which together control G1/S transition and S phase progression (Sherr and Roberts 2004).

2.2
D-Type and A-Type Cyclins

Genome-wide analysis of the cyclin family in *Arabidopsis* (Wang et al. 2004a) and in rice (La et al. 2006) revealed the absence of homologs to animal cyclin E but the presence of multiple D-type and A-type cyclins, indicating that the G1/S transition in plants is assured by D-type and A-type cyclins. *Arabidopsis* contains ten D-type cyclins that fall into seven subgroups (named CYCD1 to CYCD7). Overexpression of *Arath;CYCD1;1* or *Arath;CYCD2;1* in tobacco or in *Arabidopsis* plants increased growth rate likely by shortening the G1 phase (Cockroft et al. 2000; Masubelele et al. 2005), whereas overexpression of *Arath;CYCD3;1* led to cytokinin-independent growth and resulted in hyperplasia in leaves but not in roots (Riou-Khamlichi et al. 1999; Dewitte et al. 2003; Masubelele et al. 2005). Loss-of-function mutations of *Arath;CYCD1;1* or *Arath;CYCD4;1* delayed the onset of cell proliferation upon root protrusion during seed germination (Masubelele et al. 2005), a process by which the plant embryo resumes growth, after a period of quiescence, through G1/S transition (Barrôco et al. 2005). In suspension-cultured *Arabidopsis* and tobacco BY2 cells, overexpression of *Nicta;CYCD3;3* (Nakagami et al. 2002), *Antma;CYCD1;1* (Koroleva et al. 2004) or *Arath;CYCD3;1* (Menges et al. 2006) accelerated G1/S transition but also directly or indirectly perturbed other phases of the cell cycle. Together with previous molecular data showing that expression of CYCDs is responsive to nutrient availability and to phytohormones and that CYCD proteins bind CDKA forming active kinases (reviewed in Shen 2001; Oakenfull et al. 2002), these studies demonstrate that CYCD-CDKA kinases are essential for G1/S transition and that different CYCDs could have specific functions playing different roles in different tissues or cell types during plant development.

In contrast to animals where only a single A-type cyclin is present in invertebrates and two (with one of them expressed only in germ cells) in vertebrates, plants hold a higher complexity of A-type cyclins, comprising ten members in *Arabidopsis* that fall into three subgroups (CYCA1 to CYCA3) (Chaubet-Gigot 2000; Wang et al. 2004a). In synchronized tobacco BY2 cells, different CYCAs are expressed sequentially at different time points from late G1 till mid M phase (Reichheld et al. 1996), suggesting that plant CYCAs could exert functions spanning G1/S to G2/M transitions. Local and transient induction of *Nicta;CYCA3;2* expression in transgenic tobacco activated cell division in shoot apical meristem and leaf primordia (Wyrzykowska et al. 2002), whereas down-regulation by antisense expression induced defects in embryo formation and impaired callus regeneration in vitro from leaf disks (Yu et al. 2003). In transgenic *Arabidopsis*, Nicta;CYCA3;2 and CDKA form active kinase complex and overexpression of *Nicta;CYCA3;2* upregulated expression of an S-phase-specific histone gene but inhibited cell differentiation and endoreplication (Yu et al. 2003). Together with the fact that expression of *Nicta;CYCA3;2* occurs from late G1 and peaks at the S phase (Reichheld et al. 1996), it appears that Nicta;CYCA3;2 could function in an analogous manner as the animal cyclin E in G1/S transition (Yu et al. 2003). Interestingly, inducible overexpression of *Arath;CYCA2;3* in transgenic *Arabidopsis* resulted in a similar cellular phenotype as that observed with *Nicta;CYCA3;2* whereas null mutation of *Arath;CYCA2;3* oppositely, as expected, promoted endoreplication (Imai et al. 2006). It is worth noting that during the G1/S transition of seed germination *Arath;CYCA2;3* was classified in the same group of timely activated genes including *Arath;CYCD3;1* (Masubelele et al. 2005). In alfalfa, *Medsa;CYCA2;1* is responsive to auxin and expressed to a constant level during the cell cycle, and its antisense-expression inhibited regeneration of somatic embryos, suggesting a role in meristem function (Roudier et al. 2000; 2003). Further in supporting a role in G1/S transition, Medsa;CYCA2;1 protein binds CDKA and Rb (Retinoblastoma protein) (Roudier et al. 2000). Together these studies demonstrate that CYCAs from different plant species have important roles at the G1/S transition. Genetic study identified a role for *Arath;CYCA1;2* in meiosis during pollen formation (Wang et al. 2004b). The absence of detectable effects on mitotic division in the mutant *tam*, which is allelic to *Arath;CYCA1;2* (Wang et al. 2004b), might be explained by functional redundancy or developmental compensation by other genes.

2.3
ICKs/KRPs

Arabidopsis contains seven ICKs/KRPs but no homologs to the animal INK4 family inhibitors (Verkest et al. 2005a). In animals INK4 specifically interacts with CDK4 and CDK6 (Sherr and Roberts 2004), the simultaneous absence of INK4, CDK4 and CDK6 homologs in *Arabidopsis* further strengthens differences between plants and animals in the control of the G1/S transition. The assumption that plant ICKs/KRPs have an important function in the G1/S transition is supported by the following findings: (1) ICKs/KRPs inhibit CDK activity both in vitro and in vivo; (2) ICKs/KRPs bind CDKA and CYCDs; (3) overexpression of ICKs/KRPs reduces cell division rate and affects endoreplication in a dosage-dependent manner; (4) mutant phenotype by ICKs/KRPs-overexpression can be attenuated by co-overexpression of CYCDs (Verkest et al. 2005a; references therein). Interestingly, not only CYCD-CDK but also CYCA-CDK activity is inhibited by ICKs/KRPs in maize (Coelho et al. 2005), implying a broader role of ICKs/KRPs in the G1/S transition.

In yeast and in animals, proteasomal degradation of inhibitors which releases active cyclin-CDK complexes constitutes an important mechanism of activation of the G1/S transition. The ubiquitin ligase SCF (Skp1-Cullin/CDC53-F box protein) which is specifically responsible for such degradation is conserved in plants and essential for cell division (Shen et al. 2002; Thomann et al. 2005). The *Arabidopsis* ICK2/KRP2 was shown to be phosphorylated by both CDKA- and CDKB-type kinases, resulting in proteasome-dependent degradation (Verkest et al. 2005b). Thus, both regulated transcription and regulated proteolysis contribute to the regulation of activity of the G1/S cell-cycle machinery for an integrated cell division within plant growth and development as well as in response to environmental cues.

3
The Rb-E2F Pathway

In plants like in animals, the Rb-E2F pathway is recognized as a major mechanism bridging the activity of the cell-cycle machinery to transcription, particularly at the G1/S transition. The transcription factor E2F (the adenovirus E2-promoter-binding factor) functions as a heterodimer with a DP (Dimerization Partner) protein and controls the transcription of a wide variety of genes, among others genes involved in cell-cycle progression, DNA replication and repair. Hypophosphorylated Rb binds and inactivates E2F in transcription. Phosphorylation of Rb results in the release of E2F that activates transcription, which irreversibly commits cells to undergo DNA replication (S-phase). Many studies, in mammals as well as in plants, indicate that from late G1, in response to developmental or mitogenic cues, Rb is hyperphosphorylated by cyclin-CDK kinases that triggers G1/S transition (Fig. 1).

3.1
E2Fs

Arabidopsis contains eight proteins that can be classified by sequence homology into the E2F, DP and DEL (DP and E2F-Like) groups (Shen 2002). The three members of the E2F-group, AtE2Fa, AtE2Fb and AtE2Fc, exhibit an overall domain organization similar to the animal E2F1-5 proteins, including a highly conserved DNA-binding domain, a moderately conserved leucine-zipper dimerization domain, and a C-terminal trans-activation domain embracing a conserved Rb-binding site. While AtE2Fc, which contains a shorter C-terminal domain, was shown to function as a transcriptional repressor, to be abundant in arrested cells and to be degraded by proteasome upon cell-cycle stimulation (del Pozo et al. 2002), AtE2Fa and AtE2Fb were shown as transcriptional activators stimulating cell division (De Veylder et al. 2002; Magyar et al. 2005). The two members of the DP-group, AtDPa and AtDPb, contain a conserved DNA binding and leucine zipper dimerization domain and function, at least for AtDPa, as dimers with AtE2Fa and AtE2Fb in transcriptional activation (De Veylder et al. 2002; Magyar et al. 2005). Consistent with their hypothesized roles in the G1/S transition, AtE2Fa-DPa or AtE2Fb-DPa overexpression enhanced the S-phase and resulted in a cellular phenotype very similar to that produced by Arath;CYCD3;1, Arath;CYCA2;3 or Nicta;CYCA3;2 overexpression. The other three members belonging to the DEL-group, AtDEL1, AtDEL2 and AtDEL3, contain duplicated DNA-binding domains but lack the other conserved regions of the E2F- and DP-group. They bind to DNA as monomers and likely function as antagonists of E2F-DP dimers, probably by competition on the E2F binding site. Both AtDEL3 (also called E2Ff) and AtDEL1 are primarily involved in cell differentiation rather than in cell division (Ramirez-Parra et al. 2004; Vlieghe et al.

2005). In addition or in accordance with multiple members of the E2F family, genome-wide analysis identified a high number of E2F target genes involved in cell-cycle regulation, DNA replication and repair, chromatin dynamics, cell-wall biogenesis, and others (Ramirez-Parra et al. 2003; Vandepoele et al. 2005), implying a broad role of E2F family proteins beyond the cell-cycle control.

3.2
Rb

While only one Rb homolog exists in *Arabidopsis*, at least three different ones are present in maize. The presence of multiple RBR (Rb-Related) proteins in grasses was speculated to explain their recalcitrance to genetic transformation (Sabelli et al. 2005). Plant RBR proteins exhibit a similar domain organization as animal Rb, with highest conservation within the pocket domains A and B that bind E2F. Physical interaction between RBR and E2F has been shown in yeast two-hybrid assays (reviewed in Shen 2002) and a protein complex containing RBR, E2Fb and DPa has been demonstrated in transgenic *Arabidopsis* (Magyar et al. 2005). Loss-of-function mutations in the RBR gene in *Arabidopsis* were gametophytic lethal (Ebel et al. 2004), precluding the assessment of its precise role in cell-cycle progression. Downregulation of the RBR expression by RNAi (RNA interference) in tobacco (Park et al. 2005) and *Arabidopsis* (Wildwater et al. 2005) as well as titration of RBR protein by overexpression of the geminivirus RepA protein in tobacco and maize (Gordon-Kamm et al. 2002; Sabelli et al. 2005) and *Arabidopsis* (Desvoyes et al. 2005) stimulated cell division. Consistent with its repressor role, the reduction of RBR increased E2F activity and upregulated expression of E2F-target genes (Desvoyes et al. 2005; Park et al. 2005; Sabelli et al. 2005).

In mammals, hyperphosphorylation of Rb, sequentially by cyclin D, cyclin E and cyclin A CDK kinases, ensures the G1/S transition. In plants, both CYCA-CDK and CYCD-CDK kinases phosphorylate RBR (Nakagami et al. 1999, 2002; Roudier et al. 2000; Boniotti and Gutierrez 2001), which is thought to release the activity of RBR-bound E2F in transcription. Overall, it appears that CYCD-CDK and subsequently CYCA-CDK kinases phosphorylated RBR, resulting in the activation of the plant Rb-E2F pathway at the G1/S transition. A canonical Rb-E2F pathway appears critical in stem-cell maintenance in *Arabidopsis* roots (Wildwater et al. 2005).

On the basis of findings in mammals and conservation of regulators in plants, I previously proposed that chromatin remodelling could play a significant role in the Rb-E2F pathway (Shen 2002). In support of this assumption, Williams et al. (2003) found that the chromatin at E2F-target genes was decondensed before activation of these genes during dedifferentiation and entry into the S-phase of tobacco leaf protoplasts. Rossi et al. (2003) showed that the maize histone deacetylase ZmRpd3I binds RBR and that the two proteins additively repress expression of a reporter gene in a histone deacetylation and ZmRpd3IRBR binding dependent manner. Rb-associated MSI proteins are components of several protein-complexes involved in chromatin remodelling and play important roles in plant development (reviewed in Hennig et al. 2005). Among these complexes, one contains the SET-domain protein CLF (Curly Leaf) which itself can also directly bind RBR (Williams and Grafi 2000) and is likely involved in histone H3 lysine 27 (H3K27) methylation (Chanvivattana et al. 2004; Jullien et al. 2006), implying a possible role of H3K27 methylation in the Rb-E2F pathway. In humans, the SET-domain H3K9-methyltransferase SUV39H1 binds to Rb and functions as a corepressor of E2F activity with similarity to heterochromatin silencing (Nielsen et al. 2001). Overexpression of the SUV39H1 homolog in tobacco affected cell proliferation but did not significantly perturb expression of the E2F-target gene RNR2 (Shen and Meyer 2004; Yu et al. 2004). Plants contain a high number of homologs of SUV39H1, however, their homologies are limited to the catalytic region of the enzyme (Zhao and Shen 2004). Therefore, involvement of H3K9 methylation in the Rb-E2F pathway remains to be verified in plants.

4
DNA and Chromatin Replication

Initiation of DNA replication is licensed from multiple replication origins once and only once per origin during each cell cycle (Diffley 2004). Origin firing at the G1/S transition occurs on the prereplication complexes, composed of CDC6, CDT1, ORC (Origin Recognition Complex) and MCM (MiniChromosome Maintenance) proteins that are conserved from yeast to human (Fig. 1). In *Arabidopsis*, both *CDC6* and *CDT1* are E2F-target genes, and CDC6 and CDT1 proteins are subjected to proteolysis control by the proteasome (Castellano et al. 2001, 2004). CDT1 can be phosphorylated by CYCD-CDKA and CYCA-CDKA, which likely triggers its proteolysis (Castellano et al. 2004). Overexpression of *CDC6* or *CDT1* induced endoreplication (Castellano et al. 2001, 2004), likely because of promoted DNA replication licensing. Interestingly, down-regulation of *CDT1* by RNAi impaired S-phase progression, inhibited cell division as well as plastid division, but also increased endoreplication (Raynaud et al. 2005). This increased endoreplication could be due to elevated expression of MCM3 protein in the CDT1 RNAi transgenic plants (Raynaud et al. 2005), nevertheless its precise mechanism remains currently unknown. Loss-of-function mutations of *MCM7* or *ORC2* were embryonic lethal (Holding and Springer 2002; Collinge et al. 2004). Down-regulation by RNAi of *CDC45*, which likely plays a role in recruitment of DNA polymerase to the pre-replication complex, induced chromosome fragmentation during meiosis, resulting in defective pollen and ovule development (Stevens et al. 2004). Complete loss-of-function of DNA polymerase ε stopped cell division at an early stage of embryogenesis (Ronceret et al. 2005), whereas weak mutation alleles caused lengthening of the cell cycle and affected embryonic patterning (Jenik et al. 2005). In supporting that impaired DNA replication/S-phase progression caused mutant phenotype, treatment of wild-type embryos with the DNA replication inhibitor aphidicolin phenocopied the mutant phenotype (Jenik et al. 2005). The E2F-target gene *RNR2*, encoding ribonucleotide reductase necessary for dNTP synthesis, is also required for normal S-phase progression (Wang and Liu 2006). Several studies also revealed that DNA replication factors were involved in transcriptional gene silencing (Elmayan et al. 2005; Kapoor et al. 2005; Inagaki et al. 2006; Xia et al. 2006).

Newly synthesized DNA is immediately assembled with histones into chromatin (Fig. 2). The S-phase specific expression of histone genes is likely controlled independently of the E2F family of transcription factors (Shen and Gigot 1997; Meshi et al. 2000). Assembly of histones with DNA is mediated by different families of histone chaperones (Loyola and Almouzni 2004). Loss-of-function mutations of *FAS1* and *FAS2*, which encode subunits of the CAF1 (Chromatin Assembly Factor1) family H3/H4 chaperones, caused defects of meristems, resulting in shoot fasciation and root growth inhibition (Kaya et al. 2001). The NAP1 (Nucleosome Assembly Protein1) family genes, which encode H2A/H2B chaperones, are expressed to a relatively constant level during different phases of the cell cycle (Dong et al. 2003). Post-translational modification by phosphorylation and regulated nucleocytoplasmic shuttling of NAP1 proteins (Dong et al. 2005) likely provide

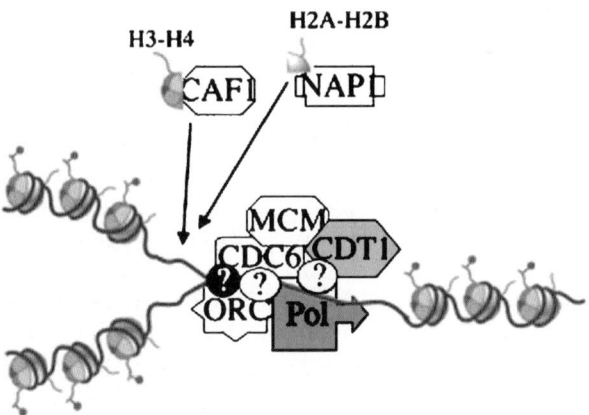

Fig. 2 Model of DNA replication and nucleosome assembly. Upon G1/S transition, transcription of the E2F-target genes CDC6 and CDT1 results in firing of the DNA replication origin, through recruitment of MCM, DNA polymerases (Pol) and other less well-characterized factors. Newly synthesized DNA is assembled with histones into nucleosome, through action of histone chaperones (CAF1 and NAP1)

mechanisms of control for the specific demand of histone chaperon activity during the S-phase. Although it is generally accepted that euchromatin is replicated in the early S-phase and heterochromatin in the late S-phase, the mechanism of this time-controlled process is still poorly studied, particularly in plants.

5
Conclusions and Perspectives

The powerful genetic approach, particularly in *Arabidopsis*, allows an exponentially increased number of genes to be examined for their role in plant growth. This likely will continue and cover the other uncharacterized genes potentially involved in the G1/S transition and the Rb-E2F pathway, as well as in DNA and chromatin replication. Many studies are starting to uncover essential functions of the G1/S transition in highly regulated cell division and cell differentiation in plant development. However, molecular and functional links among different regulators or pathways at the G1/S transition remain in many cases elusive. Furthermore, because of functional redundancy, cellular or developmental compensation as well as indirect effects, many complex aspects uncovered by the genetic approach need to be refined for mechanistic understanding. These different issues appear challenging for future research.

Acknowledgements I thank Dr. Jean Mollier for helpful comments, and apologize to colleagues whose work I cited only obliquely through references to other articles because of space limitation.

References

Barroco RM, Van Poucke K, Bergervoet JH, De Veylder L, Groot SP, Inze D, Engler G (2005) The role of the cell cycle machinery in resumption of postembryonic development. Plant Physiol 137:127–140

Boniotti MB, Gutierrez C (2001) A cell-cycle-regulated kinase activity phosphorylates plant retinoblastoma protein and contains, in *Arabidopsis*, a CDKA/cyclin D complex. Plant J 28:341–350

Boudolf V, Vlieghe K, Beemster GT, Magyar Z, Torres Acosta JA, Maes S, Van Der Schueren E, Inze D, De Veylder L (2004) The plant-specific cyclin-dependent kinase CDKB1;1 and transcription factor E2Fa-DPa control the balance of mitotically dividing and endoreduplicating cells in *Arabidopsis*. Plant Cell 16:2683–2692

Castellano MM, del Pozo JC, Ramirez-Parra E, Brown S, Gutierrez C (2001) Expression and stability of *Arabidopsis* CDC6 are associated with endoreplication. Plant Cell 13:2671–2686

Castellano MM, Boniotti MB, Caro E, Schnittger A, Gutierrez C (2004) DNA replication licensing affects cell proliferation or endoreplication in a cell type-specific manner. Plant Cell 16:2380–2393

Chanvivattana Y, Bishopp A, Schubert D, Stock C, Moon YH, Sung ZR, Goodrich J (2004) Interaction of Polycomb-group proteins controlling flowering in *Arabidopsis*. Development 131:5263–5276

Chaubet-Gigot N (2000) Plant A-type cyclins. Plant Mol Biol 43:659–675. Cockcroft CE, den Boer BG, Healy JM, Murray JA (2000) Cyclin D control of growth rate in plants. Nature 405:575–579

Coelho CM, Dante RA, Sabelli PA, Sun Y, Dilkes BP, Gordon-Kamm WJ, Larkins BA (2005) Cyclin-dependent kinase inhibitors in maize endosperm and their potential role in endoreduplication. Plant Physiol 138:2323–2336

Collinge MA, Spillane C, Kohler C, Gheyselinck J, Grossniklaus U (2004) Genetic interaction of an origin recognition complex subunit and the Polycomb group gene MEDEA during seed development. Plant Cell 16:1035–1046

del Pozo JC, Boniotti MB, Gutierrez C (2002) *Arabidopsis* E2Fc functions in cell division and is degraded by the ubiquitin-SCF(AtSKP2) pathway in response to light. Plant Cell 14:3057–3071

Desvoyes B, Ramirez-Parra E, Xie Q, Chua NH, Gutierrez C (2006) Cell type-specific role of the retinoblastoma/E2F pathway during *Arabidopsis* leaf development. Plant Physiol 140:67–80

De Veylder L, Beeckman T, Beemster GT, de Almeida Engler J, Ormenese S, Maes S, Naudts M, Van Der Schueren E, Jacqmard A, Engler G, Inze D (2002) Control of proliferation, endoreduplication and differentiation by the *Arabidopsis* E2Fa-DPa transcription factor. EMBO J 21:1360–1368

Dewitte W, Riou-Khamlichi C, Scofield S, Healy JM, Jacqmard A, Kilby NJ, Murray JA (2003) Altered cell cycle distribution, hyperplasia, and inhibited differentiation in *Arabidopsis* caused by the D-type cyclin CYCD3. Plant Cell 15:79–92

Diffley JF (2004) Regulation of early events in chromosome replication. Curr Biol 14:R778–R786

Dong A, Zhu Y, Yu Y, Cao K, Sun C, Shen WH (2003) Regulation of biosynthesis and intracellular localization of rice and tobacco homologues of nucleosome assembly protein 1. Planta 216:561–570

Dong A, Liu Z, Zhu Y, Yu F, Li Z, Cao K, Shen WH (2005) Interacting proteins and differences in nuclear transport reveal specific functions for the NAP1 family proteins in plants. Plant Physiol 138:1446–1456

Ebel C, Mariconti L, Gruissem W (2004) Plant retinoblastoma homologues control nuclear proliferation in the female gametophyte. Nature 429:776–780

Elmayan T, Proux F, Vaucheret H (2005) *Arabidopsis* RPA2: a genetic link among transcriptional gene silencing, DNA repair, and DNA replication. Curr Biol 15:1919–1925

Fulop K, Pettko-Szandtner A, Magyar Z, Miskolczi P, Kondorosi E, Dudits D, Bako L (2005) The Medicago CDKC;1-CYCLINT;1 kinase complex phosphorylates the carboxyterminal domain of RNA polymerase II and promotes transcription. Plant J 42:810–820

Gordon-Kamm W, Dilkes BP, Lowe K, Hoerster G, Sun X, Ross M, Church L, Bunde C, Farrell J, Hill P, Maddock S, Snyder J, Sykes L, Li Z, Woo YM, Bidney D, Larkins BA (2002) Stimulation of the cell cycle and maize transformation by disruption of the plant retinoblastoma pathway. Proc Natl Acad Sci USA 99:11975–11980

Gutierrez C, Ramirez-Parra E, Castellano MM, del Pozo JC (2002) G(1) to S transition: more than a cell cycle engine switch. Curr Opin Plant Biol 5:480–486

Hemerly AS, de Almeida Engler J, Bergounioux C, Van Montagu M, Engler G, Inzé D, Ferreira P (1995) Dominant negative mutants of the cdc2 kinase uncouple cell division from iterative plant development. EMBO J 14:3925–3936

Hagiwara S, Takahashi M, Yamagishi A, Zhang Y, Goto K (2001) Novel findings regarding photoinduced commitments of G1-, S- and G2-phase cells to cell-cycle transitions in darkness and dark-induced G1-, S- and G2-phase arrests in Euglena. Photochem Photobiol 74:726–733

Hennig L, Bouveret R, Gruissem W (2005) MSI1-like proteins: an escort service for chromatin assembly and remodeling complexes. Trends Cell Biol 15:295–302

Holding DR, Springer PS (2002) The *Arabidopsis* gene PROLIFERA is required for proper cytokinesis during seed development. Planta 214:373–382

Imai KK, Ohashi Y, Tsuge T, Yoshizumi T, Matsui M, Oka A, Aoyama T (2006) The A-type cyclin CYCA2;3 is a key regulator of ploidy levels in *Arabidopsis* endoreduplication. Plant Cell 18:382–396

Inagaki S, Suzuki T, Ohto MA, Urawa H, Horiuchi T, Nakamura K, Morikami A (2006) *Arabidopsis* TEBICHI, with Helicase and DNA Polymerase Domains, Is Required for Regulated Cell Division and Differentiation in Meristems. Plant Cell 18:879–892

Iwakawa H, Shinmyo A, Sekine M (2006) *Arabidopsis* CDKA;1, a cdc2 homologue, controls proliferation of generative cells in male gametogenesis. Plant J 45:819–831

Jenik PD, Jurkuta RE, Barton MK (2005) Interactions between the cell cycle and embryonic patterning in *Arabidopsis* uncovered by a mutation in DNA polymerase epsilon. Plant Cell 17:3362–3377

Jullien PE, Katz A, Oliva M, Ohad N, Berger F (2006) Polycomb group complexes selfregulate imprinting of the Polycomb group gene MEDEA in *Arabidopsis*. Curr Biol 16:486–492

Kapoor A, Agarwal M, Andreucci A, Zheng X, Gong Z, Hasegawa PM, Bressan RA, Zhu JK (2005) Mutations in a conserved replication protein suppress transcriptional gene silencing in a DNA-methylation-independent manner in *Arabidopsis*. Curr Biol 15:1912–1918

Kaya H, Shibahara KI, Taoka KI, Iwabuchi M, Stillman B, Araki T (2001) FASCIATA genes for chromatin assembly factor-1 in *Arabidopsis* maintain the cellular organization of apical meristems. Cell 104:131–142

Koroleva OA, Tomlinson M, Parinyapong P, Sakvarelidze L, Leader D, Shaw P, Doonan JH (2004) CycD1, a putative G1 cyclin from Antirrhinum majus, accelerates the cell cycle in cultured tobacco BY-2 cells by enhancing both G1/S entry and progression through S and G2 phases. Plant Cell 16:2364–2379

La H, Li J, Ji Z, Cheng Y, Li X, Jiang S, Venkatesh PN, Ramachandran S (2006) Genomewide analysis of cyclin family in rice (Oryza Sativa L.). Mol Genet Genomics 275:371–386

Leiva-Neto JT, Grafi G, Sabelli PA, Dante RA, Woo YM, Maddock S, Gordon-Kamm WJ, Larkins BA (2004) A dominant negative mutant of cyclin-dependent kinase A reduces endoreduplication but not cell size or gene expression in maize endosperm. Plant Cell 16:1854–1869

Loyola A, Almouzni G (2004) Histone chaperones, a supporting role in the limelight. Biochim Biophys Acta 1677:3–11

Magyar Z, De Veylder L, Atanassova A, Bako L, Inze D, Bogre L (2005) The role of the *Arabidopsis* E2FB transcription factor in regulating auxin-dependent cell division. Plant Cell 17:2527–2541

Masubelele NH, Dewitte W, Menges M, Maughan S, Collins C, Huntley R, Nieuwland J, Scofield S, Murray JA (2005) D-type cyclins activate division in the root apex to promote seed germination in *Arabidopsis*. Proc Natl Acad Sci USA 102:15694–15699

Menges M, Samland AK, Planchais S, Murray JA (2006) The D-type cyclin CYCD3;1 is limiting for the G1-to-S-phase transition in *Arabidopsis*. Plant Cell 18:893–906

Meshi T, Taoka KI, Iwabuchi M (2000) Regulation of histone gene expression during the cell cycle. Plant Mol Biol 43:643–657

Mourelatou M, Doonan JH, McCann MC (2004) Transition of G1 to early S phase may be required for zinnia mesophyll cells to trans-differentiate to tracheary elements. Planta 220:172–176

Nakagami H, Kawamura K, Sugisaka K, Sekine M, Shinmyo A (2002) Phosphorylation of retinoblastoma-related protein by the cyclin D/cyclin-dependent kinase complex is activated at the G1/S-phase transition in tobacco. Plant Cell 14:1847–1857

Nielsen SJ, Schneider R, Bauer UM, Bannister AJ, Morrison A, O'Carroll D, Firestein R, Cleary M, Jenuwein T, Herrera RE, Kouzarides T (2001) Rb targets histone H3 methylation and HP1 to promoters. Nature 412:561–565

Nowack MK, Grini PE, Jakoby MJ, Lafos M, Koncz C, Schnittger A (2006) A positive signal from the fertilization of the egg cell sets off endosperm proliferation in angiosperm embryogenesis. Nat Genet 38:63–67

Oakenfull EA, Riou-Khamlichi C, Murray JA (2002) Plant D-type cyclins and the control of G1 progression. Philos Trans R Soc Lond B Biol Sci 357:749–760

Park JA, Ahn JW, Kim YK, Kim SJ, Kim JK, Kim WT, Pai HS (2005) Retinoblastoma protein regulates cell proliferation, differentiation, and endoreduplication in plants. Plant J 42:153–163

Planchais S, Glab N, Tréhin C, Perennes C, Bureau JM, Meijer L, Bergounioux C (1997) Roscovitine, a novel cyclin-dependent kinase inhibitor, characterizes restriction point and G2/M transition in tobacco BY-2 cell suspension. Plant J 12:191–202

Ramirez-Parra E, Frundt C, Gutierrez C (2003) A genome-wide identification of E2Fregulated genes in *Arabidopsis*. Plant J 33:801–811

Ramirez-Parra E, Lopez-Matas MA, Frundt C, Gutierrez C (2004) Role of an atypical E2F transcription factor in the control of *Arabidopsis* cell growth and differentiation. Plant Cell 16:2350–2363

Raynaud C, Perennes C, Reuzeau C, Catrice O, Brown S, Bergounioux C (2005) Cell and plastid division are coordinated through the prereplication factor AtCDT1. Proc Natl Acad Sci USA 102:8216–8221

Reichheld JP, Chaubet N, Shen W-H, Renaudin JP, Gigot C (1996) Multiple A-type cyclins express sequentially during the cell cycle in Nicotiana tabacum BY-2 cells. Proc Natl Acad Sci USA 93:13819–13824

Riou-Khamlichi C, Huntley R, Jacqmard A, Murray JAH (1999) Cytokinin activation of *Arabidopsis* cell division through a D-type cyclin. Science 283:1541–1544

Ronceret A, Guilleminot J, Lincker F, Gadea-Vacas J, Delorme V, Bechtold N, Pelletier G, Delseny M, Chaboute ME, Devic M (2005) Genetic analysis of two *Arabidopsis* DNA polymerase epsilon subunits during early embryogenesis. Plant J 44:223–236

Rossi V, Locatelli S, Lanzanova C, Boniotti MB, Varotto S, Pipal A, Goralik-Schramel M, Lusser A, Gatz C, Gutierrez C, Motto M (2003) A maize histone deacetylase and retinoblastoma-related protein physically interact and cooperate in repressing gene transcription. Plant Mol Biol 51:401–413

Rossi V, Varotto S (2002) Insights into the G1/S transition in plants. Planta 215:345–356

Roudier F, Fedorova E, Gyorgyey J, Feher A, Brown S, Kondorosi A, Kondorosi E (2000) Cell cycle function of a Medicago sativa A2-type cyclin interacting with a PSTAIRE type cyclin-dependent kinase and a retinoblastoma protein. Plant J 23:73–83

Roudier F, Fedorova E, Lebris M, Lecomte P, Gyorgyey J, Vaubert D, Horvath G, Abad P, Kondorosi A, Kondorosi E (2003) The Medicago species A2-type cyclin is auxin regulated and involved in meristem formation but dispensable for endoreduplicationassociated developmental programs. Plant Physiol 131:1091–1103

Sabelli PA, Dante RA, Leiva-Neto JT, Jung R, Gordon-Kamm WJ, Larkins BA (2005) RBR3, a member of the retinoblastoma-related family from maize, is regulated by the RBR1/E2F pathway. Proc Natl Acad Sci USA 102:13005–13012

Shen W-H (2001) The plant cell cycle: G1/S regulation. Euphytica 118:223–232

Shen W-H (2002) The plant E2F-Rb pathway and epigenetic control. Trends Plant Sci 7:505–511

Shen W-H, Gigot C (1997) Protein complexes binding to cis-elements of the plant histone promoters: multiplicity, phosphorylation and cell cycle alteration. Plant Mol Biol 33:367–379

Shen W-H, Meyer D (2004) Ectopic expression of the NtSET1 histone methyltransferase inhibits cell expansion, and affects cell division and differentiation in tobacco plants. Plant Cell Physiol 45:1715–1719

Shen W-H, Parmentier Y, Hellmann H, Lechner E, Dong A, Masson J, Granier F, Lepiniec L, Estelle M, Genschik P (2002) Null mutation of AtCUL1 causes arrest in early embryogenesis in *Arabidopsis*. Mol Biol Cell 13:1916–1928

Sherr CJ, Roberts JM (2004) Living with or without cyclins and cyclin-dependent kinases. Genes Dev 18:2699–2711

Stevens R, Grelon M, Vezon D, Oh J, Meyer P, Perennes C, Domenichini S, Bergounioux C (2004) A CDC45 homolog in *Arabidopsis* is essential for meiosis, as shown by RNA interference-induced gene silencing. Plant Cell 16:99–113

Thomann A, Dieterle M, Genschik P (2005) Plant CULLIN-based E3s: phytohormones come first. FEBS Lett 579:3239–3245

Umeda M, Shimotohno A, Yamaguchi M (2005) Control of cell division and transcription by cyclin-dependent kinase-activating kinases in plants. Plant Cell Physiol 46:1437–1442

Umen JG, Goodenough UW Control of cell division by a retinoblastoma protein homolog in Chlamydomonas. Genes Dev 15:1652–1661

Vandepoele K, Raes J, De Veylder L, Rouze P, Rombauts S, Inze D (2002) Genome-wide analysis of core cell cycle genes in *Arabidopsis*. Plant Cell 14:903–916

Vandepoele K, Vlieghe K, Florquin K, Hennig L, Beemster GT, Gruissem W, Van de Peer Y, Inze D, De Veylder L (2005) Genome-wide identification of potential plant E2F target genes. Plant Physiol 139:316–328

Verkest A, Weinl C, Inze D, De Veylder L, Schnittger A (2005a) Switching the cell cycle. Kiprelated proteins in plant cell cycle control. Plant Physiol 139:1099–1106

Verkest A, Manes CL, Vercruysse S, Maes S, Van Der Schueren E, Beeckman T, Genschik P, Kuiper M, Inze D, De Veylder L (2005b) The cyclin-dependent kinase inhibitor KRP2 controls the onset of the endoreduplication cycle during *Arabidopsis* leaf development through inhibition of mitotic CDKA;1 kinase complexes. Plant Cell 17:1723–1736

Vlieghe K, Boudolf V, Beemster GT, Maes S, Magyar Z, Atanassova A, de Almeida Engler J, De Groodt R, Inze D, De Veylder L (2005) The DP-E2F-like gene DEL1 controls the endocycle in *Arabidopsis* thaliana. Curr Biol 15:59–63

Wang W, Chen X (2004) HUA ENHANCER3 reveals a role for a cyclin-dependent protein kinase in the specification of floral organ identity in *Arabidopsis*. Development 131:3147–3156

Wang C, Liu Z (2006) *Arabidopsis* ribonucleotide reductases are critical for cell cycle progression, DNA damage repair, and plant development. Plant Cell 18:350–365

Wang G, Kong H, Sun Y, Zhang X, Zhang W, Altman N, DePamphilis CW, Ma H (2004a) Genome-wide analysis of the cyclin family in *Arabidopsis* and comparative phylogenetic analysis of plant cyclin-like proteins. Plant Physiol 135:1084–1099

Wang Y, Magnard JL, McCormick S, Yang M (2004b) Progression through meiosis I and meiosis II in *Arabidopsis* anthers is regulated by an A-type cyclin predominately expressed in prophase I. Plant Physiol 136:4127–135

Wildwater M, Campilho A, Perez-Perez JM, Heidstra R, Blilou I, Korthout H, Chatterjee J, Mariconti L, Gruissem W, Scheres B (2005) The RETINOBLASTOMA-RELATED gene regulates stem cell maintenance in *Arabidopsis* roots. Cell 123:1337–1349

Williams L, Grafi G (2000) The retinoblastoma protein – a bridge to heterochromatin. Trends Plant Sci 5:239–240

Williams L, Zhao J, Morozova N, Li Y, Avivi Y, Grafi G (2003) Chromatin reorganization accompanying cellular dedifferentiation is associated with modifications of histone H3, redistribution of HP1, and activation of E2F-target genes. Dev Dyn 228:113–120

Wyrzykowska J, Pien S, Shen W-H, Fleming AJ (2002) Manipulation of leaf shape by modulation of cell division. Development 129:957–964

Xia R, Wang J, Liu C, Wang Y, Wang Y, Zhai J, Liu J, Hong X, Cao X, Zhu JK, Gong Z (2006) ROR1/RPA2A, a putative replication protein A2, functions in epigenetic gene silencing and in regulation of meristem development in *Arabidopsis*. Plant Cell 18:85–103

Yu Y, Dong A, Shen W-H (2004) Molecular characterization of the tobacco SET domain protein NtSET1 unravels its role in histone methylation, chromatin binding, and segregation. Plant J 40:699–711

Yu Y, Steinmetz A, Meyer D, Brown S, Shen W-H (2003) The tobacco A-type cyclin, Nicta;CYCA3;2, at the nexus of cell division and differentiation. Plant Cell 15:2763–2777

Zhao Z, Shen W-H (2004) Plants contain a high number of proteins showing sequence similarity to the animal SUV39H family of histone methyltransferases. Ann NY Acad Sci 1030:661–669

The Endoreduplication Cell Cycle: Regulation and Function

Paolo A. Sabelli · Brian A. Larkins (✉)

Department of Plant Sciences, University of Arizona, P.O. Box 210036, Tucson, AZ 85721, USA
larkins@ag.arizona.edu

Abstract The endoreduplication cell cycle is a variant of the standard mitotic cell cycle in which reiterated rounds of DNA synthesis occur in the absence of chromosome segregation and cell division. The resulting polyploid cells are frequently found in plants and often occur in tissues with a high metabolic activity that accumulate storage molecules. In this chapter we review the current understanding of the regulatory mechanisms involved in the modification of the mitotic cell cycle, based on a generally accepted template for plant cell cycle control, that bring about endoreduplication, as well as the influence of phytohormones and epigenetic factors. The regulation of endoreduplication is described in two model systems: the developing maize endosperm and the leaf trichome of *Arabidopsis*. In addition, we discuss the biological relevance and several proposed functions of this specialized cell cycle.

1
Introduction

Endoreduplication is a form of somatic nuclear polyploidization in which complete rounds of DNA synthesis occur in the absence of chromatin condensation, nuclear membrane breakdown, mitotic spindle formation, sister chromatid segregation, and cytokinesis (Edgar and Orr-Weaver 2001; Larkins et al. 2001). Because sister chromatids do not fully separate, endoreduplicated chromosomes are often larger and thicker than normal and are called polytenic (Carvalheira 2000). Endoreduplication is the most common and best characterized type of cell cycle variant leading to polyploidization. However, several other cell cycle types, such as endomitosis, result in polyploid cells and, sometimes, the distinction between them is not clear-cut (Edgar and Orr-Weaver 2001). Endoreduplication is widespread in plants; it has been estimated that it occurs in up to 90% of all angiosperm species (D'Amato 1984). In *Arabidopsis*, with the exception of the inflorescence, virtually all tissues and organs contain endoreduplicated cells (Galbraith et al. 1983), although, importantly, not all cells in a given tissue may be endoreduplicated or have undergone endoreduplication to the same extent. Cells and tissues that have been found to be endoreduplicated in plants include the seed endosperm (Larkins et al. 2001), antipodal and synergid cells of the female gametophyte, embryo suspensor, cotyledons, anther hairs and tapetum (D'Amato 1984), leaf trichomes (Schnittger and Hulskamp

2002), stem and leaf epidermis (Melaragno et al. 1993; Kudo and Kimura 2002), mesocarp of fruits (Cheniclet et al. 2005), and seedling hypocotyl (Gendreau et al. 1997).

The extent to which endoreduplication occurs varies widely among plants. For example, the average DNA content in *Arabidopsis* trichomes is 32C (where C represents the amount of unreplicated DNA in haploid cells of a given species), whereas it can reach in excess of 384C in some maize endosperms, to about 8000C in the suspensor of *Phaseolus coccineous* and 24 576C in *Arum maculatum* (equivalent to 13 endoreduplication cycles). Endoreduplication also occurs in organisms other than plants, from yeast to arthropods and, relatively infrequently, in mammals, either as part of their normal development or in response to genetic or physiological perturbations (Edgar and Orr-Weaver 2001). Endoreduplication is generally an irreversible process; that is, endoreduplicated cells cannot divide. However, in the epidermis of the *Manduca sexta* caterpillar, endoreduplicated 32C cells reenter mitosis, with a consequent reduction in their ploidy to 2C (Edgar and Orr-Weaver 2001). In addition, a recent report indicates that certain *Arabidopsis* cells, which were induced to undergo endoreduplication by the misexpression of *ICK1/KRP1*, can reenter mitosis (Weinl et al. 2005).

It has long been observed that the occurrence of endoreduplication is correlated with enlargement of differentiated cells, which are often highly active in gene expression and metabolic output. Yet exceptions to this rule have also been documented, which makes it difficult to establish a general and necessary relationship between endoreduplication, cell growth, differentiation, and the organism's development. In recent years, studies of endoreduplication in plants have made substantial progress. Thus, in this chapter we focus on current knowledge of the genetic and molecular mechanisms controlling endoreduplication in well-studied plant model systems, and discuss different theories concerning the potential functions of endoreduplication. Studies in nonplant organisms will be referred to primarily in order to provide a historical perspective or to illustrate specific issues where knowledge from plants is currently insufficient or lacking.

2
Molecular Regulation of Endoreduplication

In the mitotic cell cycle DNA synthesis and chromosome segregation are tightly coupled. However, during endoreduplication one or more rounds of DNA synthesis occur in the absence of mitosis. This observation suggests, therefore, that the factors responsible for coupling DNA synthesis to mitosis are deregulated during the endoreduplication cell cycle. Here we consider how endoreduplication may be regulated at the level of replication origin activity, the retinoblastoma/E2F pathway, the proteolysis of G1/S-phase- and G2/M-phase-specific factors, and cyclin-dependent kinase (CDK) activity.

2.1
The Control of Replication Origin Activation

Chromatin in eukaryotic cells contains an estimated 10^3–10^5 origins of replication, which typically are located 30–100 kb apart (Blow and Dutta 2005). These origins have to be activated or "fired" in a highly coordinated fashion to ensure complete and exact genome replication, with no refiring within one S-phase. Early cell fusion studies with mammalian cells demonstrated that G1 nuclei can undergo DNA synthesis, whereas G2 nuclei cannot (Rao and Johnson 1970). It is now widely accepted that G1 nuclei can proceed to S-phase because their chromatin is "licensed" for replication (Blow and Dutta 2005), whereas G2 chromatin is not. The nature of this license resides in chromosomal replication origins that become primed during late M-phase and G1 for DNA synthesis. This process involves the stepwise loading of several proteins, such as ORC, CDC6, CDT1, and the MCM2-7 complex (Sabelli et al. 1998). A drop in CDK activity, which occurs in late mitosis, is essential for the assembly of this complex, and results in a so-called prereplicative complex (pre-RC). Pre-RCs are essentially licensed origins that can be fired after a rise in CDK activity, as occurs at the G1/S-phase transition.

Firing an origin of DNA replication involves localized unwinding of the double-stranded DNA helix, and the recruitment of DNA replication enzymes at the replication fork. The firing of replication origins results in a simultaneous loss of license, as many of the key licensing proteins become excluded from chromatin and downregulated during S-phase by a range of CDK-driven mechanisms, which include specific inhibitors, nuclear export, and proteolysis. This downregulation ensures that origins are prevented from being relicensed and fired again during S-phase, and thus only one round of DNA synthesis occurs during one cell cycle. This M-phase-dependent "window of opportunity" for origin licensing, therefore, provides a crucial coupling between DNA synthesis and mitosis, which enables maintenance of genome integrity through generations of cells. In molecular terms, the anaphase-promoting complex/cyclosome (APC/C) ubiquitin ligase plays a key role during mitosis in the degradation of specific proteins that would interfere with pre-RC formation (Nakayama and Nakayama 2006) (see below). During endoreduplication, however, one or more rounds of DNA synthesis occur without intervening mitoses, suggesting that deregulation of origin licensing and firing might play an important role.

A considerable amount of information about the mechanisms that could explain the repeated rounds of DNA replication typical of endoreduplication may be provided by the study of the circumstances and causes of unscheduled replication during G2 or M-phase. For example, in *Saccharomyces cerevisiae*,

CDK activity represses rereplication within one cell cycle by three overlapping mechanisms that involve phosphorylation of ORC, reduced expression of CDC6, and nuclear exclusion of the MCM2-7 proteins (Nguyen et al. 2001). Only when all three mechanisms are inactivated does rereplication occur in G2/M, indicating that inactivation of each mechanism is not sufficient to induce rereplication. Consistent with this observation, yeast rereplication is stimulated by mutations affecting CDC2/CycB kinase (Ravid et al. 2002). Failure to induce rereplication upon CDC6 disruption in human and *Xenopus* cells supports this view. However, it is conceivable that, because ORC, CDC6, and MCM proteins work in a complex, deregulation of one type of protein could affect the activity of the others and thereby suffice to trigger rereplication, as shown for the *Schizosaccharomyces pombe* CDC6 homolog Cdc18 (Nguyen et al. 2001).

CDT1 plays a key role in regulating the licensing of DNA replication, and its overexpression results in rereplication in metazoa (DePamphilis et al. 2006) and *Arabidopsis* (Castellano et al. 2004). In mammals, alteration in the balance between CDT1 expression and its inhibitor, geminin, also results in rereplication. Recent results point to a key role played by Emi1, an inhibitor of APC/C, in the upstream regulation of both geminin/CDT1 and CycA/CDKs, which result in rereplication through chromatin-bound, active CDT1 and hypophosphorylated, and active, CDT1, CDC6, and MCMs (Machida and Dutta 2007). This work also highlights the importance of proper oscillation in APC/C activity for pre-RC formation and coupling of S-phase to M-phase (see below). However, plants do not possess Emi1 orthologs, and it remains to be seen whether they utilize this pathway.

In *Xenopus*, factors involved in nuclear import (i.e., Ran-GTP and importins) prevent rereplication by sequestering MCM proteins in a CDK-dependent manner, resulting in inhibition of pre-RCs at the G1/S-phase transition (Yamaguchi and Newport 2003). In animals, depletion of CenA induces endoreduplication in a fashion that might involve the coordinated suppression of CDC6, CDT1, and ORC1 during S-phase (Mihaylov et al. 2002).

The process leading to activation of DNA synthesis is regarded as highly conserved in eukaryotes, and plants possess orthologs of *ORC* (Gavin et al. 1995; Witmer et al. 2003), *CDC6* (de Jager et al. 2001; Ramos et al. 2001), *CDT1* (Castellano et al. 2004; Masuda et al. 2004), and *MCM* (Sabelli et al. 1996, 1999; Springer et al. 2000) genes. In *Arabidopsis*, a positive role for *CDC6* and *CDT1* in endoreduplication has been reported, at least in leaf epidermis cells (Castellano et al. 2001, 2004). Ectopic expression of *AtCDC6* stimulates endoreduplication in leaves, and the stability of CDC6 protein is enhanced in endoreduplicating cells. This observation is intriguing because it highlights the importance of context in determining the weight that certain licensing factors carry in regulating checkpoints at both G/S-phase and S-phase/G transitions during endoreduplication.

Analysis of endoreduplicating cells by pulse labeling with [^{3}H]thymidine (Lilly and Duronio 2005) or flow cytometry (Galbraith et al. 1983) shows that endoreduplication consists, similarly to mitotic cells, of DNA synthesis phases alternating with gap phases. This observation strongly suggests that the controls that normally regulate entry into S-phase and prevent rereplication in mitotic cells are largely operating during endoreduplication. Additional genetic and molecular data also support the current view that endoreduplication can be considered a derivation of the mitotic cell cycle, that endoreduplication requires more than just relaxation of the restrictions on origin licensing, and that many of the pathways involving CDK, SCF, and APC activities need reprogramming for endoreduplication to occur (DePamphilis et al. 2006).

2.2
The Retinoblastoma/E2F Pathway

The retinoblastoma (RB)/E2F pathway controls the expression of a plethora of S-phase genes and plays a pivotal role in regulating endoreduplication. RB plays a negative cell cycle role by repressing E2F/DP transcription factors (see chapter by Chen W-H, this volume). In animals, the negative role of RB in rereplication and the endoreduplication cell cycle may involve its interaction with replicated origins, the downregulation of factors required for origin licensing, or enforcing the normal expression of mitotic checkpoint proteins (Niculescu et al. 1998; Royzman et al. 1999; Bosco et al. 2001; Cobrinik 2005). Phosphorylation and consequent inactivation of RB and related (RBR) proteins, which relieves its inhibitory effect on transcription of genes involved in DNA synthesis and S-phase progression, is associated with endoreduplication in mammals and plants (Grafi et al. 1996; Harbour and Dean 2000). The direct downregulation of *RBR1* expression either using a virus-based, silencing approach in tobacco or inhibition of RBR1 protein by wheat dwarf virus (WDV) RepA expression in *Arabidopsis* results in increased expression of E2F targets and the specific stimulation of endoreduplication in differentiating organs/tissues (Park et al. 2005; Desvoyes et al. 2006). However, a rather different situation has been found in *Drosophila*, where RB/E2F complexes regulate replication origin licensing by cyclically downregulating CycE in the gap phase during endoreduplication (Weng et al. 2003). Thus, integrity of the RB/E2F pathway is required for normal endoreduplication in *Drosophila*.

S-phase gene expression and endoreduplication are also stimulated by the upregulation of RBR-repressed E2F/DP transcription factors, such as E2Fa and DPa in *Arabidopsis* (De Veylder et al. 2002) and tobacco (Kosugi and Ohashi 2003). In addition, RNAi-mediated downregulation of another RBR1-inhibited member of the E2F family, *E2Fc*, leads to decreased endoreduplication in mature leaves, possibly by controlling certain aspects of the G2/M-phase transition (del Pozo et al. 2006). Alteration in the balance within the E2F family of genes affects endoreduplication as shown by the loss of repressor-type *DEL1/E2Fe*, which promoted endoreduplication, and its overexpression, which inhibited endoreduplication in a fashion that was proposed to counteract *E2Fa/DPa* (Vlieghe et al. 2005).

That inhibition of RBR1 by phosphorylation is an important mechanism for regulating endoreduplication is indirectly confirmed by the effect of overexpressing upstream elements controlling the RBR/E2F pathway. Increased expression of *CycD3;1* and certain CKIs reduces endoreduplication (De Veylder et al. 2001; Dewitte et al. 2003; Schnittger et al. 2003; Zhou et al. 2003).

One important conclusion from several studies on the impact of the RBR/E2F pathway on endoreduplication is that it largely depends on the developmental context. In fact, downregulation of RBR (and upregulation of E2F) tends to promote cell proliferation in mitotic cells/tissues but to enhance endoreduplication in endoreduplicating/differentiated cells/tissues (Desvoyes et al. 2006).

2.3
The Role of the APC/C

Regulated protein degradation through the ubiquitin pathway plays a central role in cell cycle control. The APC/C is a cullin-based E3 ubiquitin ligase complex involved in the degradation of key cell cycle proteins, thereby regulating crucial cell cycle transition, such as the metaphase/anaphase transition, mitosis exit, and DNA replication (Nakayama and Nakayama 2006; Peters 2006). One important activator of APC/C is CDH1, which is a WD40-repeat protein required for APC/C–substrate interactions (Peters 2006). CDH1 is a conserved protein, and orthologs, such as Fizzy-related (fzr) and CSS52A, have been characterized in detail in *Drosophila* (Sigrist and Lehner 1997) and *Medicago* (Cebolla et al. 1999), respectively. Both fzr and CSS52A proteins are required for endoreduplication, most likely because they regulate the involvement of APC/C in several processes, the best studied of which is the destruction of mitotic cyclins. In *Medicago*, CSS52A plays a key role in the ploidy-dependent cell enlargement observed in nitrogen-fixing root nodules resulting from infection by the soil bacterium, *Sinorhizobium meliloti* (Vinardell et al. 2003). It also regulates endoreduplication in the giant cells found in root-knot galls, which are induced by the endoparasitic root-knot nematode, *Meloidogyne incognita* (Kondorosi and Kondorosi 2004).

Root nodules develop in *Medicago* following an indeterminate program. However, analysis of *CSS52A* orthologs in *Lotus japonicus* and *Lupinus albus*, which display determinate and intermediate types of root nodule development, respectively, has revealed that this gene has a widespread role in linking endoreduplication to nitrogen fixation in symbiotic cells, regardless of the mode of nodule morphogenesis and differentiation (Gonzalez-Sama et al. 2006). More recently, CSS52A has also been cloned from *Arabidopsis* (Tarayre et al. 2004), where it was found to be present as two distinct genes (Fulop et al. 2005). The analysis of CSS52A during root nodule development indicated that this protein plays a key role in the switch from a mitotic to an endoreduplication cell cycle as cells differentiate, rather than in sustaining or enhancing endoreduplication in differentiated cells.

The best-known function of the APC/C is to promote the separation of sister chromatids, by activating separase activity at metaphase. Since endoreduplicated chromosomes are often polytenic, it will be interesting to learn whether this aspect of APC/C activity is affected in endoreduplicating cells. Recent investigation in human cells has shown that oscillations in APC/C activity are part of a redundant pathway that regulates pre-RC formation and rereplication (Machida and Dutta 2007), primarily by the degradation of CycA and geminin in late M-phase and in G1. Plants do not have geminin orthologs, but low CycA levels are required for endoreduplication in some *Arabidopsis* cell types (Imai et al. 2006; Yoshizumi et al. 2006). Future investigation undoubtedly will focus on the role that APC/C may play in CycA downregulation in endoreduplicating plant cells.

2.4
The Role of CDKs

CDKs are the main workhorses of the cell cycle, and not surprisingly they play important roles in endoreduplication. CDKs are regulated at different levels, such as by the time and specificity of their interaction with cyclin partners (which in turn involves factors controlling the availability of cyclins, such as cyclin synthesis, degradation, and compartmentalization), by phosphorylation and dephosphorylation events, and by interaction with specific inhibitors.

A consensus view has emerged from studies of animals and plants, according to which the endoreduplication cell cycle involves sustained or upregulated S-phase CDKs and downregulation of M-phase CDKs (Sauer et al. 1995; Edgar and Orr-Weaver 2001; Larkins et al. 2001). However, specific CDKs may have distinct roles in regulating endoreduplication in different species, or even different cell types in the same organism. For example, in mammalian trophoblasts, endoreduplication involves loss of CycB expression, and a coincidental increase in CycA- and CycE-associated CDK activity that oscillates prior to and during S-phase (MacAuley et al. 1998; Hattori et al. 2000). In *Drosophila* embryos, CycA and B are present at constant levels, and cdc2 activity is constitutively high. In this system, oscillations of CycE activity and downregulation of the APC/C trigger endoreduplication. However, these differences may reflect the specific modes by which endoreduplication occurs in the two examples above, which differ significantly from the reiterated rounds of gap and complete DNA synthesis phases that define the prototypical endoreduplication cell cycle discussed in this chapter. In fact, endoreduplicating trophoblasts still display vestigial aspects of mitosis, whereas endoreduplication in *Drosophila* generally involves incomplete replication of heterochromatin (Edgar and Orr-Weaver 2001). That not withstanding, many studies have implicated CDKs in preventing rereplication in baker's yeast, fission yeast, fly, and frog. As already mentioned, in *Saccharomyces cerevisiae* CDK activity represses rereplication by three overlapping mechanisms, and rereplication is stimulated by mutations affecting CDC2/CycB kinase (Nguyen et al. 2001).

It has also been shown in *Arabidopsis* that CycA2;3 negatively regulates endoreduplication and ploidy levels (Imai et al. 2006). The involvement of CYCA2 in repressing endoreduplication was further confirmed by the observation that transcription of all *CYCA2* family members was repressed in the *increased level of polyploidy1-1D* mutant, in which endoreduplication is stimulated (Yoshizumi et al. 2006). Although CYCA2 inhibits endoreduplication, it does not seem to play a role in the mitosis/endoreduplication switch. However, the SIAMESE (SIM) protein, which interacts with and inhibits CycD and CDKA;1, appears to control the balance (switch) between mitotic and endoreduplication cell cycles (Churchman et al. 2006).

CycD expression is associated with the mitotic cell cycle but not with the endoreduplication cell cycle. Overexpression of *CYCD3;1* has an inhibitory effect on endoreduplication and cell differentiation in *Arabidopsis* leaves, suggesting that aspects of G1/S-phase regulation are distinct in mitotic and endoreduplicating cells (Dewitte et al. 2003). Likewise, trichome endoreduplication was inhibited by overexpression of *CYCB1;2*, but trichome size was maintained through compensatory stimulation of cell proliferation (Schnittger et al. 2002b). A similar result was obtained upon downregulation of the mitotic cyclin inhibitor, *CCS52A*, in *Medicago sativa* (Cebolla et al. 1999). In contrast, when the CDK inhibitor *ICK1/KRP1* was overexpressed in trichomes, it caused an inhibition of endoreduplication only (i.e., no effect on cell number) with trichome downsizing and a simplified morphology (Schnittger et al. 2003). Additional evidence involving the reduction of CDK activity by overexpressing certain CKIs, such as *NtKIS1a* (Jasinski et al. 2002) and *KRP2* (De Veylder et al. 2001), in *Arabidopsis* confirmed the important role played by CDKs in endoreduplication. Consistent with these findings, downregulation of CDKA1-associated kinase in developing maize endosperm resulted in an inhibition of the endoreduplication cell cycle (Leiva-Neto et al. 2004).

3
Endoreduplication Model Systems

3.1
Maize Endosperm

Maize endosperm is a triploid tissue that originates from the fusion of one sperm nucleus with two polar nuclei in the female gametophyte. Endosperm development progresses initially through distinct stages of acytokinetic mitosis, cellularization, and mitosis coupled to cell division. From around 8–10 days after pollination (DAP), central endosperm cells gradually switch to an endoreduplication cell cycle. By 20 DAP, ploidy levels reach and can exceed 96–192C, indicating up to six or more endoreduplication cycles (Larkins et al. 2001; Sabelli et al. 2005b) (Fig. 1). Considerable attention has been paid to the regulation of the endoreduplication phase of maize endosperm development for several reasons. First, it provides an excellent system in which to study how plant cells transition from a mitotic to an endoreduplication cell cycle. Second, this phase of endosperm development is correlated with dramatic enlargements in nuclear and cell size, the accumulation of storage compounds such as starch and storage proteins, and the rapid growth of the caryopsis. Third, endoreduplication also occurs during endosperm development in rice, wheat, and sorghum, and thus the study of endoreduplication in maize endosperm is likely to provide a blueprint that can be applied to all cereal crops.

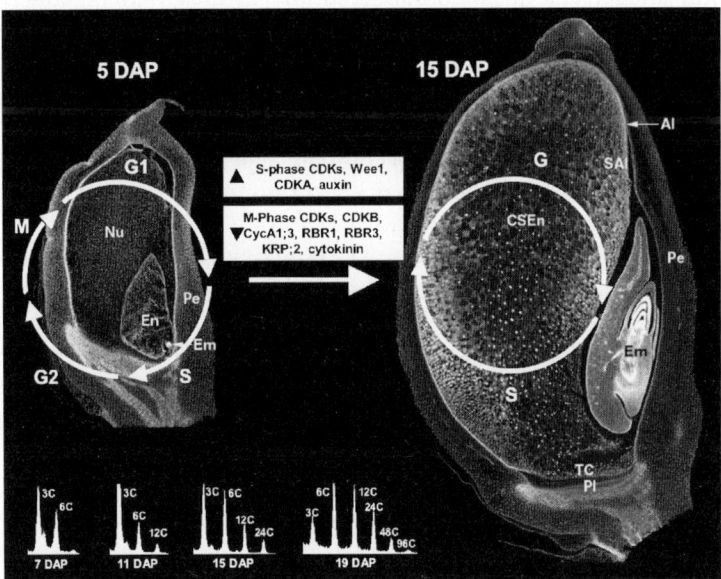

Fig. 1 Endoreduplication during maize endosperm development. Schematic diagrams of the mitotic and endoreduplication cell cycles are superimposed on micrographs of kernels at 5 and 15 days after pollination (DAP), respectively. Nuclei are stained with propidium iodide. Nuclear, cell, and endosperm sizes are clearly correlated with endoreduplication. Factors that appear up- (▲) or down- (▼) regulated as cells switch from a mitotic to an endoreduplication cell cycle are listed in boxes. The *bottom panels* show nuclear ploidy classes at different stages of endosperm development. Nu, nucellus; Em, embryo; En, endosperm; Pe, pericarp; SAl, subaleurone layer; CSEn, central starchy endosperm; Pl, placentochalaza; TC, transfer cells

Recent work and previous findings indicate that endoreduplicated chromosomes in maize endosperm are completely replicated with a loose polytene structure (Kowles and Phillips 1985; Kowles and Phillips 1988; Kowles et al. 1990) in which they appear to be tightly associated at the centromere and, unexpectedly, at the highly heterochromatic knob regions (Bauer and Birchler 2006). This observation suggested that during endoreduplication specific chromatin binding proteins, perhaps of the cohesion (cohesin) class, may be in place to ensure chromosome association with a higher concentration at centromeres and knobs. A polytene structure of chromosomes is likely to involve extensive chromatin reorganization (Zhao and Grafi 2000), although this aspect of endosperm development has been poorly investigated. However, this view is supported by the observation that interploidy crosses, which are known to perturb caryopsis development and the endoreduplication pattern, result in dramatic alteration of chromatin organization (Bauer and Birchler 2006). Chromatin reorganization has also been observed during the endoreduplication phase in the endosperm of durum wheat, but perhaps surprisingly, in this system endoreduplication seems to entail increased chromatin condensation, potentially resulting in methylation-induced gene silencing (Polizzi et al. 1998).

During the early endoreduplication period of maize endosperm development, DNA replication appears to proceed extremely rapidly. Each round of DNA synthesis is estimated to take 22–24 h, with the last round requiring an average DNA synthesis of 916 μm/second/nucleus, which is equivalent to the incorporation of 2.75×10^6 bases pairs/second/nucleus (Kowles and Phillips 1988).

Early experiments on the analysis of CDK activity typically associated with S-phase or M-phase revealed a peak in M-phase-associated CDK activity at 10 DAP, coincident with a peak in the frequency of mitotic figures (Kowles and Phillips 1988), and a peak in S-phase-associated CDK activity at 16 DAP, which is when most endosperm cells are engaged in endoreduplication (Grafi and Larkins 1995). Thus, similarly to what was observed in animal systems, a switch from a mitotic to an endoreduplication cell cycle in developing endosperm appears to involve a downregulation of M-phase-associated CDK and a sustained or upregulated S-phase CDK activity. In agreement with this view, the characterization of A- and B-type CDKs has shown that whereas CDKA protein expression is relatively constant in developing endosperm, expression of the M-phase-associated CDKB decreases in endoreduplicating endosperm (Dante 2005).

CDKA, on the other hand, plays a key role in endoreduplication. CDKA activity was specifically downregulated in transgenic endosperms by overexpressing a dominant-negative mutant form of CDKA (CDKA-DN), in which an Asp146Asn mutation abolished kinase activity, but did not interfere with cyclin binding (Leiva-Neto et al. 2004). Because CDKA-DN expression was driven by the strong 27-kD γ-zein promoter throughout the starchy endosperm, CDKA-DN out-competed the endogenous CDKA for cyclin binding, thereby effectively reducing S-phase kinase activity and inhibiting the DNA replication process. As a result, there was a 50% reduction in endoreduplication during transgenic endosperm development, with virtually no cells exceeding three cycles of endoreduplication and a maximum ploidy of 24C at 18 DAP. Nuclear size in CDKA-DN endosperm was clearly reduced compared to control, nontransgenic nuclei, especially within the larger cells in the central area of the starchy endosperm, but cell size was virtually unaffected. Only a minor decrease was measured in the transcription of genes encoding starch biosynthetic enzymes and zein storage proteins in CDKA-DN endosperms. An alternative approach involving overexpressing CDKA was inconsequential, suggesting that this kinase, thought important for endoreduplication, is not rate-limiting for the process. Investigation of the role of CDKA in maize endosperm development suggested that endoreduplication and nuclear size can be uncoupled from cell size and gene expression, thus challenging the long-standing hypothesis that endoreduplication drives cell growth and supports higher levels of gene expression.

Several A-, B-, and D-type cyclins have been characterized and found to be expressed in maize endosperm (Sun et al. 1999b; Dante 2005). CycA1;3-associated kinase was most active during the mitotic stage and then declined as the endosperm progressed into the endoreduplication stage, suggesting that CycA1;3 is more likely involved in the mitotic than the endoreduplication cell cycle. However, CycB1;3, D5;1, and D2;1 showed a peak in kinase activity at 11 DAP, a stage in which the endosperm is active in both the mitotic and endoreduplication cell cycles. The sustained (although reduced) kinase activity associated with these three cyclins throughout endosperm development suggested that they may be involved in the regulation of both types of cell cycle. This is not surprising for CycD5;1 and D2;1, as CycDs are known to be involved in S-phase entry. However, the finding that CycB1;3-associated kinase was active after mitosis ceased was unexpected and suggested a novel role for CycB1;3 in endoreduplication. Although an important role for CDKA in regulating endoreduplication in the endosperm has been established, the cyclin partner has not been identified and remains a challenge for future experiments.

Among the factors known to inhibit CDK activity, a maize Wee1 (Sun et al. 1999a) and two CKIs, KRP;1 and KRP;2 (Coelho et al. 2005) horthologs, have been investigated in developing maize endosperm. Ectopic expression of maize Wee1 in fission yeast inhibited cell division and resulted in elongated cells, consistent with an inhibitory role in CDK activity. Although specific data are lacking, the peak in Wee1 expression during endoreduplication suggests that it might contribute to inhibit cell division in endoreduplicating cells. Both *KRP;1* and *KRP;2* are expressed in the endosperm and inhibit in vitro CycA1;3- and D5;1-associated kinase activities, both of which are considered to be involved in S-phase. However, KRP;1 protein appeared constitutively expressed throughout endosperm development, whereas KRP;2 expression was downregulated after the onset of endoreduplication, suggesting different roles for these two CKIs in the mitotic and endoreduplication cycles (Coelho et al. 2005). It is possible that KRP;1 specifically plays an important role in providing the oscillation in CDK activity required to sustain endoreduplication in maize endosperm by inhibiting S-phase CDK activity after replication initiation, similarly to what has been demonstrated for the *Drosophila* endoreduplication cycle (Weiss et al. 1998).

In our laboratory, intense research is focused on the role that the RBR/E2F pathway plays in cell cycle regulation during endosperm development. Maize possesses at least two different RBR types of genes, *RBR1* (Grafi et al. 1996; Xie et al. 1996) and *RBR3* (Sabelli et al. 2005a; Sabelli and Larkins 2006). A third gene, *RBR2* (Ach et al. 1997a), belongs to the RBR1 type of sequences and encodes a protein over 90% identical to RBR1, and it is not clear whether it has a distinct function from that of RBR1. RBR3, however, is clearly distinct from RBR1 and RBR2 both in structural and functional terms. In fact, it shares only 50% sequence identity with RBR1 and is repressed by RBR1, suggesting that total pocket protein activity is determined by an intrinsic regulatory loop within the *RBR* gene family. Evidence for such a configuration of the RBR protein family has been reported in plants for only the grass family within monocots, and the significance of this finding is being actively investigated (Sabelli et al. 2005a; Sabelli and Larkins 2006). Although both RBR1 and RBR3 are expressed in developing endosperm, RBR1 expression is constitutive and actually increases during late endoreduplication stages, whereas RBR3 expression is dramatically downregulated after the onset of endoreduplication, indicating that it may be involved specifically in regulating the mitotic cell cycle but is largely dispensable (or even counterproductive) in endoreduplicating cells. The upregulation of RBR1 expression during endoreduplication might indicate a role in regulating the alternation of G- and S-phases during endoreduplication, or an involvement in inducing programmed cell death. Early results suggested that RBR1 was hyperphosphorylated (and presumably inactivated) during endoreduplication (Grafi et al. 1996), and thus it is not clear whether the total activity of this protein increases, decreases, or remains constant during endosperm endoreduplication. It is expected that forward genetic experiments aimed at down- or upregulating *RBR1* and *RBR3* expression in transgenic endosperms will provide an answer to this question.

3.2
Arabidopsis Leaf Trichome

Trichomes are unicellular, epidermal structures and endoreduplication is associated with their differentiation (Hulskamp 2004). Fully developed *Arabidopsis* trichomes average a 32C ploidy level (indicative of four rounds of endoreduplication), and typically have three or four branches. Trichome cell endoreduplication correlates with its branching pattern and size, with highly endoreduplicated trichomes being more branched and larger (Hulskamp et al. 1999) (Fig. 2). In *CDC6*- and *CDT1*-overexpressing plants, endoreduplication is enhanced in trichome-forming cells, which eventually display a more highly branched pattern (Castellano et al. 2004). RBR1 inactivation by WDV RepA protein also resulted in trichomes with increased nuclear size and larger number of branches (Desvoyes et al. 2006). The closely related *CYCB1;1* and *CYCB1;2* genes are not normally expressed in trichomes. However, ectopic expression of *CYCB1;2*, but not *CYCB1;1*, in trichomes promoted a switch to cell division resulting in a multicellular phenotype (Schnittger et al. 2002b). Interestingly, although trichome cells appeared to switch from an endoreduplication to a full mitotic cycle upon *CYCB1;2* expression, the total number of cell cycles did not change during trichome development, indicating the presence of upstream constraints on the number of cell cycle rounds. The multicellular phenotype became enhanced when *CYCB1;2* was expressed in a *siamese* mutant background (which itself results in multicellular trichomes), suggesting that *SIM* interacts genetically with *CYCB1;2* and,

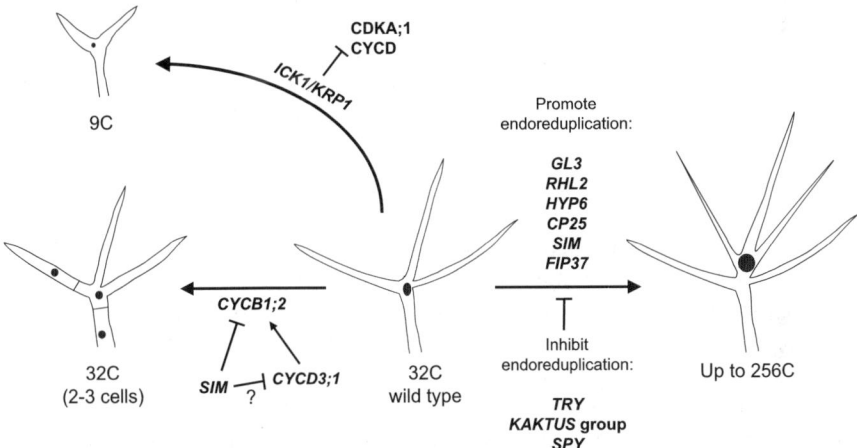

Fig. 2 Simplified diagram illustrating the relationship between endoreduplication and trichome cell size and differentiation in *Arabidopsis*. Wild-type trichomes are unicellular, have three to four branches and an average DNA content of 32C. In certain mutants endoreduplication is stimulated up to a DNA content of 256C, resulting in large cells with a complex branched morphology. These mutants identify negative regulatory genes of endoreduplication. In other mutants endoreduplication is inhibited, indicating that the corresponding genes play a positive role. In addition, expression of certain cell cycle genes can promote or repress endoreduplication, the latter generally resulting in smaller trichomes with a simplified morphology (as in the case of *ICK1/KPR1*). Expression of *CYCB1;2* and *CYCD3;1* in trichomes can induce the mitotic program, resulting in multicellular structures. The *SIAMESE* (*SIM*) gene appears to stimulate endoreduplication by repressing *CYCB1;2* and *CYCD3;1*

most likely, represses *CYCB1;2* expression in wild-type, endoreduplicating trichomes. Similarly to *CYCB1;2*, *CYCD3;1* is not expressed during normal trichome development. Its ectopic expression, however, led to multicellular trichomes, which initiated more frequently in the leaf surface than in wild-type or in *CYCB1;2* ectopically expressing plants.

In addition, the multicellular phenotype was greatly enhanced compared to *CYCB1;2*-expressing trichomes (which averaged two to three cells) (Schnittger et al. 2002b). Further experiments suggested that *CYCD3;1* induced expression of *CYCB1;2*, and that *SIM* restricts *CYCD3;1* function either in parallel or downstream. When the CDK inhibitor *ICK1/KRP1* was overexpressed in trichomes, it caused inhibition of endoreduplication (an average of two endoreduplication rounds and ploidy = 9C), with trichome downsizing and a simplified morphology with fewer branches (Schnittger et al. 2003). Co-expression experiments indicated that *ICK1/KRP1* was most likely inhibiting the G1/S-phase transition by repressing *CDKA;1*.

Trichome endoreduplication appears controlled by immunophilins. Overexpression of *FIP37*, which interacts with the immunophilin FKBP12, was shown to greatly stimulate endoreduplication specifically in trichomes, with ploidies up to 256C and enhanced trichome size and branching (Vespa et al. 2004). However, while a great deal is known about the functions of immunophilins in animals and particularly in human immune response, plant immunophilins are largely uncharacterized, and it is currently unclear how FIP37 affects the endoreduplication cycle in trichomes.

4
Factors That Affect Endoreduplication

4.1
Genetics

Genetic screens in *Arabidopsis* have identified several mutations that impact endoreduplication. Mutations affecting the switch from mitosis to endoreduplication or the extent of endoreduplication have been best characterized during the study of trichome development. The results indicate that at least five distinct pathways, such as patterning, DNA catenation, hormone response, ubiquitin-mediated proteolysis, and cell death impact the endoreduplication cell cycle in this system (Hulskamp 2004). *SIM* appears to function both as a repressor of mitosis and a stimulator of endoreduplication during the development of endoreduplicating trichomes (Walker et al. 2000). *SIM* interacts genetically with *CYCB1;2* as shown by enhanced multicellular phenotype when *CYCB1;2* was expressed in a *sim* mutant background (Schnittger et al. 2002b). *SIM* encodes a plant-specific small protein containing a motif for cyclin binding and a motif found in CKIs, similar to those that interact with both D-type cyclins and CDKA;1. Plants overexpressing *SIM* have enlarged epidermal cells with an increased ploidy resulting from enhanced endoreduplication. It was proposed that they possess a CDK inhibitory activity with a key function in the mitosis-to-endoreplication transition (Churchman et al. 2006).

Mutations at the *GLABRA3* (*GL3*) locus also reduce endoreduplication, trichome branching, and trichome cell size, whereas mutations at the *TRYPTICON* (*TRY*) and *KAKTUS* (*KAK*) loci increase endoreduplication and trichome cell size (Hulskamp et al. 1994). *KAK* encodes a protein similar to the HECT domain proteins involved in ubiquitin-mediated protein degradation (El Refy et al. 2004). In addition, a CUE protein variant that controls the endocycle leading to hypocotyl elongation was recently identified genetically (Tsumoto et al. 2006). In the *increased level of polyploidy1-1D* (*ilp1-1D*) mutant, hypocotyls showed increased polyploidy under both light and dark conditions. *ILP1* encodes a protein similar to the mammalian GC binding factor, which functions as a transcriptional repressor for all members of the *CYCA2* family. Thus ILP1 regulates endoreduplication through control of *CYCA2* (Yoshizumi et al. 2006). When the *Arabidopsis STEROL METHYLTRANSFERASE 2* gene function is mutated, such as in the *frill1* mutant, sterol composition is altered and this results in ectopic endoreduplication in petal cells (which normally do not undergo endoreduplication) and in rosette leaves (Hase et al. 2005).

Mutations in the *Arabidopsis* subunits of chromatin assembly factor CAF-1 revealed a role for this complex in limiting DNA endoreduplication (Exner et al. 2006). Additional mutations affecting endoreduplication in *Arabidopsis* have been reviewed elsewhere (Sugimoto-Shirasu and Roberts 2003).

In maize, analysis of defective kernel (*dek*) mutants, in which both the mitotic and endoreduplication cell cycle are inhibited, highlighted an elevated degree of coordination between the mitotic and endoreduplication phases of endosperm development (Kowles et al. 1992). However, mitosis and endoreduplication are also genetically uncoupled to some extent in this system, as shown by the phenotype of the *miniature1* (*mn1*) mutant (Vilhar et al. 2002), in which the amount of energy available for endosperm development is substantially decreased. In this mutant the activity of the cell wall INCW2 invertase, which is involved in providing hexose sugars to the developing endosperm, is compromised, resulting in reduced mitotic activity and cell expansion, but leaving endoreduplication largely unaffected. This suggests that endoreduplication is somewhat more resistant to certain perturbing factors than the mitotic cell cycle.

4.2
Environment and Phytohormones

Endoreduplication is affected by environmental factors and phytohormones and this subject has been recently reviewed (Barow 2006). For example, it has been shown that light has a negative effect on endoreduplication in *Arabidopsis* hypocotyls (Gendreau et al. 1998; Gendreau et al. 1999). Endoreduplication in maize endosperm is inhibited by environmental stresses, such as water deficit (Artlip et al. 1995; Setter and Flannigan 2001) and heat stress (Engelen-Eigles et al. 2000). However, the reduction in endoreduplication is more a consequence of perturbation during early mitotic development, rather than a direct result of the inhibition of the endoreduplication cell cycle per se, reinforcing the idea that a tight relationship couples the mitotic and endoreduplication phases of development mentioned above. A direct stimulatory effect of certain symbionts and parasites on endoreduplication in plant cells has been well established in the case of nitrogen-fixing root nodules of legumes infected with the bacterium *Sinorhizobium meliloti*, and in root-knot galls induced by the endoparasitic root-knot nematode *Meloidogyne incognita* (Kondorosi and Kondorosi 2004) (see Sect. 2.3).

Various plant hormones impact endoreduplication differently. For example, ethylene stimulates endoreduplication in both *Arabidopsis* (Joubes and Chevalier 2000) and cucumber hypocotyls (Dan et al. 2003). It is well established that the transition from a mitotic to an endoreduplication cell cycle in maize endosperm involves a sharp decline in the cytokinin/auxin ratio (Lur and Setter 1993; Sabelli et al. 2005b). Auxin is apparently involved in the induction of endoreduplication, but does not seem limiting for its continuation. Application of chemicals that inhibit auxin-triggered effects suggested that auxin has a rather indirect effect in stimulating endoreduplication (Lur and Setter 1993). In spite of a large body of circumstantial evidence for endosperm cell cycle control by both environmental and hormonal factors, information about the key elements linking them to the cell cycle machinery is largely missing and this is a challenge for future research.

4.3
Parent-of-Origin Effects and Epigenetic Control

It is well known that endoreduplication in maize endosperm is under maternal control, in that the endoreduplication pattern in the offspring of a cross between two inbred lines generally resembles that of the maternal parent (Kowles et al. 1997; Dilkes et al. 2002). It is possible that differential gene dosage contributes to this "maternal effect", as the genetic makeup of the endosperm involves two maternal copies and one paternal copy of the genome. An alteration to this 2 : 1 ratio is deleterious for the development of the endosperm. A maternal excess causes a shortened mitotic phase and a precocious endoreduplication period (but does not affect the extent of endoreduplication). Conversely, excess of the paternal genome delays endoreduplication through an apparent extended mitotic hiatus prior to the switch to endoreduplication, and it also inhibits DNA amplification in all endoreduplicating cells (Leblanc et al. 2002).

Genomic imprinting, which involves the differential expression of maternally and paternally derived alleles, may be involved in the maternal effect on endosperm development in general, and endoreduplication in particular (Guitton and Berger 2005). In *Arabidopsis*, several Polycomb group genes that are involved in chromatin remodeling and epigenetic regulation of gene expression, including *FIS*, *FIE*, and *MEA*, are imprinted during early seed development (Sorensen et al. 2001). After fertilization, only the maternal alleles of these genes are expressed in the endosperm, while the paternal alleles are silenced. Maize *FIE* homologs have been identified (Danilevskaya et al. 2003), but it is not known if and how they might be involved in endoreduplication. A potential pathway linking imprinting, Polycomb group genes, and endoreduplication may involve histone modification around replication-origin bound ORC (Machida et al. 2005), chromatin remodeling complexes, and interaction between RBR and Polycomb subunits (Ach et al. 1997b; Mosquna et al. 2004).

5
Functions of Endoreduplication

It has long been recognized that endoreduplication is often correlated with increased nuclear size, cell size, cell differentiation, and metabolic activity. Therefore it has been tempting to propose a general role for endoreduplication in driving growth and development. Indeed, there is strong evidence supporting this view (for an in-depth discussion on cell size control by endoreduplication, see Sugimoto-Shirasu and Roberts 2003). For example, a clear relationship between ploidy, nuclear size, and cell size was observed in floral apices of *Datura stramonium* (Satina and Blakeslee 1941), *Arabidopsis* epidermis (Melaragno et al. 1993), and in petal development in cabbage (Kudo and Kimura 2002). Endoreduplication is the first visible cellular event in trichome development prior to cell growth and branching (Hulskamp et al. 1994) and decreased ploidy content in transgenic trichomes expressing *ICK1/KRP1* results in reduced cell size (Schnittger et al. 2003). In leaf epidermis of durum wheat it promotes cell expansion and precedes cell growth (Cionini et al. 1983). Endoreduplication in the pericarp could contribute substantially to the growth of the tomato fruit (Cheniclet et al. 2005), and it is correlated with nuclear and cell size as well as starch and storage protein accumulation during endosperm development in cereals such as maize (Larkins et al. 2001) and sorghum (Kladnik et al. 2006). These (and other) observations are consistent with the "karyoplasmic ratio" theory, according to which nuclear size controls cell size through regulation of cytoplasm volume. For example, in *Arabidopsis gl3* mutant trichomes both DNA content and cell size are reduced, but the nuclear size to cell size ratio is similar to wild type.

Indeed, a convincing demonstration that ploidy directly regulates, and it is not merely correlated with, body size comes from a recent study of *Caenorhabtis elegans* (Lozano et al. 2006). This investigation also identified CycE activity as a likely effector of endoreduplication-dependent growth.

Early observations suggested that endoreduplication preferentially occurs in species with a small genome, and it was proposed that it can be viewed as an evolutionary strategy compensating for the lack of phylogenetic increase in nuclear DNA (needed to support the development and metabolism of complex organisms) (Nagl 1976; Derocher et al. 1990). However, in spite of the overwhelming evidence linking endoreduplication to increased cell size, there are instances in which the two processes can be uncoupled. For example, misexpression of CDK inhibitors and dominant-negative CDKs in *Arapidopsis* (Hemerly et al. 1993; De Veylder et al. 2001; Schnittger et al. 2003), tobacco (Jasinski et al. 2002), and maize (Leiva-Neto et al. 2004) have shown that changes in ploidy levels may not be reflected by straightforward changes in cell size. Indeed, organ and body size are often maintained through compensatory mechanisms involving changes in cell size that counteract ploidy and nuclear changes. In this respect, endoreduplication may be viewed as one important mechanism by which cell size could be attained within the developmental constraints set by higher-order control at the organ or body level. Interestingly, endoreduplication stops well before the differentiation of trichome or epidermal cells that are located on the surface of organs, arguing that it does play a role in presetting cell size. However, the size of cells that are deeply embedded in a tissue may be primarily determined by the physical and developmental constraints from neighboring cells. Thus, it is conceivable that endoreduplication allows cell size increase, which may or may not occur depending on the context (Traas et al. 1998). In this context it is important to keep in mind that plant cells are vacuolated and that there is a notable distinction between cell growth, as measured by cytoplasmic macromolecular mass, and cell expansion, which reflects cell volume increase mostly through vacuolation (Sugimoto-Shirasu and Roberts 2003). Thus, regulation of vacuolation can contribute to cell enlargement in a way that may be largely independent of endoreduplication.

How does endoreduplication impact cell growth, as defined above? One long-standing hypothesis involves an increase in gene expression in endoreduplicated cells. A twofold mechanism could be envisioned here. First, endoreduplication generates more gene templates that could support increased transcription (D'Amato 1984). Second, endoreduplicated cells often contain more (Fankhauser and Humphrey 1943) or larger (Kowles and Phillips 1988) nucleoli, potentially leading to higher rates of net protein synthesis through enhanced RNA synthesis and ribosome biogenesis. Regulation of chromatin conformation may play an important role in enhancing gene expression during endoreduplication. During the mitotic cell cycle, chromatin condenses in preparation of mitosis and this temporarily inhibits gene expression. In contrast, in endoreduplicating cells chromatin is permanently decondensed, which could favor increased gene expression by sustaining RNA synthesis. However, this hypothesis is mainly supported by correlative studies rather than by direct evidence, and recent experiments have shown that a decrease in ploidy in maize endosperm nuclei does not seem to impact transcriptional activity (Leiva-Neto et al. 2004). However, only a small set of genes was analyzed in detail, and it is not known the effect of reduced endoreduplication on a genome-wide basis. Studies on polyploid isogenic yeast strains have indicated that expression of certain genes is induced or repressed in a ploidy-dependent fashion, possibly by ploidy-related mechanisms such as transient homologous pairing (Galitski et al. 1999). Thus, endoreduplication may represent a mechanism for modulating expression of a selected pool of genes rather than indiscriminately increasing gene expression across the entire genome.

Other potential functions of endoreduplication involve a role in protecting the genome from recessive mutations, adapting to salt stress (Ceccarelli et al. 2006), or marking cells for programmed cell death (Bennett 1973; Kowles and Phillips 1988). The latter may relate to the ability of polyploid endosperm cells to provide a large pool of nucleosides and phosphorus for the development and germination of the embryo.

Is endoreduplication tied to cell differentiation? A large body of evidence suggests that endoreduplication is associated with differentiation in many cases (Edgar and Orr-Weaver 2001). One of the best examples comes from *Medicago truncatula*, in which the differentiation of nodules containing nitrogen-fixing cells in symbiotic bacteroids requires the APC/C activator *CCS52A*. Downregulation of *CCS52A* in transgenic plants resulted in lower ploidy, reduced cell size, and affected the maturation of symbiotic cells, indicating that large, endoreduplicated cells are an essential component of nodule development (Vinardell et al. 2003). Likewise, terminally differentiated cells in the maize starchy endosperm are highly endoreduplicated (Larkins et al. 2001). Although there is often a clear correlation between endoreduplication and cell differentiation in endoreduplicating tissues and organs, cells can undergo terminal differentiation without endoreduplicating, which indicates that endoreduplication and cell differentiation are not necessarily associated (Evans and Van't Hof 1975). Thus, the requirement for endoreduplication during the differentiation of certain cell types depends largely on the biological and physiological contexts.

6
Perspectives

Investigators have been aware of the occurrence of endopolyploid cells for a long time, yet until recently most studies on endoreduplication have been primarily correlative. However, the ongoing characterization of many plant cell cycle genes, coupled to the identification of mutants affecting endoreduplication patterns and the ability to perturb gene expression in transgenic plants, has resulted in substantial progress being made in recent years. Although several fundamental mechanisms controlling the endoreduplication cell cycle are evolutionarily conserved, the identities of the specific factors and pathways involved vary considerably, particularly in relation to the biological context in which they operate. This, of course, is an indication of the remarkable plasticity of cell cycle regulation in plants, which suits well the necessity of sessile organisms to respond to rapidly changing environmental conditions. Many studies have been based on increasing or decreasing expression of certain cell cycle regulators constitutively, but it is becoming increasingly apparent that in the future investigators will have to design more precise approaches to control gene expression in specific cell types and in a timely manner. Fortunately, the necessary molecular tools to do this are already available or are being developed, and there is little doubt that exciting advancements will be made in unraveling the endoreduplication cell cycle in the near future.

References

Ach RA, Durfee T, Miller AB, Taranto P, HanleyBowdoin L, Zambryski PC, Gruissem W (1997a) RRB1 and RRB2 encode maize retinoblastoma-related proteins that interact with a plant D-type cyclin and geminivirus replication protein. Mol Cell Biol 17:5077–5086

Ach RA, Taranto P, Gruissem W (1997b) A conserved family of WD-40 proteins binds to the retinoblastoma protein in both plants and animals. Plant Cell 9:1595–1606

Artlip TS, Madison JT, Setter TL (1995) Water deficit in developing endosperm of maize: cell division and nuclear DNA endoreduplication. Plant Cell Environ 18:1034–1040

Barow M (2006) Endopolyploidy in seed plants. Bioessays 28:271–281

Bauer MJ, Birchler JA (2006) Organization of endoreduplicated chromosomes in the endosperm of Zea mays L. Chromosoma 115:383–394

Bennett MD (1973) Nuclear characters in plants. Brookhaven Symp Biol 25:344–366

Blow JJ, Dutta A (2005) Preventing re-replication of chromosomal DNA. Nat Rev Mol Cell Biol 6:476–486

Bosco G, Du W, Orr-Weaver TL (2001) DNA replication control through interaction of E2F-RB and the origin recognition complex. Nat Cell Biol 3:289–295

Carvalheira GMG (2000) Plant polytene chromosomes. Genet Mol Biol 23:1043–1050

Castellano MM, del Pozo JC, Ramirez-Parra E, Brown S, Gutierrez C (2001) Expression and stability of Arabidopsis CDC6 are associated with endoreplication. Plant Cell 13:2671–2686

Castellano MM, Boniotti MB, Caro E, Schnittger A, Gutierrez C (2004) DNA replication licensing affects cell proliferation or endoreplication in a cell type-specific manner. Plant Cell 16:2380–2393

Cebolla A, Vinardell JM, Kiss E, Olah B, Roudier F, Kondorosi A, Kondorosi E (1999) The mitotic inhibitor ccs52 is required for endoreduplication and ploidy-dependent cell enlargement in plants. EMBO J 18:4476–4484

Ceccarelli M, Santantonio E, Marmottini F, Amzallag GN, Cionini PG (2006) Chromosome endoreduplication as a factor of salt adaptation in Sorghum bicolor. Protoplasma 227:113–118

Chen W-H (2007) G1/S Transition ant the Rb-E2F Pathway. Springer, Heidelberg
Cheniclet C, Rong WY, Causse M, Frangne N, Bolling L, Carde JP, Renaudin JP (2005) Cell expansion and endoreduplication show a large genetic variability in pericarp and contribute strongly to tomato fruit growth. Plant Physiol 139:1984–1994
Churchman ML, Brown ML, Kato N, Kirik V, Hulskamp M, Inze D, De Veylder L, Walker JD, Zheng Z, Oppenheimer DG, Gwin T, Churchman J, Larkin JC (2006) SIAMESE, a plant-specific cell cycle regulator, controls endoreplication onset in Arabidopsis thaliana. Plant Cell 18:3145–3157
Cionini PG, Cavallini A, Baroncelli S, Lercari B, D'Amato F (1983) Diploidy and chromosome endoreduplication in the development of epidermal cell lines in the first foliage leaf of durum wheat (Triticum durum Desf.). Protoplasma 118:36–43
Cobrinik D (2005) Pocket proteins and cell cycle control. Oncogene 24:2796–2809
Coelho CM, Dante RA, Sabelli PA, Sun YJ, Dilkes BP, Gordon-Kamm WJ, Larkins BA (2005) Cyclin-dependent kinase inhibitors in maize endosperm and their potential role in endoreduplication. Plant Physiol 138:2323–2336
D'Amato F (1984) Role of polyploidy in reproductive organs and tissues. In: Jorhri BM (ed) Embryology of angiosperms. Springer, New York, pp 523–566
Dan H, Imaseki H, Wasteneys GO, Kazama H (2003) Ethylene stimulates endoreduplication but inhibits cytokinesis in cucumber hypocotyl epidermis. Plant Physiol 133:1726–1731
Danilevskaya ON, Hermon P, Hantke S, Muszynski MG, Kollipara K, Ananiev EV (2003) Duplicated fie genes in maize: expression pattern and imprinting suggest distinct functions. Plant Cell 15:425–438
Dante RA (2005) Characterization of cyclin-dependent kinases and their expression in developing maize endosperm. Department of Plant Sciences, University of Arizona, Tucson, AZ
de Jager SM, Menges M, Bauer UM, Murray JAH (2001) Arabidopsis E2F1 binds a sequence present in the promoter of S-phase-regulated gene AtCDC6 and is a member of a multigene family with differential activities. Plant Mol Biol 47:555–568
De Veylder L, Beeckman T, Beemster GTS, Krols L, Terras P, Landrieu I, Van der Schueren E, Maes S, Naudts M, Inze D (2001) Functional analysis of cyclin-dependent kinase inhibitors of Arabidopsis. Plant Cell 13:1653–1667
De Veylder L, Beeckman T, Beemster GTS, Engler JD, Ormenese S, Maes S, Naudts M, Van der Schueren E, Jacqmard A, Engler G, Inze D (2002) Control of proliferation, endoreduplication and differentiation by the Arabidopsis E2Fa-DPa transcription factor. EMBO J 21:1360–1368
del Pozo JC, Diaz-Trivino S, Cisneros N, Gutierrez C (2006) The balance between cell division and endoreplication depends on E2FC-DPB, transcription factors regulated by the ubiquitin-SCFSKP2A pathway in Arabidopsis. Plant Cell 18:2224–2235
DePamphilis ML, Blow JJ, Ghosh S, Saha T, Noguchi K, Vassilev A (2006) Regulating the licensing of DNA replication origins in metazoa. Curr Opin Cell Biol 18:231–239
Derocher EJ, Harkins KR, Galbraith DW, Bohnert HJ (1990) Developmentally regulated systemic endopolyploidy in succulents with small genomes. Science 250:99–101
Desvoyes B, Ramirez-Parra E, Xie Q, Chua NH, Gutierrez C (2006) Cell type-specific role of the retinoblastoma/E2F pathway during Arabidopsis leaf development. Plant Physiol 140:67–80
Dewitte W, Riou-Khamlichi C, Scofield S, Healy JMS, Jacqmard A, Kilby NJ, Murray JAH (2003) Altered cell cycle distribution, hyperplasia, and inhibited differentiation in Arabidopsis caused by the D-type cyclin CYCD3. Plant Cell 15:79–92

Dilkes BP, Dante RA, Coelho C, Larkins BA (2002) Genetic analyses of endoreduplication in Zea mays endosperm: evidence of sporophytic and zygotic maternal control. Genetics 160:1163–1177

Edgar BA, Orr-Weaver TL (2001) Endoreplication cell cycles: more for less. Cell 105:297–306

El Refy A, Perazza D, Zekraoui L, Valay JG, Bechtold N, Brown S, Hulskamp M, Herzog M, Bonneville JM (2004) The Arabidopsis KAKTUS gene encodes a HECT protein and controls the number of endoreduplication cycles. Mol Genet Genomics 270:403–414

Engelen-Eigles G, Jones RJ, Phillips RL (2000) DNA endoreduplication in maize endosperm cells: the effect of exposure to short-term high temperature. Plant Cell Environ 23:657–663

Evans LS, Van't Hof J (1975) Is polyploidy necessary for tissue differentiation in higher plants? Am J Bot 62:1060–1064

Exner V, Taranto P, Schonrock N, Gruissem W, Hennig L (2006) Chromatin assembly factor CAF-1 is required for cellular differentiation during plant development. Development 133:4163–4172

Fankhauser G, Humphrey RR (1943) The relation between number of nucleoli and number of chromosome sets in animal cells. Proc Natl Acad Sci USA 29:344–350

Fulop K, Tarayre S, Kelemen Z, Horvath G, Kevei Z, Nikovics K, Bako L, Brown S, Kondorosi A, Kondorosi E (2005) Arabidopsis anaphase-promoting complexes: multiple activators and wide range of substrates might keep APC perpetually busy. Cell Cycle 4:1084–1092

Galbraith DW, Harkins KR, Maddox JM, Ayres NM, Sharma DP, Firoozabady E (1983) Rapid flow cytometric analysis of the cell cycle in intact plant tissues. Science 220:1049–1051

Galitski T, Saldanha AJ, Styles CA, Lander ES, Fink GR (1999) Ploidy regulation of gene expression. Science 285:251–254

Gavin KA, Hidaka M, Stillman B (1995) Conserved initiator proteins in eukaryotes. Science 270:1667–1671

Gendreau E, Traas J, Desnos T, Grandjean O, Caboche M, Hofte H (1997) Cellular basis of hypocotyl growth in Arabidopsis thaliana. Plant Physiol 114:295–305

Gendreau E, Hofte H, Grandjean O, Brown S, Traas J (1998) Phytochrome controls the number of endoreduplication cycles in the Arabidopsis thaliana hypocotyl. Plant J 13:221–230

Gendreau E, Orbovic V, Hofte H, Traas J (1999) Gibberellin and ethylene control endoreduplication levels in the Arabidopsis thaliana hypocotyl. Planta 209:513–516

Gonzalez-Sama A, de la Pena TC, Kevei Z, Mergaert P, Lucas MM, de Felipe MR, Kondorosi E, Pueyo JJ (2006) Nuclear DNA endoreduplication and expression of the mitotic inhibitor Ccs52 associated to determinate and lupinoid nodule organogenesis. Mol Plant Microbe Interact 19:173–180

Grafi G, Larkins BA (1995) Endoreduplication in maize endosperm: involvement of M-phase-promoting factor inhibition and induction of S-phase-related kinases. Science 269:1262–1264

Grafi G, Burnett RJ, Helentjaris T, Larkins BA, DeCaprio JA, Sellers WR, Kaelin WG (1996) A maize cDNA encoding a member of the retinoblastoma protein family: involvement in endoreduplication. Proc Natl Acad Sci USA 93:8962–8967

Guitton AE, Berger F (2005) Control of reproduction by Polycomb group complexes in animals and plants. Int J Dev Biol 49:707–716

Harbour JW, Dean DC (2000) The Rb/E2F pathway: expanding roles and emerging paradigms. Genes Dev 14:2393–2409

Hase Y, Fujioka S, Yoshida S, Sun GQ, Umeda M, Tanaka A (2005) Ectopic endoreduplication caused by sterol alteration results in serrated petals in Arabidopsis. J Exp Bot 56:1263–1268

Hattori N, Davies TC, Anson-Cartwright L, Cross JC (2000) Periodic expression of the cyclin-dependent kinase inhibitor p57(Kip2) in trophoblast giant cells defines a G2-like gap phase of the endocycle. Mol Biol Cell 11:1037–1045

Hemerly AS, Ferreira P, de Almeida Engler J, Van Montagu M, Engler G, Inze D (1993) cdc2a expression in Arabidopsis is linked with competence for cell division. Plant Cell 5:1711–1723

Hulskamp M (2004) Plant trichomes: a model for cell differentiation. Nat Rev Mol Cell Biol 5:471–480

Hulskamp M, Misera S, Jurgens G (1994) Genetic dissection of trichome cell development in Arabidopsis. Cell 76:555–566

Hulskamp M, Schnittger A, Folkers U (1999) Pattern formation and cell differentiation: trichomes in Arabidopsis as a genetic model system. Int Rev Cytol 186:147–178

Imai KK, Ohashi Y, Tsuge T, Yoshizumi T, Matsui M, Oka A, Aoyama T (2006) The A-type cyclin CYCA2;3 is a key regulator of ploidy levels in Arabidopsis endoreduplication. Plant Cell 18:382–396

Jasinski S, Riou-Khamlichi C, Roche O, Perennes C, Bergounioux C, Glab N (2002) The CDK inhibitor NtKIS1a is involved in plant development, endoreduplication and restores normal development of cyclin D3;1-overexpressing plants. J Cell Sci 115:973–982

Joubes J, Chevalier C (2000) Endoreduplication in higher plants. Plant Mol Biol 43:735–745

Kladnik A, Chourey PS, Pring DR, Dermastia M (2006) Development of the endosperm of Sorghum bicolor during the endoreduplication-associated growth phase. J Cereal Sci 43:209–215

Kondorosi E, Kondorosi A (2004) Endoreduplication and activation of the anaphase-promoting complex during symbiotic cell development. FEBS Lett 567:152–157

Kosugi S, Ohashi Y (2003) Constitutive E2F expression in tobacco plants exhibits altered cell cycle control and morphological change in a cell type-specific manner. Plant Physiol 132:2012–2022

Kowles RV, Phillips RL (1985) DNA amplification patterns in maize endosperm nuclei during kernel development. Proc Natl Acad Sci USA 82:7010–7014

Kowles RV, Phillips RL (1988) Endosperm development in maize. Int Rev Cytol 112:97–136

Kowles RV, Srienc F, Phillips RL (1990) Endoreduplication of nuclear-DNA in the developing maize endosperm. Dev Genet 11:125–132

Kowles RV, Yerk GL, Haas KM, Phillips RL (1997) Maternal effects influencing DNA endoreduplication in developing endosperm of Zea mays. Genome 40:798–805

Kudo N, Kimura Y (2002) Nuclear DNA endoreduplication during petal development in cabbage: relationship between ploidy levels and cell size. J Exp Bot 53:1017–1023

Larkins BA, Dilkes BP, Dante RA, Coelho CM, Woo YM, Liu Y (2001) Investigating the hows and whys of DNA endoreduplication. J Exp Bot 52:183–192

Leblanc O, Pointe C, Hernandez M (2002) Cell cycle progression during endosperm development in Zea mays depends on parental dosage effects. Plant J 32:1057–1066

Leiva-Neto JT, Grafi G, Sabelli PA, Woo YM, Dante RA, Maddock S, Gordon-Kamm WJ, Larkins BA (2004) A dominant negative mutant of cyclin-dependent kinase A reduces endoreduplication but not cell size or gene expression in maize endosperm. Plant Cell 16:1854–1869

Lilly MA, Duronio RJ (2005) New insights into cell cycle control from the Drosophila endocycle. Oncogene 24:2765–2775

Lozano E, Saez AG, Flemming AJ, Cunha A, Leroi AM (2006) Regulation of growth by ploidy in Caenorhabditis elegans. Curr Biol 16:493–498

Lur HS, Setter TL (1993) Role of auxin in maize endosperm development: timing of nuclear-DNA endoreduplication, zein expression, and cytokinin. Plant Physiol 103:273–280

MacAuley A, Cross JC, Werb Z (1998) Reprogramming the cell cycle for endoreduplication in rodent trophoblast cells. Mol Biol Cell 9:795–807

Machida YJ, Hamlin JL, Dutta A (2005) Right place, right time, and only once: replication initiation in metazoans. Cell 123:13–24

Machida YJ, Dutta A (2007) The APC/C inhibitor, Emi1, is essential for prevention of rereplication. Genes Dev 21:184–194

Masuda HP, Ramos GBA, de Almeida-Engler J, Cabral LM, Coqueiro VM, Macrini CMT, Ferreira PCG, Hemerly AS (2004) Genome-based identification and analysis of the pre-replicative complex of Arabidopsis thaliana. FEBS Lett 574:192–202

Melaragno JE, Mehrotra B, Coleman AW (1993) Relationship between endopolyploidy and cell size in epidermal tissue of Arabidopsis. Plant Cell 5:1661–1668

Mihaylov IS, Kondo T, Jones L, Ryzhikov S, Tanaka J, Zheng JY, Higa LA, Minamino N, Cooley L, Zhang H (2002) Control of DNA replication and chromosome ploidy by geminin and cyclin A. Mol Cell Biol 22:1868–1880

Mosquna A, Katz A, Shochat S, Grafi G, Ohad N (2004) Interaction of FIE, a Polycomb protein, with pRb: a possible mechanism regulating endosperm development. Mol Genet Genomics 271:651–657

Nagl W (1976) DNA endoreduplication and polyteny understood as evolutionary strategies. Nature 261:614–615

Nakayama KI, Nakayama K (2006) Ubiquitin ligases: cell-cycle control and cancer. Nat Rev Cancer 6:369–381

Nguyen VQ, Co C, Li JJ (2001) Cyclin-dependent kinases prevent DNA re-replication through multiple mechanisms. Nature 411:1068–1073

Niculescu AB, Chen XB, Smeets M, Hengst L, Prives C, Reed SI (1998) Effects of p21(Cip1/Waf1) at both the G(1)/S and the G(2)/M cell cycle transitions: pRb is a critical determinant in blocking DNA replication and in preventing endoreduplication. Mol Cell Biol 18:629–643

Park JA, Ahn JW, Kim YK, Kim SJ, Kim JK, Kim WT, Pai HS (2005) Retinoblastoma protein regulates cell proliferation, differentiation, and endoreduplication in plants. Plant J 42:153–163

Peters J-M (2006) The anaphase promoting complex/cyclosome: a machine designed to destroy. Nat Rev Mol Cell Biol 7:644–656

Polizzi E, Natali L, Muscio AM, Giordani T, Cionini G, Cavallini A (1998) Analysis of chromatin and DNA during chromosome endoreduplication in the endosperm of Triticum durum Desf. Protoplasma 203:175–185

Ramos GBA, Engler JD, Ferreira PCG, Hemerly AS (2001) DNA replication in plants: characterization of a cdc6 homologue from Arabidopsis thaliana. J Exp Bot 52:2239–2240

Rao PN, Johnson RT (1970) Mammalian cell fusion: studies on the regulation of DNA synthesis and mitosis. Nature 225:159–164

Ravid K, Lu J, Zimmet JM, Jones MR (2002) Roads to polyploidy: the megakaryocyte example. J Cell Physiol 190:7–20

Royzman I, Austin RJ, Bosco G, Bell SP, Orr-Weaver TL (1999) ORC localization in Drosophila follicle cells and the effects of mutations in dE2F and dDP. Genes Dev 13:827–840

Sabelli PA, Larkins BA (2006) Grasses like mammals? Redundancy and compensatory regulation within the retinoblastoma protein family. Cell Cycle 5:352–355

Sabelli PA, Burgess SR, Kush AK, Young MR, Shewry PR (1996) cDNA cloning and characterisation of a maize homologue of the MCM proteins required for the initiation of DNA replication. Mol Gen Genet 252:125–136

Sabelli PA, Burgess SR, Kush AK, Shewry PR (1998) DNA replication initiation and mitosis induction in eukaryotes: the role of MCM and CDC25 proteins. In: Francis D, Dudits D, Inze D (eds) Plant cell division. Portland, London, pp 243–268

Sabelli PA, Parker JS, Barlow PW (1999) cDNA and promoter sequences for MCM3 homologues from maize, and protein localization in cycling cells. J Exp Bot 50:1315–1322

Sabelli PA, Dante RA, Leiva-Neto JT, Jung R, Gordon-Kamm WJ, Larkins BA (2005a) RBR3, a member of the retinoblastoma-related family from maize, is regulated by the RBR1/E2F pathway. Proc Natl Acad Sci USA 102:13005–13012

Sabelli PA, Leiva-Neto JT, Dante RA, Nguyen H, Larkins BA (2005b) Cell cycle regulation during maize endosperm development. Maydica 50:485–496

Satina S, Blakeslee AF (1941) Periclinal chimeras in Datura stramonium in relation to development of leaf and flower. Am J Bot 28:862–871

Sauer K, Knoblich JA, Richardson H, Lehner CF (1995) Distinct modes of cyclin E/Cdc2c kinase regulation and S-phase control in mitotic endoreduplication cycles of Drosophila embryogenesis. Genes Dev 9:1327–1339

Schnittger A, Hulskamp M (2002) Trichome morphogenesis: a cell-cycle perspective. Philos Trans R Soc Lond B Biol Sci 357:823–826

Schnittger A, Schobinger U, Bouyer D, Weinl C, Stierhof YD, Hulskamp M (2002a) Ectopic D-type cyclin expression induces not only DNA replication but also cell division in Arabidopsis trichomes. Proc Natl Acad Sci USA 99:6410–6415

Schnittger A, Schobinger U, Stierhof YD, Hulskamp M (2002b) Ectopic B-type cyclin expression induces mitotic cycles in endoreduplicating Arabidopsis trichomes. Curr Biol 12:415–420

Schnittger A, Weinl C, Bouyer D, Schobinger U, Hulskamp M (2003) Misexpression of the cyclin-dependent kinase inhibitor ICK1/KRP1 in single-celled Arabidopsis trichomes reduces endoreduplication and cell size and induces cell death. Plant Cell 15:303–315

Setter TL, Flannigan BA (2001) Water deficit inhibits cell division and expression of transcripts involved in cell proliferation and endoreduplication in maize endosperm. J Exp Bot 52:1401–1408

Sigrist SJ, Lehner CF (1997) Drosophila fizzy-related down-regulates mitotic cyclins and is required for cell proliferation arrest and entry into endocycles. Cell 90:671–681

Sorensen MB, Chaudhury AM, Robert H, Bancharel E, Berger F (2001) Polycomb group genes control pattern formation in plant seed. Curr Biol 11:277–281

Springer PS, Holding DR, Groover A, Yordan C, Martienssen RA (2000) The essential Mcm7 protein PROLIFERA is localized to the nucleus of dividing cells during the G(1) phase and is required maternally for early Arabidopsis development. Development 127:1815–1822

Sugimoto-Shirasu K, Roberts K (2003) Big it up: endoreduplication and cell-size control in plants. Curr Opin Plant Biol 6:544–553

Sun YJ, Dilkes BP, Zhang CS, Dante RA, Carneiro NP, Lowe KS, Jung R, Gordon-Kamm WJ, Larkins BA (1999a) Characterization of maize (Zea mays L.) Wee1 and its activity in developing endosperm. Proc Natl Acad Sci USA 96:4180–4185

Sun YJ, Flannigan BA, Setter TL (1999b) Regulation of endoreduplication in maize (Zea mays L.) endosperm. Isolation of a novel B1-type cyclin and its quantitative analysis. Plant Mol Biol 41:245–258

Tarayre S, Vinardell JM, Cebolla A, Kondorosi A, Kondorosi E (2004) Two classes of the Cdh1-type activators of the anaphase-promoting complex in plants: novel functional domains and distinct regulation. Plant Cell 16:422–434

Traas J, Hulskamp M, Gendreau E, Hofte H (1998) Endoreduplication and development: rule without dividing? Curr Opin Plant Biol 1:498–503

Tsumoto Y, Yoshizumi T, Kuroda H, Kawashima M, Ichikawa T, Nakazawa M, Yamamoto N, Matsui M (2006) Light-dependent polyploidy control by a CUE protein variant in Arabidopsis. Plant Mol Biol 61:817–828

Vespa L, Vachon G, Berger F, Perazza D, Faure JD, Herzog M (2004) The immunophilin-interacting protein AtFIP37 from Arabidopsis is essential for plant development and is involved in trichome endoreduplication. Plant Physiol 134:1283–1292

Vilhar B, Kladnik A, Blejec A, Chourey PS, Dermastia M (2002) Cytometrical evidence that the loss of seed weight in the miniature1 seed mutant of maize is associated with reduced mitotic activity in the developing endosperm. Plant Physiol 129:23–30

Vinardell JM, Fedorova E, Cebolla A, Kevei Z, Horvath G, Kelemen Z, Tarayre S, Roudier F, Mergaert P, Kondorosi A, Kondorosi E (2003) Endoreduplication mediated by the anaphase-promoting complex activator CCS52A is required for symbiotic cell differentiation in Medicago truncatula nodules. Plant Cell 15:2093–2105

Vlieghe K, Boudolf V, Beemster GTS, Maes S, Magyar Z, Atanassova A, Engler JD, De Groodt R, Inze D, De Veylder L (2005) The DP-E2F-like gene DEL1 controls the endocycle in Arabidopsis thaliana. Curr Biol 15:59–63

Walker JD, Oppenheimer DG, Concienne J, Larkin JC (2000) SIAMESE, a gene controlling the endoreduplication cell cycle in Arabidopsis thaliana trichomes. Development 127:3931–3940

Weinl C, Marquardt S, Kuijt SJH, Nowack MK, Jakoby MJ, Hulskamp M, Schnittger A (2005) Novel functions of plant cyclin-dependent kinase inhibitors, ICK1/KRP1, can act non-cell-autonomously and inhibit entry into mitosis. Plant Cell 17:1704–1722

Weiss A, Herzig A, Jacobs H, Lehner CF (1998) Continuous cyclin E expression inhibits progression through endoreduplication cycles in Drosophila. Curr Biol 8:239–242

Weng L, Zhu CW, Xu JH, Du W (2003) Critical role of active repression by E2F and Rb proteins in endoreplication during Drosophila development. EMBO J 22:3865–3875

Witmer XH, Alvarez-Venegas R, San-Miguel P, Danilevskaya O, Avramova Z (2003) Putative subunits of the maize origin of replication recognition complex ZmORC1-ZmORC5. Nucleic Acids Res 31:619–628

Xie Q, SanzBurgos P, Hannon GJ, Gutierrez C (1996) Plant cells contain a novel member of the retinoblastoma family of growth regulatory proteins. EMBO J 15:4900–4908

Yamaguchi R, Newport J (2003) A role for Ran-GTP and Crm1 in blocking re-replication. Cell 113:115–125

Yoshizumi T, Tsumoto Y, Takiguchi T, Nagata N, Yamamoto YY, Kawashima M, Ichikawa T, Nakazawa M, Yamamoto N, Matsui M (2006) Increased level of polyploidy1, a conserved repressor of CYCLINA2 transcription, controls endoreduplication in Arabidopsis. Plant Cell 18:2452–2468

Zhao J, Grafi G (2000) The high mobility group I/Y protein is hypophosphorylated in endoreduplicating maize endosperm cells and is involved in alleviating histone H1-mediated transcriptional repression. J Biol Chem 275:27494–27499

Zhou Y, Wang H, Gilmer S, Whitwill S, Fowke LC (2003) Effects of coexpressing the plant CDK inhibitor ICK1 and D-type cyclin genes on plant growth, cell size and ploidy in Arabidopsis thaliana. Planta 216:604–613

**Part B
Dynamics of Chromosomes,
Cytoskeleton and Organelles During Cell Division**

Chromosome Dynamics in Meiosis

Arnaud Ronceret · Moira J. Sheehan · Wojciech P. Pawlowski (✉)

Department of Plant Breeding and Genetics, Cornell University, Ithaca, NY 14853, USA
wp45@cornell.edu

A. Ronceret and M.J. Sheehan contributed equally

Abstract Meiosis encompasses a large number of dynamic processes. Some of them are biochemical, such as formation and repair of meiotic double-strand breaks, while others are physical in nature, such as homologous chromosome segregation in anaphase I. Plants have been used as model species in meiosis studies for over 80 years. However, the past decade brought a dramatic improvement in the understanding of meiosis in plants at the mechanistic level, thanks to the adoption of genetic and molecular biology techniques in chromosome research and new microscopy methods.

1 Overview of Meiosis

Meiosis consists of two consecutive nuclear divisions (Fig. 1), a reductional division (meiosis I) and an equational division (meiosis II), without an intervening S phase between them. While meiosis II is essentially similar to a mitotic division, meiosis I is a specialized division, whose aim is to reduce the number of chromosomes in the nucleus and allow exchange of genetic material between maternal and paternal chromosomes. Based largely on chromosome dynamics, meiosis I is subdivided into four stages, prophase I, metaphase I, anaphase I, and telophase I. Meiotic prophase I, the most eventful of these stages, is further subdivided into five sub-stages: leptotene, zygotene, pachytene, diplotene, and diakinesis (Fig. 2). During **leptotene**, which follows the pre-meiotic S phase, decondensed chromatin becomes organized into chromosomes by the assemblage of a proteinaceous core. Meiotic recombination is initiated at this step by formation of double-strand breaks (DSBs) in chromosomal DNA (Pawlowski et al. 2004; Zickler and Kleckner 1999). In **zygotene**, homologous chromosomes pair. Pairing is followed by synapsis, when the central element of the synaptonemal complex (SC) is installed between the paired homologs and stabilizes pairing interactions (Page and Hawley 2004). By **pachytene**, SC formation is complete and meiotic recombination between homologs is resolved. In **diplotene**, the SC disassembles and chiasmata are visible. Chiasmata are the sites of crossovers (COs) and are responsible for holding homologous chromosomes together until their segregation in anaphase. Finally, in **diakinesis**, the chromosomes undergo the final stage of condensation.

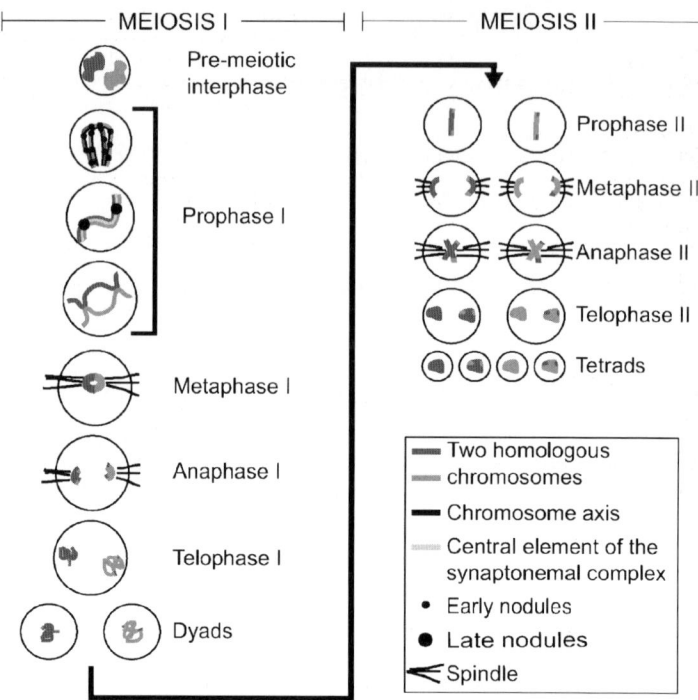

Fig. 1 A general overview of meiosis. Only one pair of chromosomes is shown and each homolog is a different shade of grey. Early and late recombination nodules are depicted as dots of different size. See Fig. 2 for a detailed list of prophase I substages and events

Recombination, pairing and synapsis are three main processes of meiotic prophase I. **Recombination** encompasses formation and repair of meiotic DSBs, including reciprocal chromosome arm exchanges. **Pairing** includes interactions between chromosomes, which involve homology recognition and lead to juxtaposition of homologs. Pairing is followed by **synapsis**, which is defined as the process of installation of the central element (CE) of SC, which binds the paired chromosomes along their entire length. These three processes are formally distinct but genetic and molecular analyses are now drawing a more complex scenario where these processes are intimately interconnected and show a great deal of coordination (Pawlowski and Cande 2005).

2
Initiation of Meiosis in Plants

The switch from the mitotic to the meiotic cell cycle in plants is preceded by a developmental pathway that assigns germ cell identity. First, archesporial

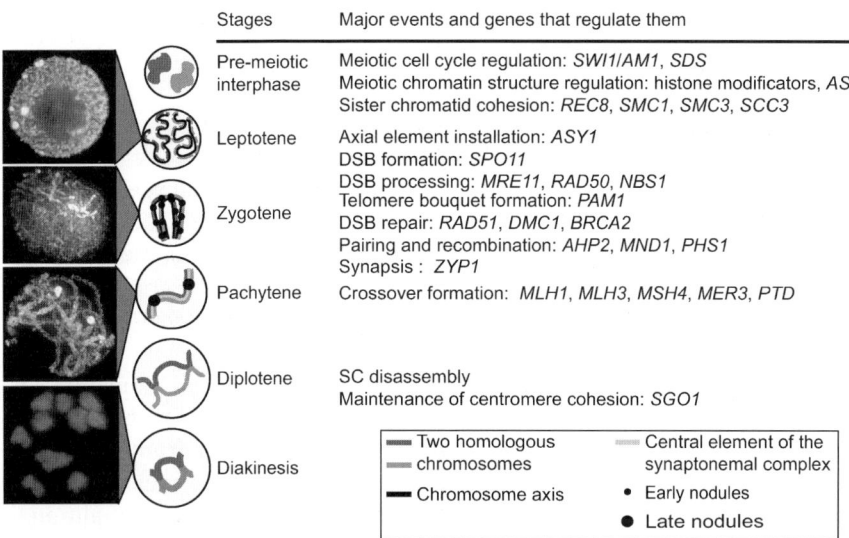

Fig. 2 Pre-meiotic interphase and meiotic prophase I in plants: substages, main events and key genes that regulate them. Only one pair of chromosomes is shown and each homolog is a different shade of grey. Early and late recombination nodules are depicted as dots of different size. *Images on the left* are chromosomes of maize male meiocytes stained with DAPI. The images are flat projections of three-dimensional image stacks collected with three-dimensional deconvolution microscopy

cells are differentiated from hypodermal cells in male and female reproductive organs. Then, the archesporial cells develop into sporocytes, which show features distinct from other cells and are destined to undergo meiosis (Ma 2005).

In contrast to the developmental events preceding the switch to meiosis, the initiation of meiosis in plants is less understood. The *ameiotic1* (*am1*) gene has been identified as a master controller of the switch from mitotic to meiotic cell cycle in maize (Golubovskaya et al. 1997, 1993) (Pawlowski et al., unpublished). Pre-meiotic cells in most *am1* mutants, instead of entering meiosis, undergo mitotic divisions. In severe cases, the progression of the cell cycle is arrested during interphase. All known specific aspects of meiosis, such as establishment of meiosis-specific chromatin structure, chromosome pairing, synapsis, recombination, and meiosis-specific chromosome dynamics require *am1*. In addition to its role in initiating meiosis, *am1* regulates the progression through the early stages of meiotic prophase (Golubovskaya et al. 1993) (Pawlowski et al., unpublished). This conclusion is based on the analysis of an unusual *am1* allele, *am1-pra1*, in which pre-meiotic cells enter meiosis but arrest during meiotic prophase.

The *Arabidopsis* homolog of *am1*, *SWITCH1* (*SWI1*), also known as *DYAD*, has been shown to regulate several meiotic prophase processes, including establishment of meiotic chromosome structure, recombination and synapsis (Agashe et al. 2002; Mercier et al. 2001, 2003). However, the phenotypes of *swi1* mutants are less obvious than those of the maize *am1* mutants. Male meiocytes in *swi1* mutants either undergo a normal meiosis or show meiotic sister-chromatid cohesion (SCC) defects. Female *swi1* meiocytes undergo an equational division. However, this division is an abnormal meiosis rather than mitosis, since the meiosis-specific cohesin REC8 is loaded onto chromosomes, and *DMC1*, a gene encoding a meiosis-specific recombinase, is expressed. None of the *swi1* mutations result in meiocytes undergoing a normal mitotic division, as is the case in *am1* mutants. This suggests that initiation at the mechanistic level differs between maize and *Arabidopsis*.

Homologs of the AM1/SWI1/DYAD proteins are confined to plants. The molecular functions of AM1 and SWI1 are not known and the pathway downstream from these proteins that results in the transition from mitotic to meiotic cell cycle is poorly understood. The decision to enter meiosis is probably made before or at the beginning of the pre-meiotic S phase, although the evidence is mostly indirect. This timing is suggested by observations that female meiocytes in several *am1* mutants arrest at interphase and that the *Arabidopsis* SWI1 shows expression exclusively during pre-meiotic G1 and S. Overall, the mechanisms of meiosis initiation in plants is likely to corroborate conclusions from yeast and mammals indicating that the signaling cascade leading to meiosis initiation shows great diversity among species while the timing of meiosis initiation is a universal feature shared by all eukaryotes (Pawlowski et al. 2007).

3
Regulation of Meiosis Progression

Several candidates for meiotic cell cycle regulators have been identified in *Arabidopsis* based on their functions and/or similarity to known cyclins. These proteins, CDC45, SOLO DANCERS (SDS), and TARDY ANSYNCHRONOUS MEIOSIS (TAM), are proposed to act at different times in meiosis. CDC45 functions during pre-meiotic S-phase (Stevens et al. 2004). SDS regulates chromosome pairing and synapsis in prophase I (Azumi et al. 2002), although the molecular mechanism of its function is not known. TAM has been proposed to regulate progression of both, meiotic prophase I and meiosis II, and its absence also leads to meiotic nuclear division becoming asynchronous with cytokinesis (Wang et al. 2004).

4
Meiotic Chromosome Structure

4.1
Dynamics of the Chromatin Structure and the Histone Code

The beginning of meiosis is associated with a dynamic re-organization of chromatin (Dawe et al. 1994; Zickler and Kleckner 1999). In fact, changes in chromosome structure and condensation have been recognized as landmarks to identify meiosis sub-stages. Specific histone H3 variant deposition (Okada et al. 2005), and histone modifications, such as methylation (Shi and Dawe 2006; Yang et al. 2006) and phosphorylation (Houben et al. 2005; Kaszas and Cande 2000) are observed in meiosis, suggesting a wide and dynamic meiotic reorganization of the histone code. In particular, phosphorylation of H3 at Ser10 correlates with the maintenance of sister chromatid cohesion (Kaszas and Cande 2000) and phosphorylation of H3 at Thr11 was found to correlate with meiotic chromosomes condensation (Houben et al. 2005). Insight into the regulation of these modifications is limited at present but an *Arabidopsis* SKP1 homolog, ASK1, was recently shown to play a role in this process (Yang et al. 2006). In hexaploid wheat, a change in chromatin conformation coincides with chromosome pairing and has been implicated in the process of homologous chromosome recognition (Prieto et al. 2004).

4.2
Chromosome Axis

The chromosome axis (axial element, AE), forming at the base of chromatin loops, contains components essential for SCC. The chromosome axis is credited with allowing preferential exchanges between homologous chromosomes rather than sister chromatids. The resulting chiasmata ensure correct segregation of chromosomes at the first meiotic division. AEs are also essential for homologous chromosome synapsis because they become, as lateral elements (LEs), components of the tripartite SC.

Loss of the ASYNAPTIC1 (ASY1) protein associated with the chromosome axis in *Arabidopsis* reduces fertility by 90% (Caryl et al. 2000; Ross et al. 1997). ASY1, proposed to be a homolog of the yeast Hop1 protein, forms foci on chromatin during pre-meiotic interphase and localizes to the entire length of chromosomes from leptotene to diplotene, except for the telomeric regions (Armstrong et al. 2002). The *Arabidopsis* genome also contains another homolog of *HOP1*, *ASY2* (At4g32200), whose function is non-redundant with *ASY1* (Caryl et al. 2000). Two *HOP1* homologs, *PAIR2* and *PAIR2c3*, were also found in rice. *PAIR2* is the ortholog of *ASY1*, since the *pair2* mutant exhibits the same phenotype as the *Arabidopsis asy1* mutant (Nonomura et al. 2004) and the PAIR2 protein in rice shows the same localization pattern as ASY1 in *Arabidopsis* (Nonomura et al. 2006).

4.3
Sister-Chromatid Cohesion

During the S-phase DNA replication, the newly copied sister chromatids are kept associated, in meiosis as well as in mitosis, with a multi-subunit cohesin complex (Revenkova and Jessberger 2006). The cohesin complex is well conserved across kingdoms. The composition of the cohesin complex differs between mitotic and meiotic cells. In mitotic cells, it contains RAD21 and SCC3 proteins associated with two Structural Maintenance of Chromosomes (SMC) proteins, SMC1 and SMC3. In meiotic cells, RAD21 is replaced by a meiosis-specific cohesin REC8. Because SMC1, SMC3, and SCC3 function in both mitosis and meiosis, mutants defective in these proteins show embryonic defects and are embryo lethal in *Arabidopsis* (Chelysheva et al. 2005; Liu et al. 2002).

In contrast to fungi and animals, plants contain more than two genes encoding the RAD21/REC8 proteins. *Arabidopsis* has four RAD21/REC8 homologs (Dong et al. 2001) but only one of them, SYN1, also known as DIF1, has a meiotic function and is the true ortholog of REC8 (Bai et al. 1999; Bhatt et al. 1999; Cai et al. 2003; Peirson et al. 1997). Immunolocalization studies showed that SYN1 is present in meiocyte nuclei from interphase to metaphase I and localizes along chromosome arms while being devoid from the centromeric region (Cai et al. 2003).

The analysis of a maize *rec8* mutant called *absence of first division1* (*afd1*) shows that the maize REC8 homolog is required for establishing the leptotene chromosome structure, the bouquet formation, as well as chromosome pairing, synapsis, and recombination (Golubovskaya et al. 2006; Pawlowski et al. 2003). AFD1 colocalizes with ASY1 from leptotene through pachytene, suggesting that both proteins are associated with the chromosome axis. ASY1 is present on chromosomes in *afd1* mutants, indicating that AFD1 is not required for ASY1 recruitment. A series of weaker *afd1* alleles revealed that the functions of REC8 in chromosome pairing, synapsis, and recombination may be independent from its function in establishing the meiotic chromosome structure (Golubovskaya et al. 2006).

5
Chromosome Dynamics of Early Meiotic Prophase—The Telomere Bouquet

The telomere bouquet is a temporary chromosome arrangement that occurs exclusively in meiosis and is found in most species examined to date (Harper et al. 2004; Zickler and Kleckner 1999). During prophase I, all telomeres cluster at a single location on the nuclear envelope (NE), generally juxtaposed with the microtubule organizing region, causing the chromosomes to resemble a bouquet of flowers. The coincidental timing of the bouquet stage with homologous chromosome pairing, recombination, and synapsis suggests a role of the bouquet in these processes.

Within the plant kingdom, the bouquet has been studied in several species, mainly grasses such as rye, wheat, and maize (Bass et al. 2000; Cowan and Cande 2002; Martinez-Perez et al. 1999; Noguchi 2002).

5.1
Function of the Bouquet

In most organisms, the bouquet forms at the beginning of, or just before, zygotene and persists until the end of zygotene (Harper et al. 2004). In rye, wheat and maize, telomeres begin to attach to the NE in leptotene and the bouquet persists until early pachytene (Bass et al. 2000; Maestra et al. 2002; Martinez-Perez et al. 1999; Noguchi 2002). In addition to coincidental timing, there is evidence supporting a close relationship between the bouquet and pairing, although the exact nature of that relationship is still unknown. In mutants with defective bouquet formation, pairing is delayed and inefficient (Golubovskaya et al. 2002; Harper et al. 2004; Niwa et al. 2000; Trelles-Sticken et al. 2000). Attachment of telomeres and confinement of chromosomes to a small area may assist the sequence-dependent homology search by limiting the physical volume within which chromosomes may interact, and by creating constructive chromosome movements (Scherthan 2001). This could enhance the efficiency of pairing (Harper et al. 2004). Telomere clustering might also eliminate or reduce ectopic pairing interactions in highly conserved syntenic regions or regions of highly repetitive sequences (Niwa et al. 2000). Furthermore, synapsis in plants initiates at the telomeres and the bouquet may help catalyze this process (Maestra et al. 2002).

5.2
Genetics of Bouquet Formation

The most complete picture of bouquet formation in a single system exists in the fission yeast, *S. pombe*, where there are a number of mutants known to cause specific bouquet defects (Davis and Smith 2006; Harper et al. 2004; Jin et al. 2002). In contrast, little is known about the bouquet in higher eukaryotes. The best-characterized plant bouquet mutant is the *pam1* (*plural abnormalities of meiosis1*) mutant of maize (Golubovskaya et al. 2002). As the name indicates, the *pam1* mutation has several obvious meiotic defects, including asynchrony of meiocytes, inhibition of telomere bouquet clustering, and abnormal synapsis (Golubovskaya et al. 2002). The initiation of meiotic recombination in the *pam1* mutant is not affected. However, later recombination stages are likely defective since it has been noticed that the RAD51 recombination protein persists on *pam1* chromosomes longer than in wild-type meiocytes.

5.3
Dynamics of Bouquet Formation

The "early bouquet" stage forms in late leptotene with the telomeres attached to the NE opposite the nucleolus, which has migrated from the center to the periphery of the nucleus. By zygotene, telomeres have migrated to coalesce with the nucleolus in the "mature bouquet" (Bass et al. 1997). Coincident with this migration, nuclear pores also migrate but in the opposite direction, away from the bouquet focus and nucleolus (Bass et al. 1997; Cowan and Cande 2002). On the basis of the low numbers of nuclei detected at early compared to late bouquet stages, Bass et al. determined that bouquet formation is a sudden and active process (Bass et al. 1997). By contrast, the termination of the bouquet seems to be a passive process with telomeres diffusing away from the focus.

The formation of the bouquet is most likely a two-stage process with telomeres first attaching to the NE, then migrating from a distal to a proximal position at or near the nucleolus (Bass et al. 1997). This model is supported by evidence from several maize meiotic mutants. In *pam1*, telomeres are able to attach to the NE in a wild-type fashion, but clustering is disrupted (Golubovskaya et al. 2002). Telomere attachment to the NE can also be disrupted as was shown in the allelic series of mutants in *afd1*, the maize homolog of *REC8* (Golubovskaya et al. 2006). In weak alleles of *afd1*, telomeres are able to attach to the NE, whereas in strong alleles, they cannot. Clustering is completely abolishing in all but the weakest *afd1* allele. Telomere attachment is not dependent on REC8 recruitment *per se* but on the formation of the AE over the entire length of the chromosome (Golubovskaya et al. 2006; Liebe et al. 2004; Trelles-Sticken et al. 2005). Evidence from mammals and fungi point to completion of DSB repair and CO formation (Liebe et al. 2006; Pandita et al. 1999; Trelles-Sticken et al. 2000) as the cue to exit from the bouquet, but there has not yet been evidence in plants for this hypothesis.

Studies indicate that telomere sequence and unknown host factors govern bouquet formation. Maize telocentric and ring chromosomes can enter the bouquet, indicating that the presence of the physical chromosome end is not necessary for bouquet formation (Carlton and Cande 2002). In another study, the meiotic behavior of an oat line carrying an addition of maize chromosome 9 was investigated (Bass et al. 2000). Normally in maize, all of the telomeres are localized in the bouquet at zygotene. In the oat-maize addition line, only about 70–90% of telomeres (both maize and oat) are localized to the bouquet during zygotene. Interestingly, the maize chromosome telomeres were observed to attach to the NE and enter the bouquet in a way similar to oat chromosomes, suggesting that telomere clustering in this line is controlled by the oat nuclei rather than by the telomeres themselves (Bass et al. 2000).

In addition to telomere sequences, there is mounting evidence that an intact cytoskeleton is required for bouquet formation. It has long been known that the microtubule (MT) depolymerizing drug colchicine applied during prophase I in plants and animals causes reduced chiasmata and impaired SC formation (Loidl 1989; Tepperberg et al. 1997). The application of colchicine to rye anthers inhibited telomere clustering but not telomere attachment to NE or nuclear pore migration, causing a resemblance to the maize *pam1* phenotype (Cowan and Cande 2002; Golubovskaya et al. 2002). Interestingly, this same effect was seen even when low levels of colchicine were applied that were insufficient to cause depolymerization of MTs. It is unknown if the target of colchicine is a cytoskeletal component on the inner NE or perhaps a membrane-associated β-tubulin (Cowan and Cande 2002; Harper et al. 2004).

5.4
Telomere Clustering in *Arabidopsis*

Arabidopsis does not form a bona fide telomere bouquet but it has been observed that *Arabidopsis* telomeres cluster around the nucleolus in pre-meiotic interphase (Armstrong et al. 2001), which could play a role equivalent to the role of the bouquet. However, this telomere clustering is also observed in mitotic interphase, and its early occurrence suggests the utilization of a chromosome arrangement existing after a previous cell division, rather than de novo clustering. In addition, a very loose bouquet-like arrangement of telomeres is observed in *Arabidopsis* in zygotene (Armstrong et al. 2001).

6
Meiotic Recombination

6.1
Formation of Meiotic DSBs and Early Recombination Steps

Meiotic recombination is universally initiated in all species by the introduction of DSBs into chromosomal DNA by SPO11, a protein belonging to the topoisomerase family (Keeney et al. 1997). In contrast to all other species, plants possess multiple SPO11 homologs. The *Arabidopsis* genome has three *SPO11* genes (Hartung and Puchta 2001), although only two of them, *AtSPO11-1* and *AtSPO11-2*, are essential for meiosis (Grelon et al. 2001; Stacey et al. 2006). Analysis of the *spo11-1* mutant demonstrated that the recombination defect is coupled with the inability of the mutant to synapse, leading to univalents, improper gamete formation, and ultimately, sterility (Grelon et al. 2001). Visualizing the presence of meiotic DSBs on maize meiotic chromosomes using the TUNEL assay showed that DSBs occur along the entire length of chromosomes, are formed in very early leptotene, and persist until early pachytene (Pawlowski et al. 2004).

Downstream from SPO11, meiotic DSBs are resected by the MRN complex, which includes MRE11, RAD50, and NBS1 (Akutsu et al. 2007; Bleuyard et al. 2004; Daoudal-Cotterell et al. 2002; Gallego et al. 2001; Puizina et al. 2004). Like *spo11-1*, the *Arabidopsis rad50* and *mre11* mutants exhibit recombination defects and are unable to synapse. However, unlike *spo11-1*, they also show fragmented and dicentric chromosomes, resulting from unrepaired DSBs (Bleuyard et al. 2004; Gallego et al. 2001; Puizina et al. 2004). An analysis of *mre11/spo11-1* double mutants indicated that this fragmentation is SPO11-dependent (Puizina et al. 2004). In addition to the meiotic defects, the *rad50* and *mre11* mutants are hypersensitive to DNA damage agents and exhibit progressive telomere shortening in mitotically dividing cells (Gallego et al. 2001; Puizina et al. 2004). This indicates that the MRN complex in plants, similarly to its counterparts in yeast and mammals, is also involved in somatic DNA repair processes.

6.2
Single-End Invasion and the Role of the RAD51 Family

Meiotic DSBs generated by SPO11 and the MRN complex are subsequently repaired by a complex of proteins containing two homologs of the bacterial RecA recombinase, RAD51 and DMC1. RAD51 and DMC1 exhibit extensive similarity at the amino acid level, but while RAD51 is expressed throughout the plant life cycle, DMC1 is only expressed during meiosis (Doutriaux et al. 1998; Jones et al. 2003; Klimyuk and Jones 1997; Li et al. 2004). RAD51 and DMC1 facilitate the single-end invasion (SEI) process, in which a single stranded DNA overhang, created by the MRN complex, invades a homologous double-stranded DNA region. Rad51 and Dmc1 in *S. cerevisiae* are known to form a complex that covers single stranded DNA ends, forming a nucleoprotein filament. This structure facilitates the recognition of the corresponding region in the homologous double helix (Neale and Keeney 2006).

RAD51 and DMC1 form distinct foci on meiotic chromosomes, presumably on the sites of DSBs (Franklin et al. 1999; Pawlowski et al. 2003; Terasawa et al. 1995). In maize, RAD51 foci first appear at the beginning of zygotene (Franklin et al. 1999; Pawlowski et al. 2003). The number of foci reaches its peak of roughly 500 per nucleus in mid zygotene, and later declines to about 10–20 in late pachytene. Observations in lily showed that most RAD51 and DMC1 foci colocalize (Terasawa et al. 1995).

The numbers and dynamics of RAD51/DMC1 foci resemble those of the early recombination nodules (EN), electron-dense structures observed in Transmission Electron Microscopy (TEM) in a number of species, including maize, lily and tomato (Anderson et al. 2003; Anderson et al. 2001). EN have been suggested to play a role in recombination and homologous chromosome pairing. RAD51 and DMC1 are indeed components of early recombination nodules, as demonstrated by immuno-gold labeling (Anderson et al. 1997).

Arabidopsis rad51 and *dmc1* mutants show defects in meiotic recombination, chromosome pairing and synapsis (Couteau et al. 1999; Li et al. 2004). The *atrad51* mutant shows also extensive SPO11-dependent chromosome fragmentation, similar to the phenotypes observed in *Arabidopsis* mutants defective in MRE11 and RAD50 (Bleuyard et al. 2004; Li et al. 2004; Puizina et al. 2004). Interestingly, fragmentation is not observed in the *Arabidopsis dmc1* mutant (Couteau et al. 1999). It is likely that, in the absence of DMC1, meiotic DSBs are repaired by RAD51 using the sister chromatid as the template. Supporting this conclusion are observations that depletion of RAD51 in the *atdmc1* mutant background does lead to chromosome fragmentation (Siaud et al. 2004).

In addition to RAD51 and DMC1, the *Arabidopsis* genome encodes five other RecA homologs, RAD51B, RAD51C, RAD51D, XRCC2, and XRCC3 (Lin et al. 2006). Of these, only RAD51C and XRCC3 have meiotic functions (Abe et al. 2005; Bleuyard and White 2004; Li et al. 2005), supporting the notion that different members of the RAD51 protein family have evolved to fulfill different requirements for meiosis and somatic DNA repair (Bleuyard et al. 2005).

The RAD51/DMC1 protein complex interacts with several other proteins in the process of DSB repair, including HOP2, MND1, and BRCA2. HOP2 and MND1 play roles in both recombination and homologous chromosome pairing and are discussed in the chromosome pairing section later in this chapter. BRCA2 is hypothesized to act in recruiting the RAD51/DMC1 complex to the sites of DSBs (Sharan et al. 2004). The *Arabidopsis* homolog of BRCA2 interacts with AtRAD51 and AtDMC1 through the conserved BRC motifs in the BRCA2 protein (Dray et al. 2006; Siaud et al. 2004). Silencing BRCA2 in *Arabidopsis* using RNAi leads to unrepaired DSB (Siaud et al. 2004).

6.3
Formation of Crossovers: Two Ways to Recombine Homologs

Only a subset of the numerous DSBs formed during meiosis produce COs, which lead to chiasmata and chromosome arm exchanges. The repair of most meiotic DSBs results in non-crossover (NCO) products (gene conversion). For example, in maize there are about 500 SEI events/nucleus that are marked by RAD51 foci, but only about 20 crossovers. Data from budding yeast indicate that COs and NCOs are made by two alternative pathways that branch very soon after DSB formation (Allers and Lichten 2001; Hunter and Kleckner 2001). In the CO pathway, SEI recombination intermediates form double Holliday junctions, which are subsequently resolved to give mostly CO products. The events that lead to NCOs are less clear, although it has been suggested that a synthesis-dependent strand-annealing mechanism is involved in their formation (Allers and Lichten 2001). Although direct evidence is still lacking, similar two pathways may exist in *Arabidopsis* (Higgins et al. 2004).

Meiotic cells must ensure that each chromosome pair has at least one CO to form chiasmata that keep the homologous chromosomes together until anaphase I. The mechanism of this "crossover assurance" is unclear. A number of proteins that specifically promote CO formation have been identified, including members of the MutL and MutS DNA mismatch-repair protein families, MLH1, MLH3, MSH4, and MSH5. The *Arabidopsis* MLH1 homolog is expressed in vegetative and reproductive tissues (Jean et al. 1999) but MLH3 is specifically expressed during meiosis (Jackson et al. 2006). The MLH1 and MLH3 proteins form foci on chromosomes and colocalize in pachytene (Franklin et al. 2006). The *Arabidopsis* MSH4 protein also has a function in meiosis. The *msh4* mutant is defective in CO formation as well as synapsis (Higgins et al. 2004). Higgins et al. observed that a residual 16% of crossovers form independent of MSH4 and these events are randomly distributed on chromosomes (Higgins et al. 2004). Similar observations were made in several other *Arabidopsis* recombination mutants, *mlh3* (Jackson et al. 2006), *mer3/rock and rollers* (*rck*) (Chen et al. 2005; Mercier et al. 2005) and *parting dancers* (*ptd*) (Wijeratne et al. 2006). These observations are consistent with the proposal that there are two classes of meiotic COs: class I COs that have a regulated distribution and are subject to crossover interference, and class II COs that occur randomly, are not subject to interference (Copenhaver et al. 2002). CO interference is a poorly understood mechanism that prevents formation of two COs close to each other. The analyses of mutants in the *MSH4*, *MLH3*, *MER3*, and *PTD* genes indicate that they all are involved in class I CO formation. The *Arabidopsis* genome contains a homolog of *MUS81*, a gene thought to be involved in class II CO formation in yeast, but its meiotic function has not been studied.

At the cytological level, CO formation corresponds to the presence of late recombination nodules (LN) (Anderson et al. 2003; Stack and Anderson 2002). LN are less numerous then EN and likely form from a subset of EN. LN in the mouse contain recombination proteins involved in CO formation, MSH4 and MLH1 (Moens et al. 2002), and it is likely these enzymes are also present in the LN in plants.

7
Homologous Chromosome Pairing

Homologous chromosome pairing is one of the least-explored processes in meiosis. In most plants, with the exception of hexaploid wheat, chromosomes enter meiosis unpaired and pair de novo during zygotene. Intuitively, chromosome pairing must occur in two steps, a step in which the homologs are brought together into a close proximity and a homology search step in which correct homologous chromosomes identify each other. The dynamic movement of chromosomes during early meiotic prophase I, including formation of the telomere bouquet, may be a part of the first step. Plants belong to a group of species, along with mammals and fungi, in which successful chromosome pairing depends on meiotic recombination (Pawlowski and Cande 2005). Consequently, it has been hypothesized that recombination plays a role in the homology search step of chromosome pairing in plants (Franklin et al. 1999; Li et al. 2004; Pawlowski et al. 2003). In addition to recombination, other mechanisms may be involved in the homology search, particularly in species with large genomes containing extensive repetitive DNA sequence families. Repetitive DNA would make identification of DNA sequence similarity insufficient alone to establish chromosome homology. However, even though existence of chromosome or chromatin-level homology recognition mechanisms may be intuitive (Stack and Anderson 2001), such mechanisms have not yet been experimentally identified.

7.1
Pairing and Recombination

Defects in chromosome pairing were observed in a number of *Arabidopsis* recombination mutants already mentioned in this chapter, including *spo11-1*, *rad50*, *mre11*, *brca2*, *rad51*, *dmc1*, *rad51c*, and *xrcc3*. In particular, members of the RecA family, DMC1, RAD51, and RAD51C, have been proposed to act in homologous chromosome pairing in addition to their roles in meiotic recombination (Franklin et al. 1999; Li et al. 2005; Pawlowski and Cande 2005; Pawlowski et al. 2003). Franklin et al. proposed a role for RAD51 in homologous pairing in maize based on the analysis of the dynamics of the distribution of chromosomal RAD51 foci in zygotene and pachytene (Franklin et al. 1999). This was supported by Pawlowski et al., who observed that, in a collection of meiotic mutants in maize, the degree of homologous pairing defects corresponded to the number of chromosomal RAD51 foci in zygotene (Pawlowski et al. 2003).

Recent years brought the identification of a small group of meiotic genes that affect both recombination and homologous chromosomes pairing and have been hypothesized to play major roles in coordinating recombination and pairing. This group contains the maize *Phs1* gene (Pawlowski et al. 2004), as well as *HOP2* and *MND1*, which were first identified in yeast (Leu et al. 1998; Tsubouchi and Roeder 2002) but recently also shown to have homologs in other species, including plants (Domenichini et al. 2006; Kerzendorfer et al. 2006; Panoli et al. 2006; Schommer et al. 2003). Mutants in these genes are defective in both pairing and recombination. The maize *phs1* mutant and the *hop2* mutant in yeast show a striking phenotype: in the absence of homologous pairing, non-homologous chromosomes associate and synapse (Leu et al. 1998; Pawlowski et al. 2004). In *phs1*, this phenotype is particularly strong and only about 5%

of chromosome associations are between proper homologs. In the maize *phs1* mutant, meiotic DSBs are formed, but the repair process is delayed and most likely proceeds through an alternative pathway, since the RAD51 protein does not form foci on chromosomes in mutant meiocytes (Pawlowski et al. 2004). Mutants in the *Arabidopsis MND1* and *AHP2* (*Arabidopsis* homolog of *HOP2*) genes also show DSB formation but the DSBs are not repaired, leading to chromosome fragmentation, even though in the *mnd1* mutant a normal number of RAD51 foci is observed (Domenichini et al. 2006; Kerzendorfer et al. 2006; Panoli et al. 2006; Schommer et al. 2003). The molecular mechanism of the PHS1 protein function is not known. In contrast, the function of MND1 and HOP2 has been extensively studied in yeast, mammals, and *Arabidopsis*. The two proteins were shown to form a heterodimer (Kerzendorfer et al. 2006). Studies in mouse and yeast indicate further that this heterodimer interacts with the Rad51/Dmc1 complex. However, the details of this interaction remain unclear. It was proposed that the Hop2/Mnd1 complex acts directly to facilitate the SEI activity of Dmc1 (Chen et al. 2004; Tsubouchi and Roeder 2002). Another hypothesis suggests that Hop2/Mnd1 influence pairing and recombination indirectly by affecting chromatin and/or higher-order chromosome structures of the homologous target (Zierhut et al. 2004).

7.2
Other Pairing Mechanisms Independent of Recombination

Polyploid species may have developed additional mechanisms that allow distinguishing between homologous and homeologous (not homologous but similar) chromosomes (Moore 2000). For example, in wheat, homeologous associations between chromosomes of different genomes are prevented by the *Ph1* locus. *Ph1* was proposed to act by regulating centromere associations that are found in meiotic and somatic cells of wheat (Martinez-Perez et al. 2001). In the presence of *Ph1*, non-homologously associated centromeres separate at the beginning of meiosis. Chromatin structure is likely to be involved in this homolog recognition process (Mikhailova et al. 1998; Prieto et al. 2005; Prieto et al. 2004). The endeavor to fine-map and clone *Ph1* has recently focused on a structure that consists of a subtelomeric heterochromatin repeat that inserted into a cluster of Cdc2 gene repeats following the polyploidization event (Griffiths et al. 2006). Presence of this structure correlates with *Ph1* activity, making it a good candidate for the *Ph1* locus.

8
Synapsis

SC forms between the paired homologs along the entire chromosome length in zygotene (Page and Hawley 2004). The tripartite SC structure is comprised of two electron-dense LEs that flank a less-dense central region. LEs correspond to the AEs, which are formed in leptotene. Between the LEs are transverse filaments (TFs) that span the central region creating a zipper-like structure. Although the SC shows a high degree of structural conservation among species, the TF proteins of different species do not show significant similarity at the amino acid sequence level, which means that they have to be identified in each species de novo. The TF protein in *Arabidopsis*, ZYP1, has recently been identified by combining bioinformatics and protein immunolocalization approaches (Higgins et al. 2005). ZYP1 is encoded by two duplicated and highly redundant genes *ZYP1a* and *ZYP1b*. ZYP1 is first visible as foci on chromatin in late leptotene and then localizes to the region between synapsed homologous chromosomes in pachytene. However, in rye, the formation of the elongated ZYP1 structures along the entire chromosome length precedes synapsis and takes place as early as leptotene (Mikhailova et al. 2006). *Arabidopsis zyp1* RNAi mutants show that the SC is not formed in absence of ZYP1 but recombination is only slightly reduced. Interestingly, the distribution of chiasmata in these plants is normal showing that in plants SC is not required for interference (Higgins et al. 2005; Osman et al. 2006). In the absence of ZYP1, recombination can occur between homologous as well as non-homologous chromosomes, suggesting that ZYP1 may also affect homologous chromosome recognition.

9
Chromosome Segregation in Meiosis I and II

After completion of prophase I, the key events in meiosis are segregation of homologous chromosomes in meiosis I, followed by segregation of sister-chromatids in meiosis II. Both events require cleavage of the cohesin complex. In addition, homologous chromosome segregation requires SC disassembly, which takes place in diplotene. Cohesion in meiosis is removed in two steps. Cohesion along chromosome arms is released prior to anaphase I. However, in the centromere region, cohesion is preserved until anaphase II by the SUGOSHIN1 (SGO1) protein (Hamant et al. 2005). In maize, SGO1 is installed in centromeric and pericentromeric chromosome regions in early leptotene and requires the presence of REC8. A specialized protease, called separase, is responsible for cohesin cleavage (Liu and Makaroff 2006).

Segregation in meiosis I and II also requires spindle formation and chromosome attachment to spindle microtubules. A large group of proteins, kinesins, is thought to be involved in spindle morphogenesis. Recently, a meiosis specific kinesin ATK1 was shown in *Arabidopsis* to be specifically required for spindle formation in male meiocytes (Chen et al. 2002).

References

Abe K, Osakabe K, Nakayama S, Endo M, Tagiri A, Todoriki S, Ichikawa H, Toki S (2005) *Arabidopsis* RAD51C gene is important for homologous recombination in meiosis and mitosis. Plant Physiol 139:896–908

Agashe B, Prasad CK, Siddiqi I (2002) Identification and analysis of *DYAD*: a gene required for meiotic chromosome organisation and female meiotic progression in *Arabidopsis*. Development 129:3935–3943

Akutsu N, Iijima K, Hinata T, Tauchi H (2007) Characterization of the plant homolog of Nijmegen breakage syndrome 1: Involvement in DNA repair and recombination. Biochem. Biophys Res Comm 353:394–398

Allers T, Lichten M (2001) Differential timing and control of noncrossover and crossover recombination during meiosis. Cell 106:47–57

Anderson LK, Doyle GG, Brigham B, Carter J, Hooker KD, Lai A, Rice M, Stack SM (2003) High-resolution crossover maps for each bivalent of Zea mays using recombination nodules. Genetics 165:849–865

Anderson LK, Hooker KD, Stack SM (2001) The distribution of early recombination nodules on zygotene bivalents from plants. Genetics 159:1259–1269

Anderson LK, Offenberg HH, Verkuijlen WMHC, Heyting C (1997) RecA-like proteins are components of early meiotic nodules in lily. Proc Natl Acad Sci USA 94:6868–6873

Armstrong SJ, Caryl AP, Jones GH, Franklin FC (2002) Asy1, a protein required for meiotic chromosome synapsis, localizes to axis-associated chromatin in *Arabidopsis* and *Brassica*. J Cell Sci 115:3645–3655

Armstrong SJ, Franklin FC, Jones GH (2001) Nucleolus-associated telomere clustering and pairing precede meiotic chromosome synapsis in *Arabidopsis thaliana*. J Cell Sci 114:4207–4217

Azumi Y, Liu D, Zhao D, Li W, Wang G, Hu Y, Ma H (2002) Homolog interaction during meiotic prophase I in *Arabidopsis* requires the SOLO DANCERS gene encoding a novel cyclin-like protein. EMBO J 21:3081–3095

Bai X, Peirson BN, Dong F, Xue C, Makaroff CA (1999) Isolation and characterization of SYN1, a RAD21-like gene essential for meiosis in *Arabidopsis*. Plant Cell 11:417–430

Bass HW, Marshall WF, Sedat JW, Agard DA, Cande WZ (1997) Telomeres cluster de novo before the initiation of synapsis: A three-dimensional spatial analysis of telomere positions before and during meiotic prophase. J Cell Biol 137:5–18

Bass HW, Riera-Lizarazu O, Ananiev EV, Bordoli SJ, Rines HW, Phillips RL, Sedat JW, Agard DA, Cande WZ (2000) Evidence for the coincident initiation of homolog pairing and synapsis during the telomere-clustering (bouquet) stage of meiotic prophase. J Cell Sci 113:1033–1042

Bhatt AM, Lister C, Page T, Fransz P, Findlay K, Jones GH, Dickinson HG, Dean C (1999) The DIF1 gene of *Arabidopsis* is required for meiotic chromosome segregation and belongs to the REC8/RAD21 cohesin gene family. Plant J 19:463–472

Bleuyard J-Y, Gallego ME, Savigny F, White CI (2005) Differing requirements for the *Arabidopsis* Rad51 paralogs in meiosis and DNA repair. Plant J 41:533–545

Bleuyard J-Y, Gallego ME, White CI (2004) Meiotic defects in the *Arabidopsis* rad50 mutant point to conservation of the MRX complex function in early stages of meiotic recombination. Chromosoma 113:197–203

Bleuyard J-Y, White CI (2004) The *Arabidopsis* homologue of Xrcc3 plays an essential role in meiosis. EMBO J 23:439–449

Cai X, Dong F, Edelmann RE, Makaroff CA (2003) The *Arabidopsis* SYN1 cohesin protein is required for sister chromatid arm cohesion and homologous chromosome pairing. J Cell Sci 116:2999–3007

Carlton PM, Cande WZ (2002) Telomeres act autonomously in maize to organize the meiotic bouquet from a semipolarized chromosome orientation. J Cell Biol 157:231–242

Caryl AP, Armstrong SJ, Jones GH, Franklin FCH (2000) A homologue of the yeast HOP1 gene is inactivated in the *Arabidopsis* meiotic mutant *asy1*. Chromosoma 109:62–71

Chelysheva L, Diallo S, Vezon D, Gendrot G, Vrielynck N, Belcram K, Rocques N, Marquez-Lema A, Bhatt AM, Horlow C, Mercier R, Mezard C, Grelon M (2005) AtREC8 and AtSCC3 are essential to the monopolar orientation of the kinetochores during meiosis. J Cell Sci 118:4621–4632

Chen C, Marcus A, Li W, Hu Y, Calzada JP, Grossniklaus U, Cyr RJ, Ma H (2002) The *Arabidopsis* ATK1 gene is required for spindle morphogenesis in male meiosis. Development 129:2401–2409

Chen C, Zhang W, Timofejeva L, Gerardin Y, Ma H (2005) The *Arabidopsis* ROCK-N-ROLLERS gene encodes a homolog of the yeast ATP-dependent DNA helicase MER3 and is required for normal meiotic crossover formation. Plant J 43:321–334

Chen YK, Leng CH, Olivares H, Lee MH, Chang YC, Kung WM, Ti SC, Lo YH, Wang AH, Chang CS, Bishop DK, Hsueh YP, Wang TF (2004) Heterodimeric complexes of Hop2 and Mnd1 function with Dmc1 to promote meiotic homolog juxtaposition and strand assimilation. Proc Natl Acad Sci USA 101:10572–10577

Copenhaver GP, Housworth EA, Stahl FW (2002) Crossover Interference in *Arabidopsis*. Genetics 160:1631–1639

Couteau F, Belzile F, Horlow C, Grandjean O, Vezon D, Doutriaux M-P (1999) Random chromosome segregation without meiotic arrest in both male and female meiocytes of a dmc1 mutant of *Arabidopsis*. Plant Cell 11:1623–1634

Cowan CR, Cande WZ (2002) Meiotic telomere clustering is inhibited by colchicine but does not require cytoplasmic microtubules. J Cell Sci 115:3747–3756

Daoudal-Cotterell S, Gallego ME, White CI (2002) The plant Rad50-Mre11 protein complex. FEBS Lett 516:164–166

Davis L, Smith GR (2006) The meiotic bouquet promotes homolog interactions and restricts ectopic recombination in Schizosaccharomyces pombe. Genetics 174:167–177

Dawe RK, Sedat JW, Agard DA, Cande WZ (1994) Meiotic chromosome pairing in maize is associated with a novel chromatin organization. Cell 76:901–912

Domenichini S, Raynaud C, Ni D-A, Henry Y, Bergounioux C (2006) Atmnd1-[Delta]1 is sensitive to gamma-irradiation and defective in meiotic DNA repair. DNA Repair 5:455–464

Dong F, Cai X, Makaroff CA (2001) Cloning and characterization of two *Arabidopsis* genes that belong to the RAD21/REC8 family of chromosome cohesin proteins. Gene 271:99–108

Doutriaux MP, Couteau F, Bergounioux C, White C (1998) Isolation and characterisation of the RAD51 and DMC1 homologs from *Arabidopsis thaliana*. Mol Gen Genet 257:283–291

Dray E, Siaud N, Dubois E, Doutriaux M-P (2006) Interaction between *Arabidopsis* Brca2 and its partners Rad51, Dmc1, and Dss1. Plant Physiol 140:1059–1069

Franklin AE, McElver J, Sunjevaric I, Rothstein R, Bowen B, Cande WZ (1999) Three-dimensional microscopy of the Rad51 recombination protein during meiotic prophase. Plant Cell 11:809–824

Franklin FC, Higgins JD, Sanchez-Moran E, Armstrong SJ, Osman KE, Jackson N, Jones GH (2006) Control of meiotic recombination in *Arabidopsis*: role of the MutL and MutS homologues. Biochem Soc Trans 34:542–544

Gallego ME, Jeanneau M, Granier F, Bouchez D, Bechtold N, White IC (2001) Disruption of the *Arabidopsis* RAD50 gene leads to plant sterility and MMS sensitivity. Plant J 25:31–41

Golubovskaya I, Avalkina N, Sheridan WF (1997) New insights into the role of the maize *ameiotic1* locus. Genetics 147:1339–1350

Golubovskaya I, Grebennikova ZK, Avalkina NA, Sheridan WF (1993) The role of the ameiotic1 gene in the initiation of meiosis and in subsequent meiotic events in maize. Genetics 135:1151–1166

Golubovskaya IN, Hamant O, Timofejeva L, Wang CJ, Braun D, Meeley R, Cande WZ (2006) Alleles of afd1 dissect REC8 functions during meiotic prophase I. J Cell Sci 119:3306–3315

Golubovskaya IN, Harper LC, Pawlowski WP, Schichnes D, Cande WZ (2002) The *pam1* gene is required for meiotic bouquet formation and efficient homologous synapsis in maize (*Zea mays* L.). Genetics 162:1979–1993

Grelon M, Vezon D, Gendrot G, Pelletier G (2001) AtSPO11-1 is necessary for efficient meiotic recombination in plants. EMBO J 20:589–600

Griffiths S, Sharp R, Foote TN, Bertin I, Wanous M, Reader S, Colas I, Moore G (2006) Molecular characterization of Ph1 as a major chromosome pairing locus in polyploid wheat. Nature 439:749–752

Hamant O, Golubovskaya I, Meeley R, Fiume E, Timofejeva L, Schleiffer A, Nasmyth K, Cande WZ (2005) A REC8-dependent plant Shugoshin is required for maintenance of centromeric cohesion during meiosis and has no mitotic functions. Curr Biol 15:948–954

Harper L, Golubovskaya I, Cande WZ (2004) A bouquet of chromosomes. J Cell Sci 117:4025–4032

Hartung F, Puchta H (2001) Molecular characterization of homologues of both subunits A (SPO11) and B of the archaebacterial topoisomerase 6 in plants. Gene 271:81–86

Higgins JD, Armstrong SJ, Franklin FC, Jones GH (2004) The *Arabidopsis* MutS homolog AtMSH4 functions at an early step in recombination: evidence for two classes of recombination in *Arabidopsis*. Genes Dev 18:2557–2570

Higgins JD, Sanchez-Moran E, Armstrong SJ, Jones GH, Franklin FC (2005) The *Arabidopsis* synaptonemal complex protein ZYP1 is required for chromosome synapsis and normal fidelity of crossing over. Genes Dev 19:2488–2500

Houben A, Demidov D, Rutten T, Scheidtmann KH (2005) Novel phosphorylation of histone H3 at threonine 11 that temporally correlates with condensation of mitotic and meiotic chromosomes in plant cells. Cytogenet Genome Res 109:148–155

Hunter N, Kleckner N (2001) The single-end invasion: An asymmetric intermediate at the double-strand break to double-Holliday junction transition of meiotic recombination. Cell 106:59–70

Jackson N, Sanchez-Moran E, Buckling E, Armstrong SJ, Jones GH, Franklin FC (2006) Reduced meiotic crossovers and delayed prophase I progression in AtMLH3-deficient *Arabidopsis*. EMBO J 25:1315–1323

Jean M, Pelletier J, Hilpert M, Belzile F, Kunze R (1999) Isolation and characterization of AtMLH1, a MutL homologue from *Arabidopsis thaliana*. Mol Gen Genet 262:633–642

Jin Y, Uzawa S, Cande WZ (2002) Fission yeast mutants affecting telomere clustering and meiosis-specific spindle pole body integrity. Genetics 160:861–876

Jones GH, Armstrong SJ, Caryl AP, Franklin FCH (2003) Meiotic chromosome synapsis and recombination in *Arabidopsis thaliana*; an integration of cytological and molecular approaches. Chromosome Res 11:205–215

Kaszas E, Cande WZ (2000) Phosphorylation of histone H3 is correlated with changes in the maintenance of sister chromatid cohesion during meiosis in maize, rather than the condensation of the chromatin. J Cell Sci 113:3217–3226

Keeney S, Giroux CN, Kleckner N (1997) Meiosis-specific DNA double-strand breaks are catalyzed by Spo11, a member of a widely conserved protein family. Cell 88:375–384

Kerzendorfer C, Vignard J, Pedrosa-Harand A, Siwiec T, Akimcheva S, Jolivet S, Sablowski R, Armstrong S, Schweizer D, Mercier R, Schlogelhofer P (2006) The *Arabidopsis thaliana* MND1 homologue plays a key role in meiotic homologous pairing, synapsis and recombination. J Cell Sci 119:2486–2496

Klimyuk VI, Jones JDG (1997) AtDMC1, the *Arabidopsis* homologue of the yeast *DMC1* gene: Characterization, transposon-induced allelic variation and meiosis-associated expression. Plant J 11:1–14

Leu JY, Chua PR, Roeder GS (1998) The meiosis-specific Hop2 protein of *S. cerevisiae* ensures synapsis between homologous chromosomes. Cell 94:375–386

Li W, Chen C, Markmann-Mulisch U, Timofejeva L, Schmelzer E, Ma H, Reiss B (2004) The *Arabidopsis* AtRAD51 gene is dispensable for vegetative development but required for meiosis. Proc Natl Acad Sci USA 101:10596–10601

Li W, Yang X, Lin Z, Timofejeva L, Xiao R, Makaroff CA, Ma H (2005) The AtRAD51C gene is required for normal meiotic chromosome synapsis and double-stranded break repair in *Arabidopsis*. Plant Physiol 138:965–976

Liebe B, Alsheimer M, Hoog C, Benavente R, Scherthan H (2004) Telomere attachment, meiotic chromosome condensation, pairing, and bouquet stage duration are modified in spermatocytes lacking axial elements. Mol Biol Cell 15:827–837

Liebe B, Petukhova G, Barchi M, Bellani M, Braselmann H, Nakano T, Pandita TK, Jasin M, Fornace A, Meistrich ML, Baarends WM, Schimenti J, de Lange T, Keeney S, Camerini-Otero RD, Scherthan H (2006) Mutations that affect meiosis in male mice influence the dynamics of the mid-preleptotene and bouquet stages. Exp Cell Res 312:3768–3781

Lin ZG, Kong HZ, Nei M, Ma H (2006) Origins and evolution of the *recA/RAD51* gene family: Evidence for ancient gene duplication and endosymbiotic gene transfer. Proc Natl Acad Sci USA 103:10328–10333

Liu C-M, McElver J, Tzafrir I, Joosen R, Wittich P, Patton D, Van Lammeren AAM, Meinke D (2002) Condensin and cohesin knockouts in *Arabidopsis* exhibit a titan seed phenotype. Plant J 29:405–415

Liu Z, Makaroff CA (2006) *Arabidopsis* separase AESP is essential for embryo development and the release of cohesin during meiosis. Plant Cell 18:1213–1225

Loidl J (1989) Colchicine action at meiotic prophase revealed by SC-spreading. Genetica 78:195–203

Ma H (2005) Molecular genetic analyses of microsporogenesis and microgametogenesis in flowering plants. Annu Rev Plant Biol 56:393–434

Maestra BN, de Jong JH, Shepherd K, Naranjo TS (2002) Chromosome arrangement and behaviour of two rye homologous telosomes at the onset of meiosis in disomic wheat-5RL addition lines with and without the Ph1 locus. Chromosome Res 10:655–667

Martinez-Perez E, Shaw P, Moore G (2001) The Ph1 locus is needed to ensure specific somatic and meiotic centromere association. Nature 411:204–207

Martinez-Perez E, Shaw P, Reader S, Aragon-Alcaide L, Miller T, Moore G (1999) Homologous chromosome pairing in wheat. J Cell Sci 112:1761–1769

Mercier R, Armstrong SJ, Horlow C, Jackson NP, Makaroff CA, Vezon D, Pelletier G, Jones GH, Franklin FC (2003) The meiotic protein SWI1 is required for axial element formation and recombination initiation in *Arabidopsis*. Development 130:3309–3318

Mercier R, Jolivet S, Vezon D, Huppe E, Chelysheva L, Giovanni M, Nogue F, Doutriaux MP, Horlow C, Grelon M, Mezard C (2005) Two meiotic crossover classes cohabit in *Arabidopsis*: one is dependent on MER3, whereas the other one is not. Curr Biol 15:692–701

Mercier R, Vezon D, Bullier E, Motamayor JC, Sellier A, Lefevre F, Pelletier G, Horlow C (2001) SWITCH1 (SWI1): a novel protein required for the establishment of sister chromatid cohesion and for bivalent formation at meiosis. Genes Dev 15:1859–1871

Mikhailova EI, Naranjo TS, Shepherd K, Wennekes-van Eden J, Heyting C, de Jong JH (1998) The effect of the wheat Ph1 locus on chromatin organisation and meiotic chromosome pairing analysed by genome painting. Chromosoma 107:339–350

Mikhailova EI, Phillips D, Sosnikhina SP, Lovtsyus AV, Jones RN, Jenkins G (2006) Molecular assembly of meiotic proteins Asy1 and Zyp1 and pairing promiscuity in rye (*Secale cereale* L.) and its synaptic mutant *sy10*. Genetics 174:1247–1258

Moens PB, Kolas NK, Tarsounas M, Marcon E, Cohen PE, Spyropoulos B (2002) The time course and chromosomal localization of recombination-related proteins at meiosis in the mouse are compatible with models that can resolve the early DNA-DNA interactions without reciprocal recombination. J Cell Sci 115:1611–1622

Moore G (2000) Cereal chromosome structure, evolution, and pairing. Annu Rev Plant Phys Plant Mol Biol 51:195–222

Neale MJ, Keeney S (2006) Clarifying the mechanics of DNA strand exchange in meiotic recombination. Nature 442:153–158

Niwa O, Shimanuki M, Miki F (2000) Telomere-led bouquet formation facilitates homologous chromosome pairing and restricts ectopic interaction in fission yeast meiosis. EMBO J 19:3831–3840

Noguchi J (2002) Homolog pairing and two kinds of bouquets in the meiotic prophase of rye, Secale cereale. Genes Genet Syst 77:39–50

Nonomura K-I, Nakano M, Eiguchi M, Suzuki T, Kurata N (2006) PAIR2 is essential for homologous chromosome synapsis in rice meiosis I. J Cell Sci 119:217–225

Nonomura KI, Nakano M, Murata K, Miyoshi K, Eiguchi M, Miyao A, Hirochika H, Kurata N (2004) An insertional mutation in the rice PAIR2 gene, the ortholog of *Arabidopsis* ASY1, results in a defect in homologous chromosome pairing during meiosis. Mol Genet Genomics 271:121–129

Okada T, Endo M, Singh MB, Bhalla PL (2005) Analysis of the histone H3 gene family in *Arabidopsis* and identification of the male-gamete-specific variant AtMGH3. Plant J 44:557–568

Osman K, Sanchez-Moran E, Higgins J, Jones G, Franklin F (2006) Chromosome synapsis in *Arabidopsis*: analysis of the transverse filament protein ZYP1 reveals novel functions for the synaptonemal complex. Chromosoma 115:212–219

Page SL, Hawley RS (2004) The genetics and molecular biology of the synaptonemal complex. Annu Rev Cell Dev Biol 20:525–558

Pandita TK, Westphal CH, Anger M, Sawant SG, Geard CR, Pandita RK, Scherthan H (1999) Atm inactivation results in aberrant telomere clustering during meiotic prophase. Mol Cell Biol 19:5096–5105

Panoli AP, Ravi M, Sebastian J, Nishal B, Reddy TV, Marimuthu MP, Subbiah V, Vijaybhaskar V, Siddiqi I (2006) AtMND1 is required for homologous pairing during meiosis in *Arabidopsis*. BMC Mol Biol 7:24

Pawlowski WP, Cande WZ (2005) Coordinating the events of the meiotic prophase. Trends Cell Biol 15:674–681

Pawlowski WP, Golubovskaya IN, Cande WZ (2003) Altered nuclear distribution of recombination protein RAD51 in maize mutants suggests involvement of RAD51 in the meiotic homology recognition. Plant Cell 8:1807–1816

Pawlowski WP, Golubovskaya IN, Timofejeva L, Meeley RB, Sheridan WF, Cande WZ (2004) Coordination of meiotic recombination, pairing, and synapsis by PHS1. Science 303:89–92

Pawlowski WP, Sheehan MJ, Ronceret A (2007) In the beginning: the initiation of meiosis. Bioessays 29:511–514

Peirson BN, Bowling SE, Makaroff CA (1997) A defect in synapsis causes male sterility in a T-DNA-tagged *Arabidopsis thaliana* mutant. Plant J 11:659–669

Prieto P, Moore G, Reader S (2005) Control of conformation changes associated with homologue recognition during meiosis. Theor Appl Genet 111:505–510

Prieto P, Shaw P, Moore G (2004) Homologue recognition during meiosis is associated with a change in chromatin conformation. Nat Cell Biol 6:906–908

Puizina J, Siroky J, Mokros P, Schweizer D, Riha K (2004) Mre11 deficiency in *Arabidopsis* is associated with chromosomal instability in somatic cells and Spo11-dependent genome fragmentation during meiosis. Plant Cell 16:1968–1978

Revenkova E, Jessberger R (2006) Shaping meiotic prophase chromosomes: cohesins and synaptonemal complex proteins. Chromosoma 115:235–240

Ross K, Fransz P, Armstrong S, Vizir I, Mulligan B, Franklin F, Jones G (1997) Cytological characterization of four meiotic mutants of *Arabidopsis* isolated from T-DNA-transformed lines. Chromosome Res 5:551–559

Scherthan H (2001) A bouquet makes ends meet. Nat Rev Mol Cell Biol 2:621–627

Schommer C, Beven A, Lawrenson T, Shaw P, Sablowski R (2003) AHP2 is required for bivalent formation and for segregation of homologous chromosomes in *Arabidopsis* meiosis. Plant J 36:1–11

Sharan SK, Pyle A, Coppola V, Babus J, Swaminathan S, Benedict J, Swing D, Martin BK, Tessarollo L, Evans JP, Flaws JA, Handel MA (2004) BRCA2 deficiency in mice leads to meiotic impairment and infertility. Development 131:131–142

Shi J, Dawe RK (2006) Partitioning of the maize epigenome by the number of methyl groups on histone H3 lysines 9 and 27. Genetics 173:1571–1583

Siaud N, Dray E, Gy I, Gerard E, Takvorian N, Doutriaux MP (2004) Brca2 is involved in meiosis in *Arabidopsis thaliana* as suggested by its interaction with Dmc1. EMBO J 23:1392–1401

Stacey NJ, Kuromori T, Azumi Y, Roberts G, Breuer C, Wada T, Maxwell A, Roberts K, Sugimoto-Shirasu K (2006) *Arabidopsis* SPO11-2 functions with SPO11-1 in meiotic recombination. Plant J 48:206–216

Stack SM, Anderson LK (2001) A model for chromosome structure during the mitotic and meiotic cell cycles. Chromosome Res 9:175–198

Stack SM, Anderson LK (2002) Crossing over as assessed by late recombination nodules is related to the pattern of synapsis and the distribution of early recombination nodules in maize. Chromosome Res 10:329–345

Stevens R, Grelon M, Vezon D, Oh J, Meyer P, Perennes C, Domenichini S, Bergounioux C (2004) A CDC45 homolog in *Arabidopsis* is essential for meiosis, as shown by RNA interference-induced gene silencing. Plant Cell 16:99–113

Tepperberg JH, Moses MJ, Nath J (1997) Colchicine effects on meiosis in the male mouse. Chromosoma 106:183–192

Terasawa M, Shinohara A, Hotta Y, Ogawa H, Ogawa T (1995) Localization of RecA-like recombination proteins on chromosomes of the lily at various meiotic stages. Genes Dev 9:925–934

Trelles-Sticken E, Adelfalk C, Loidl J, Scherthan H (2005) Meiotic telomere clustering requires actin for its formation and cohesin for its resolution. J Cell Biol 170:213–223

Trelles-Sticken E, Dresser ME, Scherthan H (2000) Meiotic telomere protein Ndj1p is required for meiosis-specific telomere distribution, bouquet formation and efficient homologue pairing. J Cell Biol 151:95–106

Tsubouchi H, Roeder GS (2002) The Mnd1 protein forms a complex with Hop2 to promote homologous chromosome pairing and meiotic double-strand break repair. Mol Cell Biol 22:3078–3088

Wang Y, Magnard JL, McCormick S, Yang M (2004) Progression through meiosis I and meiosis II in *Arabidopsis* anthers is regulated by an A-type cyclin predominately expressed in prophase I. Plant Physiol 136:4127–4135

Wijeratne AJ, Chen C, Zhang W, Timofejeva L, Ma H (2006) The *Arabidopsis thaliana* PARTING DANCERS gene encoding a novel protein is required for normal meiotic homologous recombination. Mol Biol Cell 17:1331–1343

Yang X, Timofejeva L, Ma H, Makaroff CA (2006) The *Arabidopsis* SKP1 homolog ASK1 controls meiotic chromosome remodeling and release of chromatin from the nuclear membrane and nucleolus. J Cell Sci 119:3754–3763

Zickler D, Kleckner N (1999) Meiotic chromosomes: Integrating structure and function. Annu Rev Genet 33:603–754

Zierhut C, Berlinger M, Rupp C, Shinohara A, Klein F (2004) Mnd1 is required for meiotic interhomolog repair. Curr Biol 14:752–762

Cytoskeletal and Vacuolar Dynamics During Plant Cell Division: Approaches Using Structure-Visualized Cells

Toshio Sano[1] (✉) · Natsumaro Kutsuna[1] · Takumi Higaki[1] · Yoshihisa Oda[1] · Arata Yoneda[1] · Fumi Kumagai-Sano[2] · Seiichiro Hasezawa[1]

[1]Graduate School of Frontier Sciences, The University of Tokyo, Kashiwanoha, 277-8562 Kashiwa, Japan
tsano@k.u-tokyo.ac.jp

[2]Faculty of Education, Gunma University, Aramaki, 371-8510 Maebashi, Japan

Abstract During cell cycle progression, intra-cellular cytoskeletal and membrane structures undergo dynamic changes in their form and localization, which in turn regulate further progress of the cell cycle. Despite the considerable insights into these intra-cellular structures obtained from immuno-fluorescence microscopy, the need for chemical fixation has limited the acquired images to only static ones. In contrast, more recent fluorescent protein techniques used to visualize these structures in living cell systems have allowed investigations of their dynamics. The visualization of microtubules (MTs) by using the green fluorescent protein (GFP) and the analysis of MT-associated proteins will be presented. In addition, to further understand plant cell cycle progression, dynamics of actin microfilaments (MFs) and vacuolar membranes (VMs) visualized with fluorescent proteins are also reviewed.

1
Introduction

Cell division and cell cycle progression, as well as cell elongation/expansion and differentiation, are fundamental processes to reconstitute a plant individual. From a view point of intra-cellular structures, cell division and cell cycle progression are regulated by two categories of these structures; membrane organelles, such as vacuoles, the endoplasmic reticulum (ER) and Golgi-derived vesicles, and the fibril networks of the cytoskeleton. Of these various intra-cellular structures, the plant cytoskeleton undergoes the most dynamic structural changes during cell cycle progression and cell division. Conventionally, such structures have been observed by indirect immuno-fluorescence microscopy and microinjection of fluorescent-labeled proteins, techniques that have provided a great deal of insight into the morphology of plant cells. During cell cycle progression, the microtubules (MTs), one of the major cytoskeletons of plant cells, were found to be comprised of four typical structures; the cortical microtubules, preprophase band (PPB), mitotic spindle and the phragmoplast (Zhang et al. 1990; Hasezawa et al. 1991).

However, because only static images could be obtained from these immunofluorescence microscopy studies, due to the need for chemical fixation, the bases of the structural changes in MTs during cell cycle progression could not be unequivocally demonstrated. Similarly, although microinjection allowed time-sequential observations in living cells, more detailed analyses were often limited because of the brevity of the observation period due to diffusion of the fluorescence and the applicability of the method to limited types of plant cells.

In contrast, the more recent fluorescent protein techniques allow the visualization of several intra-cellular structures in living cell systems (Hawes et al. 2001). The primary advantage of these techniques is that they allow time-sequential observations. Through the use of fluorescent proteins, the dynamics of the intra-cellular structures were observed and the sequential changes in MT configurations during cell cycle progression were clarified (Hasezawa et al. 2000; Kumagai et al. 2001). The second advantage of these techniques is that they preserve the cellular structures by avoiding the need for chemical fixation. To observe sensitive structures against chemical fixation such as actin microfilaments (MFs), these techniques could show detailed structures (Sano et al. 2005). Furthermore, as chemical fixation largely abolished membrane structures, fluorescent protein techniques will provide new insight in observing membrane organelles, such as vacuoles, ER and Golgi-derived vesicles (Hawes et al. 2001).

Herein, we present an overview of recent progress in our understanding of the dynamic changes in intra-cellular structures during cell cycle progression, as revealed by visualization of these structures by fluorescent proteins. In particular, focus is placed on MTs, MFs and the vacuolar membrane (VM) which we have analyzed using transgenic tobacco BY-2 cell lines. Starting with an introduction to the tobacco BY-2 system, the dynamics of MTs and the involvement of MT-associated proteins in cell cycle progression are described. Subsequently, results from detailed observations of MFs and VM and their dynamics are discussed in combination with our approach towards the development of 3-D reconstruction software.

2
The Tobacco BY-2 System

Here, we first introduce the tobacco BY-2 cell line as the original cell line of the transgenic lines described below. Of the numerous plant cell lines available, the tobacco BY-2 cell line is perhaps the most useful for basic plant cellular and molecular biology studies because of its rapid growth rate (Nagata et al. 1992; Nagata 2004). In addition, the establishment of an Agrobacterium-mediated transformation method (An et al. 1985) has allowed chimeric proteins, such as fusions with fluorescent proteins, to be quite easily expressed in the cell line. Furthermore, as the BY-2 cell line is the only plant cell line that can be highly synchronized, it is the main cell line used for plant cell cycle studies (Nagata et al. 1992). Synchronization in BY-2 cells is usually performed using aphidicolin, an inhibitor of DNA polymerase (Nagata and Kumagai 1999). Incubation of fully grown (usually 7-day-old) tobacco BY-2 cells for 24 h in fresh medium with 5 mg/L aphidicolin arrests the cell cycle at the border of the G1/S phase, and subsequent washing of the cell culture with fresh medium restarts the cell cycle from the S phase. After appropriate treatment, 8 to 9 h after aphidicolin release, 60 to 70% of the cells can be observed in mitosis. Sequential treatment with and subsequent release from propyzamide, a reversible inhibitor of microtubule polymerization that arrests the cell cycle at pro-metaphase and collects the cells at this stage, results in more than 95% of cells being at mitosis (Nagata and Kumagai 1999). Using these properties, the dynamics of cytoskeletons and vacuolar structures during cell cycle progression could be studied and are described below.

3
Microtubule Dynamics and Functions of MT-Associated Proteins in Plant Cell Division

3.1
Microtubule Dynamics During Cell Cycle Progression

The structure and organization of microtubules in higher plant cells have been studied mainly by immuno-fluorescence microscopy and microinjection of fluorescent-labeled tubulin (Zhang et al. 1990; Hasezawa et al. 1991; Vantard et al. 2000; Hasezawa and Kumagai 2002). The four characteristic structures of MTs that have to date been identified during cell cycle progression are as follows: (i) Cortical MTs (CMTs), which align transversely to the orientation of cell elongation, are observed from G1 through S to the G2 phase. The CMT is considered to be critical for the determination of plant cell shape, since CMT arrays determine the direction of the newly formed cellulose microfibrils in the primary cell walls (Ledbetter and Porter 1963; Giddings and Staehelin 1991; Baskin 2001; Wasteneys 2004); (ii) The preprophase band (PPB) is a bundle MT structure on the cell surface that replaces the gradually disappearing CMTs in the late G2 phase. As the PPB also disappears during mitosis, and the future cell plate fuses at the position where the PPB existed previously, the PPB is thought to leave some sort of memory to indicate the position of the future cell division plane (Mineyuki 1999); (iii) the mitotic spindle then segregates the chromosomes at mitosis; and subsequently (iv) a microtubule structure called the phragmoplast forms the cell plate.

To gain further insight into the structural changes in these different MT configurations, the development and observation of a living cell system was required. Through the use of suspension-cultured cells from *Arabidopsis* plants stably expressing GFP-tubulin (Ueda et al. 1999), the dynamics of MTs throughout cell division were firstly demonstrated (Hasezawa et al. 2000). A chimeric protein, consisting of GFP fused to the microtubule-binding domain (MBD) of the microtubule-associated protein 4 (MAP4), allowed labeling of MTs (Marc et al. 1998; Granger and Cyr 2000), but detailed observations could not be performed because of the low fluorescence of GFP-MBD. However, by establishing a transgenic BY-2 cell line stably expressing GFP-tubulin fusion proteins (BY-GT), MT dynamics throughout the cell cycle could be precisely followed by confocal laser scanning microscopy (CLSM) (Fig. 1, Kumagai et al. 2001).

Using this BY-GT cell line, the mode of CMT reorganization at the M/G1 interface was successfully clarified as follows (Kumagai et al. 2001). After disruption of the phragmoplast, the CMTs initially become organized in the perinuclear regions, and then elongate to reach the cell cortex where they form "bright spots". The first CMTs subsequently elongate from these spots and become oriented parallel to the long axis towards the distal end of the cells. Around the time when the tips of the parallel MTs reach the distal end, the formation of transverse CMTs follow in the cortex near the division site. Although the role of CMTs in cellulose microfibril deposition has been intensively discussed, recently, cellulose synthase complex visualized by a fusion of yellow fluorescent protein (YFP) with a cellulose synthase protein of CESA6 was revealed to be colocalized with CMTs, indicating a guidance of cellulose deposition by the cytoskeleton (Paredez et al. 2006). MTs were also

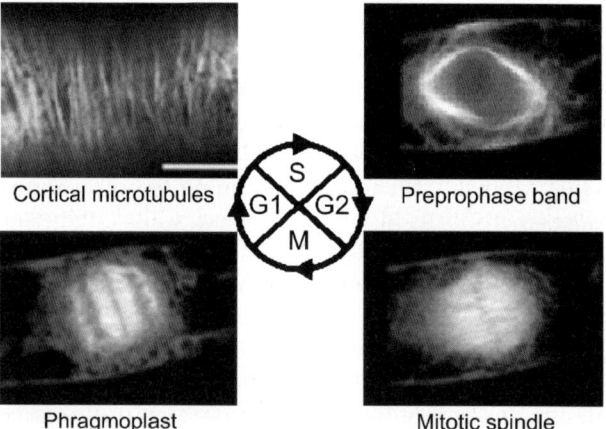

Fig. 1 Microtubule (MT) dynamics during cell cycle progression in tobacco BY-GT cells. MTs are visualized by stable expression of GFP-α tubulin fusion proteins in tobacco BY-2 cells. Typical plant MT structures observed are: cortical MTs in the G1 phase, preprophase band in the G2 phase, mitotic spindle in metaphase of mitosis, and phragmoplast in telophase in mitosis. *Bar*: 10 µm

demonstrated to be involved in secondary wall development during tracheary element differentiation in *Arabidopsis* cell culture expressing GFP-tubulin (Oda et al. 2005; Oda and Hasezawa 2006).

3.2
Involvement of MT-Associated and Regulatory Proteins in Cell Division

Microtubules (MTs) show dynamic changes during cell cycle progression as mentioned above. To regulate such MT dynamics, numerous proteins, known as MT-associated proteins and regulatory proteins, are thought to be involved (Sonobe et al. 2004), and here we describe recent findings regarding some of these, especially involved in cell cycle progression: γ-tubulin, which is a key factor of MT nucleation from the MT organizing center (MTOC); microtubule associated proteins (MAPs), which appear to have roles in MT organization and function; the 26S proteasome, which is a key component of cyclin destruction; and protein kinases, which regulate the activity of MAPs. Kinesins and other kinesin-like motor proteins that are thought to be involved in MT translocation are also described in this volume (Liu 2007, in this volume).

γ-**Tubulin.** In eukaryotic cells, MTs are thought to be nucleated and to be organized from distinct structures called MTOCs, the biochemical and structural features of which have been intensively studied using animal cell centrosomes as models (Andersen 1999). In higher plant cells, however, no distinct structures comparable to the animal centriolar MTOCs are found, and consequently the nature of the plant MTOCs remains unclear (Vaughn and Harper 1998). To explain these plant MTOC characteristics, the localization of γ-tubulin, considered a candidate protein in plant MTOCs, was examined. Gamma-tubulin is a member of the tubulin superfamily and forms γ-tubulin ring complexes (γ-TuRCs) to initiate MT assembly (Schiebel 2000; Job et al. 2003). Electron microscopy suggested that the γ-tubulin in this complex interacts directly with α-tubulin at the microtubule minus end (Keating and Borisy 2000; Wiese and Zheng 2000; Moritz et al. 2000). Although γ-tubulin is localized to a limited area of animal centrosomes and yeast spindle poles, immuno-fluorescence microscopy indicated that in higher plant cells these proteins are widely associated with MT structures of the PPB, mitotic spindle and phragmoplast, as well as with the cell cortex and nuclear surface (Liu et al. 1993, 1994, 1995).

The use of transgenic BY-2 cells stably expressing GFP-γ-tubulin not only confirmed the above localization of γ-tubulin but also revealed, through time-sequential observations, that during MT reorganization at the M/G1 interface γ-tubulin first accumulates on the daughter nuclear surfaces and then spreads onto the cell cortex along with MTs elongating from the daughter nuclei (Kumagai et al. 2003). From these observations, two possible processes can be considered in plant cells; either the MTOCs are extensively distributed as is γ-tubulin, or that the γ-tubulin proteins localize to sites other than the

MTOC. Recently, Murata et al. (2005) demonstrated that in tobacco BY-GT cells, MTs were nucleated as branches on the extant cortical MTs. In addition, at the branch points, γ-tubulin was localized to and functioned in MT nucleation. These observations suggest that in higher plant cells the MTs nucleate from the γ-tubulin complex, the localization of which is not limited to distinct structures but is rather broadly associated with MTs. The MT branching system based on the γ-tubulin complex is also involved in the expansion of phragmoplast MTs (T. Murata, personal communication). In *Arabidopsis*, depletion of γ-tubulin proteins by RNA interference or by gene disruption resulted in aberrant spindle and phragmoplast structures and affected cytokinesis (Binarová et al. 2006; Pastuglia et al. 2006), confirming the involvement of γ-tubulin in cell division through MT nucleation activity.

MAPs. Microtubule associated proteins (MAPs) play potential roles in the regulation of MT organization and function during cell cycle progression. Numerous proteins either with MT binding activity or related to proteins with MT binding activity have been identified to date, and herein, MAP65 and MOR1, which were recently cloned and biochemically characterized (Hussey et al. 2002) are described below. Readers can refer a recent review concerning plant MAPs (Hamada 2007).

MAP65 was first biochemically purified from tobacco BY-2 cells by the polymerization and depolymerization procedure of tubulin (Jiang and Sonobe 1993). The 65 kDa protein shows MT-bundling activity and forms cross-bridge structures between adjacent MTs in vitro, as confirmed by the *Arabidopsis AtMAP65-1* recombinant protein (Smertenko et al. 2004). Although the deduced amino acid sequence of the corresponding cDNA shows only a 20% homology with *Saccharomyces cerevisiae* Ase1 (anaphase spindle elongation factor) and the vertebrate PRC1 (protein regulating cytokinesis), the plant MAP65 is now proposed to be a functional equivalent of these proteins since all three proteins are components of mitotic spindles and their expression or/and localization is cell cycle dependent (Hussey et al. 2002; Mollinari et al. 2002; Schuyler et al. 2003). Immuno-fluorescence microscopy and studies with GFP fusion proteins have further revealed an overlap between the localization of MAP65 and regions of MTs, namely the PPB, mitotic spindle and phragmoplast in addition to CMTs (Smertenko et al. 2000, Mao et al. 2005), suggesting a role for MAP65 in events involving cell cycle progression. Completion of the *Arabidopsis thaliana* genome sequencing project identified nine *Arabidopsis* MAP65 genes (Hussey et al. 2002) with deduced amino acid sequences showing 28–79% identity and predicted molecular masses varying from 54 to 80 kDa. Although characterization of these plant MAP65 proteins is still incomplete, it is possible that the proteins play divergent roles in the regulation of MT organization and activity in plant cells.

MOR1 was initially identified in a screen of *Arabidopsis* mutants with aberrant microtubule patterns (Whittington et al. 2001). One mutant, *mor1*, showed temperature-sensitive cortical microtubule shortening and disorganization, and characterization of the corresponding gene revealed that the deduced amino acid sequence of MOR1 had similarity to *Xenopus* MAP215 (XMAP215) and a human homolog of TOG1p (Charrasse et al. 1998; Tournebize et al. 2000). Biochemical identification of microtubule-associated proteins from tobacco BY-2 cells revealed proteins with a molecular mass of about 200 kDa, which were found to be tobacco homologs of MOR1/XMAP215 (Yasuhara et al. 2002; Hamada et al. 2004). Purified tobacco MOR1 proteins accelerated the polymerization, lengths and numbers of MTs (Hamada et al. 2004). Although a role for MOR1 in CMT organization was initially implied by the mutant phenotype, an *Arabidopsis* mutant of *gem1-1* that affects cytokinesis and cell division patterns at pollen mitosis was found to be allelic to *mor1* (Twell et al. 2002). Recently, Kawamura et al. (2006) reported short and abnormally organized mitotic spindles and phragmoplasts in the *mor1-1* mutant, in which these MT structures persisted longer than in WT plants. Therefore, MOR1 is now implicated in mitosis and cytokinesis as well as in CMT organization.

26S proteasomes. During cell cycle progression, increased evidence indicates the rapid and timely degradation of cell cycle regulatory proteins by the ubiquitin-proteasome pathway in mammalian and yeast cells. In higher plant cells, degradation of cyclins A and B by the 26S proteasome has also been reported (Genschik et al. 1998; Criqui et al. 2000). The 26S proteasome is a highly organized protein-degrading machinery that catalyzes the ATP-dependent degradation of ubiquitinated proteins (Coux et al. 1996; Hershko and Ciechanover 1998). In tobacco BY-2 cells, the 26S proteasome was localized to nuclear envelopes and the MT structures of the PPBs, mitotic spindles and phragmoplasts (Yanagawa et al. 2002). The proteasome inhibitor, MG-132, exclusively caused cell-cycle arrest at the early stages of PPB formation at metaphase and prior to entering the G1 phase, which appears to be closely related to proteasome distribution in the cells. In BY-GT cells treated with MG-132 at metaphase, after formation and collapse of the original phragmoplast, phragmoplast-like structures appeared again (Oka et al. 2004). In such cases, the CMTs never became reorganized and the phragmoplast-like structures had the ability to form cell plates as did the original phragmoplast. Associated with phragmoplast collapse was the disappearance of the kinesin-related protein, TKRP125, which co-localized with phragmoplast microtubules, demonstrated possible microtubule translocation activity, and was implicated in the formation and/or maintenance of the bipolar structure of phragmoplasts (Asada et al. 1997). As the TKRP125 protein level remained constant following MG-132 treatment and was localized to the extra phragmoplast (Oka et al. 2004), it is possible that TKRP125 is one of the targets of the ubiquitin-proteasome degradation pathway during M/G1 transition and that degradation of MT-associated proteins is an indispensable process in cell cycle progression.

Protein kinases. MT dynamics during cell cycle progression have been shown to be regulated by several protein kinases. Mutagenesis of the consensus Cdk site in AtMAP65-1 to a non-phosphorylatable form affected the recruitment of MAP65 proteins to the central spindle at anaphase (Mao et al. 2005). During phragmoplast development, phosphorylation of MAP65-1 by a NRK1/NTF6 MAP kinase regulated MT stability and hence progression of cytokinesis (Sasabe et al. 2006). Recently, three Aurora kinase orthologs, AtAUR1, AtAUR2 and AtAUR3, were identified in *Arabidopsis* (Kawabe et al. 2005). In animal systems, the Aurora serine/threonine protein kinases regulate cell cycle events, including chromosome bi-orientation and segregation, in which the mitotic spindle MTs function (Andrews et al. 2003). GFP-tagged AtAUR proteins demonstrated AtAUR1 and 2 to be localized to the mitotic spindle and AtAUR3 to the chromosomes during cell division. Moreover, overexpression of AtAUR3 induced the disassembly of spindle MTs (Kawabe et al. 2005). These findings indicate that the characteristics and functions of plant Aurora kinases in mitotic events differ from those of other eukaryotes.

4
Dynamics of Actin Microfilaments during Cell Cycle Progression

The actin microfilaments (MFs) of higher plants are involved in various aspects of plant morphogenesis, including pollen tube growth, trichome development and the formation of stomatal complexes (McCurdy et al. 2001). During cell cycle progression, a dynamic change in the actin cytoskeleton has been demonstrated by fluorescent-phalloidin labeling after fixation or by microinjection into various plant cells (Traas et al. 1987; Hasezawa et al. 1989, Schmit and Lambert 1990, Cleary et al. 1992; Liu and Palevitz 1992). These reports described three major MF structures from the G2 phase to mitosis; the MF band in G2 phase, the so-called actin-depleted zone (ADZ) in mitosis, and a structure that overlaps the phragmoplast. The ADZ was identified as a band or zone in the middle of the cell from where cortical MF structures disappeared while adjacent cortical MF structures remained intact. This structure appeared at late G2 to mitosis at the original position of the PPB after the PPB had disappeared and has been proposed to function as a memory of the PPB position (Cleary 1995; Hoshino et al. 2003).

Recent GFP techniques have again successfully labeled plant MFs by using GFP fusions with actin-binding domains (ABD) from mouse talin or fimbrin (Kost et al. 1998; Mathur et al. 1999; Sheahan et al. 2004a; Sano et al. 2005; Voigt et al. 2005). The lower MF binding activity of the fimbrin ABD appears to successfully allow non-invading visualization of MFs (Kovar et al. 2001; Sheahan et al. 2004b) and a tobacco BY-2 cell line, stably transformed with a GFP-fimbrin ABD2 construct (BY-GF), has allowed MF dynamics to be followed during cell cycle progression (Fig. 2, Sano et al. 2005). Time-sequential observations revealed that the MF band in the late G2 phase separated to form

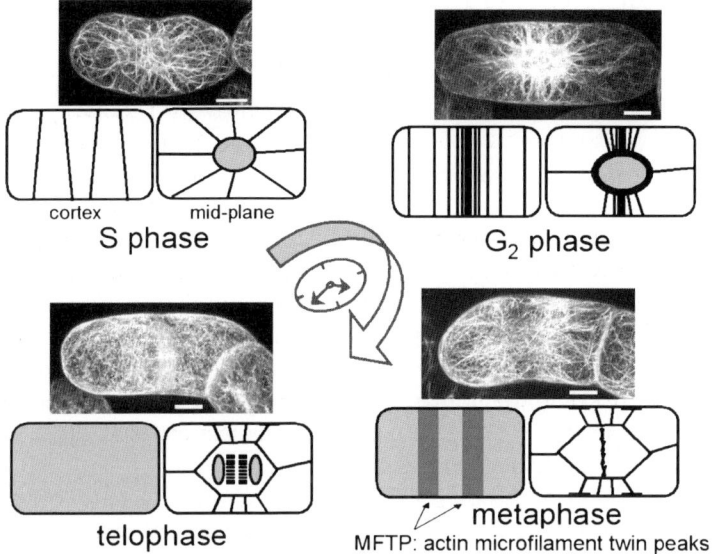

Fig. 2 Actin microfilament (MF) dynamics during cell cycle progression from S phase to telophase in tobacco BY-GF cells. MFs are visualized by stable expression of GFP-fimbrin ABD2 fusion proteins in tobacco BY-2 cells. In S phase, the MFs extend radially from the nuclear surface to the cell periphery. In G2 phase, transversely oriented cortical MFs form a band-like structure at the center of the cell. In metaphase, the cortical MF band in the G2 phase separates to form two bands at the cell cortex, which are designated as the MFTP (microfilament twin peaks). In telophase, the MFs co-localize with the MT phragmoplast. The *scale bars* represent 10 μm

a structure corresponding to the ADZ in mitosis. As measurements of fluorescent intensities of the cell cortex indicated an MF distribution that resembled two peaks, the structure was referred to as "MF twin peaks" (MFTP, Sano et al. 2005). Transition of the MF band to the MFTP may occur by a gradient of MF polymerization/depolymerization activity in this region. Time-sequential observations clearly indicated the formation of the cell plate exactly within the valley between the MFTP at cytokinesis (Sano et al. 2005), corroborating the role of the MFTP/ADZ suggested before.

In addition to clarifying MF dynamics during cell division, studies using the tobacco BY-GF line clearly revealed localization of the MFs along the cytoplasmic strands (Sano et al. 2005); a feature not previously recognized by immuno-fluorescence microscopy because of disruption of the integrity of membrane structures by fixation. Visualization of the vacuolar membrane (VM) by vital VM staining with FM4-64 in living tobacco BY-GF cells also demonstrated co-localization of MF structures and the vacuolar membrane (Higaki et al. 2006). As treatment with the actin polymerization inhibitor, bistheonellide A, disrupted the cytoplasmic strands and inhibited the reorganization of this structure at the early G1 phase, MFs have been implicated in vacuolar morphogenesis and maintenance, as well as in cytoplasmic strand formation during cell cycle progression (Higaki et al. 2006).

5
Vacuolar Dynamics and Functions in Plant Cell Division

Vacuoles constitute the largest of compartments and multifunctional organelles of plant cells (Wink 1993). During plant cell morphogenesis, vacuoles act in combination with the cell wall to generate cell turgor, in addition to their space filling properties within cells (Marty 1999). Recently, GFP fusion with the γ-tonoplast intrinsic protein (γ-TIP) or the vacuolar syntaxin-related AtVam3 protein demonstrated the detailed structure of the vacuolar membrane (Hawes et al. 2001; Saito et al. 2002; Uemura et al. 2002; Kutsuna and Hasezawa 2002). Establishment of a transgenic tobacco BY-2 cell line expressing a GFP-AtVam3 fusion protein (BY-GV) allowed us to follow vacuolar dynamics during cell cycle progression (Fig. 3, Kutsuna et al. 2003). Whereas a single, large vacuole with a rather simple shape was identified from the G1 to G2 phase, compartmented and tubular structures of the vacuolar membrane (TVMs) were found to encircle the mitotic spindle in mitosis, then to be cut off by the cell plate at cytokinesis. Although the TVMs were recognized as small ellipses on each optical section, a 3-D reconstruction, using our originally developed software, REANT (**re**constructor and **an**alyzer of **t**hree-dimensional structure) together with photobleaching analysis, revealed the topological connection between TVMs and the two large vacuoles (Fig. 3, Kutsuna et al. 2003). The REANT permitted quantification of the 3-D structures, including volume and surface area (Kutsuna and Hasezawa 2005). Following cytokinesis, the TVMs returned to form large vacuoles by expansion of their diameters and fusion with each other. Therefore, vacuolar structures changed cyclically between TVMs and large vacuoles during cell cycle progression, and the regeneration of the large vacuoles was not from the simple swelling of small vacuoles.

6
Conclusions

Herein, we have demonstrated the dynamics of the cytoskeleton and vacuoles during cell cycle progression. By establishing transgenic cell lines in which these structures could be visualized we were able to perform time-sequential observations of their structural changes in living cells during cell cycle progression. As a further analysis, by developing the original software package designated as REANT, the precise 3-D organizations of the vacuole could be understood. A new area of research, therefore, will be the

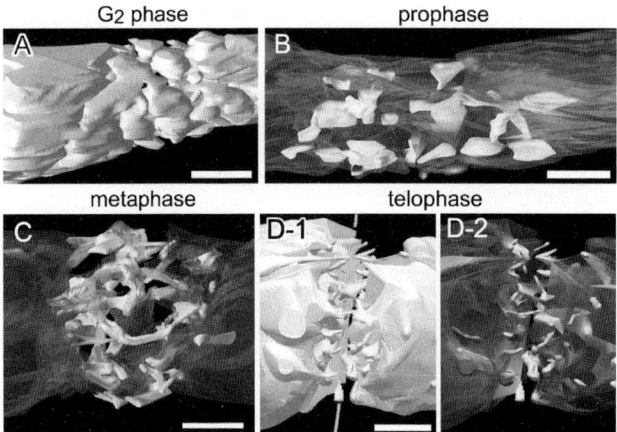

Fig. 3 Three-dimensional (3-D) surface models of vacuoles. The 3-D models are reconstructed from a series of confocal sections. **A** In the G2 phase, the large vacuoles are penetrated by thick cytoplasmic strands near the central region of the cell. **B** In prophase, the TVMs begin to elongate from the smaller compartments of the large vacuole. **C** In metaphase, the TVMs connect the two divided large vacuoles. **D** In telophase, the TVMs are cut off by the developing cell plate. The division plane is represented by the *white line* in **D-1**. To clearly visualize the TVMs, the large vacuoles are rendered transparent in panels **B**, **C** and **D-2**. *Bar*: 10 µm

development of new software for analyzing the respective images obtained by microscopy.

Although both of these techniques are expected to be applicable to other intra-structural analyses, the use of fluorescent proteins requires several points to be carefully considered. First, the fusion proteins may, because of the fused fluorescent protein, produce aberrant internal structures. Second, the fusion proteins may not always label the complete structures of interest. Third, because of the lack of appropriate plant specific promoters, and the consequent use of the cauliflower mosaic virus 35S promoter, the effects of constitutive expression of the target proteins should be carefully considered. Despite these possible limitations, we consider the techniques introduced here will provide new insights into plant intracellular structures and will be useful tools for analyzing various cellular events in higher plant and animal cells.

References

An G (1985) High efficiency transformation of cultured tobacco cells. Plant Physiol 79:568–570

Andersen SSL (1999) Molecular characteristics of the centrosome. Int Rev Cytol 187:51–109

Andrews PD, Knatko E, Moore WJ, Swedlow JR (2003) Mitotic mechanics: the auroras come into view. Curr Opin Cell Biol 15:672–683

Asada T, Kuriyama R, Shibaoka H (1997) TKRP125, a kinesin-related protein involved in the centrosome-independent organization of the cytokinetic apparatus in tobacco BY-2 cells. J Cell Sci 110:179–189

Baskin TI (2001) On the alignment of cellulose microfibrils by cortical microtubules: a review and a model. Protoplasma 215:150–171

Binarová P, Cenklova V, Prochazkova J, Doskocilova A, Volc J, Vrlik M, Bogre L (2006) γ-Tubulin is essential for acentrosomal microtubule nucleation and coordination of late mitotic events in Arabidopsis. Plant Cell 18:1199–1212

Charrasse S, Schroeder M, Gauthier-Rouviere C, Ango F, Cassimeris L, Gard DL, Larroque C (1998) The TOGp protein is a new human microtubule-associated protein homologous to the *Xenopus* XMAP215. J Cell Sci 111:1371–1383

Cleary AL (1995) F-actin redistributions at the division site in living *Tradescantia* stomatal complexes as revealed by microinjection of rhodamine-phalloidin. Protoplasma 185:152–165

Cleary AL, Gunning BES, Wasteneys GO, Hepler PK (1992) Microtubule and F-actin dynamics at the division site in living *Tradescantia* stamen hair cells. J Cell Sci 103:977–988

Coux O, Tanaka K, Goldberg AL (1996) Structure and function of the 20S and 26S proteasomes. Annu Rev Biochem 65:801–847

Criqui MC, Parmentier Y, Derevier A, Shen WH, Dong A, Genschik P (2000) Cell cycle-dependent proteolysis and ectopic overexpression of cyclin B1 in tobacco BY-2 cells. Plant J 24:763–773

Genschik P, Criqui MC, Parmentier Y, Derevier A, Fleck J (1998) Cell cycle-dependent proteolysis in plants: identification of the destruction box pathway and metaphase arrest produced by the proteasome inhibitor MG132. Plant Cell 10:2063–2075

Giddings TH, Staehelin LA (1991) Microtubule-mediated control of microfibril deposition; a re-examination of the hypothesis. In: Lloyd CW (ed) The Cytoskeletal basis of plant growth and form. Academic Press, London, pp 85–100

Granger CL, Cyr RJ (2000) Microtubule reorganization in tobacco BY-2 cells stably expressing GFP-MBD. Planta 210:502–509

Hamada T (2007) Microtubule-associated proteins in higher plants. J Plant Res 120:79–98

Hamada T, Igarashi H, Itoh TJ, Shimmen T, Sonobe S (2004) Characterization of a 200 kDa microtubule-associated protein of tobacco BY-2 cells, a member of the XMAP215/MOR1 family. Plant Cell Physiol 45:1233–1242

Hasezawa S, Hogetsu T, Syono K (1989) Changes of actin filaments and cellulose fibrils in elongating cells derived from tobacco protoplasts. J Plant Physiol 134:115–119

Hasezawa S, Marc J, Palevitz BA (1991) Microtubule reorganization during the cell cycle in synchronized BY-2 tobacco suspensions. Cell Motil Cytoskeleton 18:94–106

Hasezawa S, Ueda K, Kumagai F (2000) Time-sequence observations of microtubule dynamics throughout mitosis in living cell suspensions of stable transgenic *Arabidopsis* – direct evidence for the origin of cortical microtubules at M/G_1 interface. Plant Cell Physiol 41:244–250

Hawes C, Saint-Jore CM, Brandizzi F, Zheng H, Andreeva AV, Boevink P (2001) Cytoplasmic illuminations, in planta targeting of fluorescent proteins to cellular organelles. Protoplasma 215:77–88

Hershko A, Ciechanover A (1998) The ubiquitin system. Annu Rev Biochem 67:425–479

Higaki T, Kutsuna N, Okubo E, Sano T, Hasezawa S (2006) Actin microfilaments regulate vacuolar structures and dynamics: dual observation of actin microfilaments and vacuolar membrane in living tobacco BY-2 cells. Plant Cell Physiol 47:839–852

Hoshino H, Yoneda A, Kumagai F, Hasezawa S (2003) Roles of actin-depleted zone and preprophase band in determining the division site of higher-plant cells, a tobacco BY-2 cell line expressing GFP-tubulin. Protoplasma 222:157–165

Hussey PJ, Hawkins TJ, Igarashi H, Kaloriti D, Smertenko A (2002) The plant cytoskeleton: recent advances in the study of the plant microtubule-associated proteins MAP-65, MAP-190 and *Xenopus* MAP-215-like protein, MOR1. Plant Mol Biol 50:915–924

Jiang C-J, Sonobe S (1993) Identification and preliminary characterization of a 65 kDa higher plant microtubule-associated protein. J Cell Sci 105:891–901

Job D, Valiron O, Oakley B (2003) Microtubule nucleation. Curr Opin Cell Biol 15:111–117

Kawabe A, Matsunaga S, Nakagawa K, Kurihara D, Yoneda A, Hasezawa S, Uchiyama S, Fukui K (2005) Characterization of plant Aurora kinases during mitosis. Plant Mol Biol 58:1–13

Kawamura E, Himmelspach R, Rashbrooke MC, Whittington AT, Gale KR, Collings DA, Wasteneys GO (2006) MICROTUBULE ORGANIZATION 1 regulates structure and function of microtubule arrays during mitosis and cytokinesis in the Arabidopsis root. Plant Physiol 140:102–114

Keating TJ, Borisy GG (2000) Immunostructural evidence for the template mechanism of microtubule nucleation. Nat Cell Biol 2:352–357

Kost B, Spielhofer P, Chua N-H (1998) A GFP-mouse talin fusion protein labels plant actin filaments in vivo and visualize the actin cytoskeleton in growing pollen tubes. Plant J 16:393–401

Kovar DR, Staiger CJ, Weaver EA, McCurdy DW (2000) AtFim1 is an actin filament crosslinking protein from *Arabidopsis thaliana*. Plant J 24:625–636

Kumagai F, Yoneda A, Tomida T, Sano T, Nagata T, Hasezawa S (2001) Fate of nascent microtubules organized at the M/G_1 interface, as visualized by synchronized tobacco BY-2 cells stably expressing GFP-tubulin: time-sequence observations of the reorganization of cortical microtubules in living plant cells. Plant Cell Physiol 42:723–732

Kumagai F, Nagata T, Yahara N, Moriyama Y, Horio T, Naoi K, Hashimoto T, Murata T, Hasezawa S (2003) γ-tubulin distribution during cortical microtubule reorganization at the M/G_1 interface in tobacco BY-2 cells. Eur J Cell Biol 82:43–51

Kutsuna N, Hasezawa S (2002) Dynamic organization of vacuolar and microtubule structures during cell cycle progression in synchronized tobacco BY-2 cells. Plant Cell Physiol 43:965–973

Kutsuna N, Hasezawa S (2005) Morphometrical study of plant vacuolar dynamics in single cells using three-dimensional reconstruction from optical sections. Microsc Res Tech 68:296–306

Kutsuna N, Kumagai F, Sato MH, Hasezawa S (2003) Three-dimensional reconstruction of tubular structures of vacuolar membrane throughout mitosis in living tobacco cells. Plant Cell Physiol 44:1045–1054

Ledbetter MC, Porter KR (1963) A "microtubule" in plant cell fine structure. J Cell Biol 19:239–250

Liu B, Palevitz BA (1992) Organization of cortical microfilaments in dividing root cells. Cell Motil Cytoskel 23:252–264

Liu B, Marc J, Joshi HC, Palevitz BA (1993) A γ-tubulin-related protein associated with the microtubule arrays of higher plants in a cell-cycle dependent manner. J Cell Sci 104:1217–1228

Liu B, Joshi HC, Wilson TJ, Silflow CD, Palevitz BA, Snustad DP (1994) γ-tubulin in *Arabidopsis*: gene sequence, immunoblot, and immunofluorescence studies. Plant Cell 6:303–314

Liu B, Joshi HC, Palevitz BA (1995) Experimental manipulation of γ-tubulin distribution in *Arabidopsis* using anti-microtubule drugs. Cell Motil Cytoskel 31:113–129

Mao G, Chan J, Calder G, Doonan JH, Lloyd CW (2005) Modulated targeting of GFP-AtMAP65-1 to central spindle microtubules during division. Plant J 43:469–478

Marc J, Granger CL, Brincat J, Fisher DD, Kao T-H, McCubbin AG, Cyr RJ (1998) A *GFP-MAP4* reporter gene for visualizing cortical microtubule rearrangements in living epidermal cells. Plant Cell 10:1927–1939

Marty F (1999) Plant vacuoles. Plant Cell 11:587–599

Mathur J, Spielhofer P, Kost B, Chua N-H (1999) The actin cytoskeleton is required to elaborate and maintain spatial patterning during trichome cell morphogenesis in *Arabidopsis thaliana*. Development 126:5559–5568

McCurdy DW, Kovar DR, Staiger CJ (2001) Actin and actin-binding proteins in higher plants. Protoplasma 215:89–104

Mineyuki Y (1999) The preprophase band of microtubules: its function as a cytokinetic apparatus in higher plants. Int Rev Cytol 187:1–49

Mollinari C, Kleman JP, Jiang W, Schoehn G, Hunter T, Margolis RL (2002) PRC1 is a microtubule binding and bundling protein essential to maintain the mitotic spindle midzone. J Cell Biol 157:1175–1186

Moritz M, Braunfeld MB, Guenebaut V, Heuser J, Agard DA (2000) Structure of the γ-tubulin ring complex: a template for microtubule nucleation. Nat Cell Biol 2:365–370

Murata T, Sonobe S, Baskin TI, Hyodo S, Hasezawa S, Nagata T, Horio T, Hasebe M (2005) Microtubule-dependent microtubule nucleation based on recruitment of γ-tubulin in higher plants. Nat Cell Biol 7:961–968

Nagata T (2004) When I encountered tobacco BY-2 cells! In: Nagata T, Hasezawa S, Inze D (eds) Tobacco BY-2 cells: Biotech Agricul Forest, vol 53. Springer-Verlag, Berlin Heidelberg New York, pp 1–6

Nagata T, Kumagai F (1999) Plant cell biology through the window of the highly synchronized tobacco BY-2 cell line. Method Cell Sci 21:123–127

Nagata T, Nemoto Y, Haswzawa S (1992) Tobacco BY-2 cell line as the "HeLa" cell in the cell biology of higher plants. Int Rev Cytol 132:1–30

Oda Y, Hasezawa S (2006) Cytoskeletal organization during xylem cell differentiation. J Plant Res 119:167–177

Oda Y, Mimura T, Hasezawa S (2005) Regulation of secondary cell wall development by cortical microtubules during tracheary element differentiation in *Arabidopsis* cell suspensions. Plant Physiol 137:1027–1036

Oka M, Yanagawa Y, Asada T, Yoneda A, Hasezawa S, Sato T, Nakagawa H (2004) Inhibition of proteasome by MG-132 treatment causes extra phragmoplast formation and cortical microtubule disorganization during M/G$_1$ transition in synchronized tobacco cells. Plant Cell Physiol 45:1623–1632

Paredez AR, Somerville CR, Ehrhardt DW (2006) Visualization of cellulose synthase demonstrates functional association with microtubules. Science 312:1491–1495

Pastuglia M, Azimzadeh J, Goussot M, Camilleri C, Belcram K, Evrard JL, Schmit AC, Guerche P, Bouchez D (2006) γ-tubulin is essential for microtubule organization and development in *Arabidopsis*. Plant Cell 18:1412–1425

Saito C, Ueda T, Abe H, Wada Y, Kuroiwa T, Hisada A, Furuya M, Nakano A (2002) A complex and mobile structure forms a distinct subregion within the continuous vacuolar membrane in young cotyledons of *Arabidopsis*. Plant J 29:245–255

Sano T, Higaki T, Oda Y, Hayashi T, Hasezawa S (2005) Appearance of actin microfilament "twin peaks" in mitosis and their function in cell plate formation, as visualized in tobacco BY-2 cells expressing GFP-fimbrin. Plant J 44:595–605

Sasabe M, Soyano T, Takahashi Y, Sonobe S, Igarashi S, Itoh TJ, Hidaka M, Machida Y (2006) Phosphorylation of NtMAP65-1 by a MAP kinase down-regulates its activity of microtubule bundling and stimulates progression of cytokinesis of tobacco cells. Genes Dev 20:1004–1014

Schiebel E (2000) γ-tubulin complexes: binding to the centrosome, regulation and microtubule nucleation. Curr Opin Cell Biol 12:113–118

Schmit AC, Lambert AM (1990) Microinjected fluorescent phalloidin in vivo reveals the F-actin dynamics and assembly in higher plant mitotic cells. Plant Cell 2:129–138

Schuyler SC, Liu JY, Pellman D (2003) The molecular function of Ase1p: evidence for a MAP-dependent midzone-specific spindle matrix. Microtubule-associated proteins. J Cell Biol 160:517–528

Sheahan MB, Rose RJ, McCurdy DW (2004a) Organelle inheritance in plant cell division: the actin cytoskeleton is required for unbiased inheritance of chloroplasts, mitochondria and endoplasmic reticulum in dividing protoplasts. Plant J 37:379–390

Sheahan MB, Staiger CJ, Rose RJ, McCurdy DW (2004b) A green fluorescent protein fusion to actin-binding domain 2 of *Arabidopsis* fimbrin highlights new features of a dynamic actin cytoskeleton in live plant cells. Plant Physiol 136:3968–3978

Smertenko A, Saleh N, Igarashi H, Mori H, Hauser-Hahn I, Jiang CJ, Sonobe S, Lloyd CW, Hussey PJ (2000) A new class of microtubule-associated proteins in plants. Nat Cell Biol 2:750–753

Smertenko AP, Chang H-Y, Wagner V, Kaloriti D, Fenyk S, Sonobe S, Lloyd CW, Hauser M-T, Hussey PJ (2004) The *Arabidopsis* microtubule-associated protein AtMAP65-1: molecular analysis of its microtubule bundling capacity. Plant Cell 16:2035–2047

Sonobe S, Yokota E, Shimmen T (2004) Tobacco BY-2 cells as an ideal material for biochemical studies of plant cytoskeletal proteins. In: Nagata T, Hasezawa S, Inze D (eds) Tobacco BY-2 cells: Biotech Agricul Forest, vol 53. Springer-Verlag, Berlin Heidelberg New York, pp 98–115

Tournebize R, Popov A, Kinoshita K, Ashford AJ, Rybina S, Pozniakovsky A, Mayer TU, Walczak CE, Karsenti E, Hyman AA (2000) Control of microtubule dynamics by the antagonistic activities of XMAP215 and XKCM1 in Xenopus egg extracts. Nat Cell Biol 2:13–19

Traas JA, Doonan JH, Rawlins DJ, Shaw PJ, Watts J, Lloyd CW (1987) An actin network is present in the cytoplasm throughout the cell cycle of carrot cells and associates with the dividing nucleus. J Cell Biol 105:387–395

Twell D, Park SK, Hawkins TJ, Schubert D, Schmidt R, Smertenko A, Hussey PJ (2002) MOR1/GEM1 has an essential role in the plant-specific cytokinetic phragmoplast. Nat Cell Biol 4:711–714

Ueda K, Matsuyama T, Hashimoto T (1999) Visualization of microtubules in living cells of transgenic *Arabidopsis thaliana*. Protoplasma 206:201–206

Uemura T, Yoshimura SH, Takeyasu K, Sato MH (2002) Vacuolar membrane dynamics revealed by GFP-AtVam3 fusion protein. Gene Cell 7:743–753

Vantard M, Cowling R, Delichère C (2000) Cell cycle regulation of the microtubular cytoskeleton. Plant Mol Biol 43:691–703

Vaughn KC, Harper JDI (1998) Microtubule-organizing centers and nucleating sites in land plants. Int Rev Cytol 181:75–149

Voigt B, Timmers AC, Samaj J, Muller J, Baluska F, Menzel D (2005) GFP-FABD2 fusion construct allows in vivo visualization of the dynamic actin cytoskeleton in all cells of *Arabidopsis* seedlings. Eur J Cell Biol 84:595–608

Wink M (1993) The plant vacuole: a multifunctional compartment. J Exp Bot Suppl 44:231–246

Wasteneys GO (2004) Progress in understanding the role of microtubules in plant cells. Curr Opin Plant Biol 7:651–660

Wiese C, Zheng Y (2000) A new function for the γ-tubulin ring complex as a microtubule minus-end cap. Nat Cell Biol 2:358–364

Whittington AT, Vugrek O, Wei KJ, Hasenbein NG, Sugimoto K, Rashbrooke MC, Wasteneys GO (2001) MOR1 is essential for organizing cortical microtubules in plants. Nature 411:610–613

Yanagawa Y, Hasezawa S, Kumagai F, Oka M, Fujimuro M, Naito T, Makino T, Yokosawa H, Tanaka K, Komamine A, Hashimoto J, Sato T, Nakagawa H (2002) Cell-cycle dependent dynamic change of 26S proteasome distribution in tobacco BY-2 cells. Plant Cell Physiol 43:604–613

Yasuhara H, Muraoka M, Shogaki H, Mori H, Sonobe S (2002) TMBP200, a microtubule bundling polypeptide isolated from telophase tobacco BY-2 cells is a MOR1 homologue. Plant Cell Physiol 43:595–603

Zhang D, Wadsworth P, Hepler PK (1990) Microtubule dynamics in living dividing plant cells: confocal imaging of microinjected fluorescent brain tubulin. Proc Natl Acad Sci USA 87:8820–8824

Plant Cell Monogr (9)
D.P.S. Verma and Z. Hong: Cell Division Control in Plants
DOI 10.1007/7089_2007_126/Published online: 25 July 2007
© Springer-Verlag Berlin Heidelberg 2007

Mitotic Spindle Assembly and Function

J. Christian Ambrose · Richard Cyr (✉)

Department of Biology, The Huck Institutes of the Life Sciences,
Integrative Biosciences Graduate Degree Program, Plant Physiology Program,
The Pennsylvania State University, University Park, PA 16802, USA
rjc8@psu.edu

Abstract The ability of plant mitotic spindles to form and function with robust accuracy, in the absence of centrosomes, underscores the importance and prevalence of centrosome-independent pathways of spindle assembly. This work includes overviews of plant mitotic spindle structure and formation, microtubule-associated proteins involved in plant mitosis, and the multiple pathways used by plants to promote robust spindle assembly and function.

1
Introduction

Plant growth can be described as increases in cell numbers and cell size. Since plant cells are encased within rigid cell walls and "cemented" together, the timing, placement and direction of cell division and expansion are crucial in specifying proper tissue and organ morphogenesis. This work focuses on cell division from the point of view of the organization and functioning of the microtubule (MT) spindle apparatus and its associated proteins.

During cell division, partitioning of replicated chromosomes into daughter cells is mediated by the mitotic spindle apparatus, an array of MTs and microtubule-associated proteins (MAPs). Herein, we first describe the plant mitotic spindle during its different stages of development, follow with a survey of known MAPs involved in plant spindle function, and end with a discussion of the various pathways of spindle formation and function and how these pathways work both redundantly and synergistically in extant plants. For other reviews on various aspects of plant mitosis, see (Baskin and Cande 1990; Mineyuki 1999; Lloyd and Chan 2006).

2
Plant Mitotic Spindle Formation, Function and Morphology

2.1
General Description of the Mitotic Spindle

Although a number of variations exist between different eukaryotic kingdoms, the mitotic apparatus in its simplest form is comprised of two opposing

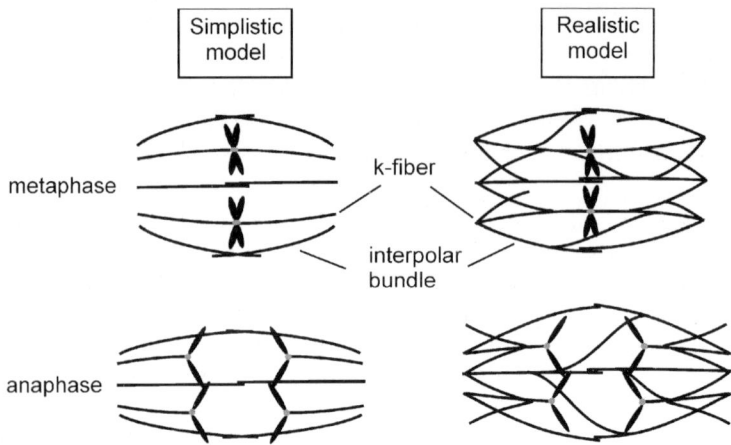

Fig. 1 Cartoons illustrating simplistic and realistic models of the higher plant mitotic spindle during metaphase and anaphase

sets of MTs oriented with their minus ends near the poles and their plus ends in the midzone, where they terminate at chromosomal kinetochores, forming kinetochore MTs (kMTs), or overlap in an antiparallel manner with MTs from the opposite half-spindle, forming interpolar MTs. Multiple kMTs associate to form kinetochore MT fibers (k-fibers), which mediate congression of chromosomes to a point equidistant from each pole during prometaphase, maintain them at the metaphase plate, and later shorten during anaphase A to pull sister chromatids to the poles (Fig. 1, left panel). Interpolar MT bundles contribute to the structural integrity of the spindle, to spindle elongation during anaphase B, and play a pivotal role in spindle formation. This simplistic description is sufficient for initially orienting the reader, however, in reality, individual k-fibers and interpolar MT bundles exhibit substantial branching and lateral interactions with one another (Fig. 1, right panel). As will be discussed, these interactions likely play a fundamental role in ensuring the integrity of spindle structure and the fidelity of chromosome segregation.

2.2
Premitotic Nuclear Migration and PPB Formation

A prominent indication of ensuing cell division in plant cells is the migration of the nucleus during S/G2 phase to the future cell division site (Sinnott and Bloch 1940). Nuclear migration is mediated via cytoplasmic strands that emanate out from the nucleus and anchor at the cortex. These strands contain actin microfilaments (MFs) and MTs, both of which are involved in nuclear migration, although the relative contributions of each vary between species and cell type (Bakhuizen et al. 1985; Venverloo and Libbenga 1987; Katsuta et al. 1990). Upon reaching the division plane, migration ceases and the nucleus becomes anchored in place, its position being more resistant to centrifugation or cytoskeletal-disrupting drugs (Venverloo and Libbenga 1987; Katsuta et al. 1990; Panteris et al. 2006). In highly vacuolated cells such as those found in *Nautilocalyx* leaves, cytoplasmic strands initially radiate out in all directions from the centralized nucleus, and then later become more or less restricted to a dense disc-like structure, the phragmosome, which defines the future division plane (Sinnott and Bloch 1940). In addition to the phragmosome, several polar cytoplasmic strands may remain, which supplement the phragmosome in nuclear positioning and division plane determination (Traas et al. 1987; Lloyd and Traas 1988). Both the polar strands and the phragmosome contain MTs and actin, the cytoskeletal elements responsible for nuclear migration and positioning. The phragmosome is thought to play a role in division plane determination, although its role may not be essential since it is not obvious in some cell types, such as the densely cytoplasmic meristematic cells of root tips. Similarly, the phragmosome itself may not play a direct role in division plane determination; instead it may be a visible manifestation of cellular polarity in large vacuolated cells, this same polarity being present also in smaller cytoplasmic cells.

In addition to the migration of the nucleus to the division site, changes in the arrangement of cortical MTs are also indicative of ensuing cell division. Concomitant with or shortly after nuclear migration, the transversely oriented MTs of the cortex begin to accumulate in a dense band encircling the nucleus, while simultaneously becoming depleted in the distal cortical regions of the cell. This preprophase band (PPB) of MTs is unique to the vegetative cells of land plants, and forecasts the future plane of cell division (Mineyuki 1999). As the cell nears prophase, MTs become more numerous and densely packed within the PPB as it narrows into a tight band (Mineyuki et al. 1991; Vos et al. 2004). For reasons that will be discussed throughout this work, it is appropriate for this cortical structure to be considered an integral component of the plant mitotic spindle.

2.3
Development of the Prophase Spindle

The first sign of the prophase spindle is the appearance of a cytoplasmic clear zone (seen with light microscopy) largely devoid of organelles, around the prophase nucleus (Bajer 1957). In narrow cells, this thin layer later becomes asymmetrically distributed into two cone-like projections along the future long axis of the spindle. These were termed "polar caps" by early light microscopists (Robyns 1929). It was later shown that the clear zone and polar caps exhibit birefringence under polarizing light microscopy (Inoue and Bajer 1961), and were subsequently shown to contain MTs (Bajer and Molè-Bajer

1969; Sakai 1969). Because of their close association with the nuclear envelope (NE), the MTs of the clear zone and polar caps will subsequently be referred to herein as perinuclear MTs. Perinuclear MTs are initially distributed randomly along the surface of the NE, radiating out into the cytoplasm in all directions (De Mey et al. 1982; Mineyuki et al. 1991). Close examination of these perinuclear MTs in *Haemanthus* endosperm reveals that most of them are organized into groups of 2–3 MTs, called MT-converging centers (MTCCs), which collect together at one end on the NE, while the other ends radiate into the cytoplasm (Bakhuizen et al. 1985; Smirnova and Bajer 1994). It was assumed that the converging ends of the MTs within MTCCs were minus ends, based on the known capacity of the nucleus to spawn MT growth (Stoppin et al. 1994). Visualization of growing MT plus ends with GFP::EB1 later confirmed that MTs grow predominately away from the nucleus at this stage (80% away, 20% toward), consistent with the minus ends being anchored at the nuclear surface (Dhonukshe et al. 2005).

Concurrent with PPB formation and narrowing, MTs become more numerous on the surface of the nucleus, radiating out into the cytoplasm and often reaching the cortical regions of the cell. In later prophase, these radiating MTs become more dynamic and shorter in all regions except in the plane of the PPB, where they continue to interact with the PPB site (Mineyuki et al. 1991; Dhonukshe et al. 2005). As the PPB narrows, MTs radiating from the nucleus and connecting to the cortex become progressively restricted to the same region of the cortex that is occupied by the PPB, appearing as spokes in a wheel from the pole view (Wick and Duniec 1983, 1984; Mineyuki and Palevitz 1990; Nogami et al. 1996). At the same time, existing perinuclear MTs become sorted asymmetrically into two poles (forming the polar caps), the axis of which defines the prophase spindle and corresponds to the future spindle axis (Schmit et al. 1983). In cells with a PPB, the axis of the prophase spindle is typically perpendicular to the plane of the PPB. MTs within the poles run parallel to the future spindle axis, often extending along the surface of the nucleus where they may interdigitate in the equatorial region; or they may extend out laterally into the cytoplasm to contact the PPB region (Pickett-Heaps and Northcote 1966; Burgess 1970; Bakhuizen et al. 1985). MTs of the prophase spindle typically converge at pointed poles, which may appear annular in structure (Wick and Duniec 1984; Marc and Gunning 1988; Liu et al. 1993). In some species, such as soybean suspension cultures, prophase spindle poles are comparatively broad (Wang et al. 1991). MTs may also become associated with protrusions or invaginations near the polar regions of the NE (Bajer and Molè-Bajer 1969; Hanzely and Schjeide 1973), or may also occasionally penetrate into the nucleus (Pickett-Heaps and Northcote 1966).

Although no discrete microtubule-organizing centers (MTOCs) have been observed in the polar caps of higher plants, extensive ER is present throughout the polar caps, and has been observed to frequently coalign and/or interact with MTs (Burgess and Northcote 1967; Bajer and Molè-Bajer 1969). Since plants lack centrosomes or spindle pole bodies, it has been suggested that the nuclear surface acts as an MTOC (Schmit 2002). Abundant evidence supports the role of the nucleus as an MTOC: (1) isolated nuclei are able to nucleate MT growth from their surfaces in vitro (Stoppin et al. 1994); (2) gamma tubulin, the universal MT nucleator localizes to the NE (Liu et al. 1993; Erhardt et al. 2002); (3) spc98, a key centrosomal component is found in plants and localizes to the nuclear surface (Erhardt et al. 2002); (4) adding antibodies to spc98 inhibits the ability of nuclei to nucleate MTs (Erhardt et al. 2002); (5) antibodies to centrosomal components frequently localize to the NE, and; (6) MTs reform on the NE after recovery from depolymerization with drugs (Falconer et al. 1988; Galatis and Apostolakos 1991).

2.4
Formation of the Prometaphase and Metaphase Spindle

Prometaphase begins at nuclear envelope breakdown (NEB) and lasts until a clear metaphase plate is formed. During prometaphase, MTs congress inward from the poles to the equator and immediately begin forming connections with kinetochores (forming k-fibers) or with other MTs from the opposite pole (forming interpolar bundles). K-fiber formation is asynchronous between sister kinetochores (Bajer 1987), and corresponds with the initiation of chromosome movements. K-fibers mediate the congression of chromosomes toward the metaphase plate and the separation of sister chromatids during anaphase, while interpolar MTs serve to stabilize the spindle along its length and width, and participate in anaphase spindle elongation. In most cases, the metaphase spindle lacks well-defined poles (as seen in animal cells), even if the preceding prophase spindle did contain focused poles (Fowke 1993). Metaphase spindles instead exhibit a barrel-shaped morphology, in contrast to animal spindles, which maintain a fusiform morphology throughout mitosis, even in the absence of centrosomes (Khodjakov et al. 2000).

Early EM studies showed that prior to NEB, MTs may penetrate the NE in the polar regions (Pickett-Heaps and Northcote 1966) and also associate with deformations of the prophase nucleus (Bajer and Molè-Bajer 1969). During NEB in *Haemanthus* endosperm, the NE appears wavy, giving the impression of "boiling", and subsequently begins fragmentation at the polar regions, with large invaginations of membrane apparently being pushed in by bundles or sheets of 50–100 MTs (Bajer and Molè-Bajer 1969). The NE fragments are later degraded, or they may persist within the spindle throughout mitosis, becoming associated with chromatin, and in some cases, becoming reincorporated into the NE of the daughter nuclei (Bajer and Molè-Bajer 1969). Electron micrographs show that most MTs invade the nucleus in the above-mentioned bundles before becoming attached to kinetochores or opposing MTs. These

MTs attach to kinetochores throughout prometaphase and metaphase, such that the number of MTs in an individual K-fiber gradually increases until the start of anaphase, at which point a drastic decrease in numbers is observed (Jensen and Bajer 1973). This phenomenon may vary between species however, since birefringence increases throughout prometaphase but then ceases abruptly at metaphase (when chromosome oscillations stop) in *Tilia americana* endosperm (Fuseler 1975).

During metaphase, a majority of spindle MTs reside within thick bundles of k-fibers or interpolar bundles, and the degree of this bundling is exacerbated by the effects of taxol (Molè-Bajer and Bajer 1982, 1983). At this stage, interpolar MT bundles have been shown to branch frequently and intermingle with k-fibers and other interpolar bundles (Jensen and Bajer 1973). The structure of an individual k-fiber resembles that of a fir tree, the base of which is at the kinetochore, and the bough is at the poles (Bajer and Molè-Bajer 1986; Palevitz 1988). Numerous MTs splay out (like branches) along the length of the fiber, becoming tapered toward the pole. Indeed these microtubule fir trees may represent a fractal manifestation of the fundamental plant MT organizational unit, the MTCC (Smirnova 1998). The k-fibers are thus higher order structures, composed of many short and overlapping MTs. Interestingly, gamma tubulin associates along the length of interphase MTs and spindle MTs, and nucleates MT branches from preexisting cortical MTs at an angle of 40 degrees (Murata et al. 2005). From examining micrographs of MTCCs, it appears that most MTs within the MTCC branch at roughly a 40 degree angle. It therefore seems possible that the nucleation of MTs occurs all along the length of k-fibers, rather than predominately at the pole regions (as is often depicted in textbooks). Similarly, chromatin may provide a source of MT nucleation, since plant spindles recovering from cold- or drug-induced MT depolymerization frequently appear as a collection of semi-autonomous "minispindles", each one comprised of an individual kinetochore of a given chromosome with its own opposing fir-trees (each fir tree emanating from a sister kinetochore), creating a diamond shape with a dark center (Bajer 1987; Cleary and Hardham 1988; Falconer et al. 1988).

Numerous ER is found throughout prometaphase and metaphase spindles, frequently as long tubular structures closely apposed to MTs (Pickett-Heaps and Northcote 1966), similar to that observed in prophase polar caps. Although possibly involved in MT stabilization or nucleation, the function of this ER is not known. Given the presence of crosslinks between this ER and spindle MTs, ER may comprise a component of the theoretical "spindle matrix", an electron-dense substance proposed to provide a structural scaffold for spindle MTs and motors (Pickett-Heaps and Northcote 1966; Pickett-Heaps et al. 1984).

One intriguing question concerning spindle function is how MTs, which are highly dynamic polymers that exhibit frequent transitions between growth and shortening, give rise to the metaphase spindle, which maintains a constant length. Poleward MT flux refers to the situation wherein tubulin subunits are incorporated at MT plus ends in the midzone, and then transported to the poles via the combined activities of motor activity and minus end depolymerization; the net result being that the spindle length remains unchanged (Mitchison 2005). Although plant spindle MTs exhibit increased dynamics compared to interphase MTs, initial experiments employing photobleached fluorescent tubulin subunits were unable to demonstrate flux (Hush et al. 1994). Using newer technology, however, our lab has detected poleward flux of bleached GFP::Tubulin subunits within BY-2 spindles (our unpublished observation). More recently, a paper by Dhonukshe et. al demonstrates this as well (Dhonukshe et al. 2006). This suggests that poleward flux may be a conserved feature of spindle functioning in eukaryotic cells, although it has not been detected to date in fungi; possibly due to technological limitations and small spindle sizes (Mallavarapu et al. 1999; Maddox et al. 2000). It is interesting to note that as early as the 1950s, one of the pioneers of live-cell microscopy, Andrew Bajer, described poleward movement of small particles, granules, and nucleoli, which he collectively termed "acentrics"—effectively demonstrating poleward flux in plant spindles before it had even been hypothesized (Bajer 1967; Allen et al. 1969; Bajer and Molè-Bajer 1972).

2.5
Anaphase

Anaphase chromosome movements in plants are derived from two different processes: (1) kinetochore-to-pole movement, which results from the shortening of k-fibers; and (2) spindle elongation, which is derived from the sliding apart of the overlapping parts of antiparallel interpolar MT bundles. In animal cells, these processes have been described as anaphase A and B, respectively, due to their sequential nature. In plants, however, the two processes appear to be largely concurrent, with varying degrees of overlap in different cell types (Ota 1961; Harris and Bajer 1965; Fuseler 1975; Ryan 1983). As k-fibers shorten during anaphase A, they become disorganized, losing their fir-tree appearance and becoming splayed out from the kinetochore into the pole region (Bajer 1968; Hard and Allen 1977). At the same time, interpolar MT bundles in the midzone become more ordered and densely packed as they elongate and slide apart, thus decreasing the degree of overlap between half-spindles (Jensen and Bajer 1973; Euteneuer and McIntosh 1980; Euteneuer et al. 1982). Formation of the cell plate often begins in mid to late anaphase, as indicated by the appearance of vesicles and cell-plate precursors near interdigitating regions of interpolar MTs (Ota 1961; Hepler and Jackson 1968; Fuseler 1975). MTs of the anaphase spindle midzone persist into telophase, giving rise to the phragmoplast during cytokinesis.

3
Microtubule-Associated Proteins in Plant Mitotic Spindles

Identification of the molecular players involved in plant mitosis has traditionally lagged behind that of animals and fungi; however, with the sequencing of the *Arabidopsis* and rice genomes, the advent of GFP technology, and an explosion in proteomic technologies, advances are being made at the forefront of the mitosis field. Since a number of excellent reviews are available on the molecular characteristics of animal mitosis (Gadde and Heald 2004; Kline-Smith and Walczak 2004), this section will emphasize the available data from plants.

Control of MTs dynamics and organization within the plant mitotic spindle is accomplished in part with a diverse ensemble of structural and regulatory microtubule-associated proteins (MAPs) (Wick 2000; Hussey et al. 2002). MAPs can be divided roughly into two main groups: structural MAPs and motor MAPs.

3.1
Structural MAPs

Structural MAPs can be further subdivided into three functional classes: (1) MAPs that affect MT dynamics (e.g. stabilize or destabilize MTs); (2) MAPs that crosslink or bundle MTs; and (3) MT-severing MAPs. It should be noted, however, that these groupings are not mutually exclusive.

MAP-65. To date, nine MAP-65 proteins have been identified or predicted in *Arabidopsis*, three in tobacco BY-2 suspension cells, and three in carrot suspension cells (Hussey et al. 2002). These proteins share homology to human PRC1 and yeast Ase1p, both of which contribute to the formation and integrity of antiparallel midzone MTs via crosslinking and stabilization (Mollinari et al. 2002; Schuyler et al. 2003; Zhu and Jiang 2005). In plants, several members of the MAP-65 family have also been localized at the midzone, suggesting they share a similar function (Muller et al. 2004; Chang et al. 2005; Mao et al. 2005). Several other members localize to the cortical arrays, where they have been implicated in MT bundling (Van Damme et al. 2004b; Mao et al. 2005); indeed, purified carrot MAP-65 bundles MTs in vitro, forming 20–25 nm crossbridges between MTs (Chan et al. 1999). In the case of AtMAP65-4, localization to the spindle is observed, but midzone enrichment is not seen (Van Damme et al. 2004b). The presence or absence of CDK phosphorylation sites between different MAP-65 family members may contribute to their differentially modulated localization dynamics. Interestingly, mutation of the MAP consensus Cdk site to an unregulated form results in the premature accumulation of AtGFP-MAP65-1 to the midzone in early prometaphase, whereas the wild-type version doesn't appear there until anaphase (Mao et al. 2005).

Mor1/Map200 (Dis1/XMAP215/TOG). The *Arabidopsis MOR1* gene was identified in a screen for plants with aberrant MT organization and was found to share homology to the Dis1/XMAP215/TOG family of MAPs (Whittington et al. 2001), which are known to facilitate polymerization of MTs (Whittington et al. 2001; Hamada et al. 2004). Although initially characterized by its disruption of cortical MT arrays at restrictive temperature, additional *mor1* defects during mitosis have been reported (Kawamura et al. 2006). At restrictive temperature, cells of *mor1* mutants frequently fail to form PPBs, and in these cells, the subsequent orientation of the spindle and phragmoplast is aberrant (Eleftheriou et al. 2005; Kawamura et al. 2006). Spindles of *mor1* mutants are also significantly shorter than normal, providing additional support for its role as an MT-polymerizing factor. This finding presents the first genetic alteration of spindle length in plants, and emphasizes the universality of MT dynamics as a key governor of spindle length; which is also becoming apparent in animal systems (Goshima et al. 2005). The tobacco MOR1 homolog, TMPB200, has been localized to all MT arrays, and purified TMBP200 has the ability to crosslink MTs in vitro, suggesting a possible role in structural organization of MT arrays, in addition to modulating dynamics (Yasuhara et al. 2002; Hamada et al. 2004). Although structural and motor proteins facilitate the integrity and functioning of spindles, alterations of MT dynamic properties generally result in the most dramatic changes in spindle length (Goshima et al. 2005).

Interestingly, another member of this family, XMAP215, from *Xenopus* acts antagonistically in mitotic spindles with the MT-destabilizing kinesin, XKCM1, in regulating MT length (Tournebize et al. 2000). A similar scenario exists in yeast, where the MT-destabilizing kinesin Kar3p counteracts the MT-stabilizing effects of another kinesin, Kip1p (Huyett et al. 1998). Whether an analogous situation exists in plants remains to be seen; although it is likely, given that the counterbalancing of MT polymerizing and depolymerizing factors is emerging as a general mechanism in regulating MT structures and fine-tuning of MT dynamics in vivo (Valiron et al. 2001).

EB1. It is the founding member of a conserved group of plus end tracking proteins (+TIPs) involved in modulation of MT dynamics and attachment of MT plus ends to cellular structures such as kinetochores and cortical division sites (Tirnauer et al. 2002; Tirnauer et al. 2004). Three *Arabidopsis* EB1 homologs have been cloned and shown to localize to mitotic spindles, although no functional data have yet been reported (Chan et al. 2003; Mathur et al. 2003; Van Damme et al. 2004a; Chan et al. 2005; Dixit et al. 2006). It will be interesting to learn the roles of plant EB1, given the presence of the novel MT arrays seen in plants.

Tangled. The maize *Tangled* gene encodes a MAP with distant homology to the animal EB-1 binding partner APC (Smith et al. 2001). Mutants in the *tangled* gene form normal mitotic arrays, but the trajectory of the phragmoplast is aberrant, leading to oblique cell plates and disorganized cell files (Cleary and Smith 1998). Although transverse PPBs appear normal, the occurrence of longitudinal PPBs is markedly reduced, suggesting a role for Tangled in division plane determination. Although Tan1 has been localized to the mitotic spindle (Smith et al. 2001), its role there remains to be determined.

Spiral. The *Arabidopsis Spiral* (*SPR1*) gene product is a novel 12 kD plant-specific MAP belonging to a family with five other *SPR*-like genes. First identified in screens for mutants with skewed root growth, *spr1* plants exhibit right-handed twisting of organs (Furutani et al. 2000; Sedbrook et al. 2004). SPR1::GFP localizes to all MT arrays and is a +TIP (Nakajima et al. 2004; Sedbrook et al. 2004). Although localized in mitotic spindles, its function also remains to be determined (Sedbrook et al. 2004).

Map70. Given the complex organization and rearrangements of plant MT arrays, a large toolbox of novel MAPs may exist that would not be identified by homology searches using known animal MAPs. The identification of the novel *Arabidopsis* protein MAP70 through proteomic screens for MT-binding proteins is one such example (Korolev et al. 2005). Identified in proteomic screens of proteins bound to taxol-stabilized MTs in *Arabidopsis* suspension cultures, MAP70 is a novel plant-specific coiled-coil protein with 5 *Arabidopsis* family members. GFP fusions to MAP70 decorate all 4 arrays (Korolev et al. 2005), although its function remains to be identified.

Map190. MAP190 is a novel plant MAP isolated from BY-2 cells based on its affinity for both MTs and MFs; and it localizes to the nucleus in interphase and to the spindle during mitosis (Igarashi et al. 2000). The presence of a MAP with actin-binding capacity within mitotic spindles suggests a possible role for actin in mitosis. Actin has been found within spindles and has been implicated in spindle positioning and orientation, but its contribution to spindle function, if any, is questionable (Mineyuki and Palevitz 1990; Staiger and Cande 1991).

3.2
Kinesins: Force-Generating MAPs

Kinesins are force-generating MAPs that use the hydrolysis of ATP to move unidirectionally along microtubules, carrying cellular cargo such as organelles, vesicles, and chromosomes (Dagenbach and Endow 2004; Miki et al. 2005). A standardized nomenclature was developed that recognizes 14 distinct kinesin families (Lawrence et al. 2004). This nomenclature will be used throughout this work. Because of the conserved nature of kinesin function within the spindle apparatus, it is helpful to give a short overview of the known functional data available from non-plant studies before describing plant mitotic kinesins.

Kinesins involved in spindle function. At least eight kinesin families contain members that are involved in some aspect of cell division (Goshima and Vale 2003; Zhu et al. 2005). Of these, five families appear to facilitate chromosome movement via direct binding to distinct regions of chromosomes. Members of the Kinesin-7 (CENP-E) and Kinesin-13 (KinI/MCAK) families associate with chromosomal kinetochores, where they mediate stable attachment of MTs to kinetochores and congression to the metaphase plate (kinesin-7), or through MT depolymerase activity, modulate spindle MT dynamics (Kinesin-13) to facilitate spindle bipolarity and length (Yen et al. 1991; Schaar et al. 1997; Goshima and Vale 2003; Sharp et al. 2005; Kapoor et al. 2006). Similar to Kinesin-13, members of the Kinesin-8 (Kip3) family also appear to act as kinetochore-localized MT depolymerases because their depletion results in elongated spindles and failed chromosome congression (West et al. 2001; Savoian et al. 2004; Goshima and Vale 2005).

Members of kinesin-4 (chromokinesin/KIF4) and kinesin-10 (Nod/Kid) families bind directly to chromosomal arms, where they facilitate motility along non-kinetochore MTs toward the metaphase plate, thus producing so-called "polar ejection forces" (Funabiki and Murray 2000; Yajima et al. 2003).

In contrast to the above kinesin families, which directly influence chromosome behavior, members of the Kinesin-5 (BimC), Kinesin-6 (MKLP1), and Kinesin-14 (C-terminal) families facilitate spindle organization and function via crosslinking MTs and sliding them relative to one another, thereby indirectly influencing chromosome motion and behavior.

Plant kinesins involved in mitotic spindle function. Of the 61 predicted kinesins in the *Arabidopsis* genome, 18 have so far been reported on, and of those, seven have been implicated directly in mitosis and genes from 23 have been shown to be up-regulated during mitosis (Vanstraelen et al. 2006). Although flowering plants lack four of the 14 kinesin families designated by Lawrence et al. (2004), they do contain representatives of each of the eight kinesin families that have so far been implicated in mitosis in animals or fungi. Furthermore, plants contain several novel plant-specific families not included in the 14-family nomenclature (Richardson et al. 2006), some of which may be involved in mitosis.

Several kinesins have been localized to the PPB and/or spindle (Liu et al. 1996; Smirnova et al. 1998; Kong and Hanley-Bowdoin 2002; Vanstraelen et al. 2004; Ambrose et al. 2005), however, functional data has only been reported for ATK1, ATK5, and KCBP (Vos et al. 2000; Marcus et al. 2003; Ambrose et al. 2005; Ambrose and Cyr 2007); therefore, the bulk of this section will focus on this functional data within the broader context of common mechanisms in eukaryotic mitosis.

Over a third of the 61 predicted kinesins in the *Arabidopsis* genome belong to the Kinesin-14 family, which comprises the sole group of minus end-directed kinesins (Reddy 2001). Members of this family typically contain a C-terminally located motor domain (although central- or N-terminal locations are also present in some plant Kinesin-14s) and function in the gathering of microtubule minus ends into poles and also in providing inward forces between overlapping MT plus ends to facilitate spindle compaction

(Walczak et al. 1998; Vos et al. 2000; Chen et al. 2002; Ambrose et al. 2005; Ambrose and Cyr 2007). In animals and fungi, these inward-directed sliding forces generated by Kinesin-14 counterbalance the outward pushing forces provided by Kinesin-5 (BimC) family members, as evidenced by the finding that loss of Kinesin-14 partially rescues the spindle-collapse phenotype found with Kinesin-5 disruption (Hoyt et al. 1993; O'Connell et al. 1993; Mountain et al. 1999; Sharp et al. 1999, 2000). Although the balance of inward forces generated by Kinesin-14 with outward forces generated by Kinesin-5 has not yet been identified in plant spindles, it probably exists based on the similarity of Kinesin-14 phenotypes (Chen et al. 2002; Marcus et al. 2003; Ambrose et al. 2005; Ambrose and Cyr 2007) and the known localization of Kinesin-5 family members to the mitotic spindle in tobacco BY-2 cells and carrot cells (Asada et al. 1997; Barroso et al. 2000).

It has been suggested that the abundance of Kinesin-14 family members in plants accounts for the absence of the other known eukaryotic minus end-directed motor, dynein (Reddy and Day 2001; Ambrose et al. 2005). While dynein-containing fungi and animal genomes generally contain only one or a few dyneins, the varied association of components of the dynein regulatory complex, dynactin, probably accounts for the regulation and modulation of the diverse cellular activities of cytoplasmic dynein. In higher plants these multiple roles may be taken on by different Kinesin-14 family members, instead of one or a few minus end motors with complex regulatory subunits. Indeed, Kinesin-14 and cytoplasmic dynein share similar functions within the spindle apparatus in acentrosomal *Xenopus* egg extracts, where the depletion of both cytoplasmic dynein and the Kinesin-14 XCTK2 leads to an exacerbation of the splayed-pole phenotype observed with inhibition of either one alone (Walczak et al. 1998).

The Kinesin-14 ATK5 is a +TIP that localizes to the mitotic spindle midzone from early prometaphase until telophase, where it generates inward forces and organizes spindle MTs (Ambrose et al. 2005; Ambrose and Cyr 2007). Spindles from *atk5-1* null mutants are elongated and bent during prometaphase, and later become abnormally broad at the midzone and poles, consistent with a loss of MT organizing activity. Additionally, although prophase spindle poles appear normal, the broadening of poles in *atk5-1* spindles appears during prometaphase, suggesting that ATK5 is a major pole determinant during prometaphase (Ambrose and Cyr 2007). The observation that poles become splayed during prometaphase, when the PPB has already disappeared, but remain tight during prophase, when the PPB is present, illustrates a redundancy between the PPB and Kinesin-14 in pole formation. Similarly, the *atk1-1* phenotype, which affects pole formation, is more pronounced in meiotic cells, which lack PPBs (Chen et al. 2002). In the next section, we discuss how different mechanisms and pathways interact and supplement one another in the formation of the plant mitotic spindle.

4
Pathways of Mitotic Spindle Assembly and Function in Higher Plants

During cell division, two general mechanistic pathways coexist to facilitate efficient and accurate segregation of chromosomes into daughter cells. The self-organizational pathway involves nucleation of MTs around condensed chromatin and subsequent sorting of these MTs into a bipolar spindle via the action of motor proteins. The search-and-capture pathway predominates in the presence of centrosomes, which provide dominant centers for nucleation of MTs that dynamically probe the cytoplasm until they encounter a kinetochore, forming stable attachments. In this section, we discuss the prevalence of these pathways in higher plants and introduce modifications to generalize the mechanistic processes, thereby including the phenomena observed in higher plants.

4.1
Evidence for the Self-Organizational Pathway in Plants

Chromatin-mediated MT nucleation involves the small GTPase Ran, which in its GTP-bound form facilitates the activation of various spindle-promoting components such as XCTK2, NuMA, TPX2, and the gamma tubulin ring complex (γTuRC), which initiates nucleation of MTs around chromatin (Wilde and Zheng 1999; Ems-McClung et al. 2004). The concentration of Ran-GTP is kept high in the vicinity of chromatin via the activity of the chromatin-bound GTP-exchange factor RCC1, and kept low in the cytoplasm via the Ran GTPase activating protein RanGAP. Thus, MT nucleation and organization are maintained in the vicinity of the chromatin.

The best evidence for chromatin-mediated MT nucleation in plants comes from drug studies where MTs are depolymerized, and then fixed and immunostained for MTs at various time points during recovery. Several studies show MTs appearing during recovery, first as kinetochore-associated dots, which then grow into small tufts and later into fir-tree structures (Cleary and Hardham 1988; Falconer et al. 1988). Similarly, small tufts of MTs form asynchronously between different granules of individual sister kinetochores during early prophase in *Haemanthus* endosperm cells (Bajer 1987). The presence of poleward flux in plant spindles also indicates probable MT plus end assembly at or near kinetochores.

4.2
Self-Organization from Random Arrays via Motor-Driven MT Sorting

When DNA-coated beads are added to *Xenopus* egg extracts, MTs are nucleated randomly around the beads via chromatin-mediated nucleation and sorted into a bipolar spindle by the action of motor proteins (Heald et al.

1996). A somewhat similar situation has been observed in plants, the difference being that the MTs are nucleated randomly about the prophase nuclear surface, instead of around chromatin or DNA-coated beads. In cells lacking PPBs or exhibiting aberrant PPB organization, the prophase spindle exhibits reduced bipolarity—with randomly organized MTs accumulating about the NE during prophase—creating apolar or multipolar spindles. These MTs later become sorted (presumably via motor proteins) during prometaphase into bipolar spindles, although establishment of a metaphase spindle takes more time compared to cells with bipolar prophase spindles (Chan et al. 2005; Yoneda et al. 2005). In the *Xenopus* extract system, sorting into a bipolar spindle is mediated by Eg5 (Kinesin-5), while Kinesin-14 and cytoplasmic dynein participate in spindle stability and pole focusing (Walczak et al. 1998). A similar situation is likely to exist in plants, given the existence of Kinesin-5 motors (Asada et al. 1997; Barroso et al. 2000), and given that spindles lacking ATK1 or ATK5 exhibit splayed spindle poles (Chen et al. 2002; Ambrose et al. 2005; Ambrose and Cyr 2007). Given the likely involvement of motors in sorting of random perinuclear MTs, the self-organizational pathway can be generalized to include both the chromatin-based MT nucleation as well as nuclear-envelope-based MT nucleation. In both cases, motor-dependent sorting of random MT arrays occurs, regardless of the nucleating structure (i.e. chromatin or NE).

4.3
Evidence for Search-and-Capture Pathways in Plants

The term "search-and-capture" was initially coined as a theoretical explanation for the functional significance of MT dynamic instability, the intrinsic property wherein individual MTs stochastically alternate between phases of growth and shortening (Mitchison and Kirschner 1984; Kirschner and Mitchison 1986). Search-and-capture of kinetochores was later demonstrated both in vitro (Mitchison and Kirschner 1985) and in vivo in animal cells (Tulu et al. 2006). Although search-and-capture has not been directly demonstrated in plant cells, its presence can be inferred from several lines of evidence. First, the bipolar organization of plant prophase spindles poises MT plus ends such that they point predominately inward, much in the same way that centrosomes do, allowing dynamic MTs to probe the three-dimensional cellular space for kinetochores after NEB. Second, GFP::EB1 moves from the poles toward the midzone during mitosis, consistent with polar nucleation and growth toward the spindle midzone, as described by the search-and-capture model (Dixit et al. 2006). Third, when the bipolar organization of prophase spindles is disrupted, the subsequent duration of metaphase spindle formation is prolonged, consistent with a less efficient search-and-capture than from pre-established spindle poles (Chan et al. 2005; Yoneda et al. 2005). Advances in live-cell imaging and GFP technology will likely reveal search-and-capture in plant mitotic spindles.

4.4
The PPB as an Equatorial Organizer of the Prophase Spindle

Substantial evidence supports the notion that the PPB facilitates the bipolar organization of the associated prophase spindle. In all reported cases where the PPB is experimentally perturbed or absent, the accompanying prophase spindle also exhibits abnormalities. Specifically, bipolarity of the prophase spindle is absent or reduced in the following cases: (1) cells that naturally lack PPBs, such as those of endosperm and meiotic tissues (Smirnova and Bajer 1992); (2) cells of the *tonneau* mutant, which lack cortical MTs and PPBs (Camilleri et al. 2002); (3) *Arabidopsis* suspension cells overexpressing EB1::GFP, which frequently lack PPBs (Chan et al. 2005); (4) cells treated with Cytochalasin D, which causes narrow PPBs to broaden (Mineyuki and Palevitz 1990; Mineyuki et al. 1991; Eleftheriou and Palevitz 1992); (5) cells treated with cycloheximide or kinase inhibitors, which also inhibit PPB narrowing (Nogami et al. 1996; Nogami and Mineyuki 1999); (6) cells exhibiting double PPBs (Yoneda et al. 2005); (7) cells treated with taxol, which interferes with PPB narrowing and causes the MTs bridging the nucleus and PPB to become more numerous and unevenly distributed (Panteris et al. 1995; Baluska et al. 1996); (8) caffeine-induced binucleate cells where one nucleus lacks a PPB (Manandhar et al. 1996); (9) cells from the *Arabidopsis atk1-1* mutant, which exhibit abnormally broad PPBs (Marcus et al. 2003); (10) asymmetrically dividing subsidiary cells of *Zea mays*, where the PPB does not encircle the nucleus (Panteris et al. 2006); and (11) meristematic cells from *Arabidopsis* root tips overexpressing the MT reporter GFP::MAP4, which frequently exhibit an absence of PPBs or lack of PPB narrowing (our unpublished observations).

Although there is little doubt that the PPB plays a role in establishing prophase spindle bipolarity, studies are lacking that suggest a clear mechanism as to how this occurs. On the basis of the available data, we propose that the cytoplasmic MTs bridging the PPB and spindle (hereafter referred to as bridge MTs) are capable of transmitting forces that facilitate the sorting of perinuclear MTs into two half-spindles, which are symmetrically mirrored about the PPB plane. These bridge MTs may exert transient tensive forces between the PPB and perinuclear MTs, thereby providing a cue for both the bipolar organization and proper orientation of the prophase spindle. Bridge MTs connect directly to the PPB, where they may become bundled with PPB MTs (Burgess 1970; Bakhuizen et al. 1985; Panteris et al. 2006). The ends of the bridge MTs that reside inside the PPB, coaligned with cortical PPB MTs, provide possible sites for anchorage or generation of force via MT-MT sliding mechanisms. Indeed crossbridges between PPB MTs, which could represent crosslinking factors or motors, have been observed (Hardham and Gunning

1978). Similarly, regions of overlap between adjacent bridge MTs within the cytoplasm between the PPB and nucleus could also provide sites for force generation via a sliding-filament mechanism.

Considerable evidence supports the presence of tension between the PPB and prophase nucleus/spindle: (1) cytoplasmic strands rapidly recoil upon laser ablation (Hoffman 1984; Goodbody et al. 1991); (2) prophase nuclei are often somewhat flattened within the PPB plane, suggesting pulling forces at the nuclear equator (Burgess 1970; Panteris et al. 1991; Granger and Cyr 2001); (3) PPBs copurify with isolated prophase nuclei, indicating a physical link between the PPB and nucleus (Wick and Duniec 1984); (4) organelle motility decreases inside mature phragmosomes (Ota 1961; Mineyuki et al. 1994) and prophase nuclei are less easily displaced by centrifugation than interphase nuclei, indicating increased gelation within mature phragmosomes (Pickett-heaps 1969; Mineyuki and Furuya 1986; Mineyuki and Palevitz 1990). Although these findings provide compelling evidence for tensile forces between the PPB and prophase spindle, the mechanism by which these interactions facilitate bipolarity along an axis perpendicular to the plane of the PPB remains speculative. We favor the idea that this is mediated, at least in part, by intervening bridge MTs.

4.5
The PPB and Centrosome Share Similar Functions

Because of their ubiquitous presence across diverse phyla and intimate association with spindle poles in animal cells, the long-standing view has been that centrosomes are essential for spindle formation and function. Although compelling, a large body of evidence challenges this notion. First, higher plants and certain animal cell types (e.g. oocytes) lack discrete MTOCs and yet still form bipolar spindles. Second, numerous studies have shown that normal bipolar spindles can still form in the absence of centrosomes, even in cells that normally contain these structures (Phalle and Sullivan 1998; Khodjakov et al. 2000; Khodjakov and Rieder 2001; Megraw et al. 2001; Wadsworth and Khodjakov 2004). Third, by disruption of key pole determinants such as NuMA, a loss of pole focusing can be induced even in cells in which centrosomes are still present (Gaglio et al. 1997).

It was initially believed that the "inside-out" or self-organizational mechanisms of spindle assembly observed in acentrosomal systems were not operative in centrosome-containing cells, however this is not the case. More recent experiments have shown these pathways are also operative in centrosome-containing cells (Maiato et al. 2004; Wadsworth and Khodjakov 2004). Similarly, when centrosomes are introduced into acentrosomal self-organizational systems, they become the dominant sites of MT nucleation and contribute to spindle formation (Heald et al. 1997). In light of these findings, the current understanding of centrosome function has shifted away from that of an MTOC essential to mitotic spindle formation, to one of an organelle that facilitates proper spindle positioning and enhances the fidelity of cell division, thus becoming important in cell survival and viability of the organism (Wadsworth and Khodjakov 2004). In animal cells, which are malleable and constantly change shape, centrosomes act to position and orient the spindle via their associated astral MTs, which attach to the cortex and anchor the spindle (Wittmann et al. 2001). In these regards, centrosomes bear several functional similarities to the PPB.

The general effects of loss or disruption of PPBs include misoriented and wobbly spindles (Chan et al. 2005), lack or reduction of prophase/prometaphase spindle bipolarity (Nogami et al. 1996; Nogami and Mineyuki 1999; Chan et al. 2003; Marcus et al. 2003), prolonged duration of prometaphase/metaphase, while MTs sort into bipolar spindles (Yoneda et al. 2005), and misguided phragmoplast trajectory during cytokinesis (Camilleri et al. 2002; Chan et al. 2005). Similarly, in animal cells wherein the centrosomes have been removed either genetically or mechanically, bipolar spindles still form, although this occurs in a slower and less efficient manner (Khodjakov et al. 2000). Furthermore, these spindles frequently become misoriented, which can lead to abortive cytokinesis (Khodjakov et al. 2000; Khodjakov and Rieder 2001). In all cases, as with loss of centrosomes in animal cells, plant cells lacking PPBs are capable of forming spindles, but the efficiency of formation and subsequent orientation of the spindles are compromised.

On the basis of these observations, we hypothesize that the PPB and centrosomes represent derived structures that act complementarily to the ancestral self-organizational pathways to facilitate efficient spindle formation and orientation/positioning, as well as to enhance the fidelity of cytokinesis. Both the PPB and centrosome facilitate bipolar organization of the prophase spindle, which poises dynamic MTs to undergo search-and-capture of kinetochores and interpolar MTs during prometaphase. In this regard, both the PPB and centrosome function within the search-and-capture pathway of spindle formation, although these two structures achieve spindle bipolarity by very different mechanisms; the centrosomes acting at the poles, and the PPB acting at the equator. In this regard, components that act in the search-and-capture pathway may be grouped into *polar organizers* (e.g. centrosomes, spindle pole bodies, plastids) and *equatorial organizers* (i.e. the PPB). All of these structures share the common function of facilitating prophase spindle bipolarity, which aids search-and-capture during prometaphase/metaphase spindle formation.

Another major function of both centrosomes and the PPB is cortical anchorage and orientation of the spindle. Centrosomes achieve this throughout mitosis via interactions of their associated astral MTs with the cortex, whereas in plants the function of astral MTs is supplanted by two cytoskeletal components: F-actin and bridge MTs. In contrast to astral MTs, the pres-

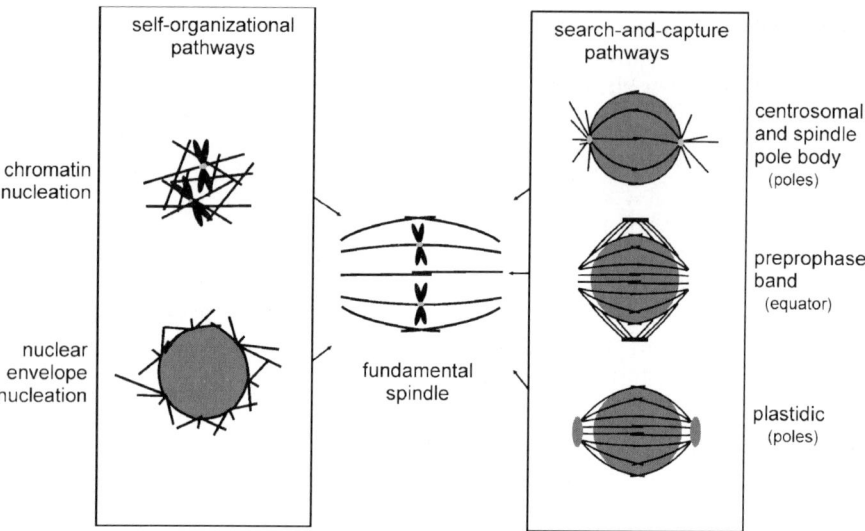

Fig. 2 Pathways of spindle assembly in eukaryotes. The self-organizational pathway involves the nucleation of MTs around chromatin or the NE and their subsequent sorting into a bipolar spindle via the action of motor proteins. The search-and-capture pathway predominates when bipolarity of the spindle is established prior to NEB, thereby predisposing MTs in a favorable orientation to establish kinetochore and interpolar MT connections. Prophase spindle bipolarity is facilitated via polar organizers or the PPB, which acts as an equatorial organizer

ence of these components is separated temporally during plant mitosis; both F-actin and bridge MTs are operative in spindle positioning and orientation during prophase, but after NEB, the bridge MTs are lost and F-actin takes over the role of spindle positioning and subsequent phragmoplast guidance during cytokinesis (Mineyuki and Palevitz 1990; Molchan et al. 2002). The idea that astral MTs in animals are analogous to bridge MTs and actin cables in plants as spindle-orienting structures is supported by the observations that each of these structures are under tension, providing a pulling force on the associated spindle in the direction of the sites of cortical anchorage (Hahne and Hoffman 1984; Aist et al. 1991; Goodbody et al. 1991). The key difference is that astral MTs act solely at the spindle poles, bridge MTs act in the phragmosomal plane, and F-actin acts at both of these sites (Traas et al. 1987; Lloyd and Traas 1988).

Figure 2 summarizes the above pathways of spindle formation in eukaryotes, with modifications accounting for the diverse mechanisms utilized within each fundamental pathway.

Acknowledgements Thanks to Deb Fisher for critical reading of this manuscript along with support from the US Department of Energy and the National Science Foundation.

References

Aist JR, Bayles CJ, Tao W, Berns MW (1991) Direct experimental evidence for the existence, structural basis and function of astral forces during anaphase B in vivo. J Cell Sci 100:279–288

Allen RD, Bajer A, Lafountain J (1969) Poleward migration of particles or states in spindle fiber filaments during mitosis in Haemanthus. J Cell Biol 43:4a

Ambrose JC, Cyr R (2007) The Kinesin ATK5 Functions in Early Spindle Assembly in Arabidopsis. Plant Cell 19:226–236

Ambrose JC, Li W, Marcus A, Ma H, Cyr R (2005) A minus-end-directed kinesin with plus-end tracking protein activity is involved in spindle morphogenesis. Mol Biol Cell 16:1584–1592

Asada T, Kuriyama R, Shibaoka H (1997) TKRP125, a kinesin-related protein involved in the centrosome-independent organization of the cytokinetic apparatus in tobacco BY-2 cells. J Cell Sci 110:179–189

Bajer A (1957) Cine-micrographic studies on mitosis in endosperm. III. The origin of the mitotic spindle. Exp Cell Res 13:493–502

Bajer A (1967) Notes on ultrastructure and some properties of transport within living mitotic spindle. J Cell Biol 33:713–720

Bajer A (1968) Behavior and fine structure of spindle fibers during mitosis in endosperm. Chromosoma 25:249–281

Bajer A, Molè-Bajer J (1969) Formation of spindle fibers, kinetochore orientation, and behaviour of the nuclear envelope during mitosis in endosperm. Fine structural and in vitro studies. Chromosoma 27:448–484

Bajer AS (1987) Substructure of the kinetochore and reorganization of kinetochore microtubules during early prometaphase in Haemanthus endosperm. Eur J Cell Biol 43:23–34

Bajer AS, Molè-Bajer J (1972) Spindle dynamics and chromosome movements. Int Rev Cytol—A Survey of Cell Biology, pp 1–255

Bajer AS, Molè-Bajer J (1986) Reorganization of microtubules in endosperm cells and cell fragments of the higher-plant Haemanthus in vivo. J Cell Biol 102:263–281

Bakhuizen R, van Spronsen PC, Sluiman-den Hertog FAJ, Venverloo CJ, Goosen-de Roo L (1985) Nuclear envelope radiating microtubules in plant cells during interphase mitosis transition. Protoplasma 128:43–51

Baluska F, Barlow PW, Parker JS, Volkmann D (1996) Symmetric reorganizations of radiating microtubules around pre- and post-mitotic nuclei of dividing cells organized within intact root meristems. J Plant Physiol 149:119–128

Barroso C, Chan J, Allan V, Doonan J, Hussey P, Lloyd C (2000) Two kinesin-related proteins associated with the cold-stable cytoskeleton of carrot cells: characterization of a novel kinesin, DcKRP120-2. Plant J 24:859–868

Baskin TI, Cande WZ (1990) The structure and function of the mitotic spindle in flowering plants. Ann Rev Plant Physiol Plant Mol Biol 41:277–315

Burgess J (1970) Interactions between microtubules and the nuclear envelope during mitosis in a fern. Protoplasma 71:77–89

Burgess J, Northcote D (1967) A function of preprophase band of microtubules in Phleum pratense. Planta 75:319–326

Camilleri C, Azimzadeh J, Pastuglia M, Bellini C, Grandjean O, Bouchez D (2002) The Arabidopsis TONNEAU2 gene encodes a putative novel protein phosphatase 2A regulatory subunit essential for the control of the cortical cytoskeleton. Plant Cell 14:833–845

Chan J, Calder GM, Doonan JH, Lloyd CW (2003) EB1 reveals mobile microtubule nucleation sites in Arabidopsis. Nat Cell Biol 5:967–971

Chan J, Calder G, Fox S, Lloyd C (2005) Localization of the microtubule end binding protein EB1 reveals alternative pathways of spindle development in Arabidopsis suspension cells. Plant Cell 17:1737–1748

Chan J, Jensen CG, Jensen LC, Bush M, Lloyd CW (1999) The 65-kDa carrot microtubule-associated protein forms regularly arranged filamentous cross-bridges between microtubules. Proc Natl Acad Sci USA 96:14931–14936

Chang HY, Smertenko AP, Igarashi H, Dixon DP, Hussey PJ (2005) Dynamic interaction of NtMAP65-1a with microtubules in vivo. J Cell Sci 118:3195–3201

Chen C, Marcus A, Li W, Hu Y, Calzada JP, Grossniklaus U, Cyr RJ, Ma H (2002) The Arabidopsis ATK1 gene is required for spindle morphogenesis in male meiosis. Development 129:2401–2409

Cleary AL, Hardham AR (1988) Depolymerization of microtubule arrays in root-tip cells by oryzalin their recovery with modified nucleation patterns. Can J Bot 66:2353–2366

Cleary AL, Smith LG (1998) The Tangled1 gene is required for spatial control of cytoskeletal arrays associated with cell division during maize leaf development. Plant Cell 10:1875–1888

Dagenbach EM, Endow SA (2004) A new kinesin tree. J Cell Sci 117:3–7

De Mey J, Lambert AM, Bajer AS, Moeremans M, De Brabander M (1982) Visualization of microtubules in interphase and mitotic plant cells of Haemanthus endosperm with the immuno-gold staining method. Proc Natl Acad Sci USA 79:1898–1902

Dhonukshe P, Vischer N, Gadella TWJ (2006) Contribution of microtubule growth polarity and flux to spindle assembly and functioning in plant cells. J Cell Sci 119:3193–3205

Dhonukshe P, Mathur J, Hulskamp M, Gadella TW Jr (2005) Microtubule plus-ends reveal essential links between intracellular polarization and localized modulation of endocytosis during division-plane establishment in plant cells. BMC Biol 3:11

Dixit R, Chang E, Cyr R (2006) Establishment of Polarity during Organization of the Acentrosomal Plant Cortical Microtubule Array. Mol Biol Cell 17:1298–1305

Eleftheriou EP, Palevitz BA (1992) The effect of cytochalasin-D on preprophase band organization in root-tip cells of Allium. J Cell Sci 103:989–998

Eleftheriou EP, Baskin TI, Hepler PK (2005) Aberrant cell plate formation in the Arabidopsis thaliana microtubule organization 1 mutant. Plant Cell Physiol 46:671–675

Ems-McClung SC, Zheng YX, Walczak CE (2004) Importin alpha/beta and Ran-GTP regulate XCTK2 microtubule binding through a bipartite nuclear localization signal. Mol Biol Cell 15:46–57

Erhardt M, Stoppin-Mellet V, Campagne S, Canaday J, Mutterer J, Fabian T, Sauter M, Muller T, Peter C, Lambert AM, Schmit AC (2002) The plant Spc98p homologue colocalizes with gamma-tubulin at microtubule nucleation sites and is required for microtubule nucleation. J Cell Sci 115:2423–2431

Euteneuer U, McIntosh JR (1980) Polarity of midbody and phragmoplast microtubules. J Cell Biol 87:509–515

Euteneuer U, Jackson WT, McIntosh JR (1982) Polarity of spindle microtubules in Haemanthus endosperm. J Cell Biol 94:644–653

Falconer M, Donaldson G, Seagull R (1988) MTOCs in higher plant cells: an immunofluorescent study of microtubule assembly sites following depolymerization by APM. Protoplasma 144:46–55

Fowke LC (1993) Microtubules in dividing root cells of the conifer Pinus radiata and the cycad Zamia furfuracea. Cell Biol Int 17:143–151

Funabiki H, Murray AW (2000) The Xenopus chromokinesin Xkid is essential for metaphase chromosome alignment and must be degraded to allow anaphase chromosome movement. Cell 102:411–424

Furutani I, Watanabe Y, Prieto R, Masukawa M, Suzuki K, Naoi K, Thitamadee S, Shikanai T, Hashimoto T (2000) The SPIRAL genes are required for directional central of cell elongation in Arabidopsis thaliana. Development 127:4443–4453

Fuseler JW (1975) Mitosis in Tilia americana endosperm. J Cell Biol 64:159–171

Gadde S, Heald R (2004) Mechanisms and molecules of the mitotic spindle. Curr Biol 14:R797–R805

Gaglio T, Dionne MA, Compton DA (1997) Mitotic spindle poles are organized by structural and motor proteins in addition to centrosomes. J Cell Biol 138:1055–1066

Galatis B, Apostolakos P (1991) Patterns of microtubule reappearance in root-cells of Vigna-sinensis recovering from a colchicine treatment. Protoplasma 160:131–143

Goodbody KC, Venverloo CJ, Lloyd CW (1991) Laser microsurgery demonstrates that cytoplasmic strands anchoring the nucleus across the vacuole of premitotic plant-cells are under tension—implications for division plane alignment. Development 113:931–939

Goshima G, Vale RD (2003) The roles of microtubule-based motor proteins in mitosis: comprehensive RNAi analysis in the Drosophila S2 cell line. J Cell Biol 162:1003–1016

Goshima G, Vale RD (2005) Cell cycle-dependent dynamics and regulation of mitotic kinesins in Drosophila S2 cells. Mol Biol Cell 16:3896–3907

Goshima G, Wollman R, Stuurman N, Scholey JM, Vale RD (2005) Length control of the metaphase spindle. Curr Biol 15:1979–1988

Granger C, Cyr R (2001) Use of abnormal preprophase bands to decipher division plane determination. J Cell Sci 114:599–607

Hahne G, Hoffman F (1984) The effect of laser microsurgery on cytoplasmic strands and cytoplasmic streaming in isolated plant protoplasts. Eur J Cell Biol 33:175–179

Hamada T, Igarashi H, Itoh TJ, Shimmen T, Sonobe S (2004) Characterization of a 200 kDa microtubule-associated protein of tobacco BY-2 cells, a member of the XMAP215/MOR1 family. Plant Cell Physiol 45:1233–1242

Hanzely L, Schjeide OA (1973) Structural and functional aspects of anastral mitotic spindle in Allium sativum root tip cells. Cytobios 7:147–162

Hard R, Allen RD (1977) Behavior of kinetochore fibers in Haemanthus katherinae during anaphase movements of chromosomes. J Cell Sci 27:47–56

Hardham AR, Gunning BES (1978) Structure of cortical microtubule arrays in plant-cells. J Cell Biol 77:14–34

Harris P, Bajer A (1965) Fine structure studies on mitosis in endosperm metaphase of Haemanthus katherinae Bak. Chromosoma 16:624–636

Heald R, Tournebize R, Habermann A, Karsenti E, Hyman A (1997) Spindle assembly in Xenopus egg extracts: Respective roles of centrosomes and microtubule self-organization. J Cell Biol 138:615–628

Heald R, Tournebize R, Blank T, Sandaltzopoulos R, Becker P, Hyman A, Karsenti E (1996) Self-organization of microtubules into bipolar spindles around artificial chromosomes in Xenopus egg extracts. Nature 382:420–425

Hepler PK, Jackson WT (1968) Microtubules and early stages of cell-plate formation in the endosperm of Haemanthus katherinae Baker. J Cell Biol 38:437–446

Hoffman GH.a.F. (1984) The effect of laser microsurgery on cytoplasmic strands and cytoplasmic streaming in isolated plant protoplasts. Eur J Cell Biol 33:175–179

Hoyt MA, He L, Totis L, Saunders WS (1993) Loss of function of Saccharomyces cerevisiae kinesin-related CIN8 and KIP1 is suppressed by KAR3 motor domain mutations. Genetics 135:35–44

Hush JM, Wadsworth P, Callaham DA, Hepler PK (1994) Quantification of microtubule dynamics in living plant-cells using fluorescence redistribution after photobleaching. J Cell Sci 107:775–784

Hussey PJ, Hawkins TJ, Igarashi H, Kaloriti D, Smertenko A (2002) The plant cytoskeleton: recent advances in the study of the plant microtubule-associated proteins MAP-65, MAP-190 and the Xenopus MAP215-like protein, MOR1. Plant Mol Biol 50:915–924

Huyett A, Kahana J, Silver P, Zeng X, Saunders WS (1998) The Kar3p and Kip2p motors function antagonistically at the spindle poles to influence cytoplasmic microtubule numbers. J Cell Sci 111:295–301

Igarashi H, Orii H, Mori H, Shimmen T, Sonobe S (2000) Isolation of a novel 190 kDa protein from tobacco BY-2 cells: Possible involvement in the interaction between actin filaments and microtubules. Plant Cell Physiol 41:920–931

Inoue S, Bajer A (1961) Birefringence in endosperm mitosis. Chromosoma 12:48–63

Jensen C, Bajer AS (1973) Spindle dynamics and arrangement of microtubules. Chromosoma 44:73–89

Kapoor TM, Lampson MA, Hergert P, Cameron L, Cimini D, Salmon ED, McEwen BF, Khodjakov A (2006) Chromosomes can congress to the metaphase plate before biorientation. Science 311:388–391

Katsuta J, Hashiguchi Y, Shibaoka H (1990) The role of the cytoskeleton in positioning of the nucleus in premitotic tobacco BY-2 cells. J Cell Sci 95:413–422

Kawamura E, Himmelspach R, Rashbrooke MC, Whittington AT, Gale KR, Collings DA, Wasteneys GO (2006) MICROTUBULE ORGANIZATION 1 regulates structure and function of microtubule arrays during mitosis and cytokinesis in the Arabidopsis root. Plant Physiol 140:102–114

Khodjakov A, Rieder CL (2001) Centrosomes enhance the fidelity of cytokinesis in vertebrates and are required for cell cycle progression. J Cell Biol 153:237–242

Khodjakov A, Cole RW, Oakley BR, Rieder CL (2000) Centrosome-independent mitotic spindle formation in vertebrates. Curr Biol 10:59–67

Kirschner M, Mitchison T (1986) Beyond self-assembly: from microtubules to morphogenesis. Cell 45:329–342

Kline-Smith SL, Walczak CE (2004) Mitotic spindle assembly and chromosome segregation: Refocusing on microtubule dynamics. Mol Cell 15:317–327

Kong LJ, Hanley-Bowdoin L (2002) A geminivirus replication protein interacts with a protein kinase and a motor protein that display different expression patterns during plant development and infection. Plant Cell 14:1817–1832

Korolev AV, Chan J, Naldrett MJ, Doonan JH, Lloyd CW (2005) Identification of a novel family of 70 kDa microtubule-associated proteins in Arabidopsis cells. Plant J 42:547–555

Lawrence CJ, Dawe RK, Christie KR, Cleveland DW, Dawson SC, Endow SA, Goldstein LS, Goodson HV, Hirokawa N, Howard J, Malmberg RL, McIntosh JR, Miki H, Mitchison TJ, Okada Y, Reddy AS, Saxton WM, Schliwa M, Scholey JM, Vale RD, Walczak CE, Wordeman L (2004) A standardized kinesin nomenclature. J Cell Biol 167:19–22

Liu B, Cyr RJ, Palevitz BA (1996) A kinesin-like protein, KatAp, in the cells of arabidopsis and other plants. Plant Cell 8:119–132

Liu B, Marc J, Joshi HC, Palevitz BA (1993) A gamma-tubulin-related protein associated with the microtubule arrays of higher plants in a cell cycle-dependent manner. J Cell Sci 104:1217–1228

Lloyd C, Chan J (2006) Not so divided: the common basis of plant and animal cell division. Nat Rev Mol Cell Biol 7:147–152

Lloyd CW, Traas JA (1988) The role of F-actin in determining the division plane of carrot suspension cells. Drug studies. Development 102:211–221

Maddox PS, Bloom KS, Salmon ED (2000) The polarity and dynamics of microtubule assembly in the budding yeast Saccharomyces cerevisiae. Nat Cell Biol 2:36–41

Maiato H, Rieder CL, Khodjakov A (2004) Kinetochore-driven formation of kinetochore fibers contributes to spindle assembly during animal mitosis. J Cell Biol 167:831–840

Mallavarapu A, Sawin K, Mitchison T (1999) A switch in microtubule dynamics at the onset of anaphase B in the mitotic spindle of Schizosaccharomyces pombe. Curr Biol 9:1423–1426

Manandhar G, Apostolakos P, Galatis B (1996) Cell division of binuclear cells induced by caffeine: Spindle organization and determination of division plane. J Plant Res 109:265–275

Mao G, Chan J, Calder G, Doonan JH, Lloyd CW (2005) Modulated targeting of GFP-AtMAP65-1 to central spindle microtubules during division. Plant J 43:469–478

Marc J, Gunning BES (1988) Monoclonal antibodies to a fern spermatozoid detect novel components of the mitotic and cytokinetic apparatus in higher plant cells. Protoplasma 142:15–24

Marcus AI, Li W, Ma H, Cyr RJ (2003) A kinesin mutant with an atypical bipolar spindle undergoes normal mitosis. Mol Biol Cell 14:1717–1726

Mathur J, Mathur N, Kernebeck B, Srinivas BP, Hulskamp M (2003) A novel localization pattern for an EB1-like protein links microtubule dynamics to endomembrane organization. Curr Biol 13:1991–1997

Megraw TL, Kao LR, Kaufman TC (2001) Zygotic development without functional mitotic centrosomes. Curr Biol 11:116–120

Miki H, Okada Y, Hirokawa N (2005) Analysis of the kinesin superfamily: insights into structure and function. Trends Cell Biol 15:467–476

Mineyuki Y (1999) The preprophase band of microtubules: Its function as a cytokinetic apparatus in higher plants. Int Rev Cytol 187:1–49

Mineyuki Y, Furuya M (1986) Involvement of colchicine-sensitive cytoplasmic element in premitotic nuclear positioning of Adiantum protenemata. Protoplasma 130:83–90

Mineyuki Y, Palevitz BA (1990) Relationship between Preprophase Band Organization, F-Actin and the Division Site in Allium - Fluorescence and Morphometric Studies on Cytochalasin-Treated Cells. J Cell Sci 97:283–295

Mineyuki Y, Marc J, Palevitz BA (1991) Relationship between the preprophase band, nucleus and spindle in dividing Allium cotyledon cells. J Plant Physiol 138:640–649

Mineyuki Y, Iida H, Anraku Y (1994) Loss of microtubules in the interphase cells of onion (Allium-cepa L) root-tips from the cell cortex and their appearance in the cytoplasm after treatment with cycloheximide. Plant Physiol 104:281–284

Mitchison T, Kirschner M (1984) Dynamic instability of microtubule growth. Nature 312:237–242

Mitchison TJ (2005) Mechanism and function of poleward flux in Xenopus extract meiotic spindles. Philos Trans R Soc Lond B Biol Sci 360:623–629

Mitchison TJ, Kirschner MW (1985) Properties of the kinetochore in vitro. II. Microtubule capture and ATP-dependent translocation. J Cell Biol 101:766–777

Molchan TM, Valster AH, Hepler PK (2002) Actomyosin promotes cell plate alignment and late lateral expansion in Tradescantia stamen hair cells. Planta 214:683–693

Molè-Bajer J, Bajer AS (1982) Modification of microtubule arrangements in the mitotic spindle of Haemanthus endosperm under the Influence of taxol. J Cell Biol 95:A308–A308

Molè-Bajer J, Bajer AS (1983) Action of taxol on mitosis— modification of microtubule arrangements and function of the mitotic spindle in Haemanthus endosperm. J Cell Biol 96:527–540

Mollinari C, Kleman JP, Jiang W, Schoehn G, Hunter T, Margolis RL (2002) PRC1 is a microtubule binding and bundling protein essential to maintain the mitotic spindle midzone. J Cell Biol 157:1175–1186

Mountain V, Simerly C, Howard L, Ando A, Schatten G, Compton DA (1999) The kinesin-related protein, HSET, opposes the activity of Eg5 and cross-links microtubules in the mammalian mitotic spindle. J Cell Biol 147:351–366

Muller S, Smertenko A, Wagner V, Heinrich M, Hussey PJ, Hauser MT (2004) The plant microtubule-associated protein AtMAP65-3/PLE is essential for cytokinetic phragmoplast function. Curr Biol 14:412–417

Murata T, Sonobe S, Baskin TI, Hyodo S, Hasezawa S, Nagata T, Horio T, Hasebe M (2005) Microtubule-dependent microtubule nucleation based on recruitment of gamma-tubulin in higher plants. Nat Cell Biol 7:961–968

Nakajima K, Furutani I, Tachimoto H, Matsubara H, Hashimoto T (2004) SPIRAL1 encodes a plant-specific microtubule-localized protein required for directional control of rapidly expanding Arabidopsis cells. Plant Cell 16:1178–1190

Nogami A, Mineyuki Y (1999) Loosening of a preprophase band of microtubules in onion (Allium cepa L.) root tip cells by kinase inhibitors. Cell Struct Func 24:419–424

Nogami A, Suzaki T, Shigenaka Y, Nagahama Y, Mineyuki Y (1996) Effects of cycloheximide on preprophase bands and prophase spindles in onion (Allium cepa L) root tip cells. Protoplasma 192:109–121

O'Connell MJ, Meluh PB, Rose MD, Morris NR (1993) Suppression of the bimC4 mitotic spindle defect by deletion of klpA, a gene encoding a KAR3-related kinesin-like protein in Aspergillus nidulans. J Cell Biol 120:153–162

Ota T (1961) Role of cytoplasm in cytokinesis of plant cells. Cytologia 26:428–447

Palevitz BA (1988) Microtubular Fir-Trees in Mitotic Spindles of Onion Roots. Protoplasma 142:74–78

Panteris E, Galatis B, Apostolakos P (1991) Patterns of cortical and perinuclear microtubule organization in meristematic root-cells of Adiantum-Capillus-Veneris. Protoplasma 165:173–188

Panteris E, Apostolakos P, Galatis B (1995) The effect of taxol on Triticum preprophase root-cells—preprophase microtubule band organization seems to depend on new microtubule assembly. Protoplasma 186:72–78

Panteris E, Apostolakos P, Galatis B (2006) Cytoskeletal asymmetry in Zea mays subsidiary cell mother cells: a monopolar prophase microtubule half-spindle anchors the nucleus to its polar position. Cell Motil Cytoskeleton 63:696–709

Phalle BD, Sullivan W (1998) Spindle assembly and mitosis without centrosomes in parthenogenetic Sciara embryos. J Cell Biol 141:1383–1391

Pickett-Heaps J (1969) Preprophase microtubules and stomatal differentiation; some effects of centrifugation on symmetrical and asymmetrical cell division. J Ultrastruct Res 27:24–44

Pickett-Heaps JD, Northcote DH (1966) Organization of microtubules and endoplasmic reticulum during mitosis and cytokinesis in wheat meristems. J Cell Sci 1:109–120

Pickett-Heaps JD, Spurck T, Tippit D (1984) Chromosome motion and the spindle matrix. J Cell Biol 99:S137–S143

Reddy AS (2001) Molecular motors and their functions in plants. Int Rev Cytol 204:97–178

Reddy AS, Day IS (2001) Kinesins in the Arabidopsis genome: a comparative analysis among eukaryotes. BMC Genomics 2:2

Richardson DN, Simmons MP, Reddy AS (2006) Comprehensive comparative analysis of kinesins in photosynthetic eukaryotes. BMC Genomics 7:18

Robyns W (1929) La figure achromatique sur materiel frais dans les divisions somatiques des phanerogames. La Cellule 39:85–117

Ryan KG (1983) Prometaphase and anaphase chromosome movements in living pollen mother cells. Protoplasma 116:24–33

Sakai A (1969) Electron microscopy of dividing cells. II. Microtubules and formation of the spindle in root tip cells of higher plants. Cytologia 34:57–70

Savoian MS, Gatt MK, Riparbelli MG, Callaini G, Glover DM (2004) Drosophila Klp67A is required for proper chromosome congression and segregation during meiosis I. J Cell Sci 117:3669–3677

Schaar BT, Chan GK, Maddox P, Salmon ED, Yen TJ (1997) CENP-E function at kinetochores is essential for chromosome alignment. J Cell Biol 139:1373–1382

Schmit AC (2002) Acentrosomal microtubule nucleation in higher plants. Int Rev Cytol 220:257–289

Schmit AC, Vantard M, de Mey J, Lambert A (1983) Aster-like microtubule centers establish spindle polarity during interphase-mitosis transition in higher plant cells. Plant Cell Rep 2:285–288

Schuyler SC, Liu JY, Pellman D (2003) The molecular function of Ase1p: evidence for a MAP-dependent midzone-specific spindle matrix. Microtubule-associated proteins. J Cell Biol 160:517–528

Sedbrook JC, Ehrhardt DW, Fisher SE, Scheible WR, Somerville CR (2004) The Arabidopsis SKU6/SPIRAL1 gene encodes a plus end-localized microtubule-interacting protein involved in directional cell expansion. Plant Cell 16:1506–1520

Sharp DJ, Mennella V, Buster DW (2005) KLP10A and KLP59C: the dynamic duo of microtubule depolymerization. Cell Cycle 4:1482–1485

Sharp DJ, Yu KR, Sisson JC, Sullivan W, Scholey JM (1999) Antagonistic microtubule-sliding motors position mitotic centrosomes in Drosophila early embryos. Nat Cell Biol 1:51–54

Sharp DJ, Brown HM, Kwon M, Rogers GC, Holland G, Scholey JM (2000) Functional coordination of three mitotic motors in Drosophila embryos. Mol Biol Cell 11:241–253

Sinnott EW, Bloch R (1940) Cytoplasmic behavior during division of vacuolate plant cells. Proc Natl Acad Sci USA 26:223–227

Smirnova EA (1998) Organization of the mitotic spindle in higher plant cells. Russ J Plant Physiol 45:165–173

Smirnova EA, Bajer AS (1992) Spindle poles in higher-plant mitosis. Cell Motil Cytoskel 23:1–7

Smirnova EA, Bajer AS (1994) Microtubule converging centers and reorganization of the interphase cytoskeleton and the mitotic spindle in higher plant Haemanthus. Cell Motil Cytoskeleton 27:219–233

Smirnova EA, Reddy AS, Bowser J, Bajer AS (1998) Minus end-directed kinesin-like motor protein, Kcbp, localizes to anaphase spindle poles in Haemanthus endosperm. Cell Motil Cytoskeleton 41:271–280

Smith LG, Gerttula SM, Han SC, Levy J (2001) TANGLED1: A microtubule binding protein required for the spatial control of cytokinesis in maize. J Cell Biol 152:231–236

Staiger CJ, Cande WZ (1991) Microfilament distribution in maize meiotic mutants correlates with microtubule organization. Plant Cell 3:637–644

Stoppin V, Vantard M, Schmit A-C, Lambert A-M (1994) Isolated plant nuclei nucleate microtubule assembly: The nuclear surface in higher plants has centrosome-like activity. Plant Cell 6:1099–1106

Tirnauer JS, Salmon ED, Mitchison TJ (2004) Microtubule plus-end dynamics in Xenopus egg extract spindles. Mol Biol Cell 15:1776–1784

Tirnauer JS, Canman JC, Salmon ED, Mitchison TJ (2002) EB1 targets to kinetochores with attached, polymerizing microtubules. Mol Biol Cell 13:4308–4316

Tournebize R, Popov A, Kinoshita K, Ashford AJ, Rybina S, Pozniakovsky A, Mayer TU, Walczak CE, Karsenti E, Hyman AA (2000) Control of microtubule dynamics by the antagonistic activities of XMAP215 and XKCM1 in Xenopus egg extracts. Nat Cell Biol 2:13–19

Traas JA, Doonan JH, Rawlins DJ, Shaw PJ, Watts J, Lloyd CW (1987) An actin network is present in the cytoplasm throughout the cell cycle of carrot cells and associates with the nucleus. J Cell Biol 105:387–395

Tulu US, Fagerstrom C, Ferenz NP, Wadsworth P (2006) Molecular requirements for kinetochore-associated microtubule formation in mammalian cells. Curr Biol 16:536–541

Valiron O, Caudron N, Job D (2001) Microtubule dynamics. Cell Mol Life Sci 58:2069–2084

Van Damme D, Bouget FY, Van Poucke K, Inze D, Geelen D (2004a) Molecular dissection of plant cytokinesis and phragmoplast structure: a survey of GFP-tagged proteins. Plant J 40:386–398

Van Damme D, Van Poucke K, Boutant E, Ritzenthaler C, Inze D, Geelen D (2004b) In vivo dynamics and differential microtubule-binding activities of MAP65 proteins. Plant Physiol 136:3956–3967

Vanstraelen M, Torres Acosta JA, De Veylder L, Inze D, Geelen D (2004) A plant-specific subclass of C-terminal kinesins contains a conserved a-type cyclin-dependent kinase site implicated in folding and dimerization. Plant Physiol 135:1417–1429

Vanstraelen M, Inze D, Geelen D (2006) Mitosis-specific kinesins in Arabidopsis. Trends Plant Sci 11:167–175

Venverloo CJ, Libbenga KR (1987) Regulation of the plane of cell division in vacuolated cells. I. The function of nuclear positioning and phragmosome formation. J Plant Physiol 131:267–284

Vos JW, Dogterom M, Emons AMC (2004) Microtubules become more dynamic but not shorter during preprophase band formation: A possible search-and-capture mechanism for microtubule translocation. Cell Motil Cytoskel 57:246–258

Vos JW, Safadi F, Reddy AS, Hepler PK (2000) The kinesin-like calmodulin binding protein is differentially involved in cell division. Plant Cell 12:979–990

Wadsworth P, Khodjakov A (2004) E pluribus unum: towards a universal mechanism for spindle assembly. Trends Cell Biol 14:413–419

Walczak CE, Vernos I, Mitchison TJ, Karsenti E, Heald R (1998) A model for the proposed roles of different microtubule-based motor proteins in establishing spindle bipolarity. Curr Biol 8:903–913

Wang H, Cutler AJ, Fowke LC (1991) Microtubule organization in Cultured Soybean and Black Spruce Cells: Interphase-mitosis Transition and Spindle Morphology. Protoplasma 162:46–54

West RR, Malmstrom T, Troxell CL, McIntosh JR (2001) Two related kinesins, klp5+ and klp6+, foster microtubule disassembly and are required for meiosis in fission yeast. Mol Biol Cell 12:3919–3932

Whittington AT, Vugrek O, Wei KJ, Hasenbein NG, Sugimoto K, Rashbrooke MC, Wasteneys GO (2001) MOR1 is essential for organizing cortical microtubules in plants. Nature 411:610–613

Wick S (2000) Plant microtubules meet their MAPs and mimics. Nat Cell Biol 2:E204–206

Wick S, Duniec J (1984) Immunofluorescence microscopy of tubulin and microtubule arrays in plant cells. II. Transition between the pre-prophase band and the mitotic spindle. Protoplasma 122:45–55

Wick SM, Duniec J (1983) Immunofluorescence microscopy of tubulin and microtubule arrays in plant cells. I. Preprophase band development and concomitant appearance of nuclear envelope-associated tubulin. J Cell Biol 97:235–243

Wilde A, Zheng YX (1999) Stimulation of microtubule aster formation and spindle assembly by the small GTPase Ran. Science 284:1359–1362

Wittmann T, Hyman A, Desai A (2001) The spindle: a dynamic assembly of microtubules and motors. Nat Cell Biol 3:E28–E34

Yajima J, Edamatsu M, Watai-Nishii J, Tokai-Nishizumi N, Yamamoto T, Toyoshima YY (2003) The human chromokinesin Kid is a plus end-directed microtubule-based motor. EMBO J 22:1067–1074

Yasuhara H, Muraoka M, Shogaki H, Mori H, Sonobe S (2002) TMBP200, a microtubule bundling polypeptide isolated from telophase tobacco BY-2 cells is a MOR1 homologue. Plant Cell Physiol 43:595–603

Yen TJ, Compton DA, Wise D, Zinkowski RP, Brinkley BR, Earnshaw WC, Cleveland DW (1991) CENP-E, a novel human centromere-associated protein required for progression from metaphase to anaphase. EMBO J 10:1245–1254

Yoneda A, Akatsuka M, Hoshino H, Kumagai F, Hasezawa S (2005) Decision of spindle poles division plane by double preprophase bands in a BY-2 cell line expressing GFP-tubulin. Plant Cell Physiol 46:531–538

Zhu C, Jiang W (2005) Cell cycle-dependent translocation of PRC1 on the spindle by Kif4 is essential for midzone formation and cytokinesis. Proc Natl Acad Sci USA 102:343–348

Zhu C, Zhao J, Bibikova M, Leverson JD, Bossy-Wetzel E, Fan JB, Abraham RT, Jiang W (2005) Functional analysis of human microtubule-based motor proteins, the kinesins and dyneins, in mitosis/cytokinesis using RNA interference. Mol Biol Cell 16:3187–3199

Cytoskeletal Motor Proteins in Plant Cell Division

Yuh-Ru Julie Lee · Bo Liu (✉)

Section of Plant Biology, University of California, Davis, CA 95616, USA
bliu@ucdavis.edu

Abstract Plant cell division involves anastral spindles with incompletely focused poles and the phragmoplast with antiparallel microtubules. The organization of the spindle microtubule array and the phragmoplast array is thought to be dependent on microtubule-based motor kinesins. Among the more than 50 kinesins encoded by a single plant genome, a number of them are either proven or predicted to be essential for cell division. Members of the Kinesin-14 subfamily play a critical role in organizing spindle poles. Some kinesins yet to be identified could be required for facilitating nuclear envelope breakdown at the end of prophase, and others for mediating the interaction between chromosomes and microtubules for spindle assembly. During anaphase, the disassembly of kinetochore fibers and the accompanying sister chromatid movement would definitely be assisted by kinesins. Similarly to fungi and animals, plants likely use Kinesin-5 for microtubule sliding, leading to spindle elongation. During cytokinesis, Kinesin-12 is required for establishing the antiparallel fashion of the phragmoplast microtubule array. Microtubule turnover in this highly dynamic array also depends on plant-specific kinesins acting on an MAP kinase cascade. More than one kinesin is predicted to deliver vesicles for cell plate assembly. Lastly, recent data also suggest that kinesins play a critical role in spatial regulation of cytokinesis. Very little has been learned about the potential roles of myosins in plant cell division. Whether myosins are also involved in vesicle transport in the phragmoplast awaits further examination. We can conclude that splendid cytoskeletal motors play splendid roles in plant cell division.

1
Introduction

In flowering plants, mitosis and meiosis involve anastral spindles with incompletely focused spindle poles. Following the segregation of genetic material, cytokinesis is brought about by a unique apparatus known as the phragmoplast, which is an evolutionary landmark found among organisms from advanced green algae to land plants (Graham et al. 2000). While plant mitosis and meiosis share many features similar to those of animal cells, plant cytokinesis emphasizes the buildup of the cell plate while fungal and animal cytokinesis emphasizes the contraction of the actomyosin ring.

The anastral spindle apparatus has a microtubule array with the kinetochore fibers focused at more than one discrete point, implicating that plant cells contain diffuse microtubule-organizing centers (MTOCs) during mitosis and meiosis (Palevitz 1993). In different plant cell types, the spindle microtubule array often exhibits flexible shapes largely due to the geometrical restriction defined by the rigid cell wall. For example, at metaphase Tradescantia generative cells have kinetochore fibers arranged along the cell axis (Liu and Palevitz 1992). Nevertheless, plant spindle, no matter how "distorted", still functions perfectly to faithfully segregate genetic material. During mitosis and meiosis, specific activities like the organization of spindle poles, attachment of kinetochores to the microtubule plus end, chromotid movement towards the poles, and microtubule polymerization/depolymerization would be dependent on activities of motor proteins acting on spindle microtubules. Compared to microtubules, during mitosis microfilaments exhibit a cortical array encaging the two half spindles (Liu and Palevitz 1992). Thus, microfilament-based motor proteins unlikely contribute to mitosis and meiosis directly in flowering plants.

The phragmoplast is organized with a framework of an anti-parallel microtubule array in which microtubule plus ends are juxtaposed in the middle. Microfilaments as well as membranous structures like the endoplasmic reticulum, vesicles, and other organelles are arranged in close proximity to microtubules (Staehelin and Hepler 1996). Microtubules are essential for assembly of the cell plate as vesicles destined to the division site are largely transported along phragmoplast microtubules. In addition to being motors for vesicle transport in the phragmoplast, microtubule-based motors also contribute to organization of phragmoplast microtubules. Microfilaments are arranged in a similar but distinct pattern as that of microtubules in the phragmoplast as they do not occupy the equatorial region (Zhang et al. 1993). Microfilaments are required for proper orientation of the cell plate although they are not essential for cell plate formation (Palevitz 1980).

The presence of both microtubule-based and microfilament-based motor proteins in plant cells was first confirmed by biochemical means around the late 1980s and early 1990s (Asada et al. 1991; Ma and Yen 1989). The splendid picture of the diversity of these motor proteins was not revealed until recently when the genomes of model plants were completely sequenced. Compared to animal species, plants have a more complicated family of microtubule motor kinesins and a simpler family of microfilament motor myosins.

1.1
Kinesin Superfamily Proteins in Plants

Directional movement along microtubules is powered by microtubule-based motors. Members of the kinesin superfamily are one of the two types of such motor proteins (Kreis and Vale 1999). The other type is dynein. However, to date completed genomes of flowering plants do not contain genes encoding the heavy chain of dynein (Lawrence et al. 2001). Thus, microtubule-based motility in plant cells is likely driven by kinesins only.

Members of the kinesin superfamily contain a conserved kinesin motor domain of approximately 350 amino acids with an ATP-binding motif and a microtubule-binding site (Vale 2003). The ATPase activity residing in this motor domain is activated upon nucleotide (ATP)-dependent binding to microtubules. The directionality of kinesins, or whether a particular kinesin moves towards the plus or minus end of microtubules, is determined by a short peptide known as the neck sequence (Endow 1999). Among kinesins, amino acid sequences of non-motor domains are not conserved. Most kinesins have a tripartite structure of the motor domain, coiled-coil domain for dimerization, and a cargo-binding domain. On the basis of phylogenetic analysis of the motor domain, a unified nomenclature has classified kinesins into Kinesin-1 to Kinesin-14 subfamilies (Lawrence et al. 2004; Miki et al. 2005). However, members of the same subfamily may have unrelated functions in different organisms. Kinesins which do not fit in these 14 subfamilies are often treated as orphan kinesins.

The *Arabidopsis thaliana* genome contains 61 genes encoding kinesins (Reddy and Day 2001b). The rice *Oryza sativa* also has more than 50 kinesin genes (Richardson et al. 2006). Unfortunately, most of these genes do not have their cDNA sequences determined. The number of plant kinesins is overwhelming especially considering mammals like mouse and human have significantly fewer kinesins (Miki et al. 2005; Vale 2003). As kinesins from other kingdoms, both Arabidopsis and rice kinesins include those with plus end-directed motor activity, and those with minus end-directed activity. Compared to kinesins from animals, plant kinesins include a large number of minus end-directed ones. For example, in *A. thaliana* 21 kinesins are predicted to be minus end-directed motors (Reddy and Day 2001b). Most of plant kinesins contain non-motor sequences not found in kinesins of other kingdoms, such as the Armadillo repeat sequence and calponin-homology domain (Reddy and Day 2001b; Richardson et al. 2006). Functions of only a few kinesins have been experimentally determined (Lee and Liu 2004).

Among the 61 kinesins in *A. thaliana*, it is intriguing how many are involved in cell division. Unfortunately, the comparison of non-motor domains of these kinesins to animal and fungal ones known to play a role in cell division usually is not informative. In a recent study, three criteria have been used to postulate that a given kinesin may play a role in mitosis: (1) the presence of a mitosis specific *cis*-acting element; (2) the presence of one or more putative cyclin-dependent kinase (CDK)-phosphorylation sites; and (3) the presence of a mitotic destruction box (Vanstraelen et al. 2006). On the basis of these criteria, 23 kinesins are designated as "mitotic kinesins", and their expression are indeed elevated during cell division. Some of the later mentioned kinesins are included in these 23 mitotic kinesins.

This article summarizes functions of kinesins in plant cell division, which have either been experimentally determined or hypothesized according to particular microtubule-based motility at certain stages during cell division. While some kinesins may only function at a certain stage of cell division, others may be active at multiple stages.

1.2
Myosin Superfamily Proteins in Plants

Motility along microfilaments is driven by myosins. The myosin motor domain of approximately 700 amino acids bears a nucleotide-binding motif and an actin-binding site. Myosins contain a special motif known as the IQ repeat for binding to calmodulin as one of their light chains. Myosins are classified into 24 subfamilies (Berg et al. 2001; Foth et al. 2006; Hodge and Cope 2000). To date, only Myosin VI has been determined as a pointed (minus) end-directed motor, and others are either determined as or predicted to be barbed (plus) end-directed motors (Berg et al. 2001). Plant cells are known for rapid cytoplasmic streaming which is powered by myosins (Shimmen and Yokota 2004). Compared to the complex figure of myosins in animals, plant myosins are grouped into only two subfamilies, Myosin VIII and Myosin XI (Reddy and Day 2001a). Plant myosins are more related to Myosin V than to others, implying that they probably play a role in membrane trafficking as animal and fungal Myosin Vs do. Because plant myosins have predicted molecular masses of larger than 150-kDa, a few of them have their cDNA sequences determined. Moreover, cellular activities of most plant myosins have not been studied experimentally.

2
Kinesins in Plant Mitosis and Meiosis

The spindle apparatus in plant cells looks very similar to that of an animal cell, except for having incompletely focused spindle poles (Fig. 1A). Therefore, the plant spindle microtubule array only consists of kinetochore microtubules linking the kinetochore to the spindle pole, and inter-polar non-kinetochore microtubules probably required for maintaining the overall structure of the spindle. Kinesins would be required for the following processes: (1) organization of microtubules to form the two spindle poles during spindle assembly; (2) nuclear envelope breakdown; (3) attachment of sister chromatids to microtubules; (4) congression of chromosomes at the metaphase plate; (5) disassembly of kinetochore fibers and movement of sister chromatid/homologous chromosomes towards the poles; and (6) spindle elongation.

2.1
Kinesins for Spindle Pole Organization

Unlike in animal cells where microtubules often originated from the centrosome prior to mitosis, plant somatic cells committed to mitosis have microtubules polymerized across the nuclear envelope, concomitantly with the

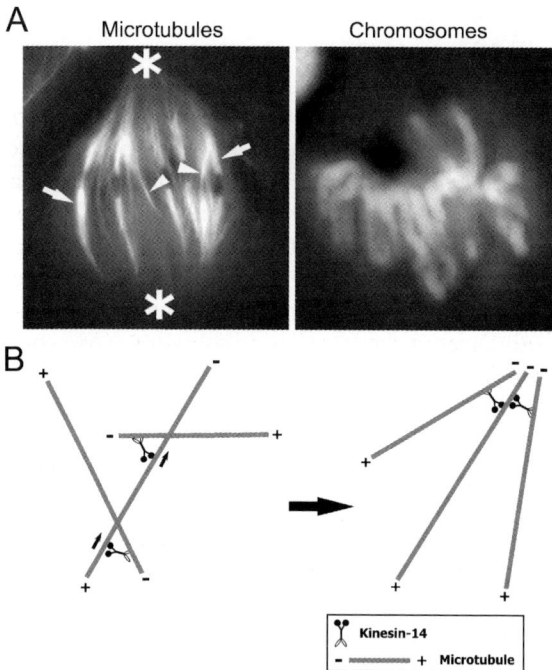

Fig. 1 Organization of spindle microtubules in plant cells. **A** Fluorescent images showing microtubules and chromosomes in an onion cell at metaphase. *Asterisks* mark the incompletely focused spindle poles. *Arrows* point at kinetochore microtubule fibers, and arrowheads at interzonal microtubules which do not end at kinetochores. **B** Proposed role of minus end-directed Kinesin-14 in converging microtubule minus ends. *Arrows* indicate the direction of kinesin motility

development of the preprophase band (PPB) (Wick and Duniec 1983). These nuclear envelope-associated microtubules are later organized into what is referred to as the "prophase spindle" with two clear poles. The formation of these spindle poles, which are the new MTOCs, involves the reorganization of microtubules with their minus ends focused at the polar region as indicated by the concentration of minus end-specific protein γ-tubulin at the poles (Liu et al. 1993). Prior to the breakdown of the nuclear envelope, microtubules form polar caps which mark the future spindle poles (Lloyd and Chan 2006). The formation of polar caps would require that microtubules get translocated so that their minus ends become converged. Such a microtubule converging phenomenon has been well documented in the endosperm cells undergoing mitosis in the African blood lily Haemanthus (Smirnova and Bajer 1998).

A microtubule gliding activity along a parallel microtubule would account for the convergence of these microtubules. Minus end-directed kinesins in the Kinesin-14 subfamily would be ideal candidates for such an activity (Fig. 1B). Members of the Kinesin-14 subfamily have their motor domain located at the C-terminus which is preceded with a minus end specific neck sequence (Henningsen and Schliwa 1997). Besides its ATP-dependent microtubule-binding site in the motor domain, Kinesin-14 also has an ATP-independent microtubule-binding site located at the N-terminus (Karabay and Walker 1999; Narasimhulu and Reddy 1998). *A. thaliana* has four kinesins structurally resembling Kinesin-14 from fungi and animals like Kar3p and NCD (Reddy and Day 2001b). Two of them, KATA/ATK1 and ATK5 demonstrate minus end-directed motility, and localize preferentially towards the plus ends of spindle microtubules (Ambrose et al. 2005; Liu et al. 1996; Liu and Palevitz 1996; Marcus et al. 2002). These kinesins are able to translocate microtubules towards the spindle poles and to form focused poles, which has been demonstrated in the presence of excess ATP (Liu et al. 1996). Genetic evidence supports the notion that KATA/ATK1 plays a role in spindle microtubule organization during mitosis and meiosis (Chen et al. 2002; Marcus et al. 2003). A mutation at this locus leads to reduced microtubule accumulation at spindle poles (Marcus et al. 2003). The meiotic spindle is dramatically affected by the loss of this kinesin as spindle poles become splayed with no obvious poles, which leads to frequent failure in meiosis (Chen et al. 2002). Plants carrying loss-of-function *atk5* mutations exhibit broad spindles in mitotic cells (Ambrose et al. 2005). Similar roles of animal Kinesin-14 in spindle pole focusing has been confirmed recently (Goshima et al. 2005). Despite the abnormal spindle morphology caused by *atk1/5* mutations, mitosis proceeds normally. Thus, other cellular factors, e.g. other members of the Kinesin-14 subfamily, may have redundant roles as these two kinesins.

2.2
Kinesins in Nuclear Envelope Breakdown

Studies in animal cells indicate that cytoplasmic dynein functions in facilitating the breakdown of the nuclear envelope by promoting nuclear envelope invaginations along astral microtubules (Salina et al. 2002). Because plant cells lack cytoplasmic dynein, such a role by dynein may have been taken over by one or more minus end-directed kinesins. Two lines of evidence would support this postulation. First, microtubules are associated with the nuclear envelope at prophase (Fig. 2). Second, kinesin(s) are present on the prophase nuclear envelope as revealed by pan kinesin antibodies raised against peptide conserved among all kinesins (Sawin et al. 1992) (Fig. 2).

A unique member of the plant Kinesin-14 subfamily, the calmodulin-binding kinesin KCBP/ZWI may play a role in nuclear envelope breakdown. KCBP/ZWI contains a C-terminal Ca^{++}/calmodulin-binding site and a talin-like domain towards the N-terminus (Reddy et al. 1996; Reddy and Reddy 1999). Otherwise, it is a minus end directed kinesin similar to KATA/ATK1 with an N-terminal ATP-independent microtubule-binding site (Narasimhulu et al. 1997; Song et al. 1997). Interestingly, its motor activity is inhibited upon

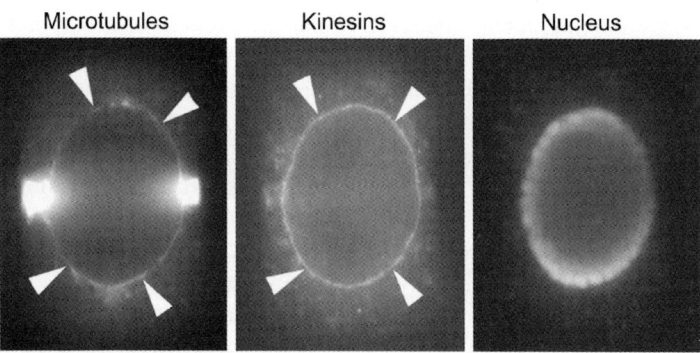

Fig. 2 Microtubules and kinesins at prophase nuclear envelope. A prophase onion cell was stained with anti-tubulin, anti-pan kinesin, and DAPI to reveal microtubules, kinesins, and DNA, respectively. Note microtubules were polymerized on the nuclear envelope (*arrowheads*) while the PPB microtubules were prominent at the cortex in the middle of the cell. Kinesins also accumulated on the nuclear envelope (*arrowheads*)

binding to Ca^{++}/calmodulin (Deavours et al. 1998; Narasimhulu et al. 1997). Conversely, its motor activity can be constitutively activated using an antibody against the calmodulin-binding peptide (Narasimhulu and Reddy 1998). It decorates microtubule arrays of the spindle and the phragmoplast (Bowser and Reddy 1997; Preuss et al. 2003; Smirnova et al. 1998). When such a protein is activated by microinjection of anti-calmodulin-binding peptide antibody in the stamen hair cells of Tradescantia, nuclear envelope breakdown is induced to allow cells to enter prometaphase precociously (Vos et al. 2000). Therefore, KCBP/ZWI's potential function in nuclear envelope breakdown may be regulated by local concentration of Ca^{++}/calmodulin at the nuclear envelope region. Other proteins may play a redundant role as KCBP/ZWI does, because null mutations at the corresponding locus in *A. thaliana* only affect trichome morphogenesis, but not cell division (Oppenheimer et al. 1997).

2.3
Kinesins Mediating the Interaction Between Chromosomes and Microtubules

In animal cells, the inventory of microtubule motors interacting with chromosomes includes cytoplasmic dynein, CENP-E/Kinesin-7, and MCAK/Kinesin-13 that appeared at the kinetochore; and chromokinesin/Kinesin-4 and Kinesin-10 along chromosome arms (Heald 2000). In addition to the absence of cytoplasmic dynein in angiosperms, plant kinesins in the Kinesin-4, -7, and -13 subfamilies appear to function in interphase cells. Null mutations at the *fra1* locus encoding a Kinesin-4 only affect cell elongation but not cell division (Zhong et al. 2002). The Arabidopsis Kinesin-13A protein is associated with the Golgi stacks, and plays a role in trichome morphogenesis, again, not in cell division (Lu et al. 2005). None of the plant Kinesin-7 members have been proved to be associated with chromosomes. It is noteworthy that the homology of plant kinesins with their animal counterparts in these subfamilies only limits to their motor domains. Functional non-motor domains found in animal Kinesin-4, -7, -10, -13 are not detected in the plant members of these subfamilies.

After the breakdown of the nuclear envelope, kinetochores exhibit a lateral interaction with microtubules prior to the formation of end-on kinetochore fibers in plant cells (Liu and Palevitz 1991). Thus, kinesins yet to be identified may be accountable for bringing kinetochores to plus ends of kinetochore fiber microtubules. To date, only a soybean kinesin decorates the kinetochore indicating that plant kinesin(s) do act at the kinetochore (Lee and Liu, unpublished data) (Fig. 3). Whether it plays a role in mitosis awaits further analysis.

Congression of chromosomes to the metaphase plate is a key step at prometaphase before cells proceed to anaphase. In animal cells, a number of

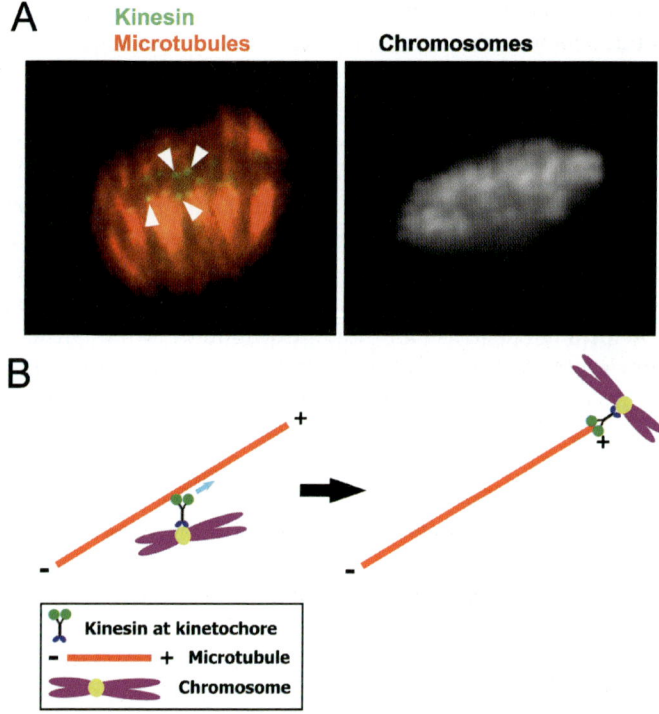

Fig. 3 Kinesin at the kinetochore. **A** A metaphase soybean cell was stained for microtubules (*red*), kinesin (*green*), and DNA (*white*). Note the kinesin was detected at the kinetochore (*arrowheads*). **B** Diagram of transition from lateral interaction between microtubules and the kinetochore to an end-on association. A plus end-directed kinesin is proposed to play a role in this process. *Arrow* indicates plus end-directed motility for bringing chromosomes to the microtubule plus end

kinesins including Kinesin-7, -10, -13 all play a non-redundant role in this process (Zhu et al. 2005). In plant cells, it is not known whether members of the Kinesin-7 subfamily are associated with kinetochore or chromosome arms. Because motor activity is required for chromosome congression, it is only a matter of time before kinesins in charge of this event are revealed experimentally in plants.

2.4
Kinesins in Kinetochore Fiber Disassembly and Sister Chromatid Movement During Anaphase

The disassembly of kinetochore fiber leads to the segregation of genetic material during anaphase among eukaryotes. A molecular model depicting motor-driven kinetochore fiber disassembly and sister chromatid segregation has been developed in animal cells. While the cytoplasmic dynein is the primary motor for the pole-ward movement, MCAK/Kinesin-13 plays an assistant role for microtubule disassembly at the kinetochore (Heald 2001). Direct evidence has been obtained in cells of the fly *Drosophila melanogaster* (Sharp et al. 2000b). Thus, ideally in plant cells a minus end-directed kinesin is needed at the kinetochore to fulfill the function of dynein in animal cells. Among 21 predicted minus end-directed kinesins encoded by the *A. thaliana* genome, to date none of the studied ones has been detected at the kinetochore. Alternatively, microtubules of the kinetochore fiber could be depolymerized solely at the spindle pole during anaphase in plant cells. An example has been shown in Drosophila cells in which a microtubule depolymerizing Kinesin-13 acting at the centrosome, and another Kinesin-13 at the kinetochore cooperatively drive sister chromatid segregation during anaphase (Rogers et al. 2004). Therefore, identification of kinesins at the kinetochore and the spindle pole would be critical for us to determine the mechanism of sister chromatid movement in plant cells.

2.5
Kinesins Acting on Midzone Microtubules for Spindle Elongation

Midzone microtubules are responsible for spindle elongation, which is the hallmark of anaphase B in animal and fungal cells. In plant meristematic cells, midzone microtubules are rather prominent although anaphase B often is not obvious (Fig. 4A) (Palevitz 1993). However, rapid polymerization and reorganization of the interpolar microtubules, usually occurring within 10 minutes, are rather intriguing and allow us to speculate that kinesins likely play a role during this process (Granger and Cyr 2000; Kumagai et al. 2001; Zhang et al. 1990). The ultimate consequence of midzone microtubule organization is the birth of the phragmoplast microtubule array. Thus, microtubule sliding activity would be responsible for the formation of this anti-parallel array.

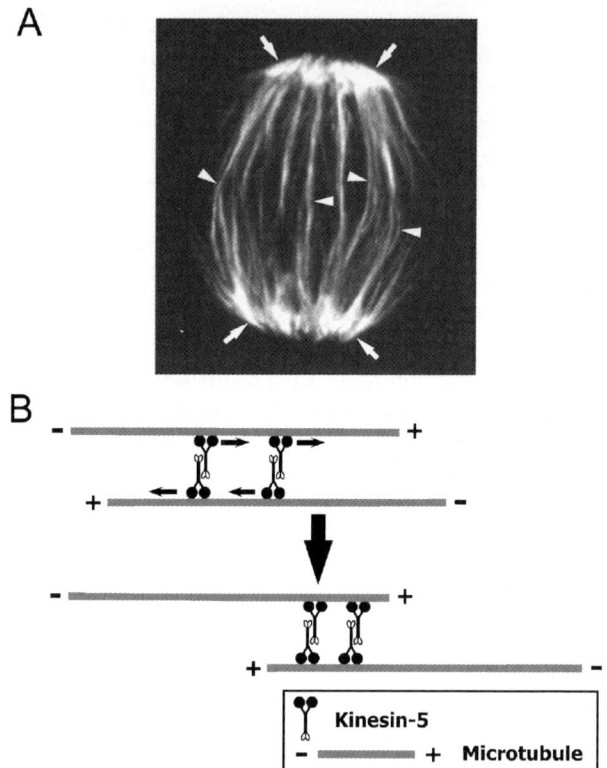

Fig. 4 Organization of spindle midzone microtubules. **A** An anti-tubulin immunofluorescent image showing microtubule organization in an anaphase onion root cell. Midzone microtubules are indicated by *arrowheads*, and remnants of shortening kinetochore microtubules are indicated by *arrows*. **B** Kinesin-5 members are proposed to play a role in sliding anti-parallel microtubules apart. Dumbbell-shaped homotetramers of Kinesin-5 are plus end directed motors, and its directionality is indicated by *arrows*

Members of the BIMC/Kinesin-5 subfamily are the key players for microtubule sliding in the spindle (Sharp et al. 2000c). BIMC was first isolated from the filamentous fungus *Aspergillus nidulans* as a motor for spindle pole body separation (Enos and Morris 1990). It contains an N-terminal motor domain followed by long coiled-coils. Towards the C-terminus of Kinesin-5, there is a signature domain known as the BIMC box with a CDK phosphorylation site, indicating its activity is regulated in a cell cycle-dependent fashion (Blangy et al. 1995). Biochemical analysis of its Drosophila homolog indicates that the native form of such kinesins is a homotetramer with two motor domains at each distal end in a dumbbell-like shape so that it is often referred to as bipolar kinesin (Kashina et al. 1996). Kinesin-5/BIMC is essential for maintaining the bipolar structure of the spindle as its inactivation leads to collapse of the bipolar spindle into a monopolar one (Blangy et al. 1995; Enos and Morris 1990).

Plant Kinesin-5 was first isolated from the tobacco BY-2 cells as TKRP125 (Asada et al. 1997). Later two similar proteins were identified in carrot suspension cells by a biochemical approach (Barroso et al. 2000). Structurally, they are very similar to their counterparts in fungi and animals. Among 61 kinesins in *A. thaliana*, four belong to Kinesin-5 with the signature BIMC box (Lee and Liu 2001, 2004). Immunolocalization data indicate that one or more Kinesin-5 decorates midzone microtubules (Asada et al. 1997; Barroso et al. 2000), suggesting they may play a role in microtubule sliding there (Fig. 4B). Unfortunately, functional characterization of Kinesin-5s in *A. thaliana* by genetic means has been lagging behind, likely due to functional redundancy of multiple homologs.

Intriguingly, the activity of Kinesin-5 is antagonized by Kar3p/NCD-like kinesins in the Kinesin-14 subfamily in both fungi and animals (Hoyt et al. 1993; O'Connell et al. 1993; Sharp et al. 2000a). The current model is that Kinesin-5 generates outward force along microtubules to allow anti-parallel microtubules to slide against each other, while Kinesin-14 acts as a brake by generating inward forces (Sharp et al. 2000c). Four members of the Kinesin-14 subfamily in *A. thaliana*, KATA/ATK1, KATB, KATC, and ATK5, structurally resemble Kar3p/NCD (Lee and Liu 2001, 2004). They decorate midzone microtubules in mitotic cells by both immunofluorescence and fluorescent protein tagging (Ambrose et al. 2005; Liu et al. 1996; Liu and Palevitz 1996; Mitsui et al. 1994). Whether these four kinesins function antagonistically against the four Kinesin-5s awaits further studies.

In summary, to our knowledge members of Kinesin-5 and Kinesin-14 may have similar functions in mitosis and meiosis as their animal and fungal counterparts. Kinesins of other subfamilies which have been proven to function in spindle operation in fungal and animal cells either do not have homologs in plant cells, or their plant counterparts do not function in mitosis or meiosis.

3
Kinesins in Plant Cytokinesis

Concomitant with the shortening of the kinetochore fibers during mitosis, midzone microtubules develop into prominent tight bundles because of new microtubule polymerization and coalescence (Zhang et al. 1993). The polarity of these midzone microtubules is later sorted out with their plus ends facing the future division site (Fig. 5). It had been speculated that phragmoplast microtubules interdigitated in the middle. However, a recent study indicates that the anti-parallel microtubules in the phragmoplast do not overlap, and are spaced by the proposed "cell plate assembly matrix" (Austin et al. 2005).

Phragmoplast microtubules serve as tracks along which Golgi-derived vesicles are transported towards microtubule plus ends for assembling the cell

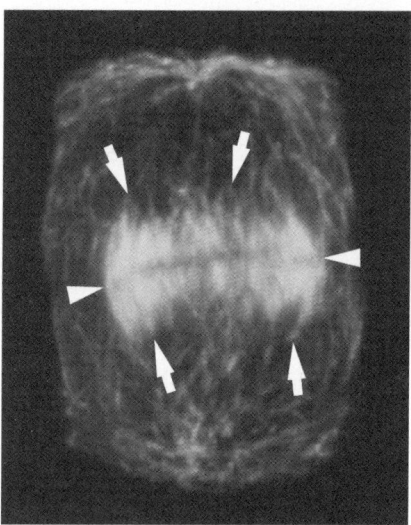

Fig. 5 Organization of microtubules in the phragmoplast. The phragmoplast microtubule array has a mirrored organization of microtubules with their plus ends in the middle (*arrowheads*) and minus ends facing daughter nuclei (*arrows*)

plate. Cell plate assembly initiates in the middle of the phragmoplast, and continues centrifugally. Upon the completion of vesicle delivery at a given area, microtubules in this area will be depolymerized from the minus end. While microtubules are depolymerized in the middle of the phragmoplast, new microtubules are being polymerized towards the periphery. Once the assembling cell plate reaches the parental plasma membrane, phragmoplast microtubules will then be completely depolymerized. One of the most intriguing questions in plant cell biology is that the cortical site where developing cell plate fuses is previous determined by the PPB before nuclear envelope breakdown in somatic cells (Wick 1991). But the molecular mechanism underlying this readout of division plane is still unknown.

In the phragmoplast, kinesins would be required for the following activities: (1) the establishment of the phragmoplast microtubule array; (2) vesicle transport along phragmoplast microtubules; (3) microtubule turnover (polymerization and depolymerization) in the phragmoplast; and (4) spatial regulation of cell plate assembly.

3.1
Kinesins for Organizing the Phragmoplast Microtubule Array

The development from midzone microtubules to the phragmoplast microtubule array is rather a rapid process (Zhang et al. 1993). The transition involves microtubule polymerization and bundling/coalescence, as well as sorting of microtubule polarity and maintenance of the mirrored array. Kinesins could be involved in all of these activities.

Although tubulins can self-polymerize into microtubules, locally emphasized polymerization in the spindle midzone would certainly be assisted by proteins that promote polymerization. Cross-linking microtubule-associated proteins in the MAP65 family are key factors for the development of midzone microtubules (Muller et al. 2004; Van Damme et al. 2004). Recent results show Kar3p/NCD-like ATK5, a member of the Kinesin-14 subfamily, decorates the plus end of midzone microtubules (Ambrose et al. 2005). The NCD kinesin has been shown to function in stabilizing microtubules, and promoting the assembly of a stable bipolar spindle in the absence of structurally defined MTOCs like the centrosome (Matthies et al. 1996). So ATK5 may function in midzone microtubule polymerization by stabilizing these microtubules. Such a stabilizing activity may be similar to that of the microtubule plus end tracking protein EB1 (Bisgrove et al. 2004; Chan et al. 2005; Dixit et al. 2006; Van Damme et al. 2004).

Midzone microtubules exhibit a phenomenon of lateral coalescence to lead the formation of the phragmoplast microtubule array (Zhang et al. 1993). Again, the Kinesin-14 member KATA/ATK1 plays a role in the lateral coalescence activity (Chen et al. 2002). In mutants lacking KATA/ATK1, midzone microtubules remain as discrete thin bundles, and often fail to develop into the phragmoplast microtubule array during meiosis. But such a phenotype is not obvious in somatic mitosis in the same mutant (Marcus et al. 2003). This may likely be because of functional redundancy of the other three Kar3p/NCD-like Kinesin-14s in *A. thaliana*. As a matter of fact, two of Kinesin-14s, KATB and KATC exhibit a cell cycle-dependent expression pattern with a peak during the M phase, which suggests potential roles in cell division (Mitsui et al. 1996).

One of the most critical steps in establishing the phragmoplast microtubule array is to sort out the polarity of midzone microtubules so that their plus ends would be facing each other at the division site. Earlier efforts have determined that microtubule motor(s) are involved in microtubule-microtubule translocation activity (Asada et al. 1991). TKRP125A of the BIMC/Kinesin-5 subfamily is a prominent phragmoplast kinesin (Asada et al. 1997; Asada and Shibaoka 1994). When TKRP125's activity was inhibited using antibodies in vitro, microtubule translocation activity was blocked (Asada et al. 1997). In tobacco suspension cells, TKRP125 decorates phragmoplast microtubules, which further supports its role in microtubule-microtubule sliding. Intriguingly, immunolocalization of a similar kinesin in carrot suspension cells indicates that a particular Kinesin-5 specifically decorates overlapping microtubules in the phragmoplast, which supports a role in sorting out the polarity of the microtubules (Barroso et al. 2000). As a bipolar motor, Kinesin-5 would be a perfect player for establishing an anti-parallel microtubule array with plus ends in the middle (Fig. 4B).

Once microtubules are arranged in two mirrored sets in the phragmoplast, they serve as tracks along which vesicles are delivered by kinesins. Early studies on microtubule dynamics in the phragmoplast indicate that tubulins are continuously added to the plus ends of phragmoplast microtubules in the middle (Asada et al. 1991; Vantard et al. 1990). The new microtubule segments would have to be pushed away. Otherwise, microtubule plus ends would be located at sites other than the future cell plate position. Since the plus end is located in the middle, a plus end-directed kinesin would be needed to continuously translocate newly added microtubule segments away from the middle of the phragmoplast. The PAKRP1 kinesin from Arabidopsis and rice exclusively decorates the plus end of phragmoplast microtubules (Lee and Liu 2000). PAKRP1 belongs to the Kinesin-12 subfamily with predicted plus end-directed motility, and uncharacterized non-motor domains (Lee and Liu 2004). In the *A. thaliana* genome, at least two genes encode Kinesin-12 with an identical spatial and temporal localization pattern in the phragmoplast (Pan et al. 2004). Inactivation of either Kinesin-12 by T-DNA insertional mutations neither alter the localization of its homolog, nor affect cytokinesis. Thus, these kinesins function redundantly during cytokinesis. It would be intriguing how microtubule organization would be affected when both kinesins are inactivated. We also ought to determine whether Kinesin-12s act as a matrix component connecting anti-parallel microtubules in the phragmoplast.

3.2
Kinesins and Microtubule Turnover in the Phragmoplast

During cell plate assembly, rapid turnover of phragmoplast microtubules includes depolymerization in the central region where cell plate materials have been delivered, and concomitant polymerization at a peripheral area where vesicle delivery is about to begin (Zhang et al. 1993). The molecular mechanism underlying this rapid turnover had been unclear until a MAP kinase cascade was connected to microtubule dynamics through serendipitous investigations (Takahashi et al. 2004). Early studies of plant MAP kinases have revealed a particular set of kinases which become localized to the division site during cytokinesis (Bogre et al. 1999; Nishihama and Machida 2001). Inactivation of this MAP kinase cascade by mutations in genes encoding individual kinases leads to incomplete cytokinesis in Arabidopsis and tobacco (Krysan et al. 2002; Nishihama et al. 2001; Soyano et al. 2003). The MAPKKK of this cascade interacts with two plant specific kinesins NACK1/HIN and STD/TES/NACK2 (Nishihama et al. 2002). These two homologous kinesins have a typical plus end-directed motor structure with the motor domain at the N-terminus (Nishihama et al. 2002; Strompen et al. 2002; Yang et al. 2003). Although they sometimes are placed inside the Kinesin-7/CENP-E subfamily, their motor domains are divergent enough from animal members of this subfamily so that they have been suggested to reside in a plant unique subfamily (Dagenbach and Endow 2004; Lawrence et al. 2004). They interact with the MAPKKK through a domain in their non-motor region (Nishihama et al. 2002). These kinesins colocalize with aforementioned MAPKKK at the cell division site (Nishihama et al. 2002; Yang et al. 2003). In *A. thaliana*, they have redundant functions in cytokinesis of all cell types as loss-of-function mutations at either loci only lead to partial failure in cytokinesis (Tanaka et al. 2004). However, NACK1/HIN functions more pronouncedly in somatic cytokinesis during embryogenesis, while STD/TES/NACK2 seems to be more critical for male meiotic cytokinesis (Hülskamp et al. 1997; Nishihama et al. 2002; Spielman et al. 1997; Strompen et al. 2002). These kinesins play a role in the localization of the MAPKKK at the division site, and consequently they are required for the expansion of the cell plate (Nishihama et al. 2002). A recent study shows that NtMAP65-1 is a substrate of this MAP kinase cascade in tobacco (Sasabe et al. 2006). Native MAP65-1 acts as a microtubule stabilizing agent by promoting microtubule polymerization and bundling, and is targeted to the spindle midzone during mitosis (Mao et al. 2005a,b; Smertenko et al. 2000). Upon phosphorylation by the MAP kinase NRK1/NTF6, MAP65-1's microtubule bundling activity is significantly down regulated, and is more restricted to the middle region of the phragmoplast where microtubule plus ends are located (Sasabe et al. 2006). Therefore, the phosphorylation of MAP65-1 promotes the turnover of phragmoplast microtubules.

Intriguingly, the depolymerization initiates at the minus ends, and progresses towards the plus end (Zhang et al. 1993). To our knowledge, none of the plant kinesins has been shown with a microtubule depolymerase activity. In animal cells, Kinesin-13 acts as the depolymerase with microtubule end-stimulated ATPase activity (Walczak 2003). In certain cells like those of the frog Xenopus, a given Kinesin-13 can act at both the plus and the minus end of microtubules (Desai et al. 1999; Walczak et al. 1996; Wordeman and Mitchison 1995). In Drosophila cells, however, one Kinesin-13 acts at the plus end, while another one acts at the minus end to depolymerize microtubules (Rogers et al. 2004). However, an Arabidopsis Kinesin-13 acts at the Golgi stacks, and does not play a role in cytokinesis (Lu et al. 2005). Whether other plant kinesins possess a depolymerase activity awaits further analysis. It will also be interesting to learn whether plant cells use the same kinesin(s) for depolymerizing microtubules in the spindle and the phragmoplast.

Concomitant with the depolymerization, new microtubules are polymerized at the peripheral region of the phragmoplast. A recent observation indicates that the Kinesin-14 ATK5 appears at the leading edge of the phragmoplast microtubule array (Ambrose et al. 2005). On the basis of its potential role in promoting polymerization as shown by its Drosophila homolog NCD, ATK5 may act to promote microtubule polymerization at the periphery of the phragmoplast.

3.3
Kinesins as Motors of Vesicle Transport for Cell Plate Assembly

One of the most conspicuous phenomena in plant cytokinesis is the formation of the cell plate. As soon as the phragmoplast microtubule array is established, along microtubules vesicles are bound towards the division site (Fig. 6). In samples prepared by rapid freezing and freeze substitution such as the one shown in Fig. 6, there is a space between the vesicle and the microtubule, suggesting that they are connected by a linker. A motor-like structure has also been shown on vesicles in the phragmoplast in an electron microscopic/tomographic study (Otegui et al. 2001). The orientation of the phragmoplast microtubules indicates that Golgi-derived vesicles are transported by plus end-directed kinesin(s) towards the plus end of these microtubules. Consequently, the fusion of these vesicles gives rise to a tubulo-

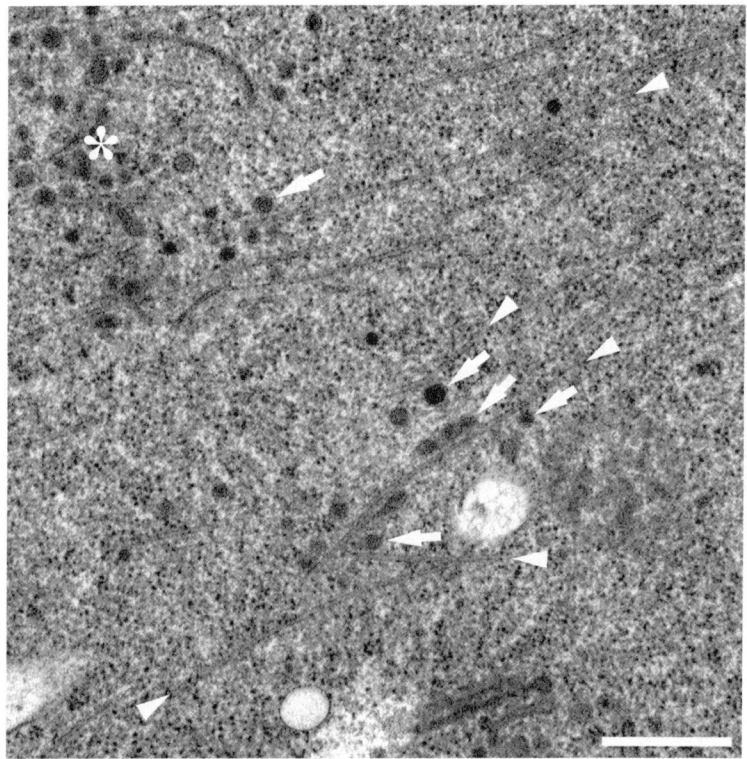

Fig. 6 Microtubules and vesicles in the phragmoplast. An *A. thaliana* cell was processed for electron microscopic examination after being fixed by rapid freezing and freeze substitution. *Arrowheads* point at microtubules, and *arrows* at vesicles bound to microtubules. *Asterisk* points at tubulo-vesicular network, formed by vesicle fusion, in the middle area of the phragmoplast. The *scale bar* represents 0.5 μm

vesicular network, and ultimately to the cell plate (Bednarek and Falbel 2002; Jürgens 2005; Otegui and Staehelin 2000; Verma 2001).

To date, very little has been known about kinesins in vesicle transport in the phragmoplast. Many questions are still standing. How many kinesins are involved in vesicle transport? Are all vesicles transported by identical kinesins? What is the mechanism of vesicle–kinesin interaction? How are such kinesins activated for vesicle transport? Are vesicles transported processively along phragmoplast microtubules? Whether, and if yes how, these kinesins are recycled during cytokinesis?

An immunocytochemical screen for kinesins in the phragmoplast has revealed a plant-specific kinesin AtPAKRP2 which is localized on vesicle-like structures in the phragmoplast (Lee et al. 2001). This N-terminal motor kinesin is very divergent from all other studied kinesins among eukaryotes, thus it is considered as an orphan kinesin. AtPAKRP2 is associated with Golgi-derived vesicles, and the association apparently is required for AtPAKRP2 to be localized along phragmoplast microtubules. This evidence suggests that AtPAKRP2 is most likely activated by its cargo (vesicle) binding. Further functional analysis is necessary for addressing how AtPAKRP2 works in the phragmoplast.

The membrane system in the phragmoplast is highly dynamic, and vesiculation is probably continuously taking place (Verma and Hong 2005). Besides relatively long distance transport along phragmoplast microtubules, there might be short distance microtubule-based transports within the phragmoplast. However, our knowledge in this aspect is vague. Obviously, more kinesins are yet to be characterized for their roles in vesicle trafficking in the phragmoplast.

3.4
Kinesins for Spatial Regulation of Cytokinesis?

One of the most intriguing questions regarding plant cell division is the forecast of the cell division plane by the PPB, whose assembly and disassembly take place before the breakdown of the nuclear envelope. Apparently, the phragmoplast is able to "read" the information left by the PPB at the cortical division site. A recent report on a predicted minus end-directed kinesin KCA1 suggests that this kinesin may be involved in the spatial regulation of cell division in *A. thaliana* (Vanstraelen et al. 2006). KCA1 and its homolog KCA2 contain typical motor and neck sequences as the Kinesin-14 member KATA/ATK1, but in addition they have an extended coiled-coil domain followed by a domain with two CDK-phosphorylation sites and a nuclear localization sequence (Vanstraelen et al. 2004). Using a heterologous expression system of tobacco cells, a GFP-KCA1 fusion protein shows a localization pattern at the cell cortex and the cell plate, but not at the cortical site which the PPB occupies (Vanstraelen et al. 2006). While such a localization pattern is dependent on the presence of the PPB, the absence of KCA1 at the PPB site is required for correct placement of the cell plate. Therefore, it would be of particular interest to identify the cargo molecule(s) of KCA1 in order to further understand how KCA1 is linked to spatial regulation of cell division.

Another unresolved mystery is how microtubules and microfilaments cross talk at the cell division site. It is known that microfilaments play a role in division plane determination, but not cell plate formation per se (Palevitz 1980). Our recent results indicate that plant kinesins containing the CH (calponin homology) domain interact with microfilaments (Preuss et al. 2004). Certain members of this plant specific kinesin group may integrate microfilaments with microtubules at the cell division site.

4
Myosins in Plant Cytokinesis: Vesicle Transport Continued

Most myosins, except for Myosin II, are implicated in vesicle transport in fungal and animal cells (Berg et al. 2001). Among them, fungal Myosin V plays a role in vesicle transport during cytokinesis/septation (Mulvihill et al. 2006). Both plant Myosin VIII and XI are more closely related to animal and fungal Myosin V than to other myosins (Lee and Liu 2004). Moreover, Myosin XI is thought to share a common ancestor with Myosin V as both contain the "dilute" domain towards their C-termini, which is considered to be involved in vesicle trafficking (Berg et al. 2001; Foth et al. 2006). But how they might be involved in cytokinesis remains to be determined. A pharmacological study with drugs disturbing myosin activities shows that certain myosin(s) may contribute to cell plate expansion and alignment in Tradescantia stamen hair cells (Molchan et al. 2002). This finding has been echoed by a recent genetic study showing that Myosin XI MYA2 plays a role in cell plate alignment in Arabidopsis (Holweg and Nick 2004).

Functional analysis of plant Myosin XI is complicated by the fact that this subfamily has many members. In *A. thaliana*, for example, 13 out of its 17 myosins are in this subfamily (Reddy and Day 2001a)! Potential functional redundancy would make functional studies more tedious and time consuming. Nevertheless, their potential role in cytokinesis will be revealed by careful genetic and cell biological studies in the near future using existing or soon to be available tools.

5
Concluding Remarks

The complexity and novelty of plant kinesin and myosin superfamilies implies that fascinating intracellular motile activities powered by these cytoskeletal motors are yet to be revealed. While concerted roles of these motors are required during cell division, their activities are regulated by means of cell-cycle dependent expression, post-translational modifications, and turnovers. Additional regulatory mechanisms include activation by cargo-binding and inhibition by Ca^{++}/calmodulin-binding. Despite many common phenomena in cell division among eukaryotes, simple recapitulation of functions of animal kinesins for plant ones in the same subfamily will certainly be an inappropriate practice. Characterizing functions of individual cytoskeletal motors in cell division would be the key for us to appreciate molecular mechanisms that regulate plant mitosis, meiosis, and cytokinesis.

Acknowledgements Our studies of plant kinesins have been generously supported by grants from the US Department of Agriculture Cooperative State Research, Education and Extension Service and the US Department of Energy Energy Biosciences programs. We wish to thank Drs. Mary Preuss and Fengli Guo for contributing pictures shown in Figs. 2 and 6.

References

Ambrose JC, Li W, Marcus A, Ma H, Cyr R (2005) A minus-end-directed kinesin with plus-end tracking protein activity is involved in spindle morphogenesis. Mol Biol Cell 16:1584–1592

Asada T, Kuriyama R, Shibaoka H (1997) TKRP125, a kinesin-related protein involved in the centrosome-independent organization of the cytokinetic apparatus in tobacco BY-2 cells. J Cell Sci 110:179–189

Asada T, Shibaoka H (1994) Isolation of polypeptides with microtubule-translocating activity from phragmoplasts of tobacco BY-2 cells. J Cell Sci 107:2249–57

Asada T, Sonobe S, Shibaoka H (1991) Microtubule translocation in the cytokinetic apparatus of cultured tobacco cells. Nature 350:238–241

Austin JR, Segui-Simarro JM, Staehelin LA (2005) Quantitative analysis of changes in spatial distribution and plus-end geometry of microtubules involved in plant-cell cytokinesis. J Cell Sci 118:3895–3903

Barroso C, Chan J, Allan V, Doonan J, Hussey P, Lloyd C (2000) Two kinesin-related proteins associated with the cold-stable cytoskeleton of carrot cells: characterization of a novel kinesin, DcKRP120-2. Plant J 24:859–868

Bednarek SY, Falbel TG (2002) Membrane trafficking during plant cytokinesis. Traffic 3:621–629

Berg J, Powell B, Cheney R (2001) A millennial myosin census. Mol Biol Cell 12:780–794

Bisgrove S, Hable W, Kropf D (2004) +TIPs and microtubule regulation. The beginning of the plus end in plants. Plant Physiol 136:3855–3863

Blangy A, Lane HA, Dherin P, Harper M, Kress M, Nigg EA (1995) Phosphorylation by $p34^{(cdc2)}$ regulates spindle association of human Eg5, a kinesin-related motor essential for bipolar spindle formation in vivo. Cell 83:1159–1169

Bogre L, Calderini O, Binarova P, Mattauch M, Till S, Kiegerl S, Jonak C, Pollaschek C, Barker P, Huskisson NS et al. (1999) A MAP kinase is activated late in plant mitosis and becomes localized to the plane of cell division. Plant Cell 11:101–113

Bowser J, Reddy ASN (1997) Localization of a kinesin-like calmodulin-binding protein in dividing cells of Arabidopsis and tobacco. Plant J 12:1429–1437

Chan J, Calder G, Fox S, Lloyd C (2005) Localization of the microtubule end binding protein EB1 reveals alternative pathways of spindle development in Arabidopsis suspension cells. Plant Cell 17:1737–1748

Chen C, Marcus A, Li W, Hu Y, Calzada J, Grossniklaus U, Cyr R, Ma H (2002) The Arabidopsis ATK1 gene is required for spindle morphogenesis in male meiosis. Development 129:2401–2409

Dagenbach EM, Endow SA (2004) A new kinesin tree. J Cell Sci 117:3–7

Deavours BE, Reddy ASN, Walker RA (1998) Ca2+/calmodulin regulation of the Arabidopsis kinesin-like calmodulin-binding protein. Cell Motil Cytoskel 40:408–416

Desai A, Verma S, Mitchison TJ, Walczak CE (1999) Kin I kinesins are microtubule-destabilizing enzymes. Cell 96:69–78

Dixit R, Chang E, Cyr R (2006) Establishment of polarity during organization of the acentrosomal plant cortical microtubule array. Mol Biol Cell 17:1298–1305

Endow SA (1999) Determinants of molecular motor directionality. Nat Cell Biol 1:E163–E167

Enos AP, Morris NR (1990) Mutation of a gene that encodes a kinesin-like protein blocks nuclear division in A. nidulans. Cell 60:1019–1027

Foth BJ, Goedecke MC, Soldati D (2006) New insights into myosin evolution and classification. Proc Natl Acad Sci USA 103:3681–3686

Goshima G, Nedelec F, Vale RD (2005) Mechanisms for focusing mitotic spindle poles by minus end-directed motor proteins. J Cell Biol 171:229–240

Graham LE, Cook ME, Busse JS (2000) The origin of plants: Body plan changes contributing to a major evolutionary radiation. Proc Natl Acad Sci USA 97:4535–4540

Granger CL, Cyr JL (2000) Microtubule reorganization in tobacco BY-2 cells stably expressing GFP-MBD. Planta 210:502–509

Heald R (2000) Motor function in the mitotic spindle. Cell 102:399–402

Heald R (2001) Chromosome movement: dynein-out at the kinetochore. Curr Biol 11:R128–R131

Henningsen U, Schliwa M (1997) Reversal in the direction of movement of a molecular motor. Nature 389:93–96

Hodge T, Cope M (2000) A myosin family tree. J Cell Sci 113:3353–3354

Holweg C, Nick P (2004) Arabidopsis myosin XI mutant is defective in organelle movement and polar auxin transport. Proc Natl Acad Sci USA 101:10488–10493

Hoyt MA, He L, Totis L, Saunders WS (1993) Loss of function of Saccharomyces cerevisiae kinesin-related Cin8 and Kip1 is suppressed by Kar3 motor domain mutations. Genetics 135:35–44

Hülskamp M, Parekh N, Grini P, Schneitz K, Zimmermann I, Lolle S, Pruitt R (1997) The STUD gene is required for male-specific cytokinesis after telophase II of meiosis in Arabidopsis thaliana. Dev Biol 187:114–124

Jürgens G (2005) Cytokinesis in higher plants. Annu Rev Plant Biol 56:281–299

Karabay A, Walker RA (1999) Identification of microtubule binding sites in the Ncd tail domain. Biochemistry 38:1838–1849

Kashina AS, Baskin RJ, Cole DG, Wedaman KP, Saxton WM, Scholey JM (1996) A bipolar kinesin. Nature 379:270–272

Kreis T, Vale R (1999) Guidebook to the Cytoskeletal and Motor Proteins. Oxford University Press, Oxford

Krysan PJ, Jester PJ, Gottwald JR, Sussman MR (2002) An Arabidopsis mitogen-activated protein kinase kinase kinase gene family encodes essential positive regulators of cytokinesis. Plant Cell 14:1109–1120

Kumagai F, Yoneda A, Tomida T, Sano T, Nagata T, Hasezawa S (2001) Fate of nascent microtubules organized at the M/G1 interface, as visualized by synchronized tobacco BY-2 cells stably expressing GFP-tubulin: time-sequence observations of the reorganization of cortical microtubules in living plant cells. Plant Cell Physiol 42:723–732

Lawrence C, Morris N, Meagher R, Dawe R (2001) Dyneins have run their course in plant lineage. Traffic 2:362–363

Lawrence CJ, Dawe RK, Christie KR, Cleveland DW, Dawson SC, Endow SA, Goldstein LSB, Goodson HV, Hirokawa N, Howard J et al. (2004) A standarized kinesin nomenclature. J Cell Biol 167:19–22

Lee Y-RJ, Giang HM, Liu B (2001) A novel plant kinesin-related protein specifically associates with the phragmoplast organelles. Plant Cell 13:2427–2439

Lee YRJ, Liu B (2000) Identification of a phragmoplast-associated kinesin-related protein in higher plants. Curr Biol 10:797–800

Lee YRJ, Liu B (2004) Cytoskeletal motors in Arabidopsis. Sixty-one kinesins and seventeen myosins. Plant Physiol 136:3877–3883

Liu B, Cyr RJ, Palevitz BA (1996) A kinesin-like protein, KatAp, in the cells of arabidopsis and other plants. Plant Cell 8:119–132

Liu B, Lee YRJ (2001) Kinesin-related proteins in plant cytokinesis. J Plant Growth Regul 20:141–150

Liu B, Marc J, Joshi HC, Palevitz BA (1993) A g-tubulin-related protein associated with the microtubule arrays of higher plants in a cell cycle-dependent manner. J Cell Sci 104:1217–1228

Liu B, Palevitz BA (1991) Kinetochore fiber formation in dividing generative cells of *Tradescantia* kinetochore reorientation associated with the transition between lateral microtubule interactions and end-on kinetochore fibers. J Cell Sci 98:475–482

Liu B, Palevitz BA (1992) Organization of cortical microfilaments in dividing root cells. Cell Motil Cytoskel 23:252–264

Liu B, Palevitz BA (1996) Localization of a kinesin-like protein in generative cells of tobacco. Protoplasma 195:78–89

Lloyd C, Chan J (2006) Not so divided: the common basis of plant and animal cell division. Nat Rev Mol Cell Biol 7:147–152

Lu L, Lee Y-RJ, Pan R, Liu B (2005) An internal motor kinesin is associated with the Golgi apparatus and plays a role in trichome morphogenesis in Arabidopsis. Mol Biol Cell 16:811–823

Ma Y-Z, Yen L-F (1989) Actin and myosin in pea tendrils. Plant Physiol 89:586–589

Mao G, Chan J, Calder G, Doonan JH, Lloyd CW (2005a) Modulated targeting of GFP-AtMAP65-1 to central spindle microtubules during division. Plant J 43:469–478

Mao T, Jin L, Li H, Liu B, Yuan M (2005b) Two microtubule-associated proteins of the Arabidopsis MAP65 family function differently on microtubules. Plant Physiol 138:654–662

Marcus A, Ambrose J, Blickley L, Hancock W, Cyr R (2002) *Arabidopsis thaliana* protein, ATK1, is a minus-end directed kinesin that exhibits non-processive movement. Cell Motil Cytoskel 52:144–150

Marcus AI, Li W, Ma H, Cyr RJ (2003) A kinesin mutant with an atypical bipolar spindle undergoes normal mitosis. Mol Biol Cell 14:1717–1726

Matthies H, McDonald H, Goldstein L, Theurkauf W (1996) Anastral meiotic spindle morphogenesis: role of the non-claret disjunctional kinesin-like protein. J Cell Biol 134:455–464

Miki H, Okada Y, Hirokawa N (2005) Analysis of the kinesin superfamily: insights into structure and function. Trends Cell Biol 15:467–476

Mitsui H, Hasezawa S, Nagata T, Takahashi H (1996) Cell cycle-dependent accumulation of a kinesin-like protein, KatB/C in synchronized tobacco BY-2 cells. Plant Mol Biol 30:177–81

Mitsui H, Nakatani K, Yamaguchishinozaki K, Shinozaki K, Nishikawa K, Takahashi H (1994) Sequencing and characterization of the kinesin-related genes KatB and KatC of *Arabidopsis thaliana*. Plant Mol Biol 25:865–876

Molchan T, Valster A, Hepler P (2002) Actomyosin promotes cell plate alignment and late lateral expansion in Tradescantia stamen hair cells. Planta (Berlin) 214:683–693

Muller S, Smertenko A, Wagner V, Heinrich M, Hussey P, Hauser M (2004) The plant microtubule-associated protein AtMAP65-3/PLE is essential for cytokinetic phragmoplast function. Curr Biol 14:412–417

Mulvihill D, Edwards S, Hyams J (2006) A critical role for the type V myosin, Myo52, in septum deposition and cell fission during cytokinesis in *Schizosaccharomyces pombe*. Cell Motil Cytoskeleton 63:149–161

Narasimhulu SB, Kao YL, Reddy ASN (1997) Interaction of *Arabidopsis* kinesin-like calmodulin binding protein with tubulin subunits: modulation by Ca2+-calmodulin. Plant J 12:1139–1149

Narasimhulu SB, Reddy ASN (1998) Characterization of microtubule binding domains in the Arabidopsis kinesin-like calmodulin binding protein. Plant Cell 10:957–965

Nishihama R, Ishikawa M, Araki S, Soyano T, Asada T, Machida Y (2001) The NPK1 mitogen-activated protein kinase kinase kinase is a regulator of cell-plate formation in plant cytokinesis. Genes Develop 15:352–363

Nishihama R, Machida Y (2001) Expansion of the phragmoplast during plant cytokinesis: A MAPK pathway may MAP it out. Curr Opin Plant Biol 4:507–512

Nishihama R, Soyano T, Ishikawa M, Araki S, Tanaka H, Asada T, Irie K, Ito M, Terada M, Banno H, et al. (2002) Expansion of the cell plate in plant cytokinesis requires a kinesin-like protein/MAPKKK complex. Cell 109:87–99

O'Connell MJ, Meluh PB, Rose MD, Morris NR (1993) Suppression of the *bimC4* mitotic spindle defect by deletion of *klpA*, a Gene Encoding a kar3-related kinesin-like protein in *Aspergillus nidulans*. J Cell Biol 120:153–162

Oppenheimer DG, Pollock MA, Vacik J, Szymanski DB, Ericson B, Feldmann K, Marks MD (1997) Essential role of a kinesin-like protein in Arabidopsis trichome morphogenesis. Proc Natl Acad Sci USA 94:6261–6266

Otegui M, Mastronarde D, Kang B, Bednarek S, Staehelin L (2001) Three-dimensional analysis of syncytial-type cell plates during endosperm cellularization visualized by high resolution electron tomography. Plant Cell 13:2033–2051

Otegui M, Staehelin LA (2000) Cytokinesis in flowering plants: more than one way to divide a cell. Curr Opin Plant Biol 3:493–502

Palevitz BA (1980) Comparatice effects of Phalloidin and Cytochalasin B on motility and morphogenesis in Allium. Can J Bot 58:773–785

Palevitz BA (1993) Morphological plasticity of the mitotic apparatus in plants and its developmental consequences. Plant Cell 5:1001–1009

Pan R, Lee YRJ, Liu B (2004) Localization of two homologous Arabidopsis kinesin-related proteins in the phragmoplast. Planta 220:156–164

Preuss ML, Delmer DP, Liu B (2003) The cotton kinesin-like calmodulin-binding protein associates with cortical microtubules in cotton fibers. Plant Physiol 132:154–160

Preuss ML, Kovar DR, Lee Y-RJ, Staiger CJ, Delmer DP, Liu B (2004) A plant-specific kinesin binds to actin microfilaments and interacts with cortical microtubules in cotton fibers. Plant Physiol 136:3945–3955

Reddy A, Day I (2001a) Analysis of the myosins encoded in the recently completed *Arabidopsis thaliana* genome sequence. Genome Biol 2:7

Reddy ASN, Day IS (2001b) Kinesins in the Arabidopsis genome: a comparative analysis among eukaryotes. BMC Genomics 2:2

Reddy ASN, Safadi F, Narasimhulu SB, Golovkin M, Hu X (1996) A novel plant calmodulin-binding protein with a kinesin heavy chain motor domain. J Biol Chem 271: 7052–7060

Reddy VS, Reddy ASN (1999) A plant calmodulin-binding motor is part kinesin and part myosin. Bioinformatics 15:1055–1057

Richardson DN, Simmons MP, Reddy AS (2006) Comprehensive comparative analysis of kinesins in photosynthetic eukaryotes. BMC Genomics 7:18

Rogers G, Rogers S, Schwimmer T, Ems-McClung S, Walczak C, Vale R, Scholey J, Sharp D (2004) Two mitotic kinesins cooperate to drive sister chromatid separation during anaphase. Nature 427:364–370

Salina D, Bodoor K, Eckley D, Schroer T, Rattner J, Burke B (2002) Cytoplasmic dynein as a facilitator of nuclear envelope breakdown. Cell 108:97–107

Sasabe M, Soyano T, Takahashi Y, Sonobe S, Igarashi H, Itoh TJ, Hidaka M, Machida Y (2006) Phosphorylation of NtMAP65-1 by a MAP kinase down-regulates its activity of microtubule bundling and stimulates progression of cytokinesis of tobacco cells. Gene Dev 20:1004–1014

Sawin K, Mitchison T, Wordeman L (1992) Evidence for kinesin-related proteins in the mitotic apparatus using peptide antibodies. J Cell Sci 101:303–313

Sharp DJ, Brown HM, Kwon M, Rogers GC, Holland G, Scholey JM (2000a) Functional coordination of three mitotic motors in Drosophila embryos. Mol Biol Cell 11: 241–253

Sharp DJ, Rogers GC, Scholey JM (2000b) Cytoplasmic dynein is required for poleward chromosome movement during mitosis in Drosophila embryos. Nat Cell Biol 2:922–930

Sharp DJ, Rogers GC, Scholey JM (2000c) Microtubule motors in mitosis. Nature 407:41–47

Shimmen T, Yokota E (2004) Cytoplasmic streaming in plants. Curr Opin Cell Biol 16:68–72

Smertenko A, Saleh N, Igarashi H, Mori H, Hauser-Hahn I, Jiang C-J, Sonobe S, Lloyd CW, Hussey PJ (2000) A new class of microtubule-associated proteins in plants. Nat Cell Biol 2:750–753

Smirnova EA, Bajer AS (1998) Early stages of spindle formation and independence of chromosome and microtubule cycles in *Haemanthus* endosperm. Cell Motil Cytoskel 40:22–37

Smirnova EA, Reddy ASN, Bowser J, Bajer AS (1998) Minus end-directed kinesin-like motor protein, KCBP, localizes to anaphase spindle poles in Haemanthus endosperm. Cell Motil Cytoskel 41:271–280

Song H, Golovkin M, Reddy ASN, Endow SA (1997) In vitro motility of AtKCBP, a calmodulin-binding kinesin protein of Arabidopsis. Proc Natl Acad Sci USA 94:322–327

Soyano T, Nishihama R, Morikiyo K, Ishikawa M, Machida Y (2003) NQK1/NtMEK1 is a MAPKK that acts in the NPK1 MAPKKK-mediated MAPK cascade and is required for plant cytokinesis. Genes Dev 17:1055–1067

Spielman M, Preuss D, Li F, Browne W, Scott R, Dickinson H (1997) TETRASPORE is required for male meiotic cytokinesis in *Arabidopsis thaliana*. Development 124:2645–2657

Staehelin LA, Hepler PK (1996) Cytokinesis in higher plants. Cell 84:821–824

Strompen G, El Kasmi F, Richter S, Lukowitz W, Assaad FF, Jurgens G, Mayer U (2002) The Arabidopsis HINKEL gene encodes a kinesin-related protein involved in cytokinesis and is expressed in a cell cycle-dependent manner. Curr Biol 12:153–158

Takahashi Y, Soyano T, Sasabe M, Machida Y (2004) A MAP kinase cascade that controls plant cytokinesis. J Biochem 136:127–132

Tanaka H, Ishikawa M, Kitamura S, Takahashi Y, Soyano T, Machida C, Machida Y (2004) The AtNACK1/HINKEL and STUD/TETRASPORE/AtNACK2 genes, which encode functionally redundant kinesins, are essential for cytokinesis in Arabidopsis. Genes Cells 9:1199–1211

Vale RD (2003) The molecular motor toolbox for intracellular transport. Cell 112:467–480

Van Damme D, Bouget F, Van Poucke K, Inze D, Geelen D (2004) Molecular dissection of plant cytokinesis and phragmoplast structure: a survey of GFP-tagged proteins. Plant J 40:386–398

Van Damme D, Van Poucke K, Boutant E, Ritzenthaler C, Inze D, Geelen D (2004) In vivo dynamics and differential microtubule-binding activities of MAP65 proteins. Plant Physiol (Rockville) 136:3956–3967

Vanstraelen M, Inze D, Geelen D (2006) Mitosis-specific kinesins in Arabidopsis. Trends Plant Sci 11:167–175

Vanstraelen M, Torres Acosta J, De Veylder L, Inze D, Geelen D (2004) A plant-specific subclass of C-terminal kinesins contains a conserved a-type cyclin-dependent kinase site implicated in folding and dimerization. Plant Physiol 135:1417–1429

Vanstraelen M, Van Damme D, De Rycke R, Mylle E, Inze D, Geelen D (2006) Cell cycle-dependent targeting of a kinesin at the plasma membrane demarcates the division site in plant cells. Curr Biol 16:308–314

Vantard M, Levilliers N, Hill AM, Adoutte A, Lambert AM (1990) Incorporation of paramecium axonemal tubulin into higher plant cells reveals functional sites of microtubule assembly. Proc Natl Acad Sci USA 87:8825–8829

Verma DPS (2001) Cytokinesis and building of the cell plate in plants. Annu Rev Plant Physiol Plant Mol Biol 52:751–784

Verma DPS, Hong Z (2005) The ins and outs in membrane dynamics: tubulation and vesiculation. Trends Plant Sci 10:159–165

Vos JW, Safadi F, Reddy ASN, Hepler PK (2000) The kinesin-like calmodulin binding protein is differentially involved in cell division. Plant Cell 12:979–990

Walczak CE (2003) The Kin I kinesins are microtubule end-stimulated ATPases. Mol Cell 11:286–288

Walczak CE, Mitchison TJ, Desai A (1996) XKCM1: A Xenopus kinesin-related protein that regulates microtubule dynamics during mitotic spindle assembly. Cell 84:37–47

Wick SM (1991) Spatial aspects of cytokinesis in plant cells. Curr Opin Cell Biol 3:253–260

Wick SM, Duniec J (1983) Immunofluorescence microscopy of tubulin and microtubule arrays in plant cells. I. Preprophase band development and concomitant appearance of nuclear envelope-associated tubulin. J Cell Biol 97:235–243

Wordeman L, Mitchison T (1995) Identification and partial characterization of mitotic centromere-associated kinesin, a kinesin-related protein that associates with centromeres during mitosis. J Cell Biol 128:95–104

Yang C, Spielman M, Coles J, Li Y, Ghelani S, Bourdon V, Brown R, Lemmon B, Scott R, Dickinson H (2003) TETRASPORE encodes a kinesin required for male meiotic cytokinesis in Arabidopsis. Plant J 34:229–240

Zhang D, Wadsworth P, Hepler PK (1993) Dynamics of microfilaments are similar, but distinct from microtubules during cytokinesis in living, dividing plant cells. Cell Motil Cytoskel 24:151–155

Zhang DH, Wadsworth P, Hepler PK (1990) Microtubule dynamics in living dividing plant cells – confocal imaging of microinjected fluorescent brain tubulin. Proc Natl Acad Sci USA 87:8820–8824

Zhong R, Burk DH, Morrison W Herbert III, Ye Z-H (2002) A kinesin-like protein is essential for oriented deposition of cellulose microfibrils and cell wall strength. Plant Cell 14:3101–3117

Zhu C, Zhao J, Bibikova M, Leverson J, Bossy-Wetzel E, Fan J, Abraham R, Jiang W (2005) Functional analysis of human microtubule-based motor proteins, the kinesins and dyneins, in mitosis/cytokinesis using RNA interference. Mol Biol Cell 16:3187–3199

Organelle Dynamics During Cell Division

Andreas Nebenführ

University of Tennessee,
Department of Biochemistry and Cellular and Molecular Biology,
Knoxville, TN 37996-0840, USA
nebenfuehr@utk.edu

Abstract Many organelles in plant cells show a more or less random distribution in the interphase cell but assume very specific positions during mitosis and/or cytokinesis. Most prominent among these is the Golgi apparatus which is thought to provide the majority of raw materials for the assembly of the forming cell plate. However, the localization of other organelles also seems to indicate specific functions during cell division. In addition, organelle positioning mediated by the actin cytoskeleton has been implicated in equal inheritance of organelles by the daughter cells. This review summarizes the current knowledge of dynamic organelle positioning during mitosis and cytokinesis and discusses the mechanisms responsible for the observed localizations.

1
Introduction

Plant cells, like those of all eukaryotes, are compartmentalized into a number of membrane-bound organelles that carry out specialized functions within the cell (Lunn 2006). An important aspect of cell division is the distribution of these organelles into the daughter cells to ensure proper functioning of these progeny cells (Warren 1993). This concept is immediately obvious for organelles that originate from fission of preexisting organelles, such as mitochondria and plastids. Once a cell has lost either of these organelles, it cannot create new copies of them since their genetic information is lost.

However, the issue of organelle inheritance also applies to compartments that do not contain their own genome. For example, it is not clear how a cell would regenerate a new endoplasmic reticulum (ER) should it ever lose this important biosynthetic organelle. In other cases, an organelle may be crucial for the process of cell division itself and may be required in both daughter cells for its successful completion. The Golgi apparatus which provides the raw material for cell plate assembly can serve as an example for this class of organelles.

To ensure faithful inheritance of organelles during cell division, a number of different approaches can be envisioned that fall broadly into two categories: regulated and random (Warren and Wickner 1996). An extreme example for regulated inheritance is the mitotic division of the nucleus itself. In this case individual chromosomes are tightly attached to a bipolar spindle to ensure that their duplicated products, the chromatids, are reliably partitioned into the two daughter cells. A similar approach also applies to those organelles that are present in small copy numbers such as chloroplasts and sometimes mitochondria in some algae. Many organelles of flowering plant cells, on the other hand, exist in high copy numbers and a random distribution throughout the cytoplasm should suffice to ensure inheritance of at least some copies by both daughters (Sheahan et al. 2004).

In reality, most organelle inheritance schemes fall somewhere between these extreme cases of total control and pure chance. In particular, those organelles that function in some aspect of cell division may not be randomly distributed in order to allow for efficient cell division to occur. As a result, any deviation from a purely random distribution of high copy number organelles can be used to infer that this organelle may play a role in mitosis or cytokinesis. This reasoning formed the basis of a number of studies that have examined the positioning of organelles during various stages of cell division. This work will review these studies and highlight some of the conclusions derived from these observations.

2
Organelle Positioning During Mitosis and Cytokinesis

2.1
Endoplasmic Reticulum

One of the first organelles to be examined in detail for its distribution during mitosis and cytokinesis was the ER. Staining of *Haemanthus* endosperm cells with chlorotetracycline, which marks high Ca^{2+} concentrations, revealed intense signals in the mitotic spindle (Wolniak et al. 1980). Since it is well known that the ER serves as an intracellular store of Ca^{2+} ions it was proposed that this staining represented the presence of ER in the mitotic spindle (Fig. 1). This conclusion was confirmed by electron microscopy (EM) of Os-FeCN stained barley and lettuce root tip cells (Hepler 1980, 1982). In both species a high density of ER elements was detected at the spindle poles as well as intermixed and associated with spindle microtubules (MTs). Newer studies employing ER-targeted GFP or immunofluorescence of HDEL-carrying proteins further confirmed these results for tobacco suspension cultured cells (Gupton et al. 2006; Nebenführ et al. 2000) and gymnosperms and pteridophytes (Zachariades et al. 2003). It has been proposed that the spindle ER plays a role in regulating local cytoplasmic Ca^{2+} concentrations that in turn are crucial for spindle function (Hepler 1980).

Prior to mitosis, the ER was found to align with the pre-prophase band (PPB) of MTs in *Pinus* (Zachariades et al. 2001). During cytokinesis, the

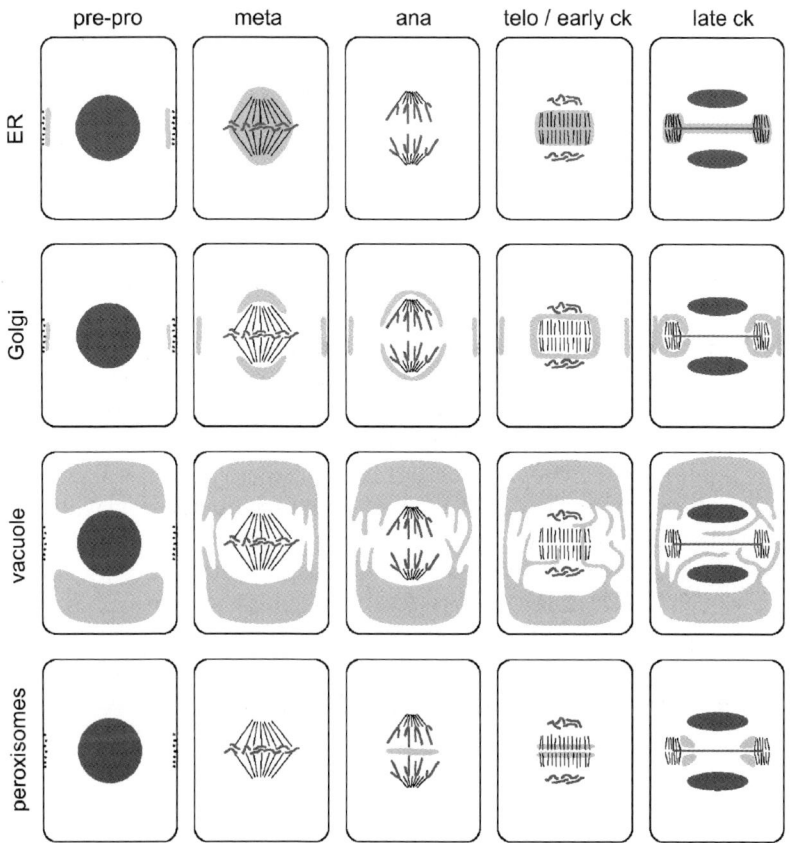

Fig. 1 Several organelles accumulate in specific regions during various stages of cell division. *Light gray* regions identify regions where ER, Golgi stacks, and peroxisomes preferentially accumulate at the given mitotic phase. The vacuole is largely confined to the periphery of the cell but sends tubular extensions into the phragmosome. For details see text. pre-pro, preprophase; meta, metaphase; ana, anaphase; telo/early ck, telophase to early cytokinesis; late ck, late cytokinesis

ER is prominently present within the phragmoplast where tubular elements seem to align with the MTs when viewed by fluorescence microscopy (Gupton et al. 2006; Nebenführ et al. 2000; Zachariades et al. 2003). ER also associates closely with the cell plate leading to the appearance of a brightly fluorescing line at the division plane (Gupton et al. 2006; Nebenführ et al. 2000; Zachariades et al. 2003). The function of this cell-plate associated ER is not fully established. Initially, it has been proposed that the ER may form a "cage" within which Golgi-derived vesicles fuse to form the cell plate (Hepler 1982). However, a more recent study employing three-dimensional EM tomography revealed that the ER is not present during the earliest stages of vesicle fusion and approaches the cell plate only later (Seguí-Simarro et al. 2004). This study speculates that the close proximity of ER and cell plate is required to facilitate exchange of membrane lipids between these compartments during cell plate maturation (Seguí-Simarro et al. 2004).

2.2
Golgi Apparatus

The Golgi apparatus assumes a special position among the organelles of plant cells in that its activity is directly necessary for cell plate formation. This special function has been postulated for the first time based on the unusual arrangement of Golgi stacks in the vicinity of the growing cell plate in maize root tips (Whaley and Mollenhauer 1963). In fact, until recently it has been assumed that the cell plate is formed exclusively from Golgi-derived vesicles (e.g., Staehelin and Hepler 1996). However, recent evidence suggests that endocytosed material from the maternal plasma membrane may also contribute to the new dividing structure (Bolte et al. 2004; Dettmer et al. 2006; Dhonukshe et al. 2006). Irrespective of these new findings, it is clear that Golgi products are required for cell plate formation, a conclusion that is further supported by recent studies on Golgi stack partitioning.

Golgi stacks are randomly distributed throughout the cytoplasm during interphase (Nebenführ et al. 2000; Seguí-Simarro and Staehelin 2006) and double in number during G2 phase (Seguí-Simarro and Staehelin 2006). In larger cells that contain large numbers of Golgi stacks, such as the highly vacuolated tobacco BY-2 cells, these stacks start to accumulate in an equatorial ring underneath the thinning PPB (Dixit and Cyr 2002; Nebenführ et al. 2000). This accumulation, termed the "Golgi belt" (Fig. 1), fully develops during pro-metaphase and is specific for this organelle since mitochondria do not accumulate in this region (Nebenführ et al. 2000). The Golgi belt is not found in meristematic cells of the *Arabidopsis* shoot meristem (Seguí-Simarro and Staehelin 2006) which contain fewer stacks and may also impose spatial constraints on organelle distribution due to their smaller size. The Golgi belt continues to mark the future division site after disappearance of the PPB which has led to the speculation that these stacks are involved in preparing the cortical division site for insertion of the cell plate (Nebenführ et al. 2000). Contrary to this prediction it was found that disruption of Golgi stacks with brefeldin A (BFA) did not inhibit insertion of the cell plate in this area (Dixit and Cyr 2002). However, it has to be noted that cell-plate insertion is also possible at non-division sites (Mineyuki and Gunning 1990), in other words, secretion from Golgi stacks to the PM at the division site may not be necessary for cell division, but may only facilitate some aspect such as maturation of the division wall. The unusual positioning of Golgi stacks in the Golgi belt at this stage of cell division clearly deserves further study to elucidate its function.

During metaphase, another accumulation of Golgi stacks becomes evident at the opposing poles of the metaphase spindle (Nebenführ et al. 2000). The density of stacks in close proximity to the spindle of BY-2 cells is approximately twice as high as in the rest of the cytoplasm. The function of this accumulation is again unknown, but has been linked to the presence of Golgi vesicles and products in the metaphase spindle (Hepler 1980; Sonobe et al. 2000) and a priming of the division area with cell plate building blocks (Nebenführ et al. 2000). During anaphase, Golgi stacks begin to appear in the interzone between the separating chromosomes but are excluded from the forming phragmoplast (Nebenführ et al. 2000).

Golgi stacks are prominently associated with the growing cell plate and accumulate around the phragmoplast as demonstrated by conventional EM in root tips (Whaley and Mollenhauer 1963), fluorescence microscopy of GFP-labeled stacks in BY-2 cells (Nebenführ et al. 2000) and EM serial sections in shoot meristems (Seguí-Simarro and Staehelin 2006). These stacks do not display a preferred orientation relative to the phragmoplast (Seguí-Simarro and Staehelin 2006), but are most likely involved in the delivery of secretory products to the phragmoplast. This association of Golgi stacks with the phragmoplast is not static but allows dynamic repositioning of the organelle as the cell plate expands (Nebenführ et al. 2000). It is not clear how long individual stacks remain in close proximity to the cytokinetic machinery. Golgi stacks are also often seen close to the maturing cell plate in late stages of cytokinesis and following completion of the cell wall (Kawazu et al. 1995; Nebenführ et al. 2000).

Golgi products have been tracked with anti-xyloglucan antibodies to fluorescently label the presence of hemicelluloses within BY-2 cells (Sonobe et al. 2000). Although direct evidence from EM labeling is not available, it is assumed that the signals generated in this way represent post-Golgi secretory vesicles. The random distribution of small spots found in interphase cells is first broken in metaphase when a diffuse staining of the metaphase spindle is detected. During anaphase, the signal appears as a broad band in the interzone between the chromosomes that gradually narrows until it finally, in telophase, is confined to a narrow line that corresponds to the cell plate (Sonobe et al. 2000). This distribution is consistent with the expected movement of Golgi vesicles along phragmoplast MTs to the division plane. Notably, the presence of hemicelluloses in the metaphase spindle supports the idea that Golgi stacks already produce cell plate precursors prior to cytokinesis and deliver these precursors to the spindle region where they are available for cell plate assembly as soon as the phragmoplast forms (Nebenführ et al. 2000).

2.3
Endosomes and Prevacuolar Compartments

As mentioned above, there is recent evidence that endocytic membrane traffic may contribute to cell plate formation. This is mostly based on the endocytic tracer FM4-64, a lipophilic dye that partitions into the plasma membrane and is thought to be endocytosed and eventually delivered to the vacuole (Aniento and Robinson 2005; Bolte et al. 2004). Interestingly, FM4-64 prominently labels the region of the forming cell plate in cytokinetic cells (Bolte et al. 2004; Dettmer et al. 2006; Dhonukshe et al. 2006) suggesting that endocytic traffic in these cells is redirected to the forming cell plate. This conclusion appears to be supported by immuno-labeling of PM proteins and cell wall polysaccharides in the cell plate as well as small punctate structures that seem to colocalize with FM4-64 spots (Dhonukshe et al. 2006).

Curiously, FM4-64 label is found not only in small spots that likely represent endosomes but also throughout the metaphase spindle (Dhonukshe et al. 2006). It is not clear which membrane compartment is labeled in this case, although the ER shows a similar distribution. Some of the FM4-64 spots were also labeled with GFP-RabF2b (= Ara7) near the cell plate (Dhonukshe et al. 2006), a Rab protein that has been implicated in Golgi-to-vacuole traffic and localizes to prevacuolar compartments (PVCs) (Kotzer et al. 2004). It is unclear whether this partial colocalization of FM4-64 with a PVC marker represents a bifurcation of the endocytic pathway between a vacuolar and a cell plate branch. This finding, however, would suggest that PM-to-vacuole traffic is not completely blocked in dividing cells. Interestingly, PVCs or more specifically GFP-RabF2b-positive structures were also labeled with a YFP-2xFYVE construct that binds to phosphatidylinositol-3-phosphate (PI3P) in membranes (Vermeer et al. 2006). In dividing BY-2 cells this marker was found in areas that correspond to the phragmoplast (Vermeer et al. 2006).

Multi-vesicular bodies (MVBs) have been identified previously as PVCs (Tse et al. 2004), although it is not clear whether all MVBs are PVCs, or whether all PVCs are MVBs. However, their unique morphology makes it possible to identify MVBs unambiguously in the EM and this was exploited in a 3D reconstruction from serial sections of *Arabidopsis* meristem cells (Seguí-Simarro and Staehelin 2006). It was found that during interphase MVBs often are located in small clusters near Golgi stacks and during cytokinesis are also associated with the cell plate. The size of MVBs increases during cell division such that their volume quadruples. This volume increase coincides with the appearance of clathrin coated buds and vesicles on the cell plate, suggesting that these cell-plate associated MVBs are involved in targeting excess membrane to the lytic vacuole (Seguí-Simarro and Staehelin 2006).

It should be cautioned that the identity of fluorescently labeled compartments is not always known but only inferred from colocalization data at the LM level. Similarly, the composition and function of membrane compartments identified in the EM is often unclear and only predicted based on morphological similarity to known structures. An additional difficulty in this part of the endomembrane system is that all compartments are only temporary containers and can change their composition and hence identity as they mature. A better understanding of the endocytic/post-Golgi/pre-vacuolar trafficking pathways is needed before we can assign definitive functions to any of these labeled structures and come to final conclusions about their role in cell plate formation.

2.4
Vacuole

The vacuole is, by volume, the largest organelle in mature plant cells and at the same time displays an enormous variability in shapes and structures (Higaki et al. 2006). In most dividing cells, the vacuole is much smaller and even decreases in volume during division (Seguí-Simarro and Staehelin 2006), but nevertheless is a prominent component of the cell. In effect, the absence of the vacuole from the center of the cell defines the "phragmosome", the continuous cytoplasmic domain within which mitosis and cytokinesis occurs. Two recent studies have pursued complementary approaches to follow vacuole dynamics in two very different dividing cells. In the first study, fluorescent labeling of the tonoplast, the vacuolar membrane, was used to visualize vacuoles in 3D confocal reconstructions (Kutsuna et al. 2003). The second study used EM serial sections to visualize vacuole structure in *Arabidopsis* shoot meristem cells (Seguí-Simarro and Staehelin 2006).

Interestingly, both cell types displayed the formation of tubular extensions of the vacuole that surrounded the mitotic apparatus and connected the two parts of the vacuole across the phragmosome (Kutsuna et al. 2003; Seguí-Simarro and Staehelin 2006; Fig. 1). In meristematic cells the vacuole breaks down into smaller units around metaphase (Seguí-Simarro and Staehelin 2006), a feature that was not evident in BY-2 cells presumably due to the higher degree of vacuolation in the latter. However, during telophase the vacuoles of both cell types projected tubular extensions into the region surrounding the cell plate. These parts of the vacuole may be participating in the degradation of material that was removed from the maturing cell plate (see above).

2.5
Other Organelles: Peroxisomes, Mitochondria, and Plastids

Organelles outside the endomembrane system have received relatively little attention with respect to their dynamics during cell division. The most thorough study was conducted on immuno-labeled peroxisomes in dividing cells of the onion root tip and leek leaf epidermis (Collings et al. 2003). In this case it was found that peroxisomes, which are randomly distributed during interphase and up to metaphase, start to accumulate in the division plane in anaphase (Fig. 1). Interestingly, this accumulation preceded formation of a clear phragmoplast but coincided with an accumulation of actin filaments (Collings et al. 2003). This cluster of peroxisomes is then split into two layers by the forming cell plate. As the ring-phragmoplast expands outward, the peroxisome clusters seem to trail behind and continue to remain closely associated with the maturing cell plate (Collings et al. 2003). The authors speculate that this localization of peroxisomes indicates the production of hydrogen peroxide radicals in the cell plate or a role of these organelles in membrane lipid recycling (Collings et al. 2003). However, it has to be cautioned that not all species accumulate peroxisomes near the cell plate to the same extent as onions and leek. In particular, the peroxisome accumulation is not as prominent in BY-2 cells and not detectable at all in *Arabidopsis* root tip cells (Collings et al. 2003). This suggests that organelle accumulations during cell division may reflect species-specific adaptations.

The distribution of mitochondria and plastids in dividing cells has not been studied in detail. Staining of these organelles with a fluorescent dye in tobacco BY-2 cells revealed that they accumulate in the phragmosome but largely are relegated to the periphery and don't approach the spindle apparatus closely (Nebenführ et al. 2000). This is particularly true for the larger plastids which are found mostly close to the vacuolar membrane. This pattern persists also during cytokinesis when some mitochondria can be found in close proximity of the phragmoplast but plastids are mostly confined to the area behind the re-forming daughter nuclei (Nebenführ et al. 2000). This pattern is also seen in EM images of the dividing cell (e.g., Seguí-Simarro and Staehelin 2006). The proximity of mitochondria to the cytokinetic apparatus may reflect the energy requirements of this machinery, while the plastids (at least in the non-photoautotrophic BY-2 cells) do not contribute to the cell division process.

3
Mechanisms of Organelle Positioning

While considerable attention has been paid to the role of the cytoskeleton in moving cell plate precursors through the phragmoplast to the division plane (Vanstraelen et al. 2006), relatively little information has been garnered on the mechanisms that lead to the specific positioning of organelles surrounding the mitotic and cytokinetic machinery. It is reasonable to assume that the cytoskeleton plays a major role in getting the organelles to their correct positions. In interphase cells, all organelles discussed so far can move along the actin cytoskeleton with the help of myosin motor proteins (Collings et al. 2002; Higaki et al. 2006; Kwok and Hanson 2003; Nebenführ et al. 1999; Runions et al. 2006; Ruthardt et al. 2005; van Gestel et al. 2002). However, all these rapid, saltatory motions come to a standstill during mitosis and all movements in dividing cells are much slower (Mineyuki et al. 1984). In addition, MTs have been found to bind organelles in interphase (Sonobe et al. 2000; van Gestel et al. 2002) and MT-based motors are known to be associated with various organelles (e.g., Lu et al. 2005; Romagnoli et al. 2003). Thus, it is not possible to a priori predict which cytoskeletal system is being used to mediate the observed distribution of the different organelles.

The presence of tubular ER elements in the metaphase spindle and the phragmoplast is suggestive of an interaction of this organelle with MTs. Close apposition of ER membranes with spindle MTs (Hepler 1980) seems to support this notion. Disruption of MTs with oryzalin, unlike cytochalasin D treatment, does indeed lead to a redistribution of the ER in mitotic cells (Zachariades et al. 2003) but in addition leads to a dramatic rearrangement of cellular elements that make interpretation difficult.

The specific accumulation of Golgi stacks in certain regions of metaphase cells could not be disrupted by actin depolymerizing drugs (Nebenführ et al. 2000) suggesting that the actin cytoskeleton is not involved in anchoring them in specific areas. However, it should be pointed out that this kind of experiment does not rule out a role for actin tracks during delivery of Golgi stacks to these positions. Disruption of MTs with propyzamide also did not result in a loss of Golgi accumulation, although the segregation of mitochondria and plastids from Golgi stacks could be disrupted by additional mechanical force, i.e. by shaking of the treated cells (Nebenführ et al. 2000). This might indicate that MTs provide anchoring points for Golgi stacks. As for the ER, these results are difficult to interpret since removal of the metaphase spindle leads to a complete loss of normal cell architecture and structural integrity of the phragmosome.

A clear involvement of the actin cytoskeleton has been found for the positioning of two organelles, the vacuole and peroxisomes. In these cases, treatment with various actin-disrupting drugs (bistheonellide A, latrunculin B, or cytochalasin D) broke up the tubular extensions of the vacuoles seen during mitosis (Kutsuna et al. 2003) and prevented accumulation of peroxisomes in the division plane (Collings et al. 2003), respectively. In the latter case, the same result was obtained with the myosin inhibitor 2,3-butanedione monoxime (BDM) indicating that the accumulation of peroxisomes at the cell plate depends on myosin-driven movements (Collings et al. 2003).

The actin cytoskeleton also seems to play a role on a global scale in ensuring even distribution of organelles into the two daughter cells. Inheritance of mitochondria, chloroplasts, and ER is normally very even in cells derived from tobacco mesophyll protoplasts (Sheahan et al. 2004). However, disruption of actin filaments with latrunculin B resulted in many cells receiving only a small fraction of some organelles while disruption of MTs with oryzalin did not yield this effect (Sheahan et al. 2004). Thus, the actin cytoskeleton seems to be needed to position roughly equal numbers of organelles on both sides of the division plane. A complementary interpretation could be that actin is required for proper positioning of the division plane so that the cells are divided evenly. This interpretation is supported by the observation that alignment of the forming cell plate with the cortical division site in *Tradescatia* stamen hair cells depends on the acto-myosin system (Molchan et al. 2002). Current data seem to suggest that both effects occur at the same time (Sheahan et al. 2004). However, a detailed spatio-temporal analysis of organelle distributions in actively dividing cells in the presence and absence of latrunculin B will be required to determine their relative contributions.

4
Outlook

Progress in recent years has demonstrated that many organelles assume non-random positions during cell division in a wide variety of plant cells. These specific accumulations indicate that these organelles are performing specific functions at these positions, which in some cases has led to new insights in the process of cell division. For example, the proximity of peroxisomes, multivesicular bodies, and vacuolar elements near the forming cell plate has resulted in new hypotheses that can now be tested experimentally. More research will also be necessary to elucidate the mechanisms that mediate this organelle positioning. In particular, precise surgical interventions will be required since the wholesale disruption of all MTs or all actin filaments have often proven too crude to identify specific localization mechanisms. Knowledge of these mechanisms then should allow us to disrupt positioning of specific organelles which in turn will directly address the function of these organelles during cell division.

Acknowledgements Research in my lab is supported by the National Science Foundation, grant MCB-0416931.

References

Aniento F, Robinson DG (2005) Testing for endocytosis in plants. Protoplasma 226:3–11
Bolte S, Talbot C, Boutte Y, Catrice O, Read ND, Satiat-Jeunemaitre B (2004) FM-dyes as experiemntal probes for dissecting vesicle trafficking in living plant cells. J Microsc 214:159–173
Collings DA, Harper JDI, Marc J, Overall RL, Mullen RT (2002) Life in the fast lane: actin-based motility of plant peroxisomes. Can J Bot 80:430–441
Collings DA, Harper JDI, Vaughn KC (2003) The association of peroxisomes with the developing cell plate in dividing onion root cells depends on actin microfilaments and myosin. Planta 218:204–216
Dettmer J, Hong-Hermesdorf A, Stierhof Y-D, Schumacher K (2006) Vacuolar H+-ATPase is required for endocytic and secretory trafficking in *Arabidopsis*. Plant Cell 18:715–730
Dhonukshe P, Baluska F, Schlicht M, Hlavacka A, Samaj J, Friml J, Gadella TWJ (2006) Endocytosis of cell surface material mediates cell plate formation during plant cytokinesis. Devel Cell 10:137–150

Dixit R, Cyr R (2002) Golgi secretion is not required for marking the preprophase band site in cultured tobacco cells. Plant J 29:99–108

Gupton SL, Collings DA, Allen NS (2006) Endoplasmic reticulum targeted GFP reveals ER organization in tobacco NT-1 cells during cell division. Plant Physiol Biochem 44:95–105

Hepler PK (1980) Membranes in the mitotic apparatus of barley cells. J Cell Biol 86:490–499

Hepler PK (1982) Endoplasmic reticulum in the formation of the cell plate and plasmodesmata. Protoplasma 111:121–133

Higaki T, Kutsuna N, Okubo E, Sano T, Hasezawa S (2006) Actin microfilaments regulate vacuolar structures and dynamics: dual observation of actin microfilaments and vacuolar membrane in living tobacco BY-2 cells. Plant Cell Physiol 47:839–852

Kawazu T, Kawano S, Kuroiwa T (1995) Distribution of Golgi apparatus in the mitosis of cultured tobacco cells as revealed by $DiOC_6$ fluorescence microscopy. Protoplasma 186:183–192

Kotzer AM, Brandizzi F, Neumann U, Paris N, Moore I, Hawes C (2004) AtRabF2b (Ara7) acts on the vacuolar trafficking pathway in tobacco leaf epidermal cells. J Cell Sci 117:6377–6389

Kutsuna N, Kumagai F, Sato MH, Hasezawa S (2003) Three-dimensional reconstruction of tubular structure of vacuolar membrane throughout mitosis in living tobacco cells. Plant Cell Physiol 44:1045–1054

Kwok EY, Hanson MR (2003) Microfilaments and microtubules control the morphology and movement of non-green plastids and stromules in *Nicotiana tabacum*. Plant J 35:16–26

Lu L, Lee Y-RJ, Pan R, Maloof JN, Liu B (2005) An internal motor kinesin is associated with the Golgi apparatus and plays a role in trichome morphogenesis in *Arabidopsis*. Mol Biol Cell 16:811–823

Lunn JE (2006) Compartmentation in plant metabolism. J Exp Bot 58:35–47

Mineyuki Y, Gunning BES (1990) A role for preprophase bands of microtubules in maturation of new cell walls, and a general proposal on the function of preprophase band sites in cell division in higher plants. J Cell Sci 97:527–537

Mineyuki Y, Takagi M, Furuya M (1984) Changes in organelle movement in the nuclear region during the cell cycle of *Adiantum* protonemata. Plant Cell Physiol 25:297–308

Molchan TM, Valster AH, Hepler PK (2002) Actomyosin promotes cell plate alignment and late lateral expansion of *Tradescantia* stamen hair cells. Planta 214:683–693

Nebenführ A, Frohlick JA, Staehelin LA (2000) Redistribution of Golgi stacks and other organelles during mitosis and cytokinesis in plant cells. Plant Physiol 124:135–151

Nebenführ A, Gallagher L, Dunahay TG, Frohlick JA, Masurkiewicz AM, Meehl JB, Staehelin LA (1999) Stop-and-go movements of plant Golgi stacks are mediated by the acto-myosin system. Plant Physiol 121:1127–1141

Romagnoli S, Cai G, Cresti M (2003) In vitro assays demonstrate that pollen tube organelles use kinesin-related motor proteins to move along microtubules. Plant Cell 15:251–269

Runions J, Brach T, Kühner S, Hawes C (2006) Photoactivation of GFP reveals protein dynamics within the endoplasmic reticulum membrane. J Exp Bot 57:43–50

Ruthardt N, Gulde N, Spiegel H, Fischer R, Emans N (2005) Four-dimensional imaging of transvacuolar strand dynamics in tobacco BY-2 cells. Protoplasma 205–215

Seguí-Simarro JM, Austin JR, White EA, Staehelin LA (2004) Electron tomographic analysis of somatic cell plate formation in meristematic cells of Arabidopsis preserved by high-pressure freezing. Plant Cell 16:836–856

Seguí-Simarro JM, Staehelin LA (2006) Cell cycle-dependent changes in Golgi stacks, vacuoles, clathrin-coated vesicles and multivesicular bodies in meristematic cells of *Arabidopsis thaliana*: A quantitative and spatial analysis. Planta 223:223–236

Sheahan MB, Rose RJ, McCurdy DW (2004) Organelle inheritance in plant cell division: the actin cytoskeleton is required for unbiased inheritance of chloroplasts, mitochondria and endoplasmic reticulum in dividing protoplasts. Plant J 37:379–390

Sonobe S, Nakayama N, Shimmen T, Sone Y (2000) Intracellular distribution of subcellular organelles revealed by antibody against xyloglucan during cell cycle in tobacco BY-2 cells. Protoplasma 213:218–227

Staehelin LA, Hepler PK (1996) Cytokinesis in higher plants. Cell 84:821–824

Tse YC, Mo B, Hillmer S, Zhao M, Lo SW, Robinson DG, Jiang L (2004) Identification of multivesicular bodies as prevacuolar compartments in *Nicotiana tabacum* BY-2 cells. Plant Cell 16:672–693

van Gestel K, Köhler RH, Verbelen J-P (2002) Plant mitochondria move on F-actin, but their positioning in the cortical cytoplasm depends on both F-actin and microtubules. J Exp Bot 53:659–667

Vanstraelen M, Inzé D, Geelen D (2006) Mitosis-specific kinesins in *Arabidopsis*. Trends Plant Sci 11:167–175

Vermeer JEM, Leeuwen Wv, Tobeña-Santamaria R, Laxalt AM, Jones DR, Divecha N, Gadella TWJ, Munnik T (2006) Visualization of PtdIns3P dynamics in living plant cells. Plant J 47:687–700

Warren G (1993) Membrane partitioning during cell division. Ann Rev Biochem 62:323–348

Warren G, Wickner W (1996) Organelle inheritance. Cell 84:395–400

Whaley WG, Mollenhauer HH (1963) The Golgi apparatus and cell plate formation: a postulate. J Cell Biol 17:216–221

Wolniak SM, Hepler PK, Jackson WT (1980) Detection of the membrane-calcium distribution during mitosis in *Haemanthus* endosperm with chlorotetracycline. J Cell Biol 87:23–32

Zachariades M, Quader H, Galatis B, Aposolakos P (2003) Organization of the endoplasmic reticulum in dividing cells of the gymnosperms *Pinus brutia* and *Pinus nigra*, and the pteridophyte *Asplenium nidus*. Cell Biol Int 27:31–40

Zachariades M, Quader H, Galatis B, Apostolakos P (2001) Endoplasmic reticulum prepohase band in dividing root-tip cells of *Pinus brutia*. Planta 213:824–827

Open Mitosis: Nuclear Envelope Dynamics

Annkatrin Rose

Department of Biology, Appalachian State University, 572 Rivers Street, Boone, NC 28608, USA
rosea@appstate.edu

Abstract The nuclear envelope separating the cell nucleus from the cytoplasm is a common feature of all eukaryotic cells, but its origin is still an enigma. Its early evolution appears closely linked with the evolution of the mitotic spindle apparatus. Many regulatory proteins are playing critical dual roles in spindle assembly as well as nuclear envelope and nuclear pore complex formation. During the evolution of higher eukaryotes, open mitosis evolved independently in the plant and animal lineages, leading to a marked diversification of nuclear envelope compositions and roles in mitosis. Unique features of the plant nuclear envelope include its function as mitotic spindle organizing center and the lack of nuclear lamins and associated proteins. Nuclear envelope dynamics observed during mitosis appear to be similar between plants and animals. The nuclear envelope is absorbed into the endoplasmic reticulum after breakdown and reformed from the endoplasmic reticulum membrane pool after mitosis. In addition, nuclear envelope material contributes to the newly forming cell plate in plant cells. Plant and animal cells might use the same underlying molecular signals for nuclear envelope reassembly, but modified as variations of a common theme.

1
Introduction

In all eukaryotic cells, the nucleus containing the genetic material is separated from the cytoplasm by a double-membrane structure termed the nuclear envelope. The outer nuclear membrane is continuous with the endoplasmic reticulum and presents a surface for the organization of cytoskeletal components. The inner nuclear membrane serves similarly as anchoring surface for chromatin. Both membranes are connected by nuclear pore complexes forming channels to allow for the exchange of macromolecules between the nucleoplasm and the cytoplasm.

The compartmentalization of the genetic material within a membrane-bound nucleus poses a challenge to the cell during mitosis: The cell has to segregate the nuclear material before cytokinesis to ensure the equal distribution of the genetic material onto the daughter cells. How do eukaryotic cells solve this problem?

The answer is two-pronged:

1. To achieve segregation of the genetic material, the eukaryotic cell utilizes a dynamic microtubule array, the mitotic spindle, attached to specialized

chromosome regions, the centromers, to separate the condensed chromosomes
2. To achieve the separation of one membrane-bound nucleus into two identical daughter nuclei, a variety of mechanisms has evolved ranging from karyokinesis in closed mitosis to complete breakdown and reassembly of the nuclear envelope in open mitosis

The evolution of both processes appears to be tightly interlocked, with the nuclear envelope playing crucial roles in mitotic checkpoint control and spindle assembly. This chapter takes a closer look at the evolution of open mitosis in the "green" lineage and the dynamics and functional roles of the nuclear envelope in plant cells compared to the other branches of the eukaryotic domain.

2
Evolution of the Nuclear Envelope

The exact events leading to the development of the nucleus as an organelle are subject to controversial discussion. Two basic models for the formation of the nuclear envelope have been proposed:

1. *Karyogenic hypothesis*: formation of the nuclear envelope through invagination of the plasma membrane in a proto-eukaryote
2. *Endokaryotic hypothesis*: formation of the nucleus (and centrosomes) by endosymbiosis, with the nuclear envelope a remnant of a phagocytic event

The possibility of analyzing fully sequenced genomes to determine likely prokaryotic origins of eukaryotic signature proteins has renewed the debate. While sequence data seems to support the theory of endosymbiosis of an archaebacterium by a eubacterial host (Horiike et al. 2002) or endosymbiosis of a number of archaea and bacteria by a postulated "chronocyte" (Hartman and Fedorov 2002), the endokaryotic hypothesis remains controversial (Martin 2005). Another hypothesis argues that not endosymbiosis, but the formation of a chimera between an archaebacterium and motile eubacteria led to the development of the karymastigont, a nuclear structure coupled with a microtubular flagellum that is found in amitochondriate protists (Margulis et al. 2000, 2006).

A common premise of these hypotheses is the coupling of the emergence of a nuclear envelope to the presence of an endoskeleton and its co-evolution with mitosis (Cavalier-Smith 2002; Dolan et al. 2002). In the compartmentalization model, the appearance of the nucleus would have been predated by the evolution of an endoskeleton allowing for motility and vesicle transport and laying the basis for the development of the mitotic apparatus (Martin 1999). Membrane enclosure of the nucleus might have been triggered by the need to protect the genetic material from digestive enzymes used during phagocytosis (Mitchison 1995; Roos 1984). Phagocytosis also would have been a requirement for endosymbiosis, favoring the "chrononcyte" hypothesis (Hartman and Fedorov 2002). In the karyomastigont hypothesis, the DNA of the chimera recombined and remained attached to the membrane as well as the kinetosome of its motile component, resulting in the formation of a membrane-bound nucleus linked to cytoskeletal structures giving rise to the kinetosome/centrosome microtubule organizing center (MTOC) facilitating mitosis (Dolan et al. 2002; Margulis et al. 2000, 2006).

2.1
Origin of Nuclear Envelope Membranes and Proteins

The nuclear membranes are thought to have evolved from or together with the endoplasmic reticulum (ER) in the evolution of eukaryotic cells (Cavalier-Smith 2002; Lopez-Garcia and Moreira 2006; Mans et al. 2004). This is supported by the fact that the outer nuclear envelope membrane is continuous with the ER. It has been suggested that the radiation of the Ras-family of signaling GTPases, leading to the development of the Ran cycle, played a pivotal role in the development of a functional nuclear envelope. This hypothesis is supported by several observations:

- The Ran cycle components are highly conserved throughout the eukaryotic domain and are in fact the most strikingly conserved nuclear pore components in a comparative genomics analysis, suggesting that they preexisted in the eukaryotic ancestor (Mans et al. 2004)
- While ER vesicles show a propensity for association with chromatin to form a rudimentary nucleus, the fusion of such vesicles docked to chromatin to form a continuous nuclear envelope requires the Ran cycle (Hetzer et al. 2000, Newport and Dunphy 1992)
- The nuclear pore assembly likewise requires the action of the Ran cycle (Ryan et al. 2003, Walther et al. 2003)
- The Ran cycle plays an essential role in nucleocytoplasmic transport across the nuclear envelope (Steggerda and Paschal 2002)

Nug-type GTPases involved in ribosomal subunit export are similarly highly conserved, suggesting that a primitive nuclear pore complex depending on Ran and Nug GTPase-driven essential transport functions was part of the ancestral nuclear envelope in early eukaryotic cells. Based on comparative genomics, it has been estimated that these early nuclear pores were already highly complex structures composed of about 20 proteins. The primordial nuclear import machinery likely included the karyopherins importin α and β, which have additional functions in spindle formation and in nuclear envelope and pore complex assembly, respectively (Askajer et al. 2002; Harel et al. 2003; Mans et al. 2004; Zhang et al. 2002).

The nuclear envelope and nuclear pore complexes subsequently gained complexity and functionality through diversification of existing protein families, such as the karyopherins, and the incorporation of novel proteins in the different eukaryotic lineages. Some of the domains in these proteins can be traced back to bacterial origins (endosymbiotic horizontal gene transfer), while others are paralogs of eukaryotic proteins that have gained new functions. The nuclear lamina, a prominent feature of the metazoan nuclear envelope, evolved only in that lineage, possibly as an adaptation to the greater mechanical stress, in the form of contractile forces, in animal cells as compared to fungi and plants (Mans et al. 2004). As a result of this divergent evolution, the nuclear envelopes found in the three kingdoms of higher eukaryotes show distinctive differences in composition as well as function, despite their common origin.

The evolutionarily late acquisition of novel functions in nuclear envelope proteins is also supported by the finding of distinct targeting mechanisms for Ran GTPase-activating protein (RanGAP). RanGAP is targeted to the nuclear pore in animal as well as plant cells, but not in yeast. This subcellular localization is achieved through additional nuclear envelope targeting domains present in the plant and animal versions of the protein. However, these domains differ from each other, are attached to opposite termini of the catalytic core of the protein, and are not functionally interchangeable (Jeong et al. 2005; Rose and Meier 2001). This suggests that they evolved after the divergence of the major eukaryotic kingdoms and utilize kingdom-specific nuclear envelope components as anchors. The extension of animal and plant RanGAP with a specific targeting domain changing its subcellular localization is an example for the common principle of "tinkering together" existing protein domains to gain functionality during the evolution of the nuclear envelope and nuclear pore complex (Mans et al. 2004). It has been postulated that the targeting of RanGAP to the nuclear envelope constitutes an adaptation to the development of open mitosis in which the nuclear envelope disassembles (Rose and Meier 2001).

2.2
From Closed Mitosis to Open Mitosis

In most unicellular eukaryotes, the nuclear envelope stays intact during mitosis. In this closed mitosis, an intranuclear spindle separates the chromosomes before karyokinesis occurs. In fenestral mitosis, the spindle forms outside the nucleus and penetrates the nuclear envelope; however, the nuclear envelope does not disassemble completely. During the evolution of multicellular organisms, open mitosis evolved where the nuclear envelope disintegrates before or during spindle formation and reforms around the two daughter nuclei after the separation of the chromosomes. Open mitosis occurs in the plant and animal kingdoms and possibly some fungi of the phylum *basidiomycota*. As it is not found in more basal lineages, open mitosis likely evolved independently in the different kingdoms and is a result of convergent evolution of mitotic mechanisms rather than divergence from a common ancestor (see Fig. 1).

When did open mitosis arise in the plant lineage? The evolutionary ancestors of the higher plants are the green algae. Modes of mitosis found in green algae include the proposed archetype of eukaryotic mitosis via kary-

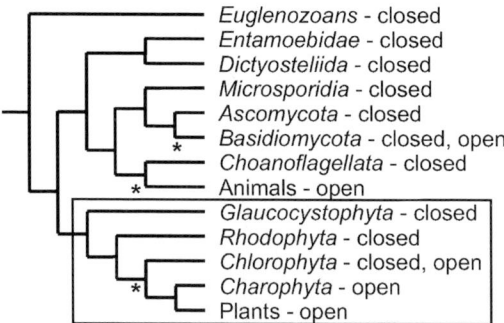

Fig. 1 Occurrence of closed and open mitosis in eukaryotes after mitochondrial endosymbiosis (phylogeny based on Hedges 2002). *Asterisks* mark independent appearances of open mitosis during evolution. The *boxed* phyla represent organisms with plastids originating from photosynthetic endosymbiosis (Moreira et al. 2000)

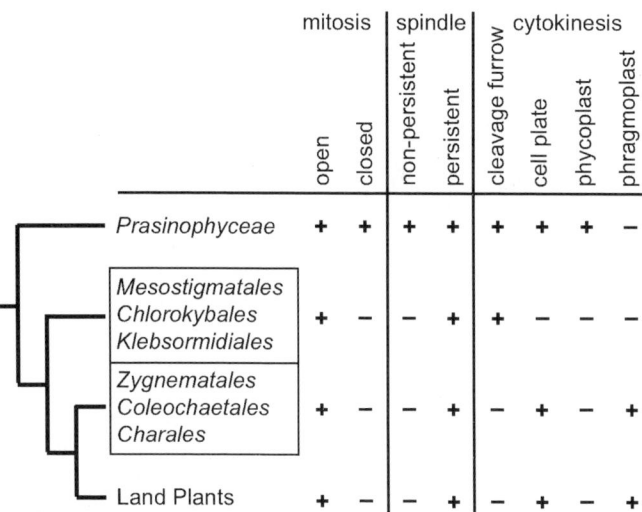

Fig. 2 Variations in mitotic features present in the groups of green algae implicated in the ancestry of land plants. Presence or absence of features listed on top are marked by *plus* and *minus*, respectively. The *box* indicates *Charophyta* (Based on van den Hoek et al. 1995; McCourt et al. 2004)

omastigont duplication, for example in the model organism *Chlamydomonas* (Brugerolle and Mignot 2003; Coss 1974), and a large variety of diverse mechanisms (summarized in Fig. 2). In addition to the occurrence of open and closed mitosis as well as cytokinesis via cleavage furrow or cell plate, two fundamentally different modes of microtubule organization can be observed at the end of mitosis. Members of the *Chlorophyceae* and some *Prasinophyceae*

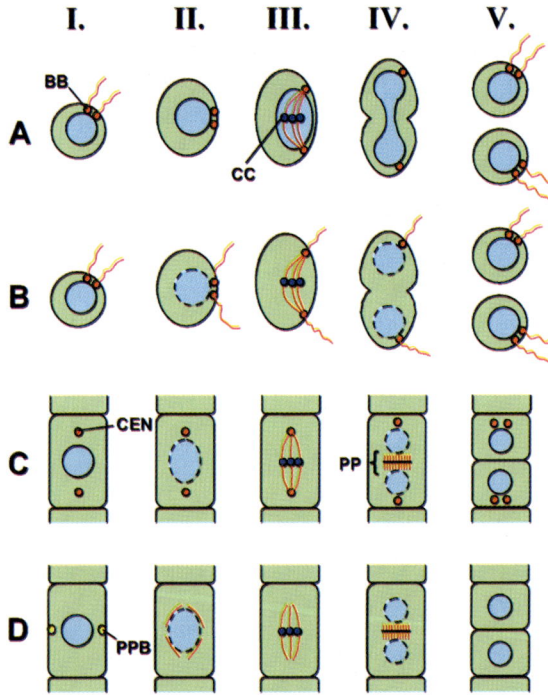

Fig. 3 Schematic representation of mitotic stages during the development of open mitosis in the plant lineage. *A Chlamydomonas*-type mitosis: *Chlamydomonas* loses its flagella before undergoing karyomastigont-type karyokinesis followed by cytokinesis via cleavage furrow. *B Pyramimonas (Prasinophyceae)*-type mitosis: *Pyramimonas* contains four basal bodies with flagella, but only two are shown for simplification. Mitosis is open and followed by cytokinesis via cleavage furrow. *C Coleochaete*-type mitosis: Mitosis is open and utilizes cytoplasmic centrioles as microtubule organizing centers. Cytokinesis is achieved by a phragmoplast and centrifugal cell plate formation. *D* Higher plant mitosis: Microtubules form a preprophase band at the cell periphery and are organized around the nuclear envelope before it breaks down. Cytokinesis is achieved by a phragmoplast and centrifugal cell plate formation. *I* Preprophase, *II* prophase, *III* metaphase, *IV* telophase, *V* interphase. Nuclei are shown in *light blue* and nuclear envelope disassembly and reassembly is represented by *broken lines*. Microtubules are represented by *red lines*. *BB* membrane-bound basal bodies, *CC* condensed chromosomes, *CEN* cytoplasmic centrioles, *PP* phragmoplast with new forming cell plate, *PPB* preprophase band of microtubules

contain a phycoplast where the spindle apparatus collapses and microtubules align parallel to the plane of cell division in telophase. In contrast, the *Zygnematales*, *Coleochaetales*, and *Charales* contain a phragmoplast where the spindle persists into telophase and microtubules align perpendicular to the plane of cell division (McCourt et al. 2004). These characteristics coincide with the division of the green algae into *Charophyta* and other *Chlorophyta*. Outside the *Charophyta*, only a few species of algae in the *Ulvophycean* clade utilize phragmoplasts as well (Lopez-Bautista et al. 2003).

The *Charophyta*, and within this group the *Charales* (stoneworts), are considered the closest relatives to the land plants based on their mitotic behavior as well as phylogeny (Battacharya and Medlin 1998; Karol et al. 2001; Lewis and McCourt 2004). Morphological and sequence-based data suggest that the ancestor of the charophytes belonged to the green algae group *Prasinophyceae* (Battacharya and Medlin 1998). Cell division within the *Prasinophyceae* is diverse and includes open and closed mitosis with a persistent spindle and cleavage furrow (*Pyramimonas* and *Nephroselmis*, respectively) and closed mitosis with a non-persistent spindle and cell plate formation via phycoplast (*Tetraselmis*).

Putting phylogeny and differences and similarities in mitosis together, the following order of events might have led to the development of the phragmoplast-type open mitosis of higher plants:

- Spindle persists into telophase (some *Prasinophyceae*, all *Charophyta*)
- Nuclear envelope breaks down during mitosis (some *Prasinophyceae*, all *Charophyta*)
- Telophase spindle becomes more prominent (all *Charophyta*)
- Phragmoplast with cell plate forms (some *Charophyta*)
- Centrioles are lost in higher plants

Schematic examples of proposed stages along the development of higher plant type open mitosis are shown in Fig. 3.

3
Role of Nuclear Envelope Components in Mitosis and Meiosis

The nuclear envelope plays a role in several processes during cell division, such as:

- Chromosome condensation
- Telomere attachment and homologous chromosome synapsis in meiois
- Spindle assembly and orientation
- Kinetochore assembly and function
- Mitotic checkpoint control
- Cell plate formation in higher plants

3.1
Spindle and Kinetochore Assembly

Most prominently, components of the nuclear envelope are tightly linked with mitotic spindle assembly. In yeast and some protists, the spindle pole bodies are embedded into the nuclear envelope. Additional nuclear envelope proteins are involved in the regulation of spindle pole body formation. For example, the yeast nuclear pore protein MLP2 has been found to promote spindle pole body assembly (Niepel et al. 2005). Proteins of the nuclear pore complexes and the nucleocytoplasmic transport machinery also play crucial roles during mitotic spindle assembly in animals. In *C. elegans*, several nucleoporins are required for proper spindle orientation (Schetter et al. 2006).

In mammalian cells, the nuclear pore proteins RanBP2/Nup358 and RanGAP redistribute to the kinetochores during mitosis where they are essential for kinetochore assembly and function (Joseph et al. 2004; Salina et al. 2003). The depletion of RanBP2/Nup358 from mammalian cells causes mitotic arrest, suggesting a function in mitotic checkpoint control (Salina et al. 2003). Another protein shuttling between the nuclear envelope and mitotic kinetochores, MEL-28/ELYS, is required for structural and functional integrity of the nuclear envelope as well as chromosome condensation and kinetochore and spindle assembly in *C. elegans* egg cells (Fernandez and Piano 2006; Galy et al. 2006). MEL-28/ELYS has been implicated in postmitotic nuclear pore complex assembly, mirroring the mitotic roles of RanBP2/Nup358 (Franz et al. 2007; Rasala et al. 2006).

3.2
Mitotic Functions of the Ran Cycle

In animal cells, the Ran cycle and karyopherins, facilitators of nucleocytoplasmic transport during interphase, play important roles in spindle assembly as well as vesicle fusion and nuclear envelope reassembly (Askajer et al. 2002; Di Fiore et al. 2004). Disturbances of the proper ratios and localization of proteins within the Ran cycle interfere with mitotic check point control (Arnaoutov and Dasso 2003; Li et al. 2003; Quimby et al. 2000). In plant cells, the nuclear envelope protein and Ran cycle component RanGAP is redistributed to the mitotic spindle and subsequently to the growing cell plate of the phragmoplast, linking nuclear envelope components to cytokinesis (Jeong et al. 2005; Pay et al. 2002). The functional significance of this localization is unclear, but it is likely that the plant Ran cycle plays roles in mitotic microtubule assembly and vesicle fusion in analogy to its functions found in animal cells. In the fungus *Aspergillus nidulans*, the nuclear pore complexes disassemble partially and the nuclear envelope becomes "leaky", allowing cytoplasmic RanGAP access to the nucleus during a short time window preceding nuclear division (De Souza et al. 2004). Yeast RanGAPs contain nuclear export signals, suggesting that they are capable of shuttling between the nucleus and the cytoplasm. In *Schizosaccharomyces pombe*, RanGAP appears to be involved in heterochromatin assembly, centromeric silencing and mitotic chromosome segregation (Kusano et al. 2004; Nishijima et al. 2006). Together, these data emphasize the dual role of components of the nucleocytoplasmic transport machinery during mitotic events.

3.3
The Plant Nuclear Envelope as Microtubule Organizing Center

One key difference between higher plants and other organisms is the absence of centrosomes or spindle pole bodies as microtubule organizing centers. With the development from single cell, motile algae to multicellular, sessile plants, plant cells have lost their centrioles and therefore depend on alternative means for microtubule nucleation during spindle formation. This function is taken over by the nuclear envelope serving as a site for microtubule nucleation at the onset of mitosis (Bakhuizen et al. 1985; Stoppin et al. 1994; reviewed by Canaday et al. 2000). Microtubule-associated proteins (MAPs) modulate the microtubule nucleating activity of the plant nuclear envelope (Stoppin et al. 1996).

The minimal core of the microtubule nucleation machinery consists of the proteins Spc98/Spc97 and γ-tubulin. Together, these form the γ-tubulin ring complex (γ-TuRC) to initiate microtubule nucleation (reviewed by Job et al. 2003). Homologs of all three components of the γ-TuRC can be identified in plant genomes. Consistent with its function and the position of microtubule nucleation sites in plant cells, the plant homolog of Spc98 has been found to co-localize with γ-tubulin at the plant nuclear envelope (Erhardt et al. 2002). It is unclear how the attachment of the γ-TuRC to the nuclear envelope is achieved in plant cells. In yeast, the spindle pole body components Spc72 and Spc110 serve as anchoring sites on the cytoplasmic and nucleoplasmic surfaces of the spindle pole body, respectively (Knob and Schiebel 1997, 1998). No homologs of these proteins exist in plants. However, they share the common feature of long coiled-coil domains, a common protein interaction motif. Long coiled-coil proteins have also been identified at the plant nuclear envelope, for example the carrot nuclear matrix constituent protein NMCP1 (Masuda et al. 1997). It is possible that these proteins might be involved in anchoring the microtubule nucleation machinery to the plant nuclear envelope.

3.4
Meiotic Telomere Tethering at the Nuclear Envelope

Another function of the nuclear envelope becomes evident in meiosis. Homologous chromosome synapsis during meiotic prophase is accompanied by the clustering of telomeres at the nuclear envelope in a so-called bouquet arrangement (reviewed by Scherthan 2001). In animal and yeast cells, this clustering occurs adjacent to the microtubule organizing center, the centrosome or spindle pole body, respectively. Despite the lack of these organelles in plant cells, telomere bouquets at the nuclear envelope can also be observed in some plant species (Bass et al. 1997; Cowan et al. 2001; Martinez-Perez et al. 1999). However, not all plant species follow this pattern. In Arabidopsis, telomeres cluster inside the nucleus and associate with the nucleolus (Armstrong et al. 2001). How telomeres are attached to the nuclear or nucleolar periphery is unknown. In the maize *pam1* mutant, bouquet formation is delayed or abolished, suggesting that PAM1 protein is required for meiotic bouquet formation (Golubovskaya et al. 2002). In Arabidopsis, the Skp1-like protein ASK1 is essential for the release of chromatin from the nuclear and nucleolar periphery (Yang et al. 2006). Since ASK1 is thought to be involved in ubiquitin-mediated proteolysis, this suggests that nuclear chromatin release in plants might require the degradation of proteins linking chromatin to the nuclear envelope.

4
Nuclear and Nuclear Envelope Dynamics in Plants

Plant nuclei are highly mobile organelles. Their movement appears to be mediated by actin rather than tubulin in Arabidopsis (Chytilova et al. 2000; Ketelaar et al. 2002). The nuclear envelope itself exhibits a high degree of plasticity and can contain extensive grooves and invaginations. In onion and tobacco cells, these structures were observed having cytoplasmic cores with actin bundles supporting cytoplasmic streaming and vesicle movement (Collings et al. 2000). Taken together, these findings suggest a role for actin in nuclear mobility and nuclear envelope architecture at least during interphase in plant cells, in contrast to microtubule-dynein based nuclear position and migration mechanisms in other eukaryotes. Mitotic roles of actin and its motor myosin appear to concentrate on cell plate formation during later stages of the cell cycle (Hepler et al. 2002; Sano et al. 2005; Yoneda et al. 2004).

4.1
Markers for the Plant Nuclear Envelope

Marker proteins for the nuclear envelope are valuable tools for studying the dynamics of the nuclear envelope during mitosis in more detail. In animal cells, lamins and lamina-associated inner nuclear membrane proteins are frequently used as markers for the nuclear envelope. Since plant cells do not possess homologs of these proteins, it is surprising that antibodies directed against lamins are capable of detecting antigens in plant cells (Li and Roux 1992; McNulty and Saunders 1992; Minguez and Morena Diaz de la Espina

1993). Attempts to identify plant lamin-like antigens implicate coiled-coil proteins as likely candidates, suggesting that this structural feature shared by lamins is responsible for the cross-reaction with anti-lamin antibodies (Blumenthal et al. 2004). The precise targets of these antisera in plant cells are unknown.

Several other antisera directed against mammalian proteins have been used successfully in labeling the plant nuclear envelope. One example is a monoclonal antibody against calf thymus centrosomes, which might cross-react with microtubule organizing components at the plant nuclear envelope (Schmit et al. 1994). Another example is a peptide antibody raised against LCA1, a SERCA-type ATPase of mammalian ER. This antibody localizes to nuclear envelopes in tomato cells with high specificity both at light microscopic as well as electron microscopic levels and was found to relocalize to spindle poles during mitosis (Downie et al. 1998). An antiserum against chicken spectrin also decorated the plant nuclear envelope in pea (de Ruijter et al. 2000). Spectrins are nuclear envelope proteins specific to animals with no homologs in plants and are involved in the attachment of actin filaments at the nuclear membrane.

While immunological studies provide snapshots of nuclear envelope behavior in fixed cells, tagged proteins can provide dynamic data of mitotic processes in living cells. A heterologous nuclear envelope protein suitable for use as such a marker in plant cells is the mammalian lamin B receptor (LBR). This protein is typically only found in animal cells, but not in plants. LBR is an integral membrane protein of the inner nuclear membrane and binds to chromatin via its N-terminus (Pyrasopoulou et al. 1996; Ye and Worman 1994). An adjacent region and the first transmembrane domain are responsible and sufficient for the nuclear envelope targeting of the protein (Smith and Blobel 1993; Soullam and Worman 1993, 1995). While mammalian LBR is tethered inside the nucleus via interaction with lamin B, it appears that its chromatin-binding capabilities are sufficient to ensure similar tethering at the inner nuclear membrane in plant cells that do not possess lamins. This phenomenon has been utilized to visualize the plant nuclear envelope during mitosis in live cells by using the N-terminal 238 amino acids of LBR fused to green fluorescent protein (Irons et al. 2003).

Several plant nuclear envelope proteins have been identified and utilized for plant nuclear envelope dynamics studies (see Table 1). These include NMCP1 from carrot (Masuda et al. 1999), the tomato protein MAF1 (Dixit and Cyr 2002), and the Ran cycle component RanGAP (Jeong et al. 2005).

NMCP1 was the first nuclear envelope protein cloned from plant cells. It is a 134-kDa protein found exclusively at the nuclear periphery during interphase in immunostaining and relocates to the spindle poles during mitosis (Masuda et al. 1997, 1999). NMCP1 staining of the nuclear envelope disperses during late prophase, metaphase, and anaphase with a concentration in the spindle region. In late anaphase, staining becomes visible around the polar re-

Table 1 Protein markers for the study of nuclear envelope dynamics in plant mitosis

Protein	Description	Reference
NMCP1	Carrot coiled-coil protein	Masuda et al. 1997, 1999;
MAF1, WPPs	Tomato WPP domain protein and Arabidopsis homologs	Dixit and Cyr 2002; Gindullis et al. 1999; Patel et al. 2004
RanGAP1	Arabidopsis Ran GTPase activating protein	Jeong et al. 2005; Pay et al. 2002; Rose and Meier 2001
LBR	N-terminal targeting domain of mammalian lamin B receptor	Irons et al. 2003

gion of the daughter nuclei and circles the nuclei completely during telophase (Masuda et al. 1999). Similar to animal lamins, NMCP1 is predicted to have a primarily coiled-coil structure, making it the best candidate for a lamin-like protein in plants.

MAF1 and RanGAP share a homologous targeting domain, the WPP domain (Rose and Meier 2001). This domain is responsible for targeting green fluorescent protein fusions to the nuclear envelope during interphase (see Fig. 4) and to the cell plate of the phragmoplast in cytokinesis (Jeong et al. 2005; Patel et al. 2004). Three homologs of MAF1 exist in Arabidopsis. Interestingly, they are differentially targeted to the nuclear envelope. Only two members of the family, WPP1 and WPP2, concentrate at the nuclear envelope in meristematic root cells while the third one, WPP3, does not. Their targeting is abolished once the cells reach the zone of differentiation, but is reestablished after inducing root tissue to form callus. The reduction of these proteins in roots via RNAi is associated with reduced mitotic activity, functionally linking WPP proteins to mitosis (Patel et al. 2004).

In addition to nuclear envelope proteins, the Golgi protein *N*-acetyl-glucosaminyl transferase (Nag) has been found to label the plant nuclear

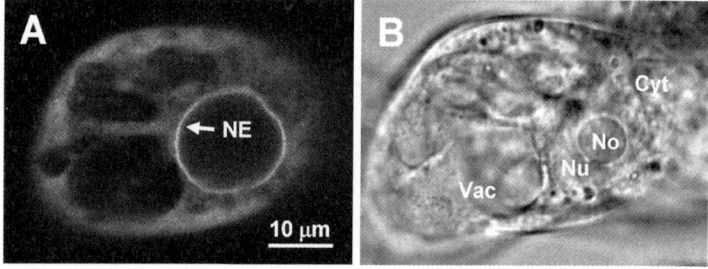

Fig. 4 Arabidopsis RanGAP1 marks the nuclear envelope in tobacco BY-2 cells. **a** Fluorescence of RanGAP1-GFP fusion protein transiently expressed in a BY-2 cell as imaged with a laser scanning confocal microscope. **b** Corresponding transmitted light image. *NE* nuclear envelope, *Nu* nucleus, *No* nucleolus, *Cyt* cytoplasm, *Vac* vacuole

envelope (Dixit and Cyr 2002). This suggests a connection between these two membrane systems in plants cells, which is further strengthened by the finding that MAF1 also exhibits a dual targeting to the nuclear envelope as well as the Golgi (Patel et al. 2005). Likewise, ER-targeted green fluorescent protein fusions have been used to visualize the nuclear envelope during cell division (Gupton et al. 2006).

4.2
Nuclear Envelope Dynamics During Mitosis

Microscopy studies suggest that the plant nuclear envelope joins the ER after it breaks down during open mitosis and reforms from the ER membrane pool during telophase. In higher plant mitosis, the nuclear envelope becomes indistinguishable from tubular and cisternal ER surrounding and invaginating the spindle apparatus. Based on observations of green fluorescent protein-labeled ER through cell division, the nuclear envelope appears to accumulate ER material shortly after chromosome condensation, but before nuclear envelope breakdown, and subsequently elongates with an increased accumulation of ER at the spindle poles. It dissipates into a network of tubules aligning in the orientation of the spindle microtubules (Gupton et al. 2006). These tubular ER/nuclear envelope strands appear to form channels for the separation of the daughter chromatids (Hawes et al. 1981; Hepler 1980). During later stages of mitosis, the tubular ER/nuclear envelope material provides the vesicles for cell plate formation in the phragmoplast (Gupton et al. 2006).

Nuclear envelope labeled with the more specific marker LBR fused to green fluorescent protein disperses during metaphase into an ER-like meshwork. Later, it accumulates in tubular structures at the division plate between the two daughter nuclei, indicating the prevalence of nuclear envelope material at the newly forming cell plate. The nuclear envelopes around the daughter nuclei become visible during phragmoplast assembly, and ER fluorescence decreases, indicating a migration of material from the ER back to the nuclear envelope (Irons et al. 2003).

4.2.1
Disassembly of the Nuclear Envelope

At the onset of open mitosis in higher eukaryotes, the nuclear envelope breaks down, but where does it go? Two theories have been proposed:

1. Nuclear envelope vesiculation
2. Absorption into the endoplasmic reticulum

Evidence for the first theory stems from in vitro experiments on *Xenopus* cell-free extracts. When disrupted, nuclear membranes formed vesicles and reformed around chromatin (Vigers and Lohka 1991). However, the alternative model of absorption into the ER is favored by strong evidence from in vivo experiments. While the outer nuclear envelope membrane is continuous with the ER even in interphase, the inner nuclear envelope membrane, visualized through immunostaining or with fluorescent markers, also shows continuity with mitotic ER membranes (Ellenberg et al. 1997; Irons et al. 2003; Yang et al. 1997). No vesiculation could be observed, suggesting that the nuclear envelope integrates with the endoplasmic reticulum in both animal and plant mitosis.

What triggers the disassembly of the nuclear envelope? Since the process of open mitosis has evolved independently in plants, animals, and certain fungi, it would be reasonable to assume that the mechanisms of nuclear envelope breakdown and reassembly differ. However, a common theme is emerging in the form of mechanical forces exerted onto the nuclear membrane by the cytoskeleton. Alternatively, nuclear pore complex disassembly has been implicated in creating a "leaky" envelope and might constitute the first step in nuclear envelope disassembly.

In animal cells, phosphorylation of the proteins of the nuclear lamina has been suggested as the starting point for nuclear envelope breakdown (Foisner and Gerace 1993). More recently, a model has emerged that suggests a crucial role for the physical tearing of the nuclear envelope by microtubules as the first step in nuclear envelope breakdown (Beaudouin et al. 2002; Lenart and Ellenberg 2003). Early spindle microtubules cause folds and invaginations in the nuclear envelope as early as an hour prior to the onset of mitosis. The nuclear envelope finally ruptures under increasing tension, allowing kinases to enter the nuclear interior and in turn release the nuclear membrane from the lamina and chromatin by phosphorylating inner nuclear membrane proteins and their ligands (Gerace and Foisner 1994; Goldberg et al. 1999; Takano et al. 2002).

This process of physical tearing is facilitated by dynein and dynactin concentrated at the nuclear envelope in animal cells (Salina et al. 2002). A similar mechanism apparently exists in fungi. In the basidiomycete *Ustilago maydis*, the nuclear envelope is stripped off the chromosomes via dynein-mediated nuclear migration (Straube et al. 2005).

Is this mechanical disruption of the nuclear envelope applicable to plant cells as well? Plants lack both dyneins and nuclear lamins, therefore other components must be involved in nuclear envelope disassembly in plants. While the breakdown process appears similar in plant and animal open mitosis on a microscopic level, some features are unique to plants. As mentioned earlier, the plant nuclear envelope serves as a microtubule organizing center for spindle formation at the end of the G2 phase of the cell cycle. The resulting microtubule array around the plant nuclear envelope has been suggested to play a role in moving and anchoring the nucleus in the division plane (Bakhuizen et al. 1985). The microtubules align outside the nuclear membrane, forming a bipolar spindle before the nuclear envelope breaks down.

Electron microscopy studies revealed that in the charophyte alga *Spirogyra*, nuclear envelope disruption starts around the spindle equator in metaphase, suggesting a connection between spindle formation and nuclear membrane disassembly early in the plant lineage (Ueda et al. 1986).

The orientation of the spindle and position of the division plane in a plant cell is determined by the preprophase band (PPB), a ring of microtubules concentrated below the plasma membrane. Disintegration of the PPB is closely linked with mitotic breakdown of the nuclear envelope, suggestive of a coregulation of these two events. Evidence for this mechanism stems from dual labeling experiments with tubulin and the Golgi/nuclear envelope-marker Nag. When fused to green and red fluorescent proteins, respectively, these markers showed that nuclear envelope breakdown in tobacco cells preceded the disappearance of the PPB microtubules by roughly 2 min. Both processes appeared to stimulate each other as nuclear envelope breakdown was accompanied by a ruffling of the nuclear envelope in an area adjacent to the PPB, whereas the PPB was disassembled most rapidly in a region closest to the disrupted nuclear envelope (Dixit and Cyr 2002).

Since higher plants lack dyneins, an intriguing candidate for a plant-specific microtubule motor functioning during mitosis is the kinesin-like calmodulin-binding protein (KCBP). KCBP has been found only in the green algae and plants (Abdel-Ghany et al. 2005) and has been implicated in mitotic functions in *Chlamydomonas* and higher plant cells (Dymek et al. 2006; Preuss et al. 2003). Constitutive activation of KCBP via antibodies directed against the calmodulin-binding region during late prophase has been shown to induce nuclear envelope breakdown within minutes of injection (Vos et al. 2000). This suggests a role for Ca^{2+}/calmodulin-mediated activation of KCBP in the control of nuclear envelope breakdown in plants. Interestingly, the plant nuclear envelope markers mentioned above included LCA1, a Ca^{2+} pump in tomato (Downie et al. 1998). It is tempting to speculate that Ca^{2+} is stored in the nuclear envelope lumen and could be readily available as a signal in KCBP-mediated nuclear envelope breakdown in plants. Alternatively, the ER–PPB, a membrane structure underlying the microtubular PPB, has been proposed as a possible storage site for Ca^{2+} as a signal at the onset of mitosis (Zachariadis et al. 2001).

An alternative to the mechanical force model for nuclear envelope breakdown has been proposed, based on observations of nuclear pores in starfish oocytes (Terasaki et al. 2001). Nuclear envelope breakdown in animal cells consists of two distinct phases of permeabilization. Using dextran entry as a marker for "leaky" nuclear envelopes, it has been shown that permeabilization occurs prior to visible disruption of the nuclear membranes. This first permeabilization step coincides with the partial disassembly of nuclear pore complexes and is followed by a second wave of rapidly spreading fenestration of the nuclear envelope (Lenart et al. 2003; Terasaki et al. 2001). Partially disassembled nuclear pores have also been observed in *Drosophila* (Kiseleva et al. 2001) and in *Aspergillus nidulans*, a fungus exhibiting an intermediate between closed and open mitosis (De Souza et al. 2004).

4.2.2
Reassembly of the Nuclear Envelope

Reformation of the nuclear envelope occurs during late anaphase and telophase and involves association of membranes containing nuclear envelope proteins with the separated chromosomes. Microscopic studies in the charophyte alga *Spirogyra* show that vesicles remain associated with the spindle and chromosomes until they fuse to form the envelopes of the daughter nuclei during early anaphase (Ueda et al. 1986). In higher plant cells, the tubular ER/nuclear envelope network interweaving the spindle apparatus gives rise to the newly forming nuclear envelopes as well as the cell plate during formation of the phragmoplast. Intense staining of the phragmoplast region has been observed using an ER marker (Gupton et al. 2006). During the reassembly of the daughter nuclei, the condensed chromosomes are surrounded by ER/nuclear envelope material. Chromosomes lagging in the transition from metaphase to anaphase can cause the trapping of such material within the nucleus, leading to the formation of nuclear grooves and invaginations observed in plant cells (Gupton et al. 2006).

What drives nuclear envelope reassembly? In vitro reconstitution experiments suggest that the principles of plant and animal nuclear envelope assembly are roughly the same. *Xenopus* sperm chromatin, when presented with carrot or tobacco cell extracts, is capable of recruiting plant components for nuclear envelope formation, suggesting that the underlying mechanisms are ubiquitous to animals and plants (Lu and Zhai 2001; Zhao et al. 2000). Conversely, plant sperm chromatin triggered the nuclear reconstitution from animal cell extracts, including the nuclear lamina (Lu et al. 2001).

Since animal and plant systems seem compatible in terms of nuclear envelope assembly, it is feasible to postulate a common regulatory mechanism for the initiation of nuclear envelope assembly around chromatin. How is this process regulated on a molecular level?

Increasing evidence has been accumulated to suggest a crucial role for the Ran cycle and karyopherins in nuclear envelope reassembly. Beads coated with Ran have been shown to induce the formation of nuclear envelopes complete with nuclear pore complexes capable of transport (Zhang and Clarke 2000, 2001). Depletion of Ran by RNAi results in failure of nuclear envelope assembly in *C. elegans* (Askajer et al. 2002; Bamba et al. 2002). This process requires both the exchange of RanGDP to RanGTP by the nucleotide exchange factor RCC1 as well as hydrolysis of RanGTP to RanGDP (Hetzer et al. 2000; Zhang and Clarke 2000). The correct balance of RCC1 and Ran binding protein 1 is critical for nuclear envelope formation (Nicolas et al. 1997; Pu and Dasso 1997). The Ran GTPase cycle is also involved in nuclear pore complex assembly in yeasts and animals (Ryan et al. 2003; Walther et al. 2003). Both nuclear envelope and nuclear pore assembly mediated by the Ran cycle are inhibited by importin β in a RanGTP-reversible manner (Harel et al. 2003).

While none of these mechanisms have been confirmed in plants yet, the major components of the Ran cycle have been identified (Ach and Gruissem 1994; Merkle et al. 1994), and the localization of plant RanGAP to the mitotic spindle and the growing cell plate are suggestive of similar roles for the plant Ran cycle to its mammalian and yeast counterparts in regulating spindle assembly and vesicle fusion events (Jeong et al. 2005; Pay et al. 2002).

Mammalian nuclear envelope components are recruited sequentially to the newly forming daughter nuclei with the nucleoporin Nup153, RanBP2/Nup358, LBR, and emerin being among the first proteins to associate with chromatin (Bodoor et al. 1999; Chaudhary and Courvalin 1993; Haraguchi et al. 2000; Schwartz 2005). In *Drosophila*, membrane vesicles containing nuclear envelope proteins associate with chromatin depending on the presence of lamins (Ulitzur et al. 1997). The association of A-type lamins to chromatin does not require the presence of membranes (Burke 1990; Glass and Gerace 1990). The exact order of recruiting events leading to the reassembly of the nuclear envelope and pores in plant cells is not known, partly due to a lack of knowledge about the corresponding plant cell components. None of the above mentioned animal proteins has a homolog in plants.

5
Outlook

While the plant nuclear envelope shares similar functions and mitotic behavior with the animal nuclear envelope, distinct differences exist in protein compositions and specialized functions, such as microtubule nucleation in plants. Comparatively few nuclear envelope proteins have been identified so far in plants, and many are still functionally uncharacterized. While a role during mitosis can be inferred for conserved components of the Ran cycle and it is tempting to speculate that plant and animal cells might use the same underlying molecular signals for nuclear envelope assembly, experimental evidence from plant systems corroborating this hypothesis is still lacking. Further studies need to be done to determine the exact events of nuclear envelope disassembly and reassembly and their regulation in plant cells. With increasing knowledge about plant nuclear envelope proteins and the availability of specific plant nuclear envelope markers, it is becoming possible to address these questions in more detail. Future studies may be directed towards the identification of integral membrane proteins of the plant nuclear envelope and towards protein and membrane trafficking during the plant cell cycle.

References

Abdel-Ghany SE, Day IS, Simmons MP, Kugrens P, Reddy ASN (2005) Origin and evolution of kinesin-like calmodulin-binding protein. Plant Physiol 138:1711–1722

Ach RA, Gruissem W (1994) A small nuclear GTP-binding protein from tomato suppresses a *Schizosaccharomyces pombe* cell-cycle mutant. Proc Natl Acad Sci USA 91:5863–5867

Armstrong SJ, Franklin CH, Jones GH (2001) Nucleolus-associated telomere clustering and pairing precede meiotic chromosome synapsis in *Arabidopsis thaliana*. J Cell Sci 114:4207–4217

Arnaoutov A, Dasso M (2003) The Ran GTPase regulates kinetochore function. Dev Cell 5:99–111

Askjaer P, Galy V, Hannak E, Mattaj IW (2002) Ran GTPase cycle and importins α and β are essential for spindle formation and nuclear envelope assembly in living *Caenorhabditis elegans* embryos. Mol Biol Cell 13:4355–4370

Backhuizen R, van Spronsen PC, Sluiman-den Hertog FAJ, Venverloo CJ, Goosen-de Roo L (1985) Nuclear envelope radiating microtubules in plant cells during interphase mitosis transition. Protoplasma 128:43–51

Bamba C, Bobinnec Y, Fukuda M, Nishida E (2002) The GTPase Ran regulates chromosome positioning and nuclear envelope assembly in vivo. Curr Biol 12:503–507

Bass HW, Marshall WF, Sedat JW, Agard DA, Cande WZ (1997) Telomeres cluster de novo before the initiation of synapsis: a three-dimensional spatial analysis of telomere positions before and during meiotic prophase. J Cell Biol 137:5–18

Battacharya D, Medlin L (1998) Algal phylogeny and the origin of land plants. Plant Physiol 116:9–15

Beaudouin J, Gerlich D, Daigle N, Eils R, Ellenberg J (2002) Nuclear envelope breakdown proceeds by microtubule-induced tearing of the lamina. Cell 108:83–96

Blumenthal SS, Clark GB, Roux SJ (2004) Biochemical and immunological characterization of pea nuclear intermediate filament proteins. Planta 218:965–975

Bodoor K, Shaihk SA, Salina D, Raharjo WH, Bastos R, Lohka MJ, Burke B (1999) Sequential recruitment of NPC proteins to the nuclear periphery at the end of mitosis. J Cell Sci 112:2253–2264

Brugerolle G, Mignot JP (2003) The rhizoplast of chrysomonads, a basal body-nucleus connector that polarizes the dividing spindle. Protoplasma 222:13–21

Burke B (1990) On the cell-free association of lamins A and C with metaphase chromosomes. Exp Cell Res 186:169–176

Canaday J, Stoppin-Mellet V, Mutterer J, Lambert AM, Schmit AC (2000) Higher plant cells: γ-tubulin and microtubule nucleation in the absence of centrosomes. Microsc Res Tech 49:487–495

Cavalier-Smith T (2002) The phagotropic origin of eukaryotes and phylogenetic classification of protozoa. Int J Syst Evol Microbiol 52:297–354

Chaudhary N, Courvalin JC (1993) Stepwise reassembly of the nuclear envelope at the end of mitosis. J Cell Biol 122:295–306

Chytilova E, Macas J, Sliwinska E, Rafelski SM, Lambert GM, Galbraith DW (2000) Nuclear dynamics in *Arabidopsis thaliana*. Mol Biol Cell 11:2733–2741

Collings DA, Carter CN, Rink JC, Scott AC, Wyatt SE, Allen SN (2000) Plant nuclei can contain extensive grooves and invaginations. Plant Cell 12:2425–2439

Coss RA (1974) Mitosis in *Chlamydomonas reinhardtii* basal bodies and the mitotic apparatus. J Cell Biol 63:325–329

Cowan CR, Carlton PM, Cande WZ (2001) The polar arrangement of telomeres in interphase and meiosis. Rabl organization and the bouquet. Plant Physiol 125:532–538

De Ruijter NCA, Ketelaar T, Blumenthal SS, Emons AM, Schel JHN (2000) Spectrin-like proteins in plant nuclei. Cell Biol Int 24:427–438

De Souza CPC, Osmani AH, Hashmi SB, Osmani SA (2004) Partial nuclear pore complex disassembly during closed mitosis in *Aspergillus nidulans*. Curr Biol 14:1973–1984

Di Fiore B, Ciciarello M, Lavia P (2004) Mitotic functions of the Ran GTPase network: the importance of being in the right place at the right time. Cell Cycle 3:305–313

Dixit R, Cyr RJ (2002) Spatio-temporal relationship between nuclear-envelope breakdown and preprophase band disappearance in cultured tobacco cells. Protoplasma 219:116–121

Dolan MF, Melnitsky H, Margulis L, Kolnicki R (2002) Motility proteins and the origin of the nucleus. Anat Rec 268:290–301

Downie L, Priddle J, Hawes C, Evans DE (1998) A calcium pump at the higher plant nuclear envelope? FEBS Lett 429:44–48

Dymek EE, Goduti D, Kramer T, Smith EF (2006) A kinesin-like calmodulin-binding protein in *Chlamydomonas*: evidence for a role in cell division and flagellar functions. J Cell Sci 119:3107–3116

Ellenberg J, Siggia ED, Moreira JE, Smith CL, Presley JF, Worman HJ, Lippincott-Schwartz J (1997) Nuclear membrane dynamics and reassembly in living cells: targeting of an inner nuclear membrane protein in interphase and mitosis. J Cell Biol 138:1193–1206

Erhardt M, Stoppin-Mellet V, Campagne S, Canaday J, Mutterer J, Fabian T, Sauter M, Muller T, Peter C, Lambert AM, Schmit AC (2002) The plant Spc98p homologue colocalizes with γ-tubulin at the nucleation sites and is required for microtubule nucleation. J Cell Sci 115:2423–2431

Fernandez AG, Piano F (2006) MEL-28 is downstream of the Ran cycle and is required for nuclear-envelope function and chromatin maintenance. Curr Biol 16:1757–1763

Foisner R, Gerace L (1993) Integral membrane proteins of the nuclear envelope interact with lamins and chromosomes, and binding is modulated by mitotic phosphorylation. Cell 73:1267–1279

Franz C, Walczak R, Yavuz S, Santarella R, Gentzel M, Askjaer P, Galy V, Hetzer M, Mattaj IW, Antonin W (2007) MEL-28/ELYS is required for the recruitment of nucleoporins to chromatin and postmitotic nuclear pore complex assembly. EMBO Rep 8:165–172

Galy V, Askjaer P, Franz C, Lopez-Iglesias C, Mattaj IW (2006) MEL-28, a novel nuclear-envelope and kinetochore protein essential for zygotic nuclear-envelope assembly in *C. elegans*. Curr Biol 16:1748–1756

Gerace L, Foisner R (1994) Integral membrane proteins and dynamics of the nuclear envelope. Trends Cell Biol 4:127–131

Gindullis F, Peffer NJ, Meier I (1999) MAF1, a novel plant protein interacting with matrix attachment region binding protein MFP1, is located at the nuclear envelope. Plant Cell 11:1755–1767

Glass JR, Gerace L (1990) Lamins A and C bind and assemble at the surface of mitotic chromosomes. J Cell Biol 111:1047–1057

Goldberg M, Harel A, Gruenbaum Y (1999) The nuclear lamina: molecular organization and interaction with chromatin. Crit Rev Eukaryot Gene Exp 9:285–293

Golubovskaya IN, Harper LC, Pawlowski WP, Schichnes D, Cande WZ (2002) The *pam1* gene is required for meiotic bouquet formation and efficient homologous synapsis in maize (*Zea mays* L.). Genetics 162:1979–1993

Gupton SL, Collings DA, Allen DS (2006) Endoplasmic reticulum targeted GFP reveals ER organization in tobacco NT-1 cells during cell division. Plant Physiol Biochem 44:95–105

Haraguchi T, Koujin T, Hayakawa T, Kaneda T, Tsutsumi C, Imamoto N, Akazawa C, Sukegawa J, Yoneda Y, Hiraoka Y (2000) Live fluorescence imaging reveals early recruitment of emerin, LBR, RanBP2, and Nup153 to reforming functional nuclear envelopes. J Cell Sci 113:779–794

Harel A, Chan RC, Lachish-Zalait A, Zimmerman E, Elbaum M, Forbes DJ (2003) Importin β negatively regulates nuclear membrane fusion and nuclear pore complex assembly. Mol Biol Cell 14:4387–4396

Hartman H, Fedorov A (2002) The origin of the eukaryotic cell: a genomic investigation. Proc Natl Acad Sci USA 99:1420–1425

Hawes CR, Juniper BE, Horne JC (1981) Low and high-voltage electron-microscopy of mitosis and cytokinesis in maize roots. Planta 152:397–407

Hedges SB (2002) The origin and evolution of model organisms. Nat Rev Genet 3:838–849

Hepler PK (1980) Membranes in the mitotic apparatus of barley cells. J Cell Biol 86:490–499

Hepler PK, Valster A, Molchan T, Vos JW (2002) Roles for kinesin and myosin during cytokinesis. Phil Trans R Soc Lond B 357:761–766

Hetzer M, Bilbao-Cortes B, Walther TC, Gruss OJ, Mattaj JW (2000) GTP hydrolysis by Ran is required for nuclear envelope assembly. Mol Cell 5:1013–1024

Horiike T, Hamada K, Shinozawa T (2002) Origin of eukaryotic cell nuclei by symbiosis of archaea in bacteria supported by the newly clarified origin of functional genes. Genes Genet Syst 77:369–376

Irons SL, Evans DE, Brandizzi F (2003) The first 238 amino acids of the human lamin B receptor are targeted to the nuclear envelope in plants. J Exp Bot 54:1–8

Jeong SY, Rose A, Joseph J, Dasso M, Meier I (2005) Plant-specific mitotic targeting of RanGAP requires a functional WPP domain. Plant J 42:270–282

Job D, Valiron O, Oakley B (2003) Microtubule nucleation. Curr Opin Cell Biol 15:111–117

Joseph J, Liu ST, Jablonski SA, Yen TJ, Dasso M (2004) The RanGAP1–RanBP2 complex is essential for microtubule-kinetochore interactions in vivo. Curr Biol 14:611–617

Karol KG, McCourt RM, Cimino MT, Delwiche CF (2001) The closest living relatives of land plants. Science 294:2351–2353

Ketelaar T, Faivre-Moskalenko C, Esseling JJ, de Ruijter NCA, Grierson CS, Dogterom M, Emons AMC (2002) Positioning of nuclei in Arabidopsis root hairs: an actin-regulated process of tip growth. Plant Cell 14:2941–2955

Kiseleva E, Rutherford S, Cotter LM, Allen TD, Goldberg MW (2001) Steps of nuclear pore complex disassembly and reassembly during mitosis in early *Drosophila* embryos. J Cell Sci 114:3607–3618

Knob M, Schiebel E (1997) Spc98p and Spc97p of the yeast γ-tubulin complex mediate binding to the spindle pole body via their interaction with Spc110p. EMBO J 16:6985–6995

Knob M, Schiebel E (1998) Receptors determine the cellular localization of a γ-tubulin complex and thereby the site of microtubule formation. EMBO J 17:3952–3997

Kusano A, Yoshioka T, Nishijima H, Nishitani H, Nishimoto T (2004) Schizosaccharomyces pombe RanGAP homolog, SpRNA1, is required for centromeric silencing and chromosome segregation. Mol Biol Cell 15:4960–4970

Lenart P, Ellenberg J (2003) Nuclear envelope dynamics in oocytes: from germinal vesicle breakdown to mitosis. Curr Opin Cell Biol 15:88–95

Lenart P, Rabut G, Daigle N, Hand AR, Terasaki M, Ellenberg J (2003) Nuclear envelope breakdown in starfish oocytes proceeds by partial NPC disassembly followed by a rapidly spreading fenestration of nuclear membranes. J Cell Biol 160:1055–1068

Lewis LA, McCourt RM (2004) Green algae and the origin of land plants. Am J Bot 91:1535–1556

Li H, Roux SJ (1992) Casein kinase II protein kinase is bound to nuclear lamina-matrix and phosphorylates lamin-like protein in isolated pea nuclei. Proc Natl Acad Sci USA 89:8434–8438

Li HY, Cao K, Zheng Y (2003) Ran in the spindle checkpoint: a new function for a versatile GTPase. Trends Cell Biol 13:553–557

Lopez-Bautista JM, Waters DA, Chapman RL (2003) Phragmoplastin, green algae and the evolution of cytokinesis. Int J Syst Evol Microbiol 53:1715–1718

Lopez-Garcia P, Moreira D (2006) Selective forces for the origin of the eukaryotic nucleus. Bioessays 28:525–533

Lu P, Ren M, Zhai ZH (2001) Nuclear reconstitution of plant (*Orychophragmus violaceus*) demembranated sperm in cell-free extracts from animal (*Xenopus laevis*) eggs. J Struct Biol 136:89–95

Lu P, Zhai ZH (2001) Nuclear assembly of demembranated *Xenopus* sperm in plant cell-free extracts from *Nicotiana* ovules. Exp Cell Res 270:96–101

Mans BJ, Anantharaman V, Aravind L, Koonin EV (2004) Comparative genomics, evolution and origins of the nuclear envelope and nuclear pore complex. Cell Cycle 3:1612–1637

Margulis L, Chapman M, Guerrero R, Hall J (2006) The last eukaryotic common ancestor (LECA): acquisition of cytoskeletal motility from aerotolerant spirochetes in the proterozoic eon. Proc Natl Acad Sci USA 103:13080–13085

Margulis L, Dolan MF, Guerrero R (2000) The chimeric eukaryote: origin of the nucleus from the karymastigont in amitochondriate protists. Proc Natl Acad Sci USA 97:6954–6959

Martin W (1999) A briefly argued case that mitochondria and plastids are descendants of endosymbionts, but that the nuclear compartment is not. Proc R Soc Lond B 266:1387–1395

Martin W (2005) Archaebacteria (archaea) and the origin of the eukaryotic nucleus. Curr Opin Microbiol 8:630–637

Martinez-Perez E, Shaw P, Reader S, Aragon-Alcaide L, Miller T, Moore G (1999) Homologous chromosome pairing in wheat. J Cell Sci 112:1761–1769

Masuda K, Haruyama S, Fujino K (1999) Assembly and disassembly if the peripheral architecture of the plant nucleus during mitosis. Planta 210:165–167

Masuda K, Xu ZJ, Takahashi S, Ito A, Ono M, Nomura K, Inoue M (1997) Peripheral framework of carrot cell nucleus contains a novel protein predicted to exhibit a long α-helical domain. Exp Cell Res 232:173–181

McCourt RM, Delwiche CF, Karol KG (2004) Charophyte algae and land plant origins. Trends Ecol Evol 19:661–666

McNulty AK, Saunders MJ (1992) Purification and immunological detection of pea nuclear intermediate filaments: evidence for plant nuclear lamins. J Cell Sci 103:407–414

Merkle T, Haizel T, Matsumoto T, Harter K, Dailmann G, Nagy F (1994) Phenotype of the fission yeast cell cycle regulatory mutant *pim1-46* is suppressed by a tobacco cDNA encoding a small, Ran-like GTP-binding protein. Plant J 6:555–565

Minguez A, Morena Diaz de la Espina S (1993) Immunological characterization of lamins in the nuclear matrix of onion cells. J Cell Sci 106:431–439

Mitchison TJ (1995) Evolution of a dynamic cytoskeleton. Phil Trans R Soc Lond B 349:299–304

Moreira D, Le Guyader H, Phillipe H (2000) The origin of red algae and the evolution of chloroplasts. Nature 405:69–72

Newport J, Dunphy W (1992) Characterization of membrane binding and fusion events during nuclear envelope assembly using purified components. J Cell Biol 116:295–306

Nicolas FJ, Zhang C, Hughes M, Goldberg MW, Watton SJ, Clarke PR (1997) *Xenopus* Ran-binding protein 1: molecular interactions and effects on nuclear assembly in *Xenopus* egg extracts. J Cell Sci 110:3019–3030

Niepel M, Strambio-de-Castillia C, Fasolo J, Chait BT, Rout MP (2005) The nuclear pore complex-associated protein, Mlp2p, binds to the yeast spindle pole body and promotes its efficient assembly. J Cell Biol 170:225–235

Nishijima H, Nakayama J, Yoshioka T, Kusano A, Nishitani H, Shibahara K, Nishimoto T (2006) Nuclear RanGAP is required for the heterochromatin assembly and is reciprocally regulated by histone H3 and Clr4 histone methyltransferase in *Schizosaccharomyces pombe*. Mol Cell Biol 17:2524–2536

Patel S, Brkljacic J, Gindullis F, Rose A, Meier I (2005) The plant nuclear envelope protein MAF1 has an additional location at the Golgi and binds to a novel Golgi-associated coiled-coil protein. Planta 222:1028–1040

Patel S, Rose A, Meulia T, Dixit R, Cyr RJ, Meier I (2004) Arabidopsis WPP-domain proteins are developmentally associated with the nuclear envelope and promote cell division. Plant Cell 16:3260–3273

Pay A, Resch K, Frohnmeyer H, Fejes E, Nagy F, Nick P (2002) Plant RanGAPs are localized at the nuclear envelope in interphase and associated with microtubules in mitotic cells. Plant J 30:699–709

Preuss ML, Delmer DP, Liu B (2003) The cotton kinesin-like calmodulin-binding protein associates with cortical microtubules in cotton fibers. Plant Physiol 132:154–160

Pu RT, Dasso M (1997) The balance of RanBP1 and RCC1 is critical for nuclear assembly and nuclear transport. Mol Biol Cell 8:1955–1970

Pyrasopoulou A, Meier J, Maison C, Simos G, Georgatos SD (1996) The lamin B receptor (LBR) provides essential chromatin docking sites at the nuclear envelope. EMBO J 15:7108–7119

Quimby BB, Wilson CA, Corbett AH (2000) The interaction between Ran and NTF2 is required for cell cycle progression. Mol Biol Cell 11:2617–2629

Rasala BA, Orjalo AV, Shen Z, Briggs S, Forbes DJ (2006) ELYS is a dual nucleoporin/kinetochore protein required for nuclear pore assembly and proper cell division. Proc Natl Acad Sci USA 103:17801–17806

Roos UP (1984) From proto-mitosis to mitosis – an alternative hypothesis on the origin and evolution of the mitotic spindle. Orig Life 13:183–193

Rose A, Meier I (2001) A domain unique to plant RanGAP is responsible for its targeting to the plant nuclear rim. Proc Natl Acad Sci USA 98:15377–15382

Ryan KJ, McCaffery JM, Wente SR (2003) The Ran GTPase cycle is required for yeast nuclear pore complex assembly. J Cell Biol 160:1041–1053

Salina D, Bodoor K, Eckley DM, Schroer TA, Rattner JB, Burke B (2002) Cytoplasmic dynein is a facilitator of nuclear envelope breakdown. Cell 108:97–107

Salina D, Enarson P, Rattner JB, Burke B (2003) Nup358 integrates nuclear envelope breakdown with kinetochore assembly. J Cell Biol 162:991–1001

Sano T, Higaki T, Oda Y, Hayashi T, Hasezawa S (2005) Appearance of actin microfilament "twin peaks" in mitosis and their function in cell plate formation, as visualized in tobacco BY-2 cells expressing GFP-fimbrin. Plant J 44:595–605

Scherthan H (2001) A bouquet makes ends meet. Nat Rev 2:621–627

Schetter A, Askjaer P, Piano F, Mattaj I, Kemphues K (2006) Nucleoporins NPP-1, NPP-3, NPP-4, NPP-11 and NPP-13 are required for proper spindle orientation in *C. elegans*. Dev Biol 289:360–371

Schmit AC, Stoppin V, Chevrier V, Job D, Lambert AM (1994) Cell-cyle dependent distribution of a centrosomal antigen at the perinuclear MTOC or at the kinetochores of higher-plant cells. Chromosoma 103:343–351

Schwartz TU (2005) Modularity within the architecture of the nuclear pore complex. Curr Opin Struct Biol 15:221–226

Smith S, Blobel G (1993) The first membrane spanning region of the lamin B receptor is sufficient for sorting to the inner nuclear-membrane. J Cell Biol 120:631–637

Soullam B, Worman HJ (1993) The amino-terminal domain of the lamin B receptor is a nuclear-envelope targeting signal. J Cell Biol 120:1093–1100

Soullam B, Worman HJ (1995) Signals and structural features involved in integral membrane-protein targeting to the inner nuclear-membrane. J Cell Biol 130:15–27

Steggerda SM, Paschal BM (2002) Regulation of nuclear import and export by the GTPase Ran. Int Rev Cytol 217:41–91

Stoppin V, Lambert AM, Vantard M (1996) Plant microtubule-associated proteins (MAPs) affect microtubule nucleation and growth at plant nuclei and mammalian centrosomes. Eur J Cell Biol 69:11–23

Stoppin V, Vantard M, Schmit AC, Lambert AM (1994) Isolated plant nuclei nucleate microtubule assembly: the nuclear surface in higher plants has centrosome-like activity. Plant Cell 6:1099–1106

Straube A, Weber I, Steinberg G (2005) A novel mechanism of nuclear envelope breakdown in a fungus: nuclear migration strips off the envelope. EMBO J 24:1674–1685

Takano M, Takeuchi M, Ito H, Furukawa K, Sugimoto K, Omata S, Horigome T (2002) The binding of lamin B receptor to chromatin is regulated by phosphorylation in the RS region. Eur J Biochem 269:943–953

Teresaki M, Campagnola P, Rolls MM, Stein PA, Ellenberg J, Hinkle B, Slepchenko B (2001) A new model for nuclear envelope breakdown. Mol Biol Cell 12:503–510

Ueda K, Abhayavardhani P, Noguchi T (1986) Formation of the nuclear envelope during mitotic anaphase in *Spirogyra*. J Plant Res 99:301–308

Ulitzur N, Harel A, Goldberg M, Feinstein N, Gruenbaum Y (1997) Nuclear membrane vesicle targeting to chromatin in a *Drosophila* embryo cell-free system. Mol Biol Cell 8:1439–1448

Van den Hoek C, Mann DG, Jahns HM (1995) Algae: an introduction to phycology. Cambridge University Press, Cambridge NY

Vigers GPA, Lohka MJ (1991) A distinct vesicle population targets membranes and nuclear pore complexes to the nuclear-envelope in *Xenopus* cells. J Cell Biol 112:545–556

Vos JW, Safadi F, Reddy ASN, Hepler PK (2000) The kinesin-like calmodulin-binding protein is differentially involved in cell division. Plant Cell 12:979–990

Walther TC, Askajer P, Gentzel M, Habermann A, Griffiths G, Wilm M, Mattaj JW, Hetzer M (2003) RanGTP mediates nuclear pore complex assembly. Nature 424:689–694

Yang L, Guan T, Gerace L (1997) Integral membrane proteins of the nuclear envelope are dispersed throughout the endoplasmic reticulum during mitosis. J Cell Biol 137:1199–1210

Yang X, Timofejeva L, Ma H, Makaroff CA (2006) The Arabidopsis SKP1 homolog ASK1 controls meiotic chromosome remodeling and release of chromatin from the nuclear membrane and nucleolus. J Cell Sci 119:3754–3763

Ye Q, Worman HJ (1994) Primary structure-analysis and lamin-b and DNA-binding of human LBR, an integral protein of the nuclear-envelope inner membrane. J Biol Chem 269:11306–11311

Yoneda A, Akatsuka M, Kumagai F, Hasezawa S (2004) Disruption of actin microfilaments causes cortical microtubule disorganization and extra-phragmoplast formation at M/G1 interface in synchronized tobacco cells. Plant Cell Physiol 45:761–769

Zachariadis M, Quader M, Galatis B, Apostolakos P (2001) Endoplasmic reticulum preprophase band in dividing root-tip cells in *Pinus brutia*. Planta 213:824–827

Zhang C, Hutchins JR, Mühlhäusser P, Kutay U, Clarke PR (2002) Role of importin β in the control of nuclear envelope assembly by Ran. Curr Biol 12:498–502

Zhang CM, Clarke PR (2000) Chromatin-independent nuclear envelope assembly induced by Ran GTPase in *Xenopus* egg extracts. Science 288:1429–1432

Zhang CM, Clarke PR (2001) Roles of Ran-GTP and Ran-GDP in precursor vesicle recruitment and fusion during nuclear envelope assembly in a human cell-free system. Curr Biol 11:208–212

Zhao Y, Liu X, Wu M, Tao W, Zhai Z (2000) In vitro nuclear reconstitution could be induced by a plant cell-free system. FEBS Lett 480:208–212

Part C
Cell Plate Formation and Cytokinesis

… plant Cell Monogr (9)

MAP Kinase Signaling During M Phase Progression

Michiko Sasabe · Yasunori Machida (✉)

Division of Biological Science, Graduate School of Science, Nagoya University, Chikusa-ku, 464-8602 Nagoya, Japan
yas@bio.nagoya-u.ac.jp

Abstract NPK1 (nucleus- and phragmoplast-localized protein kinase 1) from tobacco was the first isolated mitogen-activated protein kinase kinase kinase (MAPKKK) from plants. In tobacco, the *NPK1* gene is transcribed in meristematic tissues or immature organs, and, at a cellular level, the protein is expressed from S to M phase and decreases thereafter. We have recently revealed the involvement of a mitogen-activated protein kinase (MAPK) cascade including NPK1 in mitosis. This is related, at least in part, to its control of microtubule turnover. In this review, we summarize the results of recent investigations on the mechanism of plant cytokinesis, focusing on the role of the MAPK cascade.

1
MAPK Cascades in Plants

The MAPK family is highly conserved in eukaryotes, including yeast, animals, and plants, and it is one of the major kinase families involved in transducing extracellular and intracellular signals to cellular responses (Robinson and Cobb 1997; English et al. 1999; Chen et al. 2001). The classic MAPK cascade consists of three sequential protein kinase activation steps and is initiated when the first member, MAPKKK, is activated. Activation of MAPKKK results in continuous phosphorylation and activation of MAPK kinase (MAPKK). This is followed by activation of the specific MAPK, which is often mediated by a series of protein-protein interactions. The *Arabidopsis* genome sequencing project has revealed the existence of approximately 60 MAPKKKs, 10 MAPKKs, and 20 MAPKs, and it has served as the basis for standard annotation of these gene families in all plants (Ichimura et al. 2002).

There is considerable divergence in the primary structures of the kinase domains for the plant MAPKKKs. Although biochemical and genetic characterization is necessary to determine their position in the various signaling cascades, these kinases can be classified into two large subgroups, MEKK-type and the Raf-like kinases (Ichimura et al. 2002; Nakagami et al. 2005). NPK1 and its *Arabidopsis* homologs (ANP1, ANP2, and ANP3) are MEKK-type MAPKKKs. The plant MAPKKKs are serine/threonine kinases that activate MAPKKs by phosphorylating the consensus sequence $S/T-X_5-S/T$ in the putative kinase active site, whereas the consensus sequence for mammalian

MAPKKs is S/T-X$_4$-S/T. Activated MAPKKs also activate MAPKs by phosphorylating the conserved amino acid motif TDY or TEY, which is structurally similar to the typical motif in animals and yeast. The large number of MAPKKKs may be due to their role in processing a wide variety of various external and internal stimuli.

Plant MAPK cascades are activated by a wide variety of stimuli, including wounding, cold stress, and osmotic stress and participate in defensive responses to pathogens, cellular responses to hormones, and regulation of the cell cycle (Chang 2003; Chinnusamy et al. 2004; Guo and Ecker 2004; Šamaj et al. 2004; Takahashi et al. 2004; Pedley and Martin 2005; Boudsocq and Lauriere 2005; Nakagami et al. 2005). Some MAPK cascades also appear to be involved in the growth and differentiation of plant organs. For example, SIMK MAPK is required for the formation of root hairs in alfalfa (Šamaj et al. 2002), and YODA MAPKKK regulates the first cell fate decision and stomatal development and patterning in *Arabidopsis* (Lukowitz et al. 2004; Bergmann et al. 2004). Further studies, however, are needed to identify all of the components of the MAPK cascades and the mechanisms regulating these protein kinases.

Here, we review our recent studies on a MAPK cascade that is involved in the regulation of cytokinesis in plant cells. This MAPK cascade includes NPK1 MAPKKK, NQK1 MAPKK, and NRK1 MAPK, and it is activated by the NACK kinesin-like proteins (KLPs) in tobacco. All of the components of this cascade are both structurally and functionally conserved in tobacco and *Arabidopsis* (Fig. 1; Hulskamp et al. 1997; Spielman et al. 1997; Strompen et al. 2002; Nishihama et al. 1997, 2001, 2002; Soyano et al. 2003; Yang et al. 2003; Tanaka et al. 2004). In addition, we recently identified a microtubule-associated protein (MAP), NtMAP65, as a downstream target of this cascade in tobacco.

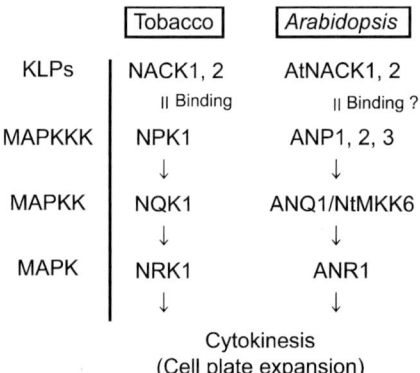

Fig. 1 The NACK-PQR pathway that controls plant cytokinesis. The components of the NACK-PQR pathways in tobacco and *Arabidopsis* are shown

2
The MAPK Cascade Regulating Plant Cytokinesis

2.1
NPK1 MAPKKK Is Required for Cell Plate Expansion

In yeast and animal cells, MAPK cascades are involved in various aspects of the cell cycle, including entry into the cell cycle, transition from G2 to M phase, and the spindle assembly checkpoint (Pages et al. 1993; Minshull et al. 1994; Takenaka et al. 1998; Wright et al. 1999). Some MAPKs have been reported to localize to the midzone of the spindle and midbody (Shapiro et al. 1998; Zecevic et al. 1998; Willard and Crouch 2001) or to the phragmoplast midzone (Calderini et al. 1998; Bögre et al. 1999). These findings suggest that a MAPK cascade regulates cytokinesis.

The *NPK1* gene from tobacco encodes a member of the MAPKKK family, and the kinase domain of NPK1 can replace the functions of several yeast MAPKKKs (Banno et al. 1993, Machida et al. 1998). The *NPK1* gene is transcribed in meristematic tissues and immature organs but not in mature organs (Nakashima et al. 1998). *Arabidopsis* homologs (*ANP1*, *ANP2*, and *ANP3*) of *NPK1* are also preferentially transcribed in organs that contain proliferating cells (Nishihama et al. 1997). These results suggest that NPK1 MAPKKKs play a role in the signaling pathway regulating plant cell division, and further biochemical and immunological analysis in tobacco (see below) have suggested that they are involved in formation of the cell plate (Nishihama et al. 2001).

Structure of NPK1. Figure 2A shows a schematic representation of the domain and motif organization of NPK1. NPK1 consists of a carboxy-terminal regulatory domain and an amino-terminal kinase domain related to that of MAPKKKs. Deletion of the carboxy-terminal domain increases its kinase activity, indicating that this domain is a negative regulator of NPK1 (Banno et al. 1993; Machida et al. 1998). This domain also has a nuclear localization signal, a coiled-coil structure, and consensus sequences for phosphorylation by cyclin-dependent protein kinases (CDKs) (Nishihama et al. 1997). The carboxy-terminal region was later identified as the binding site for the NACK1 KLP, which regulates the activity and localization of NPK1 (Nishihama et al. 2002; Ishikawa et al. 2002). Thus, it appears that phosphorylation of the regulatory domain by CDKs or its interaction with other factors mediates the activation of NPK1.

Characterization of the kinase activity and the localization of NPK1 in tobacco cells. In BY-2 tobacco cells, the kinase activity of NPK1 specifically increases late in the M phase of the cell cycle (Fig. 3), following transcription of the *NPK1* gene and accumulation of NPK1 protein at the initiation of the M phase (Nishihama et al. 2001). NPK1 is localized in the nucleus at interphase and prophase prior to breakdown of the nuclear envelope, whereas it

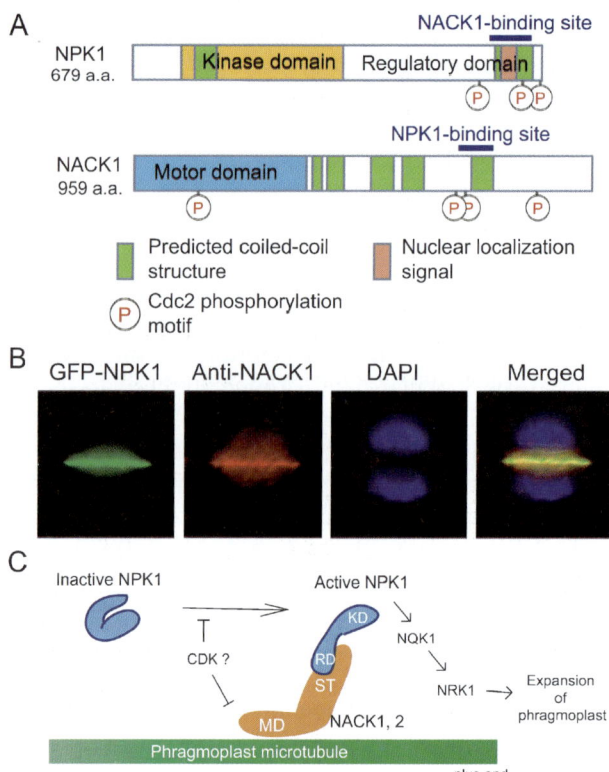

Fig. 2 The direct interaction of NPK1 and NACK1 at the equator of the phragmoplast in telophase. **A** Domain organization in the NPK1 and NACK1 proteins. The coiled-coil structures (*green*) were predicted by the COILS program (Lupas et al. 1991), and the nuclear localization signal (*red*) and NACK1/NPK1-binding sites (*bars*) were determined in our previous studies (Ishikawa et al. 2002). **B** Colocalization of NPK1 and NACK1 at the equator of the phragmoplast. BY-2 cells expressing GFP-NPK1 (*green*) were double-stained with rabbit antibodies against NACK1 (*red*) and 4'-6-diamidino-2-phenylindole (DAPI) for nuclei (*blue*). **C** The model for the activating mechanism of the NACK1/NPK1 complex followed by the activation of the MAPK cascade during plant cytokinesis. KD, kinase domain; RD, regulatory domain; MD, motor domain; ST, stalk domain. See text for details

is localized in the cytoplasm at metaphase (Nishihama et al. 2001). During cytokinesis, when the kinase activity of NPK1 increases, NPK1 shifts to the leading edge of the equatorial zone of the phragmoplast (Fig. 2B; Nishihama et al. 2001). Because of this pattern of localization, this kinase was designated nucleus- and phragmoplast-localized protein kinase 1 (NPK1).

Function of NPK1. In BY-2 cell lines overexpressing a kinase-defective mutant of NPK1 (NPK1:KW), a significant fraction of cells are binucleate or multinucleate and unusually large (Nishihama et al. 2001). Staining with calcofluor, which binds β-glucans in cell walls, or aniline blue, which binds

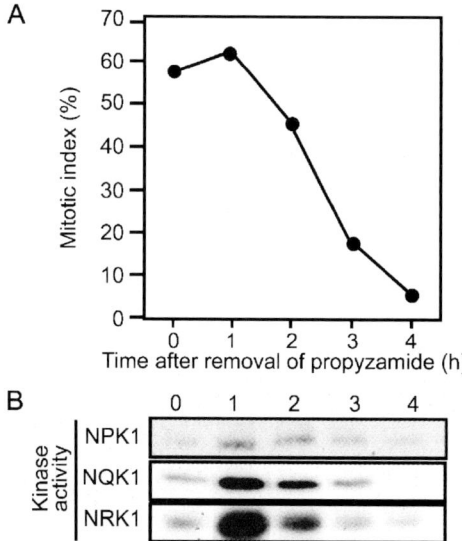

Fig. 3 Specific activation of NPK1, NQK1, and NRK1 at late M phase. **A** The graph shows a plot of the mitotic indices of BY-2 cells synchronized at M phase. The cell cycle was arrested at prometaphase by propyzamide after release from an aphidicolin block. **B** The protein kinase activities of NPK1, NQK1, and NRK1 were determined by immunocomplex kinase assays using recombinant kinase-negative NQK1, kinase-negative NRK1, and myelin basic proteins as substrates, respectively

callose, a polysaccharide synthesized at the initial stage of vesicle fusion, reveals the formation of incomplete cell plates in these multinucleated cells (Nishihama et al. 2001). Such multinucleated cells could also be observed in transgenic tobacco plants harboring NPK1:KW. The cotyledons of these NPK1:KW plants develop with a rough surface, and the stomata of the cotyledons consistently contained many binucleated guard cells and cells without a nucleus (Nishihama et al. 2001). Thus, expression of NPK1:KW inhibits the lateral expansion of the cell plate in both tobacco cells and plants.

Arabidopsis homologs of NPK1 (ANP1, ANP2, and ANP3) also appear to be involved in cytokinesis. Loss of function of two of the three homologs of NPK1 (ANP2 and ANP3) causes defects in cytokinesis, specifically, the formation of multinucleated cells with incomplete cell walls (Krysan et al. 2002). These results suggest that the MAPK cascade including the NPK1 MAPKKK family is necessary for the progression of cytokinesis in plants.

2.2
Activator of NPK1 MAPKKK: NACK1, a KLP

As described above, deletion of the regulatory domain in the carboxy-terminal half of NPK1 increases kinase activity, indicating that this domain negatively regulates NPK1 (Banno et al. 1993). Therefore we speculated that there are activators of NPK1. Animals and yeast have several proteins that can regulate the MAPKKKs via protein-protein interaction. For example, the small G protein Ras (Farrar et al. 1996; Luo et al. 1996) and the 14-3-3 protein regulate Raf MAPKKK (Irie et al. 1994); TAB1 regulates TAK1 MAPKKK (Shibuya et al. 1996); and SSK1 regulates SSK2 MAPKKK (Posas and Saito 1998).

To isolate the activators of NPK1 MAPKKK, we took advantage of a cloning strategy using a functional yeast genetic system based on the mating pheromone-responsive MAPK cascade, which consists of STE11 MAPKKK, STE7 MAPKK, and FUS3 MAPK (Irie et al. 1994). This system is based on suppression of a mutation in the *STE11* gene for MAPKKK by a MAPKKK and its potential activator introduced into the yeast cells from a heterologous organism. We transformed yeast *ste11* mutant cells with tobacco *NPK1* cDNA and then introduced a tobacco cDNA library into the transformed cells. Using this system, we isolated the tobacco cDNAs for two proteins that stimulate the activity of NPK1 MAPKKK (Machida et al., 1998, Nishihama et al. 2002). These cDNAs encoded two KLPs, which we designated NACK1 and NACK2 for **N**PK1-**ac**tivating **k**inesin-like proteins 1 and 2.

Structure and biochemical characterization of NACK proteins. Coexpression experiments in yeast cells revealed that NACK1 associates with NPK1 and increases the activity of NPK1 (Nishihama et al. 2002). In tobacco BY-2 cells, the *NACK1* and *NACK2* mRNAs and NACK1 protein accumulate only at the M phase of the cell cycle, which is consistent with the increase of NPK1 kinase activity (Nishihama et al. 2002).

The amino-terminal halves of NACK1 and NACK2 contain sequences that are similar to that of the MT-based motor domain, which are also conserved among various KLPs, whereas the carboxy-terminal halves have typical stalk domains with coiled-coil structures (Fig. 2A; Nishihama et al. 2002). Yeast two-hybrid and in vitro immunoprecipitation assays using recombinant proteins have shown that the stalk domain of NACK1 binds directly to the regulatory domain of NPK1 via these predicted coiled-coil structures (Ishikawa et al. 2002). This direct interaction with NACK1 seems to regulate the subcellular localization as well as the activity of NPK1 (Fig. 2C). During late anaphase and telophase, NACK1 is consistently localized to the equatorial zone of the phragmoplast, similar to NPK1 (Fig. 2B), whereas the deletion of the regulatory domain of NPK1, which contains the NACK1-binding site, eliminates its localization to the equator of the phragmoplast (Nishihama et al. 2002). This suggests that NACK1 is responsible for the proper localization and activation of NPK1 (Fig. 2C).

The motor and stalk domains of NACK1 also have several consensus sequences for phosphorylation by CDKs (Fig. 2A). Recently, Weingartner et al. (2004) reported that overexpression of the constitutively active form cyclin B1 disrupts the proper localization of NACK1 on phragmoplast MTs during cytokinesis. Whether NACK proteins directly bind to MTs remains to be determined, but it will be interesting to investigate whether the binding of NACK1 to NPK1 and MTs is controlled by cyclin/CDK complexes (Fig. 2C).

Function of NACK proteins. We postulated that overexpression of a truncated form of NACK1 (NACK:ST) that contained the stalk (ST) region might have a phenotypic effect similar to overexpression of a kinase-defective NPK1 on cytokinesis if the direct binding of NACK1 is required for the localization and activation of NPK1. As expected, both BY-2 cells and tobacco plants overexpressing NACK:ST have many multinucleated cells with incomplete cell plates (Fig. 4A; Nishihama et al. 2002). In these BY-2 cells, time-lapse observations of cytokinesis reveals that expansion of cell plates is markedly suppressed (Nishihama et al. 2002). We therefore concluded that NACK1 regulates plant cytokinesis by activating NPK1 MAPKKK.

The homologs of the *NACK1* and *NACK2* genes in *Arabidopsis* are designated *AtNACK1* and *AtNACK2* and are identical to *HINKEL* (*HIK*) and *STUD*

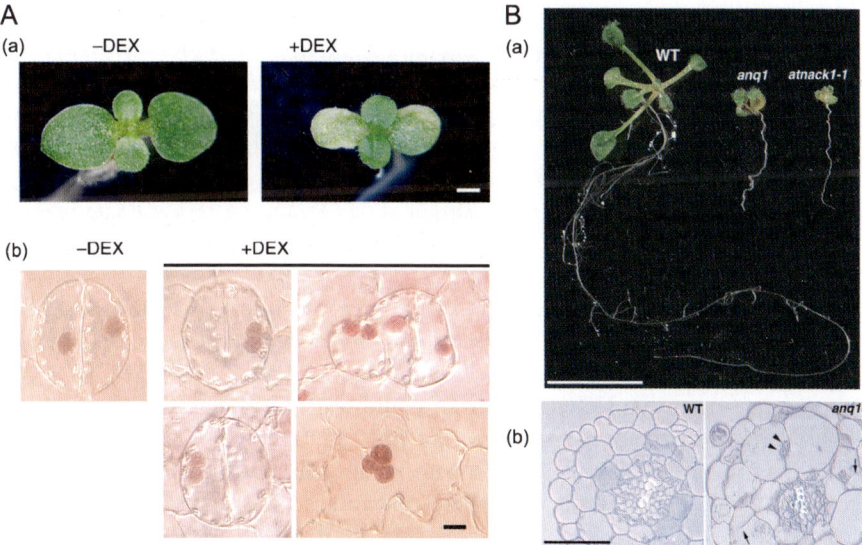

Fig. 4 The involvement of the NACK-PQR pathway in plant cytokinesis. **A** Generation of bi- or multi-nucleate stomata and pavement cells upon expression of a motor domain-less NACK1 (+DEX) in tobacco plants (*bottom panels*). Bar, 10 μm. Seeds of a plant transformed with the dexamethasone (DEX)-inducible NACK:ST construct were allowed to germinate on solid medium with DEX (+ DEX) or without DEX (-DEX). *Top panels* show the surface of cotyledons of 10-day-old seedlings. Bar, 10 mm. **B** Effects of the knockout of the *ANQ1* gene on the development and cell division of *Arabidopsis*. **a** 11-day-old plants of wild-type (WT; Wassiliewskija; *left*), *anq1* (*center*) and *atnack1-1* (*right*). Bar, 1 cm. **b** Sections of roots of 12-day-old plants of WT (*left*) and *anq1* (*right*) stained with toluidine *blue*. *Arrows* indicate incomplete cell walls. *Arrowheads* indicate nuclei in multinucleate cells. Bar, 50 μm

(*STD*)/*TETRASPORE* (*TES*), respectively (Nishihama et al. 2002; Strompen et al. 2002; Yang et al. 2003). Loss-of-function mutations in *AtNACK1/HIK* and *STD/TES/AtNACK2* result in the occasional failure of somatic and male-meiotic cytokinesis, respectively (Hülskamp et al. 1997; Spielman et al. 1997; Strompen et al. 2002; Nishihama et al. 2002; Yang et al. 2003). Recently, it has been shown that these genes have redundant functions and are essential for cytokinesis during both male and female gametogenesis (Tanaka et al. 2004).

2.3
Identification of Other Components of the MAPK Cascade Activated by the NACK/NPK1 Complex

To isolate downstream factors of NPK1, we used a yeast genetic system that is based on the yeast osmosensing MAPK cascade, which includes Ssk2/22p and Ste11p MAPKKK, Pbs2p MAPKK, and Hog1p MAPK (Brewster et al. 1993; Maeda et al. 1994, 1995). Expression of *NQK1* (*Nicotiana kinase next to NPK1*) cDNA replaced the function of the yeast MAPKK in the presence of both NPK1 and NACK1 (Soyano et al. 2003). NRK1 (*Nicotiana* kinase next to NQK1) MAPK was isolated as a binding partner of NQK1 using a yeast two-hybrid system (Soyano et al. 2003).

Characterization and function of NQK1 MAPKK. The cDNA cloned by the yeast genetic system encoded a putative polypeptide that had high homology to members of the MAPKK family, and the amino acid sequence of NQK1 was identical to that of NtMEK1 (Calderini et al. 2001). Of the 10 MAPKKs in *Arabidopsis*, AtMKK6 (Ichimura et al. 2001) has the highest amino acid sequence homology with NQK1/NtMEK1.

To examine the biochemical and biological function of NQK1/NtMEK1, we prepared a DNA construct for expression of a kinase-defective mutant of NQK1 in which the deduced ATP-binding site was mutated (NQK1:KW). This kinase-defective NQK1 construct could not complement the osmosensing MAPK of yeast cells that have the *pbs2* mutation, which was used for the screening of *NQK1* cDNA (Soyano et al. 2003), suggesting that NQK1 acts as a MAPKK. Recombinant NQK1 protein is phosphorylated by the active form of NPK1 and is activated (Soyano et al. 2003). Although NPK1 and NACK1 proteins rapidly disappear after M phase, NQK1 proteins accumulate throughout the cell cycle. The activity of NQK1 in tobacco cells, however, increases at the late M phase of the cell cycle and decreases thereafter. The pattern of NQK1 activity is similar to that of NPK1 as well as to the pattern of NACK1 accumulation (Fig. 3; Nishihama et al. 2001, 2002; Soyano et al. 2003). Green fluorescent protein (GFP) fusions of NQK1 and AtMKK6 also localize to the phragmoplast equator, similar to NPK1 and NACK1 (our unpublished results). These results suggest that activation of NPK1 MAPKKK by NACK1 binding causes the activation of NQK1 MAPKK at the equator of the phragmoplast during late M-phase.

The requirement of NQK1 for the expansion of the phragmoplast and for the formation of cell plates was demonstrated by overexpressing a kinase-defective mutant of NQK1 (NQK1:KW) in BY-2 cells and tobacco plants. The cells overexpressing NQK1:KW are multinucleated and form incomplete cell plates (Soyano et al. 2003). Also, the seedlings of the NQK1:KW transgenic plants are poorly developed, and its surfaces have bi- or un-nucleated guard and pavement cells, similar to those of plants overexpressing NACK:ST. We observed consistent abnormalities in the cells in which the kinase-defective NQK1 was overexpressed; therefore, NQK1 MAPKK seems to promote progression of cytokinesis downstream of NPK1 MAPKKK.

This idea is further supported by observation of the mutant of *ANQ1/AtMKK6*, which is the *Arabidopsis* homolog of *NQK1*. Like the *atnack1* mutant of *Arabidopsis*, the *anq1* homozygous mutant of *Arabidopsis* exhibits a severe dwarf phenotype (Fig. 4B; Soyano et al. 2003). The *anq1* mutants have typical defects in cytokinesis, large cells with incomplete cell walls, and multiple nuclei in leaves and roots. In addition to the growth defect, *anq1* mutants produce abnormal pollen grains that exhibit a tetrad structure (Soyano et al. 2003). This phenotype of pollen grains is similar to those from *tetraspore*, *stud*, and *atnack2* mutants, which have allelic mutations (Hülskamp et al. 1997; Spielman et al. 1997; Tanaka et al. 2004) and cause defects in male meiotic cytokinesis. Thus, ANQ1/AtMKK6 may also be involved in mitotic and meiotic cytokinesis under the regulation of NACK.

Identification of NRK1 MAPK that acts downstream of NQK1 MAPKK. NRK1 MAPK, which may act downstream of NQK1, was identified by yeast two-hybrid screening using NQK1:KW as the bait (Soyano et al. 2003). The predicted amino acid sequence of NRK1 is almost identical to that of NTF6 (Calderini et al. 2001). We found that recombinant NRK1 and NQK1 proteins bind each other in vitro (Soyano et al. 2003), which is consistent with a report by Calderini et al. (2001). Active recombinant NQK1 MAPKK phosphorylates and activates NRK1 MAPK (Soyano et al. 2003). Calderini et al. (1998) reported that NTF6 localizes to the midzone of phragmoplasts in anaphase tobacco cells. Using NRK1-specific antibodies, we also recently found that NRK1 localizes to the equator of phragmoplasts at telophase (our unpublished observation). Although NRK1 protein is present throughout cell cycle, the activity of NRK1 is activated in parallel with NQK1 at late M phase (Fig. 3; Soyano et al. 2003), suggesting that NQK1 phosphorylates NRK1 in vivo. We designated the kinase cascade that includes NPK1, NQK1, NRK1, NACK1, and NACK2, the NACK-PQR pathway (Fig. 1; Soyano et al. 2003).

Although the activation of NRK1 is tightly coupled to the activation of NPK1 and NQK1, involvement of NRK1 in cytokinesis has not yet been experimentally demonstrated. The preliminary observations in our laboratory indicate that the loss-of-function mutants of some MAPKs in Group B in *Arabidopsis* are defective in cytokinesis (our unpublished observations). These results indicate the NACK-PQR pathway is conserved and promotes cytokinesis in tobacco and *Arabidopsis*. In *Arabidopsis*, the pathway includes AtNACK1 and AtNACK2; ANP1, ANP2, and ANP3; and ANQ1/AtMKK6 (Fig. 1).

3
Factors Downstream of the NACK-PQR Pathway

In eukaryotes, the last stage of cell division is cytokinesis, which leads to the appropriate distribution and separation of the sets of chromosomes and cell components. Plant cytokinesis occurs at the late anaphase step of cell division in the phragmoplast, which is a cytokinetic apparatus composed mainly of MTs and actin filaments. Cytokinesis consists of four main processes: (1) formation of the phragmoplast; (2) transport, accumulation, and fusion of Golgi-derived vesicles (see the works by Sanderfoot, Nebenführ, in this volume); (3) synthesis and maturation of cell walls (see the work by Verma and Hong, in this volume); and (4) expansion of the phragmoplast (see the work by Seguí-Simarro, Otegui, Austin and Staehelin, in this volume). The NACK1/NPK1 complex and the subsequent MAPK cascade (the NACK-PQR pathway) appears to control cytokinesis of plant cells by regulating phragmoplast expansion, which, as described above, is the final step; however, what kind of molecules and how they are regulated downstream of the NACK-PQR pathway has not been established. Recently, a MAP was identified as a substrate of NRK1/NTF6 MAPK (Sasabe et al. 2006). This supports the hypothesis that the NACK-PQR pathway might regulate an event related to the dynamics of phragmoplast MTs. We propose a novel regulatory mechanism for the progression of cytokinesis, namely, via phragmoplast expansion directed by the NACK-PQR pathway.

3.1
The Molecular Roles of the NACK-PQR Pathway during Cytokinesis

Cytokinesis is executed by the cytokinesis-specific apparatus, the central spindle, or midbody in animals and the phragmoplast in plants (Nishihama and Machida 2001; Otegui et al. 2005). These structures develop from late anaphase to telophase between the two daughter nuclei and consist of two bundles of antiparallel MTs. As cytokinesis proceeds, the central spindle becomes compacted in animal cells, and in plant cells, the phragmoplast expands centrifugally. In addition, new membranes and/or cell walls are generated inside or outside the midzone of the central spindle or the phragmoplast. These dynamic processes appear to be mediated by the turnover of MTs, which involves the depolymerization of MTs and the polymerization of tubulins at the plus-ends (Shelden and Wadsworth 1990; Asada et al. 1991; Hush et al. 1994; Straight and Field 2000). Thus, despite the opposite directions of cytokinesis in plant and animal cells, the cytoskeletal structures and the mode of MT dynamics appear to be largely conserved. Notably, however, plant cytokinesis includes the synthesis of the cell wall in addition to the fusion of cell membranes.

The phenotypes generated by overexpression of the dominant-negative mutants of NACK1, NPK1, and NQK1 are similar to those of cells treated with taxol, a compound that blocks the depolymerization of MTs (Yasuhara et al. 1993). This suggests that MT disassembly is required for phragmoplast expansion. In cells from *atnack1/hik* mutant plants of *Arabidopsis*, phragmoplast MTs persist in the center of the division plane, suggesting that the disassembly of phragmoplast MTs is inhibited in these cells (Strompen et al. 2002). Phragmoplast MTs usually disappear from the division plane after formation of the cell plate. Thus, the activation of the NACK-PQR pathway appears to be required for the reorganization of phragmoplast MTs in expansion of the phragmoplast during cell plate formation. Factors acting downstream of the NACK-PQR pathway may therefore control MT dynamics.

3.2
MAP65 Protein is a Downstream Factor of NRK1/NTF6

Identification of NtMAP65-1 as a substrate for NRK1/NTF6 MAPK. In animals, several components belonging to the KLP or microtubule-associated protein (MAP) families have been shown to regulate the formation or dynamics of the central spindle (Glotzer 2005). In plants, however, few factors involved in phragmoplast dynamics have been identified.

We found that several MAPs purified from tobacco BY-2 cells can be phosphorylated by active NRK1 in vitro. One of these candidate substrates is NtMAP65-1a, a protein belonging to the MAP65/Ase1/PRC1 family (Sasabe et al. 2006). This family of proteins is conserved among a variety of organisms and includes Ase1p (anaphase spindle elongation factor) in yeast (Pellman et al. 2002), PRC1 (protein regulator of cytokinesis 1) in mammals (Jiang et al. 1998), SPD1 (spindle defective1) in *C. elegans* (Verbrugghe and White 2004), and Feo (Fascetto) in *Drosophila* (Vernì et al. 2004). Although members of the MAP65/Ase1/PRC1 protein family have low amino acid sequence similarities, the secondary structures in their predicted coiled-coil regions are highly conserved (Schuyler et al. 2003). These MAPs localize to the cytokinetic apparatuses, and most of them are involved in cytokinesis.

Tobacco has at least three members of the MAP65 subfamily (NtMAP65-1a, -1b and -1c). The amino acid sequences of these proteins are more than 85% identical (Hussey et al. 2002), and when produced as recombinant proteins in *Escherichia coli*, they are effectively phosphorylated by active NRK1 (our unpublished observation). In vitro, NRK1 phosphorylates NtMAP65-1a at a single site, Thr-579, in the carboxy-terminal region. Specific antibodies against Thr-579-phosphorylated NtMAP65-1 have revealed that NtMAP65-1 is phosphorylated at this site in vivo. Because PRC1 (a mammalian MAP65 homolog) is phosphorylated by Cdc2 (Jiang et al. 1998), we also examined whether NtMAP65-1a can be also phosphorylated by CDKs from BY-2 cells in vitro. We found that the sites phosphorylated by CDKs were different from that phosphorylated by NRK1 (Fig. 5A; Sasabe et al. 2006).

Characterization of NtMAP65-1 phosphorylated by MAPK. In synchronized BY-2 cells, NtMAP65-1 phosphorylated at Thr-579 accumulates at late M phase, although the total amount of NtMAP65-1 does not change. Such a pattern of phosphorylation is consistent with the pattern of NACK1 accumulation and NPK1, NQK1/NtMEK1, and NRK1/NTF6 activation. We have found that a GFP-NtMAP65-1a fusion protein localizes on all kinds of MT structures, including cortical MTs, the preprophase band, spindles, and the phragmoplast (our unpublished observation). Immunostaining with NtMAP65-1 antibodies also revealed NtMAP65-1 on various MT structures throughout the cell cycle. Interestingly, NtMAP65-1 phosphorylated on Thr-579 is concentrated at the equator of the phragmoplast along with other components of the NACK-PQR pathway, although NtMAP65-1 can be found throughout the entire phragmoplast (Fig. 5B; Sasabe et al. 2006). These findings suggest that Thr-579 of NtMAP65-1 is phosphorylated by NRK1/NTF6 at the phragmoplast midzone during cytokinesis.

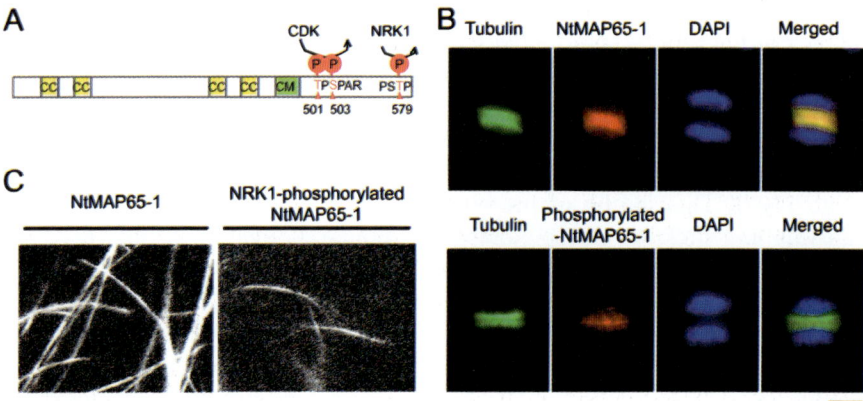

Fig. 5 NtMAP65-1 acts downstream of the NACK-PQR pathway. **A** A schematic representation of NtMAP65-1a. The coiled-coil motif (CC) and the motif that is conserved in the MAP65 family (CM) are described. Sites of phosphorylation (P) by NRK1 and CDKs are also indicated. **B** Subcellular localization of NtMAP65-1 (*top*) and NtMAP65-1 phosphorylated on Thr-579 (*bottom*) at telophase in BY-2 cells. BY-2 cells were triple-stained with mouse antibodies against a-tubulin (*green*), rabbit antibodies against NtMAP65-1 or NtMAP65-1 phosphorylated on Thr-579 (*red*), and DAPI (*blue*). **C** Dark-field micrographs of MTs bundled in vitro by NtMAP65-1 (*left*) and NRK1-phosphorylated NtMAP65-1 (*right*). The MT-bundling activity of NtMAP65-1 was decreased by phosphorylation with NRK1

Function of phosphorylated NtMAP65-1 during M phase. In BY-2 cells, overexpression of mutant NtMAP65-1a proteins that cannot be phosphorylated by NRK1 delays M phase progression and phragmoplast expansion, whereas mutants that cannot be phosphorylated by CDKs do not affect either process (Sasabe et al. 2006). Similar delays in expansion are often observed in cells overexpressing dominant-negative mutants of NACK1, NPK1, and NQK1 (Nishihama et al. 2001, 2002; Soyano et al. 2003). Further analysis has revealed that the cortical MTs and phragmoplast MTs in cells overexpressing an NRK1-non-phosphorylatable mutant are resistant to the MT-depolymerizing drug propyzamide, suggesting that the MTs are stabilized by the non-phosphorylatable mutant of NtMAP65-1a.

It is well known that the MAP65/Ase1p/PRC1 protein family can bind and bundle MTs. Recombinant NtMAP65-1a also has MT binding and bundling activity in vitro. On the other hand, phosphorylation of NtMAP65-1a by NRK1 reduces its bundling activity (Fig. 5C). This is not due to the phosphorylation of CDKs, which suggests that the stability of MT-containing structures depends on the MT-bundling activity of MAP65. Recently, three groups have reported that the MTs bundled by MAP65 proteins have increased resistance to some chemicals or environmental stresses in vivo and in vitro (Wicker-Planquart et al. 2004; Van Damme et al. 2004b; Mao et al. 2005). The phenotypes of various organisms with loss-of-function mutations of MAP65/Ase1/PRC1 imply that these proteins maintain the distribution of the spindle or phragmoplast MTs (Jiang et al. 1998; Smertenko et al. 2000; 2004; Mollinari et al. 2002; Schuyler et al. 2003; Verbrugghe and White 2004; Vernì et al. 2004; Müller et al. 2004). The activity of MT-bundling seems to be concerned with the stability of MTs. Thus, the simplest model to explain the molecular effect of phosphorylation of NtMAP65 by NRK1/NTF6 is that, during cytokinesis, it suppresses the stability of the phragmoplast at the midzone, resulting in increased MT turnover and, therefore, phragmoplast expansion. Whether the phosphorylation of MAP65 by CDKs is important during the progression of M phase in plants remains to be determined. In animals, the phosphorylation of PRC1 by CDK before metaphase appears to be important for progression from metaphase to anaphase and may suppress MT bundling in the mitotic spindle (Mollinari et al. 2002). In addition, mammalian PRC1 interacts separately with many other proteins that are involved in the formation of the midzone of the central spindle, and its interaction appears to be involved in the phosphorylation of CDK (Ban et al. 2004; Kurasawa et al. 2004; Zhu and Jiang 2005). These reports suggest that regulation of the dynamics of MT-containing structures must be very complex. Further investigations of MAP65-interacting proteins and the NACK-PQR pathway-related factors are required if we are to fully understand the regulation of mitosis.

4
Conclusions and Perspectives

Plant cytokinesis involves the formation of a cell plate. This event includes many plant-specific steps and is accomplished with the help of the phragmoplast. The phragmoplast acts as a conduit for vesicle transport and must expand to maintain phragmoplast MTs at the leading edge of the cell plate as it grows. Recent research has identified potential regulators of microtubule dynamics such as MAP65 and MOR1/GEM1 as well as a MAPK cascade in the control of phragmoplast expansion (Smertenko et al. 2000; Whittington et al. 2001; Nishihama et al. 2001, 2002; Twell et al. 2002; Soyano et al. 2003; Müller et al. 2004; Van Damme et al. 2004a; Chang et al. 2005). Finally, this MAPK cascade controls phragmoplast expansion in tobacco cells via regulation of MAP65-1 function (Sasabe et al. 2006).

Many components of the cytokinetic machinery including MAPs have been identified (Jürgens 2005), and, intriguingly, despite different modes of cytokinesis in the various kingdoms, most of these proteins are conserved (Otegui et al. 2005). This indicates that the system for regulating the cytoskeletal structures involved in cytokinesis may be conserved between animals and plants. The involvement of MAPKs in cytokinesis has not been reported in non-plant organisms. Plant cytokinesis includes (1) the dynamics of MTs, (2) the synthesis of cell membranes, and (3) the synthesis of cell walls. These subcellular processes might be coordinately controlled. How MAPKs do this is an interesting question that warrants further investigation.

Acknowledgements This work was supported in part by a grant from the Program for Promotion of Basic Research Activities for Innovative Biosciences, by a Grant-in-Aid for Scientific Research on Priority Areas (no. 14036216), and by a Grant-in-Aid for the 21st Century COE Program (System Bioscience) from the Ministry of Education, Culture, Sports, Science, and Technology of Japan.

References

Asada T, Sonobe S, Shibaoka H (1991) Microtubule translocation in the cytokinetic apparatus of cultured tobacco cells. Nature 350:238–241

Ban R, Irino Y, Fukami K, Tanaka H (2004) Human mitotic spindle-associated protein PRC1 inhibits MgcRacGAP activity toward Cdc42 during the metaphase. J Biol Chem 279:16394–16402

Banno H, Hirano K, Nakamura T, Irie K, Nomoto S, Matsumoto K, Machida Y (1993) NPK1, a tobacco gene that encodes a protein with a domain homologous to yeast BCK1, STE11, and Byr2 protein kinases. Mol Cell Biol 13:4745–4752

Bergmann DC, Lukowitz W, Somerville CR (2004) Stomatal development and pattern controlled by a MAPKK kinase. Science 304:1494–1497

Bögre L, Calderini O, Binarova P, Mattauch M, Till S, Kiegerl S, Jonak C, Pollaschek C, Barker P, Huskisson NS, Hirt H, Heberle-Bors E (1999) A MAP kinase is activated late in plant mitosis and becomes localized to the plane of cell division. Plant Cell 11:101–113

Boudsocq M, Lauriere C (2005) Osmotic signaling in plants: multiple pathways mediated by emerging kinase families. Plant Physiol 138:1185–1194

Brewster JL, de Valoir T, Dwyer ND, Winter E, Gustin MC (1993) An osmosensing signal transduction pathway in yeast. Science 259:1760–1763

Calderini O, Bogre L, Vicente O, Binarova P, Heberle-Bors E, Wilson C (1998) A cell cycle regulated MAP kinase with a possible role in cytokinesis in tobacco cells. J Cell Sci 111:3091–3100

Calderini O, Glab N, Bergounioux C, Heberle-Bors E, Wilson C (2001) A novel tobacco mitogen-activated protein (MAP) kinase kinase, NtMEK1, activates the cell cycle-regulated p43Ntf6 MAP kinase. J Biol Chem 276:18139–18145

Chang C (2003) Ethylene signaling: the MAPK module has finally landed. Trends Plant Sci 8:365–368

Chang HY, Smertenko AP, Igarashi H, Dixon DP, Hussey PJ (2005) Dynamic interaction of NtMAP65-1a with microtubules in vivo. J Cell Sci 118:3195–3201

Chen Z, Gibson TB, Robinson F, Silvestro L, Pearson G, Xu B, Wright A, Vanderbilt C, Cobb MH (2001) MAP kinases. Chem Rev 101:2449–2476

Chinnusamy V, Schumaker K, Zhu JK (2004) Molecular genetic perspectives on cross-talk and specificity in abiotic stress signalling in plants. J Exp Bot 55:225–236

English J, Pearson G, Wilsbacher J, Swantek J, Karandikar M, Xu S, Cobb MH (1999) New insights into the control of MAP kinase pathways. Exp Cell Res 253:255–270

Farrar MA, Alberol-Ila J, Perlmutter RM (1996) Activation of the Raf-1 kinase cascade by coumermycin-induced dimerization. Nature 383:178–181

Glotzer M (2005) The molecular requirements for cytokinesis. Science 307:1735–1739

Guo H, Ecker JR (2004) The ethylene signaling pathway: new insights. Curr Opin Plant Biol 7:40–49

Hülskamp M, Parekh NS, Grini P, Schneitz K, Zimmermann I, Lolle SJ, Pruitt RE (1997) The STUD gene is required for male-specific cytokinesis after telophase II of meiosis in Arabidopsis thaliana. Dev Biol 187:114–124

Hush JM, Wadsworth P, Callaham DA, Hepler PK (1994) Quantification of microtubule dynamics in living plant cells using fluorescence redistribution after photobleaching. J Cell Sci 107:775–784

Hussey PJ, Hawkins TJ, Igarashi H, Kaloriti D, Smertenko A (2002) The plant cytoskeleton: recent advances in the study of the plant microtubule-associated proteins MAP-65, MAP-190 and the Xenopus MAP215-like protein, MOR1. Plant Mol Biol 50:915–924

Irie K, Gotoh Y, Yashar BM, Errede B, Nishida E, Matsumoto K (1994) Stimulatory effects of yeast and mammalian 14-3-3 proteins on the Raf protein kinase. Science 265:1716–1719

Ishikawa M, Soyano T, Nishihama R, Machida Y (2002) The NPK1 mitogen-activated protein kinase kinase kinase contains a functional nuclear localization signal at the binding site for the NACK1 kinesin-like protein. Plant J 32:789–98

Jiang W, Jimenez G, Wells NJ, Hope TJ, Wahl GM, Hunter T, Fukunaga R (1998) PRC1: a human mitotic spindle-associated CDK substrate protein required for cytokinesis. Mol Cell 2:877–885

Jurgens G (2005) Plant cytokinesis: fission by fusion. Trends Cell Biol 15:277–283

Krysan PJ, Jester PJ, Gottwald JR, Sussman MR (2002) An Arabidopsis mitogen-activated protein kinase kinase kinase gene family encodes essential positive regulators of cytokinesis. Plant Cell 14:1109–1120

Kurasawa Y, Earnshaw WC, Mochizuki Y, Dohmae N, Todokoro K (2004) Essential roles of KIF4 and its binding partner PRC1 in organized central spindle midzone formation. EMBO J 23:3237–3248

Lukowitz W, Roeder A, Parmenter D, Somerville C (2004) A MAPKK kinase gene regulates extra-embryonic cell fate in Arabidopsis. Cell 116:109–119

Luo Z, Tzivion G, Belshaw PJ, Vavvas D, Marshall M, Avruch J (1996) Oligomerization activates c-Raf-1 through a Ras-dependent mechanism. Nature 383:181–185

Lupas A, Van Dyke M, Stock J (1991) Predicting coiled coils from protein sequences. Science 252:1162–1164

Machida Y, Nakashima M, Morikiyo K, Soyano T, Nishihama R (1998) MAPKKK-related protein kinase NPK1: Involvement in the regulation of the M phase of plant cell cycle. J Plant Res 111:243–246

Maeda T, Takekawa M, Saito H (1995) Activation of yeast PBS2 MAPKK by MAPKKKs or binding of an SH3-containing osmosenser. Science 269:554–558

Maeda T, Wurgler-Murphy SM, Saito H (1994) A two-component system that regulates an osmosensing MAP kinase cascade in yeast. Nature 369:242–245

Mao T, Jin L, Li H, Liu B, Yuan M (2005) Two microtubule-associated proteins of the Arabidopsis MAP65 family function differently on microtubules. Plant Physiol 138:654–662

MAPK Group (Ichimura K, Shinozaki K, Tena G, Sheen J, Henry Y, Champion A, Kreis M, Zhang S, Hirt H, Wilson C, Heberle-Bors E, Ellis BE, Morris PC, Innes RW, Ecker JR, Scheel D, Klessig DF, Machida Y, Mundy J, Ohashi Y, Walker JC) (2002) Mitogen-activated protein kinase cascades in plants: a new nomenclature. Trends Plant Sci 7:301–308

Minshull J, Sun H, Tonks NK, Murray AW (1994) A MAP kinase-dependent spindle assembly checkpoint in Xenopus egg extracts. Cell 79:475–486

Mollinari C, Kleman JP, Jiang W, Schoehn G, Hunter T, Margolis RL (2002) PRC1 is a microtubule binding and bundling protein essential to maintain the mitotic spindle midzone. J Cell Biol 157:1175–1186

Müller S, Smertenko A, Wagner V, Heinrich M, Hussey PJ, Hauser MT (2004) The plant microtubule-associated protein AtMAP65-3/PLE is essential for cytokinetic phragmoplast function. Curr Biol 14:412–417

Nakagami H, Pitzschke A, Hirt H (2005) Emerging MAP kinase pathways in plant stress signalling. Trends Plant Sci 10:339–346

Nakashima M, Hirano K, Nakashima S, Banno H, Nishihama R, Machida Y (1998) The expression pattern of the gene for NPK1 protein kinase related to mitogen-activated protein kinase kinase kinase (MAPKKK) in a tobacco plant: correlation with cell proliferation. Plant Cell Physiol 3:690–700

Nishihama R, Banno H, Kawahara E, Irie K, Machida Y (1997) Possible involvement of differential splicing in regulation of the activity of Arabidopsis ANP1 that is related to mitogen-activated protein kinase kinase kinases (MAPKKKs). Plant J 12:39–48

Nishihama R, Ishikawa M, Araki S, Soyano T, Asada T, Machida Y (2001) The NPK1 mitogen-activated protein kinase kinase kinase is a regulator of cell-plate formation in plant cytokinesis. Genes Dev 15:352–363

Nishihama R, Machida Y (2001) Expansion of the phragmoplast during plant cytokinesis: a MAPK pathway may MAP it out. Curr Opin Plant Biol 4:507–512

Nishihama R, Soyano T, Ishikawa M, Araki S, Tanaka H, Asada T, Irie K, Ito M, Terada M, Banno H, Yamazaki Y, Machida Y (2002) Expansion of the cell plate in plant cytokinesis requires a kinesin-like protein/MAPKKK complex. Cell 109:87–99

Otegui MS, Verbrugghe KJ, Skop AR (2005) Midbodies and phragmoplasts: analogous structures involved in cytokinesis. Trends Cell Biol 15:404–413

Pages G, Lenormand P, L'Allemain G, Chambard JC, Meloche S, Pouyssegur J (1993) Mitogen-activated protein kinases p42mapk and p44mapk are required for fibroblast proliferation. Proc Natl Acad Sci USA 90:8319–8323

Pedley KF, Martin GB (2005) Role of mitogen-activated protein kinases in plant immunity. Curr Opin Plant Biol 8:541–547

Pellman D, Bagget M, Tu YH, Fink GR, Tu H (1995) Two microtubule-associated proteins required for anaphase spindle movement in Saccharomyces cerevisiae. J Cell Biol 130:1373–1385

Posas F, Saito H (1998) Activation of the yeast SSK2 MAP kinase kinase kinase by the SSK1 two-component response regulator. EMBO J 17:1385–1394

Robinson MJ, Cobb MH (1997) Mitogen-activated protein kinase pathways. Curr Opin Cell Biol 9:180–186

Samaj J, Baluska F, Hirt H (2004) From signal to cell polarity: mitogen-activated protein kinases as sensors and effectors of cytoskeleton dynamicity. J Exp Bot 55:189–198

Samaj J, Ovecka M, Hlavacka A, Lecourieux F, Meskiene I, Lichtscheidl I, Lenart P, Salaj J, Volkmann D, Bogre L, Baluska F, Hirt H (2002) Involvement of the mitogen-activated protein kinase SIMK in regulation of root hair tip growth. EMBO J 21:3296–3306

Sasabe M, Soyano T, Takahashi Y, Sonobe S, Igarashi H, Itoh TJ, Hidaka M, Machida Y (2006) Phosphorylation of NtMAP65-1 by a MAP kinase down-regulates its activity of microtubule bundling and stimulates progression of cytokinesis of tobacco cells. Genes Dev 20:1004–1014

Schuyler SC, Liu JY, Pellman D (2003) The molecular function of Ase1p: evidence for a MAP-dependent midzone-specific spindle matrix. Microtubule-associated proteins. J Cell Biol 160:517–528

Shapiro PS, Vaisberg E, Hunt AJ, Tolwinski NS, Whalen AM, McIntosh JR, Ahn NG (1998) Activation of the MKK/ERK pathway during somatic cell mitosis: direct interactions of active ERK with kinetochores and regulation of the mitotic 3F3/2 phosphoantigen. J Cell Biol 142:1533–1545

Shelden E, Wadsworth P (1990) Interzonal microtubules are dynamic during spindle elongation. J Cell Sci 97:273–281

Shibuya H, Yamaguchi K, Shirakabe K, Tonegawa A, Gotoh Y, Ueno N, Irie K, Nishida E, Matsumoto K (1996) TAB1: an activator of the TAK1 MAPKKK in TGF-beta signal transduction. Science 272:1179–1182

Smertenko A, Saleh N, Igarashi H, Mori H, Hauser-Hahn I, Jiang CJ, Sonobe S, Lloyd CW, Hussey PJ (2000) A new class of microtubule-associated proteins in plants. Nat Cell Biol 2:750–753

Smertenko AP, Chang HY, Wagner V, Kaloriti D, Fenyk S, Sonobe S, Lloyd C, Hauser MT, Hussey PJ (2004) The Arabidopsis microtubule-associated protein AtMAP65-1: molecular analysis of its microtubule bundling activity. Plant Cell 16:2035–2047

Soyano T, Nishihama R, Morikiyo K, Ishikawa M, Machida Y (2003) NQK1/NtMEK1 is a MAPKK that acts in the NPK1 MAPKKK-mediated MAPK cascade and is required for plant cytokinesis. Genes Dev 17:1055–1067

Spielman M, Preuss D, Li FL, Browne WE, Scott RJ, Dickinson HG (1997) TETRASPORE is required for male meiotic cytokinesis in Arabidopsis thaliana. Development 124:2645–2657

Straight AF, Field CM (2000) Microtubules, membranes and cytokinesis. Curr Biol 10:R760–R770

Strompen G, El Kasmi F, Richter S, Lukowitz W, Assaad FF, Jurgens G, Mayer U (2002) The Arabidopsis HINKEL gene encodes a kinesin-related protein involved in cytokinesis and is expressed in a cell cycle-dependent manner. Curr Biol 12:153–158

Takahashi Y, Soyano T, Sasabe M, Machida Y (2004) A MAP kinase cascade that controls plant cytokinesis. J Biochem 136:127–132

Takenaka K, Moriguchi T, Nishida E (1998) Activation of the protein kinase p38 in the spindle assembly checkpoint and mitotic arrest. Science 280:599–602

Tanaka H, Ishikawa M, Kitamura S, Takahashi Y, Soyano T, Machida C, Machida Y (2004) The AtNACK1/HINKEL and STUD/TETRASPORE/AtNACK2 genes, which encode functionally redundant kinesins, are essential for cytokinesis in Arabidopsis. Genes Cells 9:1199–1211

Twell D, Park SK, Hawkins TJ, Schubert D, Schmidt R, Smertenko A, Hussey PJ (2002) MOR1/GEM1 has an essential role in the plant-specific cytokinetic phragmoplast. Nat Cell Biol 4:711–714

Van Damme D, Bouget FY, Van Poucke K, Inzé D, Geelen D (2004a) Molecular dissection of plant cytokinesis and phragmoplast structure: a survey of GFP-tagged proteins. Plant J 40:386–398

Van Damme D, Van Poucke K, Boutant E, Ritzenthaler C, Inze D, Geelen D (2004b) In vivo dynamics and differential microtubule-binding activities of MAP65 proteins. Plant Physiol 136:3956–3967

Verbrugghe KJ, White JG (2004) SPD-1 is required for the formation of the spindle midzone but is not essential for the completion of cytokinesis in *C. elegans* embryos. Curr Biol 14:1755–1760

Vernì F, Somma MP, Gunsalus KC, Bonaccorsi S, Belloni G, Goldberg ML, Gatti M (2004) Feo, the Drosophila homolog of PRC1, is required for central-spindle formation and cytokinesis. Curr Biol 14:1569–1575

Weingartner M, Criqui MC, Meszaros T, Binarova P, Schmit AC, Helfer A, Derevier A, Erhardt M, Bogre L, Genschik P (2004) Expression of a nondegradable cyclin B1 affects plant development and leads to endomitosis by inhibiting the formation of a phragmoplast. Plant Cell 16:643–657

Whittington AT, Vugrek O, Wei KJ, Hasenbein NG, Sugimoto K, Rashubrooke MC, Wasteneys GO (2001) MOR1 is essential for organizing cortical microtubules in plants. Nature 411:610–613

Wicker-Planquart C, Stoppin-Mellet V, Blanchoin L, Vantard M (2004) Interactions of tobacco microtubule-associated protein MAP65-1b with microtubules. Plant J 39:126–134

Willard FS, Crouch MF (2001) MEK, ERK, and p90RSK are present on mitotic tubulin in Swiss 3T3 cells: a role for the MAP kinase pathway in regulating mitotic exit. Cell Signal 13:653–664

Wright JH, Munar E, Jameson DR, Andreassen PR, Margolis RL, Seger R, Krebs EG (1999) Mitogen-activated protein kinase kinase activity is required for the G(2)/M transition of the cell cycle in mammalian fibroblasts. Proc Natl Acad Sci USA 96:11335–11340

Yang CY, Spielman M, Coles JP, Li Y, Ghelani S, Bourdon V, Brown RC, Lemmon BE, Scott RJ, Dickinson HG (2003) TETRASPORE encodes a kinesin required for male meiotic cytokinesis in Arabidopsis. Plant J 34:229–240

Yasuhara H, Sonobe S, Shibaoka H (1993) Effects of taxol on the development of the cell plate and of the phragmoplast in tobacco BY-2 cells. Plant Cell Physiol 34:21–29

Zecevic M, Catling AD, Eblen ST, Renzi L, Hittle JC, Yen TJ, Gorbsky GJ, Weber MJ (1998) Active MAP kinase in mitosis: localization at kinetochores and association with the motor protein CENP-E. J Cell Biol 142:1547–1558

Zhu C, Jiang W (2005) Cell cycle-dependent translocation of PRC1 on the spindle by Kif4 is essential for midzone formation and cytokinesis. Proc Natl Acad Sci USA 102:343–348

Plant Cytokinesis – Insights Gained from Electron Tomography Studies

José M. Seguí-Simarro[1] (✉) · Marisa S. Otegui[2] · Jotham R. Austin II[3] · L. Andrew Staehelin[4]

[1]Instituto para la Conservación y Mejora de la Agrodiversidad Valenciana (COMAV), Universidad Politécnica de Valencia, Camino de Vera s/n, edificio 9B, 46022 Valencia, Spain
seguisim@btc.upv.es

[2]Department of Botany, University of Wisconsin, 430 Lincoln Drive, Madison, WI 53706, USA

[3]MLK 007D, University of Chicago, 910 E. 58th st, Chicago, IL 60637, USA

[4]MCD Biology, University of Colorado, UCB 347, Boulder, CO 80309-0347, USA

Abstract Cytokinesis is the final step of the cell division sequence. During cytokinesis, the plant cell completes its partition into two equal daughter cells by the transient formation of the cytokinetic apparatus, a complex structural scaffold that assists in the formation of a new cell wall. In the last few years, technical advances in sample processing and three-dimensional reconstruction and modeling have permitted the analysis of the dramatic structural changes undergone by the plant cell during cytokinesis at an unprecedented level of spatial resolution. The main purpose of this chapter is to summarize the contribution of dual-axis electron tomography of cells cryopreserved by high-pressure freezing and freeze-substitution (HPF-FS) to the understanding of the mechanisms underlying cytokinesis in angiosperms. We focus on the new structural and functional insights into the complex process of assembly of both somatic- and syncytial-type cell plates, the architecture of the different cell plate membrane intermediates, the structural and functional properties of the recently characterized Cell Plate Assembly Matrix (CPAM), and the relationship between the CPAM and phragmoplast microtubule dynamics.

1
Introduction

Plant cytokinesis is the final event in the plant cell division cycle that yields two daughter nuclei separated by a new cell wall. During this process, the plant cell divides its whole cell mass, including its genetic material and its organelles, into two daughter cells. Typically, cytokinesis (the separation of the daughter cytoplasms) is initiated once the partition and segregation of the nuclear components (caryokinesis) is completed and the daughter nuclei are being rebuilt. When caryokinesis and cytokinesis are temporally and mechanistically coupled, the process is known as somatic-type cytokinesis, whereas the term syncytial-type cytokinesis is used to denote systems where caryokinesis and cytokinesis are both temporarily and structurally separate events (Otegui and Staehelin 2000a).

Our current concept of the plant cytokinetic machinery includes three distinct structures, each of which is an essential component of the cytokinetic apparatus: the phragmoplast, the cell plate assembly matrix (CPAM), and the cell plate. The phragmoplast is a cytoskeletal array of anti-parallel microtubules (MTs) and actin filaments. The principal function of the MTs is to deliver both the Golgi-derived vesicles and the CPAM molecules to the cell plate assembly region. A major function of the actin filaments is to guide the cell plate to the proper site of fusion of the cell plate with the cell wall (Gunning 1982; Whalen et al. 2001; Molchan et al. 2002). The CPAM is a transient, membrane-less cytoplasmic domain that can be recognized in electron micrographs of cryofixed and freeze-substituted cells as a ribosome-excluding zone that encompasses the fusing vesicles of growing cell plates (Samuels et al. 1995; Otegui et al. 2001; Otegui and Staehelin 2004; Seguí-Simarro et al. 2004). It is brought to the cell plate assembly region together with the Golgi-derived, cell plate-forming vesicles, and it appears to consist of scaffolding proteins and enzymatic, structural and regulatory proteins that are required for cell plate assembly. The cell plate itself is a transient membranous organelle that arises within the CPAM from the fusion of Golgi-derived vesicles during late anaphase and telophase, and progresses through a defined sequence of maturational steps to give rise to the new cell wall (Samuels et al. 1995; Otegui et al. 2001; Otegui and Staehelin 2004; Seguí-Simarro et al. 2004).

The focus of this review is on the structural events associated with the assembly of both somatic- and syncytial-type cell plates as seen in electron tomograms of cells preserved by high pressure freezing and freeze-substitution (HPF-FS) techniques. In particular, we review the structure of the membrane assembly intermediates associated with the different maturational stages of cell plate assembly, as well as discuss the concomitant changes in the organization and function of the CPAM and how the CPAM influences MT (+)-end dynamics.

2
Techniques: High Pressure Freezing and Electron Tomography Have Ushered in a New Era in Plant Cytokinesis Research

The cytokinetic apparatus and growing cell plates are highly dynamic molecular assemblies that undergo changes in organization in the time range of seconds and even fractions of seconds. Preserving the assembly intermediates of the cell plate for structural analysis has been a central goal of electron microscopists for half a century (reviewed in Seguí-Simarro and Staehelin 2006b). Classical chemical fixatives, such as glutaraldehyde and osmium tetroxide, are multifunctional reagents that preserve cell ultrastructure by crosslinking subsets of cellular molecules, thereby preserving their spatial organization. However, because chemical fixation takes seconds to minutes to immobilize cellular processes, and because different cellular components are fixed at different rates, chemical fixatives are not ideal tools for preserving cell ultrastucture (reviewed in Gilkey and Staehelin (1986)). For structural studies of cytokinesis, the most limiting aspect of chemical fixatives is their inability to preserve short-lived structural intermediates of cell plate assembly and phragmoplast MT dynamics in a reliable manner. Furthermore, due to the selective crosslinking activity of chemical fixatives, which leaves many types of molecules in an unfixed state, many cellular structures can continue to undergo structural changes long after the sample has been "fixed".

Most of these limitations of chemical fixatives can be overcome by employing ultra-rapid freezing methods such as high pressure freezing (HPF) to stabilize the cells, due to their ability to immobilize all cellular molecules within milliseconds (Gilkey and Staehelin 1986). Samples preserved in this manner can then be freeze-substituted (FS) at -80 to 90 °C, which increases the probability of viewing even the most labile cellular structures.

The second limitation of classical electron microscopy, even when cryofixed and freeze-substituted samples are examined, is that conclusions have to be drawn from essentially 2-dimensional (2-D) samples (most frequently, thin sections). Since the thin sections are typically 60–80 nm thick, this means that whereas the resolution in the x and y axes of the sections is $\sim$4 nm, the resolution in the z-axis is only 100–150 nm, which greatly limits the ability of researchers to determine the complete 3-D architecture of most cellular organelles and cytoskeletal systems at high resolution. The solution to this latter problem is the use of dual-axis electron tomography (McIntosh et al. 2005).

Electron tomography is analogous to the various tomographic techniques used in modern medicine, which rely on X-rays, magnetic resonance or ultrasound. By taking multiple images of a specimen (typically a 150–300 nm thick plastic section) as it is systematically tilted from +60° to -60° along two orthogonal axes, it is possible to record information from the z-axis of a sample, and by computing the back-projection of these images, it is then possible to produce a tomogram, which is a 3-D block of data that is represented as an array of volume elements (voxels). This 3-D data block, in turn, can be used to display 2-D slice images of different z-axis levels of the sample. These images resemble thin section images (compare Fig. 1A and B), but because they are only $\sim$2 nm thick, versus 60–80 nm for thin sections, many more details of cellular architecture can be resolved. Furthermore, by tracing the observed structures in serial slices in a process called hand-segmentation or modeling, it is possible to produce reconstructions with a 3-D resolution of 5 to 8 nm (Fig. 1C). The overall size of the reconstructions can be further increased by producing tomograms and models of serial sections (Fig. 1D), and

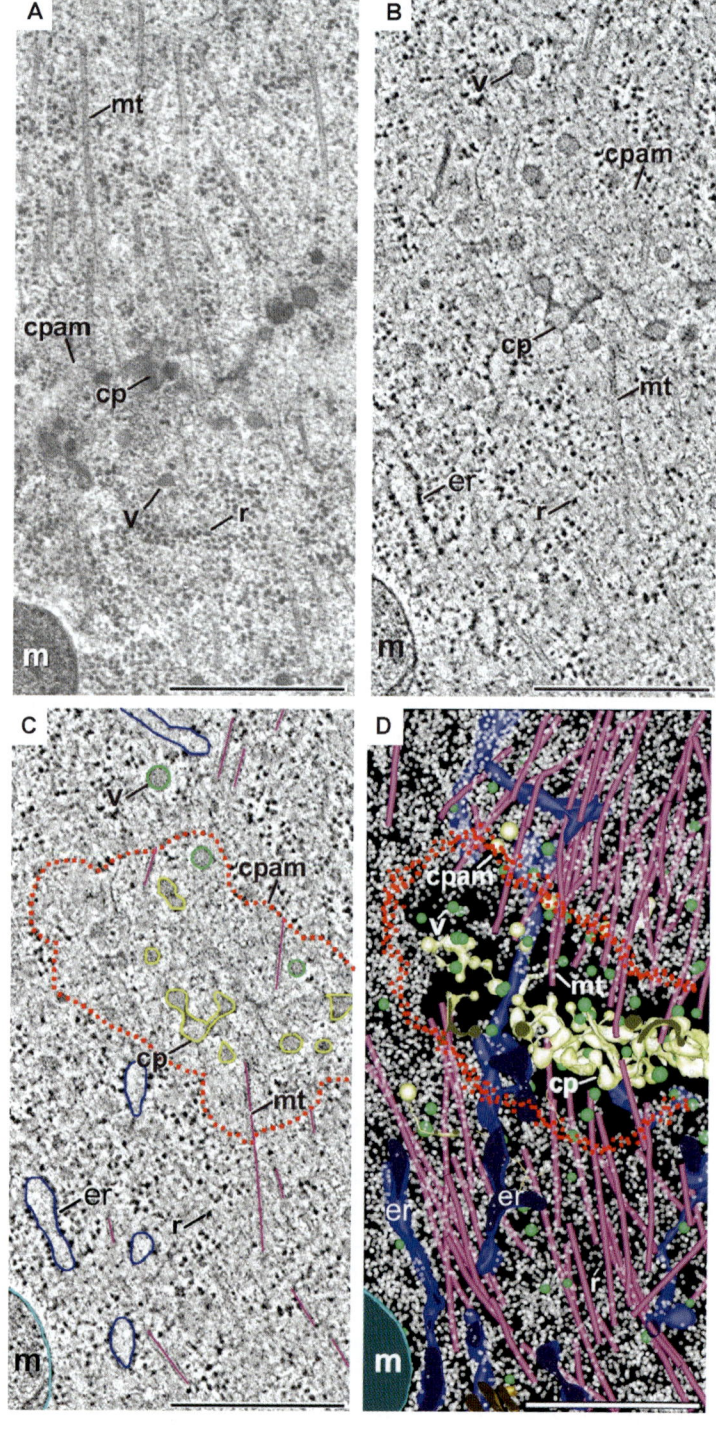

Fig. 1 Different steps in the tomographic reconstruction and modeling process, compared with conventional transmission electron micrographs. **A** Electron micrograph of a 100 nm-thick section of an Arabidopsis meristem cell at the solid phragmoplast stage of cytokinesis. Large organelles such as mitochondria (m), or structures with simple geometries such as microtubules (mt) are clearly resolved, but irregular membranous structures such as the cell plate (cp) often appear as dark, blurred spots. Due to the section thickness, the cell plate assembly matrix (cpam) can only be distinguished at certain regions around the cell plate. r: ribosome. **B** Two-nm-thick virtual slice of a tomogram from the same cell. The overall image is sharper, and the cell plate (cp) membranes, Golgi-derived vesicles (v) and endoplasmic reticulum (er) membranes are better resolved. The ribosome-excluding CPAM is now more clearly seen around the cell plate. **C** Modeling of the same tomographic slice. Each object within the tomographic volume is given a color, and the contours are manually traced around them across the whole set of tomographic slices. Each contour is then connected by a mesh to reconstruct the virtual surface of each object in the model. **D** Tomographic model. After object modeling and rendering, the whole set of objects can be visualized, giving a 3-D picture of their actual size, distribution and associations between them. The *scale bars* represents 500 nm

by then extending the reconstructions from one section to the next (Otegui and Austin 2006). Most of the tomography-based models presented in this review are based on serial section reconstructions.

3
Structure and Function of the Phragmoplast Cytoskeleton

3.1
Phragmoplast Microtubules Undergo Changes in Organization During Cell Plate Formation

Microtubules are the most visible component of the cytoskeletal elements involved in cytokinesis, such as the premitotic pre-prophase band (Fig. 2A) and the phragmoplast. They have been the subject of hundreds of studies of dividing plant cells, both at the light and the electron microscope level of analysis. Here we only review contributions made by electron microscopy to the study of the phragmoplast and cell plate of cryofixed cells.

Figures 2 and 3 illustrate the organization of the MT arrays associated with phragmoplast formation and maturation in *Arabidopsis* meristem cells (somatic-type cytokinesis). The two opposing sets of MTs in Fig. 3 are shown in different colors so that their spatial organization can be more readily discerned. The comparison between the late anaphase organization of the mitotic spindle MTs and the organization of the MTs at the time of phragmoplast initial assembly during very late anaphase (Fig. 3A and B) highlights how the phragmoplast initials arise from clusters of opposing sets of spindle MTs. The distinction between the two types of arrays is based on the presence of multiple CPAMs in the equatorial plane of the phragmoplast initials

Fig. 2 The different stages of somatic-type plant cytokinesis (see text for further details). **A** G2-M stage (premitotic). **B** Late mitotic anaphase – phragmoplast initials stage of cytokinesis. **C** Early telophase stage of caryokinesis – solid phragmoplast stage of cytokinesis. **D** Mid telophase stage of caryokinesis – transitional phragmoplast stage of cytokinesis. **E** Late telophase stage of caryokinesis – ring phragmoplast stage of cytokinesis. **F** G1 stage in the newly formed daughter cells. chr: chromosome; cpam: cell plate assembly matrix; cw: cell wall; db: dumbbell-shaped intermediate; gs: golgi stack; mt: microtubule; mvb: multivesicular body; n: nucleus; ne: nuclear envelope; nu: nucleolus; pd: plasmodesmata; pfs: planar fenestrated sheet cell plate; pgz: peripheral cell plate growth zone; pm: plasma membrane; ppb: pre-prophase band; tn: tubular network cell plate; tvn: tubulo-vesicular network cell plate; v: golgi-derived vesicle

(Fig. 2B; Seguí-Simarro et al. 2004). Formation of the initials starts when the chromosomes reach the pole regions and the nuclear envelope begins to reassemble. The largest MT arrays are associated with the solid phragmoplast stage of cell plate formation (Figs. 2C, 3C). Based on tomographic data, the

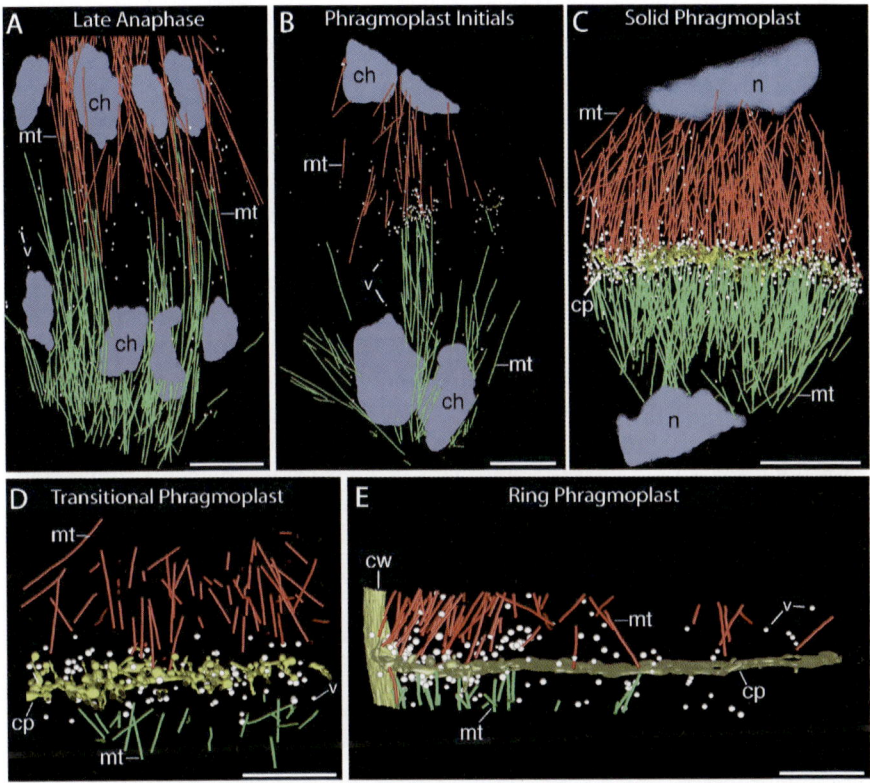

Fig. 3 Large volume tomographic reconstructions of different stages of mitosis and cytokinesis in Arabidopsis meristem cells. Based on Austin et al. (2005). **A** Late-anaphase (volume = $5.9 \times 5.9 \times 0.8$ mm^3), which shows two opposing sets of spindle microtubules (mt), *red* and *green*, above and between the migrating sister chomatids (ch), and vesicles (v). **B** Phragmoplast initials stage (vol = $5.9 \times 5.9 \times 0.8$ mm^3), in which MT arrays arrange parallel to the spindle axis, between the decondensing chromatin masses. Individual phragmoplast initials arise in the cell equator surrounded by clouds of cell plate forming vesicles. **C** Solid phragmoplast stage (vol = $4.3 \times 4.3 \times 1.2$ mm^3), which displays two dense sets of opposing MTs (mt) between the re-forming daughter nuclei (n). The growing cell plate (cp) is surrounded by numerous vesicles and is sandwiched between the two sets of MTs. **D** Transitional phragmoplast stage (vol = $2.8 \times 2.8 \times 0.8$ mm^3), which lacks a CPAM and displays many short MTs on both sides of the maturing cell plate. **E** Ring phragmoplast stage (vol = $5.9 \times 2.8 \times 0.9$ mm^3). Two dense sets of opposing MTs flank the growing edge of the cell plate. The more mature central cell plate region has fewer interacting MTs. The *scale bars* represents 1 μm

solid phragmoplasts in *Arabidopsis* meristem cells are comprised of 800 to 1,000 MTs (Austin and Seguí-Simarro, unpublished results).

A controversy has arisen recently over the question as to whether the phragmoplast MTs overlap in the equatorial plane or not. The unambiguous answer to this question, based on the analysis of cells preserved by HPS-FS methods, is that there is no systematic overlap between the MT (+)-ends in somatic-type cytokinesis phragmoplast arrays, but there is overlap between the MTs in the mini-phragmoplasts of endosperm cells (Otegui and Staehelin 2000b; Austin et al. 2005). The reason for this difference is unknown, but it could be related to the fact that the mini-phragmoplasts formed during syncytial-type cytokinesis contain only $\sim 2 \times 10$ MTs versus $\sim 2 \times 400$ to 500 in the solid phragmoplasts formed during somatic-type cytokinesis. Interdigitation of the MT ends might be required for the stabilization of the mini-phragmoplast MT assemblies, which have to remain operative during the simultaneous assembly of all of the syncytial cell walls, a process that would be expected to take more time than the assembly of a single cell plate and cell wall during somatic-type cytokinesis (Otegui and Staehelin 2000b).

Upon completion of the tubulo-vesicular stage of cell plate formation (see below), the solid phragmoplast MT array breaks down in conjunction with the disassembly of the CPAM. This stage is called the transitional stage (Figs. 2D, 3D), because it signals the transition from a solid phragmoplast to a ring phragmoplast type of MT organization. Formation of the ring phragmoplast MT arrays follows the reformation of the CPAM around the edges of the centrifugally expanding cell plate (Figs. 2E, 3E). The few MTs and secondary CPAMs that are seen to interact with the central cell plate region at this stage of development are focused on the remaining pores in the fenestrated sheet-type cell plate region (Seguí-Simarro et al. 2004), which have to be closed to produce a new cell wall (Fig. 2F).

3.2
Transport of Cell Plate-Forming Vesicles to the Future Site of Cell Plate Assembly Starts during Metaphase

In addition to illustrating the organization of the MTs in *Arabidopsis* meristem cells during the anaphase stage of mitosis and at different stages of somatic-type cytokinesis, Fig. 3 shows the positions of all of the Golgi-derived, cell plate-forming vesicles in the same samples. One of the unexpected discoveries to result from the analysis of these tomograms was that the transport of the vesicles towards the equatorial plane starts already during mitosis, i.e., much earlier than the onset of cytokinesis. Indeed, Golgi stacks accumulate at the equatorial plane in a belt-shaped manner as soon as in prometaphase (Seguí-Simarro and Staehelin 2006a). Initiation of this vesicle trafficking starts at metaphase and is largely completed by the end of the solid phragmoplast stage of cell plate formation. This explains how it is possible for a meristematic cell to produce a cell plate from $\sim 20\,000$ vesicles within less than 5 minutes of the onset of phragmoplast assembly.

In this context, the following simple calculations provide interesting insights into the rate of vesicle production and cisternal turnover by individual Golgi stacks during cytokinesis. With a diameter of $\sim 9\,\mu m$, a new cross wall in an *Arabidopsis* meristem cell requires the deposition of ~130 μm² of plasma membrane. Taking into account the dimensions of the Golgi-derived, cell plate-forming vesicles (~50 nm; Seguí-Simarro et al. 2004) and the mean number of Golgi stacks per cell (~35 during interphase and ~68 during G2 and mitosis; Seguí-Simarro and Staehelin 2006a), the assembly of a new cross wall would require that each (mitotic stage) Golgi stack produce ~240 vesicles for the new cell plate. However, because ~30% of the cell plate membrane is recycled via clathrin-coated vesicles during maturation (Seguí-Simarro et al. 2004), the actual number of Golgi vesicles required is ~310. Since each TGN cisterna fragments into ~35 secretory-type (cell plate forming) vesicles (unpublished results), the 310 vesicles correspond to ~8.8 cisterna equivalents. According to the live cell measurements of Ueda et al. (2003), it takes *Arabidopsis* cells 6 to 8 minutes to progress from metaphase through the solid phragmoplast stage of cytokinesis. This means that each Golgi has to deliver ~300 vesicles in <8 minutes to the cell plate forming region, which amounts to ~37 vesicles or 1 TGN cisterna equivalent per minute. In conclusion, this rough calculation suggests that during cytokinesis, each Golgi stack produces approximately one new cisterna per minute.

4
Structure and Function of the Cell Plate Assembly Matrix (CPAM)

4.1
The CPAM Defines the Site of Cell Plate Assembly

The discovery of the CPAM is an example of a cellular structure whose discovery can be traced to the use of HPF-FS techniques to preserve cells undergoing cytokinesis for electron microscope analysis. Although early electron microscope studies of chemically fixed dividing cells provided hints of a cell plate-associated matrix (reviewed in Gunning 1982), the first definitive evidence for the presence of a ribosome-excluding zone around growing regions of cell plates and associated MT ends was provided in an investigation of tobacco BY-2 cells preserved by HPF-FS (Samuels et al. 1995). In that study, a prominent CPAM was seen to encompass the tubulo-vesicular network-type cell plates, and this ribosome-excluding material was reported to abruptly disappear when the tubulo-vesicular cell plate matured into a tubular network-type of cell plate. A similar ribosome-excluding matrix was subsequently recognized in endosperm cells of *Arabidopsis* undergoing syncytial-type cell plate formation (Otegui and Staehelin 2000b). In particular, it was shown that the formation of the first membrane tubules from cell plate forming vesicles always occurred within a ribosome-free matrix, and that this matrix persisted until the wide tubular stage cell plate network was converted into a fenestrated sheet-type of cell plate.

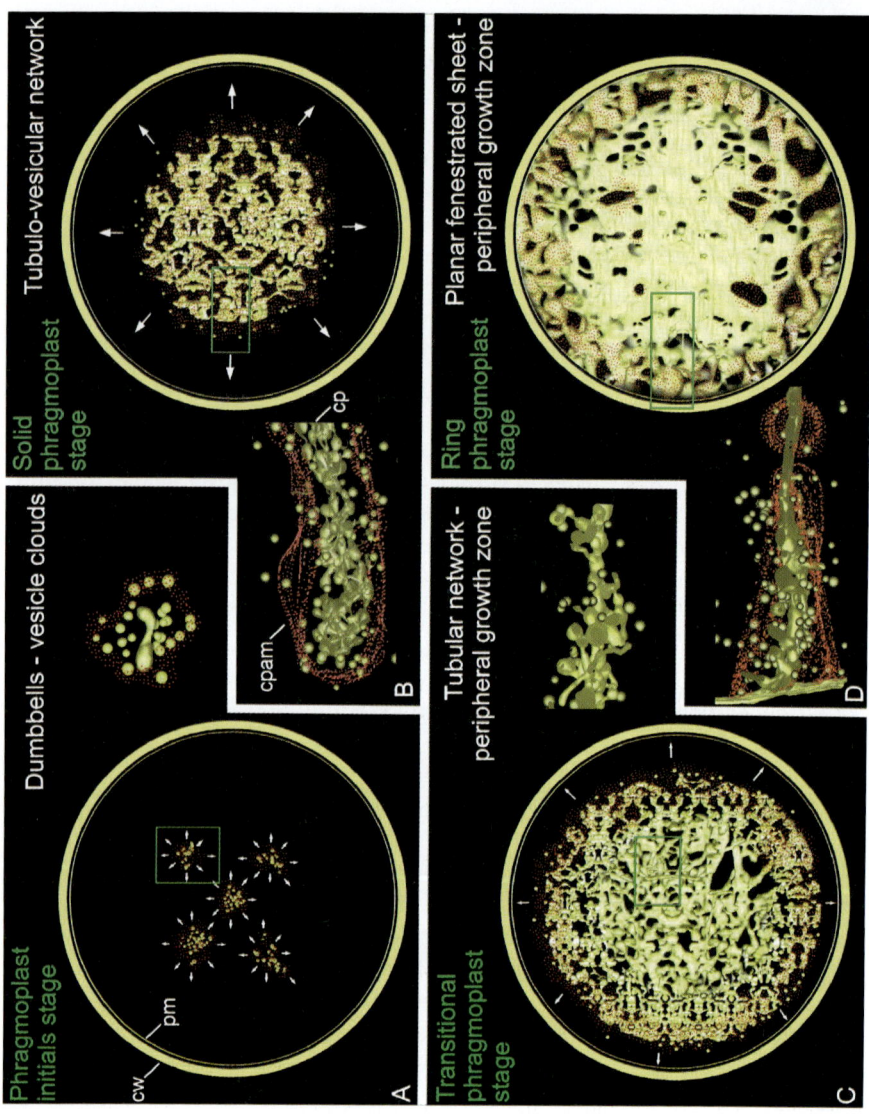

However, the full extent of the CPAM system and its importance for cell plate formation only became evident when the cryofixed dividing cells were analyzed by means of electron tomography (Otegui et al. 2001; Otegui and Staehelin 2004; Seguí-Simarro et al. 2004; Austin et al. 2005). Thus, as mentioned above, the definition of the phragmoplast initials during somatic type cytokinesis includes the formation of multiple CPAMs in the equatorial plane within which vesicles accumulate and the first vesicle fusion intermediates can be observed (Figs. 2B, 4A). The lateral enlargement of the phragmoplast

◄ **Fig. 4** Diagrams illustrating the structural changes in cell plate membrane and CPAM organization during somatic-type cytokinesis based on tomographic models. The cell plate membranes are depicted in *yellow*, the CPAM is outlined with *red dots*, and the cell plate is shown in a face-on view. The microtubules have been omitted to facilitate viewing of the cell plate and CPAM. **A** Phragmoplast initials stage. Different vesicle clouds concentrate within individual CPAMs at the central region of the cell equator. Within them, the first dumbbell-shaped cell plate intermediates are detected. Arrows indicate the direction of expansion of the phragmoplast initials. **B** Solid phragmoplast stage. The isolated early intermediates have now merged into a coherent tubulo-vesicular network cell plate (cp) surrounded by a single, cocoon-like CPAM (cpam). **C** Transitional phragmoplast stage. The cell plate transforms into a CPAM-free tubular network at its central domain, whereas the peripheral region actively expands outwards (*arrows*) surrounded by a ring-shaped CPAM. **D** Ring phragmoplast stage. The central cell plate region matures into a planar fenestrated sheet whereas the peripheral growth zone merges with the parental plasma membrane within the ring-shaped CPAM. Small, secondary CPAMs are created de novo around the larger perforations (fenestrae) of the planar sheet central cell plate. The *green boxes* outline the region of the cell plate shown in *insets* in cross-sectional views. cp: cell plate; cpam: cell plate assembly matrix; cw: cell wall; pm: plasma membrane

initials that yields the solid phragmoplast is paralleled by an expansion and merging of the individual CPAMs of the initials into a coherent CPAM across the entire width of the solid phragmoplast MT array (Seguí-Simarro et al. 2004), encompassing the nascent tubulo-vesicular cell plate (Figs. 2C, 4B). The breakdown and subsequent reassembly of the CPAM during the transitional and the ring phragmoplast stages of cell plate formation (Figs. 2D,E, 4C,D) highlights further the close structural and functional relationship between the CPAM, the distribution of MTs, and the growing, but not the maturing, cell plate domains.

Structurally, the CPAM can be defined as a fine filamentous matrix that excludes ribosomes to a distance of ∼160 nm from the cell plate in somatic cells, and ∼100 nm in syncytial cells. The first insights into the composition and the functional properties of the CPAM have come from a combination of electron tomography and immunolocalization studies. A list of putative components of the CPAM is presented in Table 1, and the functional importance of these molecules for cell plate formation is discussed in the following sections.

4.2
The CPAM Stabilizes MT (+)-Ends

Phragmoplast MTs originate from MT organizing centers (MTOCs) located in the vicinity of the nuclei (Vantard et al. 1990; Baskin 2000). This leads to the (+)-ends of the MTs being oriented towards the cell plate (Euteneuer et al. 1982). Prior to the discovery of the CPAM, it was reported that the region surrounding the cell plate, where the MT (+)-ends terminate, is critical for

Table 1 Putative components of the cell plate assembly matrix (CPAM)

Family/Type	Structures observed	Candidate molecule(s)	Localization	Function	References
Plant exocyst	Y/L-shaped tethers	AtSec5, AtSec6, AtSec15, AtExo70, AtExo84	Golgi-derived vesicles	Vesicle tethering	(Cvrckova et al., 2001).
Plant exocyst	Y-shaped tethers	AtSec3	CPAM	Vesicle tethering	(Seguí-Simarro et al., 2004; Bednarek and Falbel, 2002; Falbel et al., 2003).
Rab GTPases	—	???	CPAM matrix and cell plate	Vesicle tethering	
Rops (plant Rho-GTPases)	—	AtRop4, Rop1	CPAM matrix and cell plate	Vesicle tethering, formation of short actin filaments, regulation of callose synthase complexes	(Hong et al., 2001a; Molendijk et al., 2001; Bednarek and Falbel, 2002).
Syntaxins	—	KNOLLE, SNAP33, NSPN11, KEULE	Fusing vesicles	Vesicle fusion	(Lauber et al., 1997; Assaad et al., 2001; Zheng et al., 2002).
DRP1	Rings & springs	DRP1A, phragmoplastin	Cell plate intermediates	Membrane squeezing, stretching and stabilization	(Gu and Verma, 1996, 1997; Otegui et al., 2001; Kang et al., 2003; Seguí-Simarro et al., 2004).
DRP2	Rings & springs	DRP2A	Clathrin-coated buds	Formation of recycling clathrin-coated vesicles	(Verma, 2001; Konopka et al., 2006).
EB1	MT–cell plate linkers	AtEB1, AtMAP65-3	CPAM-embedded MT (+) ends	Anchoring of MTs to the cell plate	(Mathur et al., 2003; Muller et al., 2004; van Damme et al., 2004).
actin	Fine, 6 nm filaments	F-actin	CPAM matrix and cell plate	Multiple roles (see text)	(Endle et al., 1998; Otegui et al., 2001; Bednarek and Falbel, 2002; Otegui and Staehelin, 2004).
Long coiled-coil proteins	—	???	CPAM matrix	CPAM scaffold	Kang and Staehelin, submitted

MT stability (Inoué 1964). In particular, it was shown that irradiation of this proximal phragmoplast region with UV light not only led to MT depolymerization but also prevented reformation of a stable phragmoplast. In contrast, UV irradiation of the distal zone of the phragmoplast, which also caused MT breakdown, did not prevent the eventual reformation of the phragmoplast MT array (Inoué 1964). Further evidence for MT-cell plate interactions came from studies with isolated phragmoplasts in which the MTs were shown to remain connected to cell plate membranes throughout the cell disruption and the phragmoplast isolation and purification processes (Kakimoto and Shibaoka 1992).

Most cytoplasmic MTs are highly dynamic structures that can form de novo, grow and then rapidly break down (Baskin 2000). Based on fluorescence redistribution after photobleaching experiments, phragmoplast MTs have a turnover rate of $t_{1/2} = 60$ s (Hush et al. 1994). In vitro studies of growing and shrinking MTs in cryogenically fixed samples that were examined in the frozen state in a cryoelectron microscope have led to the correlation of MT (+)-end morphology with the dynamic state of the MT (Mandelkow et al. 1991; Chretien et al. 1995; Carvalho et al. 2003). In particular, extended MT ends are typical of growing MTs, ramshorn-like ends are seen on disassembling MTs, and blunt ends are characteristic of MTs in a metastable state.

The demonstration that the same types of MT-end morphologies can be visualized in intact plant cells preserved by HPF-FS has opened up the possibility of analyzing phragmoplast MT dynamics in such cells by quantifying the number of MTs exhibiting a specific type of MT (+)-end geometry. This approach has been employed to analyze the effects of CPAM association on MT (+)-end dynamics during cytokinesis of meristem cells of *Arabidopsis* (Austin et al. 2005). The highlights of that study were that most of the (+)-ends of the solid phragmoplast MTs terminate within the CPAM, and that of those MTs 50–70% exhibit a blunt geometry, which corresponds to a metastable state. In contrast, only 20% of the (+)-ends of MTs that were located outside

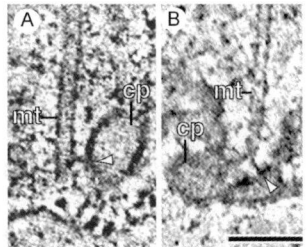

Fig. 5 Microtubule – cell plate linkers. ~30 nm long linker-like structures (*arrowheads*) physically connecting the microtubule (mt) blunt (+)-ends with cell plate membranes (cp), as revealed in properly oriented tomographic reconstructions. The *scale bar* represent 0.1 µm

of the CPAM were blunt ended. During the transitional phragmoplast stage, where the CPAM is absent from the central region of the cell plate, 80% of the (+)-ends exhibited a horned morphology, which is typical of disassembling MTs. These results suggest that the CPAM contains factors that stabilize MT (+)-ends and thereby prolong the life of the CPAM-associated MTs compared to those that terminate outside the CPAM. This stabilization, in turn, increases the efficiency of vesicle transport along the MTs to the growing cell plate regions.

To further characterize the spatial relationship between MT (+)-ends and the cell plate, we used our tomographic database to determine the distance between the ends of blunt MTs and the closest cell plate membrane (Austin et al. 2005). This quantitative analysis demonstrated that blunt-ended MTs often terminate 30 nm from the surface of a cell plate membrane, whereas no such relationship was detected for the ends of horned and extended MTs. This finding suggested that the (+)-ends of the metastable MTs could be physically connected through specific linkers to the cell plate membranes, and that these linkers could contribute to the stabilization of the blunt-ended MTs. As shown in Fig. 5A and B, the postulated 30 nm linkers can even be recognized in suitably oriented tomographic slice views. Yet to be determined is the molecular identity of the observed linkers, but two candidate proteins have been identified. In *Arabidopsis*, the AtEB1 protein is a (+)-end MT-associated protein (van Damme et al. 2004), which has been localized to the phragmoplast midline (Mathur et al. 2003). Since evidence from other systems indicates that EB1 can stabilize MT (+)-ends (Tirnauer et al. 2002), a similar role for AtEB1 in stabilizing cell plate-associated MTs seems possible. A second candidate is AtMAP65-3, an EB1-like protein (van Damme et al. 2004), which, when mutated, causes a widening of the MT-depleted midline region of the phragmoplast (Muller et al. 2004).

4.3
The CPAM Scaffold is Probably Composed of Long Coiled-Coil Proteins

As shown in Fig. 4, the CPAM completely encompasses the growing cell plate regions and is absent from the maturing cell plate domains (Seguí-Simarro et al. 2004). In meristem cells, the ribosome-excluding CPAM zone that encompasses the tubulo-vesicular cell plate membranes of the solid phragmoplast (Fig. 4B) is ~160 nm wide (Seguí-Simarro et al. 2004), whereas the CPAM surrounding the wide tubular network of syncytial-type cell plates is ~100 nm (Otegui et al. 2001). The identity of the scaffolding proteins of the CPAM is unknown, but in analogy to the composition of the ribosome-excluding COPII vesicle scaffold and the Golgi matrix (Kang and Staehelin, submitted), the most likely candidates are long coiled-coil-type proteins similar to Uso1, p115, and golgins (Gillingham and Munro 2003; Latijnhouwers et al. 2005). The *Arabidopsis* genome contains a total of 252 long coiled-coil proteins (Rose et al. 2004), and three of these proteins – AtUso1 (Kang and Staehelin, submitted), AtGRIP (Gilson et al. 2004), AtCASP (Renna et al. 2005) – have been shown to localize to budding COPI vesicles and/or to Golgi stacks (Golgi matrix). Yet to be determined is which type of long-coiled coil of protein produces the scaffolding structure of the CPAM.

In other systems, long coiled-coil proteins are recruited to specific membrane systems through activated GTP-binding proteins and lipid species, and this recruitment has been postulated to help each membrane compartment create and express its own identity (Munro 2004). Once assembled, the scaffold can recruit additional cytosolic proteins that facilitate the binding of specific vesicles to this structure, and catalyze the vesicle budding and vesicle fusion reactions that accompany the membrane transformation processes characteristic of each type of membrane compartment.

4.4
The CPAM is an Affinity Matrix that Assembles and Organizes the Molecules Required for Vesicle Tethering and Vesicle Fusion

The hypothesis that the CPAM recruits Golgi-derived vesicles, enzymes and regulatory molecules to the cell plate assembly site is based on the finding that specific types of molecular assemblies involved in cell plate formation localize exclusively to this structure. For example, it has been reported that many cell plate-forming vesicles both inside and outside the CPAM carry ~35 nm long, L-shaped appendages, whereas paired vesicles held together by Y-shaped linkers of the same length can only be detected within the CPAM region (Fig. 6; Seguí-Simarro et al. 2004). Based on their structural and functional similarities with the exocyst complexes of mammalian and yeast cells (Terbush et al. 1996, 2001; Hsu et al. 1998), both of which are involved in polarized secretion, it has been postulated that that the Y-shaped complexes are completed vesicle tethering complexes, and the L-shaped precursors that lack the final protein(s) needed for the tethering function (Seguí-Simarro et al. 2004). Since the completed tethers are seen only inside the CPAM, it is likely that the protein(s) needed to complete the tethers are also found only inside this matrix. Furthermore, because Sec3 is an exocyst complex protein that in other systems is recruited to target membranes (Finger et al. 1998), the plant homologue of Sec3 could be the subunit that actually gives rise to the Y-shaped tethering complexes within the CPAM. Once tethered by an exocyst complex, the vesicles are induced to fuse through interactions between proteins associated with v- and t-SNARE complexes (Waizenegger et al. 2000; see also Chap. 13). Several SNARE proteins have also been localized to the cell plate forming regions of phragmoplasts (Jürgens 2005a).

Putative homologues of all of the exocyst protein subunits, including Sec3, have been identified in the *Arabidopsis* genome (Cvrckova et al. 2001; Jurgens and Geldner 2002; Elias et al. 2003; Seguí-Simarro et al. 2004; Cole et al. 2005).

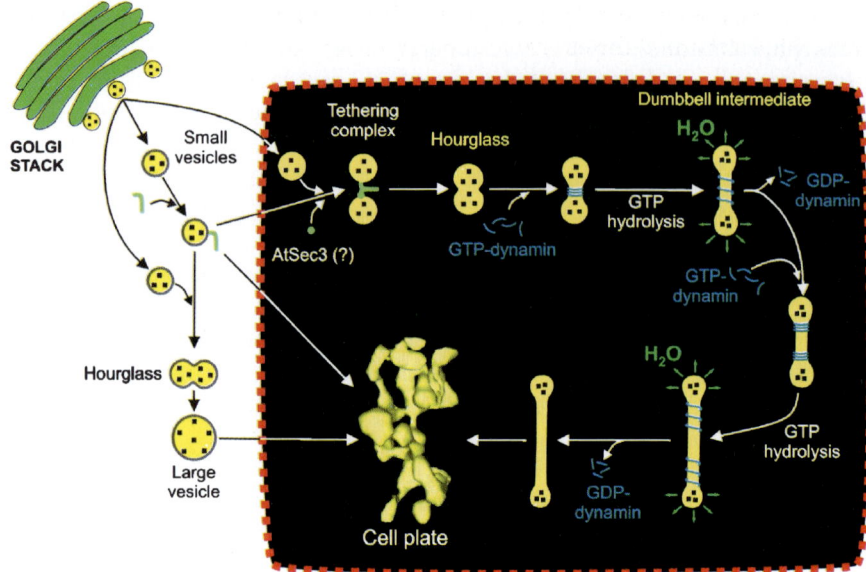

CELL PLATE ASSEMBLY MATRIX (CPAM)

Fig. 6 Diagrammatic model of the putative mechanisms of CPAM-mediated vesicle tethering and squeezing during cell plate growth. Based on Seguí-Simarro et al. (2004). Tethering of the Golgi-derived vesicles to each other within the CPAM is mediated by the assembly of Y-shaped, putative exocyst complexes. Once tethered and fused, dynamin (DRP) spirals assemble around the waist of the unstable hourglass intermediates, thereby stabilizing the hourglass configuration. Hydrolysis of the GTP bound to the DRP subunits causes the spirals to expand and squeeze the constricted domain of the hourglass-shaped vesicles, thereby converting them into dumbbell-shaped intermediates. The concomitant loss of vesicle volume is most likely achieved through water expulsion from the dumbbell lumen. After several rounds of assembly-expansion disassembly of the DRP springs, the volume of the dumbbells is reduced by up to 50%, and the concomitant loss of water appears to cause the contents of the vesicles (pectic polysaccharides and xyloglucans) to gel. This gelling, in turn, appears to mechanically stabilize the elongated shape of the dumbbells. When new vesicles fuse with the ends of the stabilized dumbbells, the latter do not swell, i.e., they retain their shape, thereby giving rise to increasingly complex vesicular-tubular-type cell plate intermediates. In contrast, when the same types of vesicles fuse outside of the CPAM, no DRP springs assemble around the necks of the hourglass intermediates and the hourglass structures are rapidly transformed into larger, round vesicles. These larger vesicles may also enter the CPAM at a later stage and fuse with the growing cell plate

Very recently, two studies have demonstrated not only the existence of an exocyst-like complex in plants, but also a functional role in polar exocytosis in two distantly related plant species, *Arabidopsis* (Cole et al. 2005) and maize (Wen et al. 2005). The maize rth1/sec3 mutant exhibits defects in root hair elongation as well as delayed plant growth and flowering (Wen et al. 2005). In

Arabidopsis, pollen from AtSec8 mutants display reduced pollen tube growth rates, while retaining the ability to complete both post-meiotic and pollen cytokinesis (Cole et al. 2005). These phenotypes are consistent with a role for Sec3 and Sec8 in polarized exocytosis not only during root hair and pollen tube expansion, but also during cell plate assembly and growth, since the expected phenotype for a mutated exocyst complex would be a loss of specificity in vesicle tethering and a concomitant slowdown in vesicle fusion, but not an inhibition of the fusion process itself.

In yeast and mammalian cells, regulation of the exocyst tethering complexes is dependent on Rho and Rab GTPases, and on short actin filaments (Hsu et al. 1999; Guo et al. 2001; Lipschutz and Mostov 2002; Wang et al. 2003). In plants, Rops (Rho-related GTPases of plants) have been localized to cell plates by means of GFP-tagged AtRop4 (Molendijk et al. 2001). Short actin filaments, in turn, have been shown to be present at the growing edges of cell plates (Endle et al. 1998), which is also where the CPAM is seen. Although there is currently no direct evidence localizing Rab-GTPases to cell plates, their importance for cell plate formation has been inferred from other studies (Bednarek and Falbel 2002; Falbel et al. 2003). Based on these results, we propose that the CPAM is the site where plant exocyst complex subunits, SNARE proteins, and molecules involved in the regulation of vesicle fusion are stored.

4.5
The CPAM is Enriched in Dynamin-Related Proteins that Assist in the Shaping of Cell Plate Assembly Intermediates, and the Budding of Clathrin-coated Vesicles

The first microscopic evidence of membrane fusion within the CPAM is the appearance of vesicles with a bilobed or hourglass-type of morphology (Fig. 6; Otegui and Staehelin 2000b; Seguí-Simarro et al. 2004). However, whereas these early, hourglass-shaped vesicle fusion intermediates are rarely encountered (3 out of >1000 vesicles examined; Seguí-Simarro et al. 2004), indicating that this is a short-lived intermediate membrane configuration, other more dumbbell-shaped vesicles with longer connecting necks are more frequently observed (Fig. 6; Samuels et al. 1995; Rensing et al. 2002; Seguí-Simarro et al. 2004). A distinguishing feature of these dumbbell-shaped vesicles is that the neck region is often surrounded by a more densely stained layer, which in the best tomographic samples can be resolved into 40 to 45 nm wide rings and spirals (Fig. 7A,A′; Otegui et al. 2001; Seguí-Simarro et al. 2004). Similar rings and spirals have been found to pinch the tubules of the wide tubular network of syncytial-type cell plates (Otegui et al. 2001), and to be present around the narrow tubular domains of the tubulo-vesicular network in somatic-type cell plates (Fig. 7B; Seguí-Simarro et al. 2004). Furthermore, the spirals exhibit two configurations, they can be tightly wound or they can be expanded (Fig. 6). These rings and spirals label with antibod-

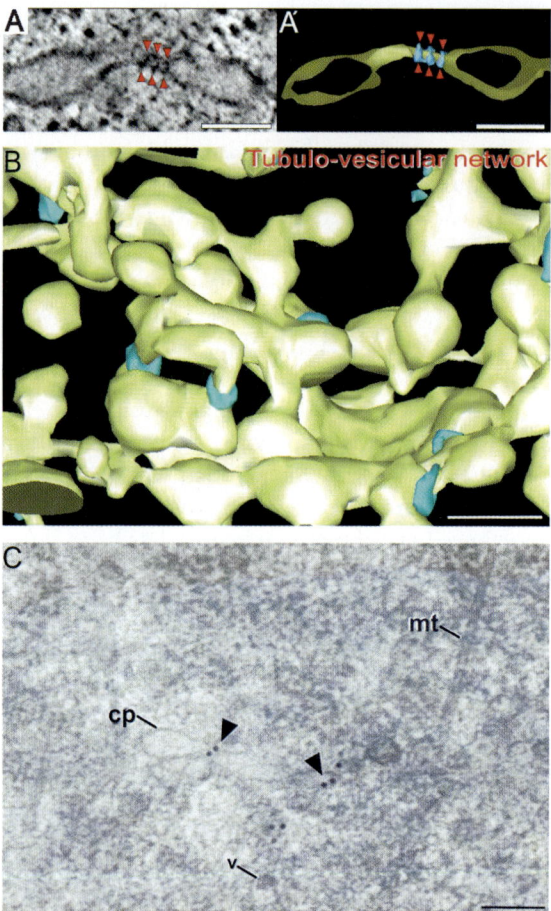

Fig. 7 DRP rings and spirals. Tomographic slice **A** and model **A'** of a spiral in an extended conformation, around the thin tubule that connects two vesicular cell plate domains. **B** Tomographic model of a tubulo-vesicular cell plate (solid phragmoplast stage) of Arabidopsis somatic cells showing several compacted rings around thin tubules. **C** Immunogold labeling with anti-DRP1A antibodies decorating a tubulo-vesicular somatic cell plate. Note that particles are mainly present at the narrow regions (likely tubules) of the cell plate. The *scale bars* represents 100 nm

ies against AtDRP1a (ADL1a), a dynamin-related protein that is present in *Arabidopsis* endosperm (Otegui et al. 2001) and meristem cells (Fig. 7C). Two GTP-binding, dynamin-related proteins (DRPs), phragmoplastin and DRP1a, have been identified in cell plates, and especially in the growing regions of the cell plates of both soybean and *Arabidopsis* (Gu and Verma 1996; Kang et al. 2001, 2003; Verma 2001). None of the adl1 mutants produced to date have shown any impairment in cytokinesis, most likely due to gene redundancy (Bednarek and Falbel 2002; Konopka et al. 2006).

In in vitro studies, dynamin has been shown to form spiral polymers around lipid bilayer tubules, with GTP-dynamin producing tight spirals and GDP-dynamin expanded ones (Stowell et al. 1999; Zhang et al. 2000). Based on these observations, it has been proposed that GTP hydrolysis causes the dynamin spirals to expand like a spring, going from a tight to an expanded spiral configuration (Fig. 6). The energy released during this conformational change appears to be used by cells to shape or fission membranes (Stowell et al. 1999; Hong et al. 2001b). For example, the dynamin spirals that are assembled around the necks of budding clathrin-coated vesicles have been postulated to break the neck of the vesicle buds and thereby create free vesicles (Stowell et al. 1999). In the cellularizing endosperm, the dynamin spirals within the CPAM wrap around and pinch but do not sever the wide cell plate tubules, a process that appears to be used to prevent the tubules from prematurely expanding into sheet-like structures (Otegui et al. 2001). In contrast, during somatic type cytokinesis, dynamin spirals are employed to convert the hourglass-shaped fused vesicles into dehydrated and mechanically stable, dumbbell-shaped vesicles (Fig. 6), which serve as the building blocks of somatic-type cell plates (Seguí-Simarro et al. 2004). The rationale for this type of vesicle transformation is discussed below. When the cell plate-forming vesicles fuse outside the CPAM, they appear to give rise to larger round vesicles and not to dumbbell-shaped vesicle intermediates (Seguí-Simarro et al. 2004). Only when vesicle fusion occurs inside the CPAM are dumbbell-shaped vesicles produced. This suggests that the CPAM serves as a scaffold for concentrating the dumbbell vesicle-forming DRPs in the vicinity of the cell plate, much like the accumulation of the proteins that complete the formation of the vesicle-tethering exocyst complexes.

The maturation of both syncytial and somatic-type cell plates also involves the budding of large numbers of clathrin-coated vesicles (Otegui et al. 2001; Otegui and Staehelin 2004; Seguí-Simarro et al. 2004; Seguí-Simarro and Staehelin 2006a), and this budding process too seems to be mediated by DRP rings and spirals. In fact, similar rings have been observed constricting the clathrin-coated vesicle buds that remove membrane from the maturing cell plates (Seguí-Simarro et al. 2004; Seguí-Simarro and Staehelin 2006a). The molecular nature of these latter rings is unknown, but DRP2A, another dynamin-related protein that has also been localized to cell plates is a likely candidate (Konopka et al. 2006). The presence of a pleckstrin homology domain and of a proline-rich motif (Hong et al. 2003) relates DRP2A to animal dynamins involved in membrane trafficking processes mediated by clathrin-coated vesicles.

5
Membrane Events Associated with Cell Plate Formation

5.1
The Generation of a Cell Plate From Vesicles and its Transformation into a Cell Wall Requires Many Coordinated Cellular Activities

Cell plates are the precursors of cell walls, and understanding how new cell walls are produced from Golgi-derived vesicles is central to the understanding of plant cell cytokinesis. Somatic-type cytokinesis starts with the appearance of the cell plate initials during late anaphase (Figs. 2B, 4A). During the subsequent growth and transformation of these initials, the cell plate accumulates not only new membrane, which is destined to give rise to new plasma membrane, but also large quantities of polysaccharides and proteoglycans, structural and enzymatic proteins, and diverse types of lipidic compounds to create a new cell wall (Whaley and Mollenhauer 1963; Verma 2001). Whereas some of the polysaccharides, such as pectins and hemicelluloses, are produced by enzymes in the Golgi apparatus and delivered in vesicles to the cell plate (Staehelin and Moore 1995; Sorensen et al. 2000), others like callose and cellulose are produced by enzymes located in the cell plate membrane and deposited directly into the cell plate lumen to give rise to the new cell wall (Samuels et al. 1995; Verma 2001). Upon fusion of a cell plate with the mother cell wall, mobile cellulose synthase rosette complexes translocate laterally within the plasma membrane from the old to the new plasma membrane regions while spinning out new cellulose microfibrils, which, together with the matrix polysaccharides, firmly anchor the new cross wall into the cell wall system as a whole (Brown et al. 1996).

During this carefully orchestrated cell plate assembly and maturation process, many logistic, mechanistic and biochemical problems have to be solved, including the timely delivery of the precursor molecules to the site of cell plate formation (Ueda et al. 2003), the control of the vesicle fusion processes to ensure that a planar cell plate and not a vacuole-like balloon is created (Seguí-Simarro et al. 2004), and the regulation of the timing of the onset of polysaccharide (e.g., callose) synthesis (Samuels et al. 1995) and of membrane recycling (Seguí-Simarro and Staehelin 2006a). All of these changes coincide with the appearance of characteristic cell plate intermediates, which are discussed in greater detail in the following sections.

5.2
Cell Plate Formation Requires Vesicular Building Blocks that do not Balloon when Fused Together

The fusion of spherical cytoplasmic vesicles typically leads to the formation of larger spherical vesicles. However, if a dividing cell wants to form a new cell wall, the cell plate-forming vesicles have to be coaxed into forming a planar membrane structure. The mechanism by which this is achieved involves exploiting the mechanical force of GTP-driven dynamin springs to deform and dehydrate freshly-fused, hourglass-shaped vesicles as well as the tubulo-vesicular membranes of growing cell plates (Figs. 6, 7; Seguí-Simarro et al. 2004). As mentioned above, the dynamin-like proteins appear to be confined to the CPAM, where all cell plate growth is seen to occur. The evidence supporting this hypothesis is as follows. Dumbbell-shaped vesicles with dynamin springs around their necks typically have a surface area that is equivalent to the surface area of two small cell plate-forming vesicles. Thus, the change in vesicle shape does not involve a change in membrane surface area. However, as the neck region of a dumbbell-shaped vesicle is extended, the diameter and the surface area of the bulbous ends is reduced. The simultaneous reduction in volume can amount to 50%, and this change in volume is most likely achieved by the "squeezing out" of water from the bulbous ends as the membrane is transiently stretched (Fig. 6). The concomitant dehydration of the vesicle contents cell wall forming pectic polysaccharides, hemicelluloses and arabinogalactan proteins (Staehelin and Moore 1995) – most likely causes gelling of the polysaccharides, especially of the esterified pectins (Zhang and Staehelin 1992; Thakur et al. 1997). Such a gelling reaction would be expected to mechanically stabilize the elongated vesicles, the new cell plate building blocks. Supporting the idea of mechanically stabilized building blocks is the observation that even as new vesicles fuse with the ends of the dumbbell vesicles, the fused structures do not balloon. The continued formation of dynamin springs around the tubular regions of the tubulo-vesicular cell plate membrane networks, and the resulting continual compression of the cell plate contents, appears to ensure that the cell plate retains a planar configuration during the rapid vesicle fusion stage of cell plate assembly.

5.3
Assembly of the Tubulo-Vesicular Cell Plate Membrane Network Occurs Exclusively Within the CPAM

Expansion and merging of the CPAM islands of the phragmoplast initials leads to the formation of a large and continuous CPAM that spans the width of the solid phragmoplast (Figs. 2C, 4B). This large CPAM defines where the phragmoplast MT ends are stabilized, where vesicle delivery occurs, where dumbbell-shaped vesicles are produced, and where cell plate growth occurs. The initial cell plate has a tubulo-vesicular-type membrane architecture (Figs. 2C, 4B, 7B). It is generated by the joining of the expanding tubulo-vesicular membrane structures that arise from the fusion of spherical vesicles with the dumbbell-shaped vesicles of the phragmoplast initials. Growth of the tubulo-vesicular network involves both fusion of individual vesicles with the network and the generation of new dumbbell-shaped vesicles that serve as new nucleators for cell plate assembly. The resulting structure is a plate-like, labyrinthic membranous network consisting of vesicular and cisternal domains interconnected by thin, short tubules (Figs. 4B, 7B).

One of the valuable aspects of electron tomography is the ability to quantify the structures that have been reconstructed. In the Seguí-Simarro et al. (2004) study, the tomographic data were used to calculate both the unit cell plate surface area and the unit cell plate volume changes during the different stages of cell plate assembly. Since publication of that study, we have been informed that there was an error in the algorithm used to calculate the volume of the objects with irregular morphologies (D. Mastronarde, personal communication). We have now recalculated all of the volumetric data of our tomographic database using the corrected algorithm (Fig. 8A). Compared to the original data (Fig. 11 of Seguí-Simarro et al. 2004), all of the recalculated cell plate volumes per μm^3 of tomogram volume are threefold higher. In terms of understanding cell plate assembly, the only significant change relates to the events associated with the assembly of the tubulo-vesicular network. In particular, the final tubulo-vesicular cell plate has a volume similar to the volume of all of the vesicles that contributed to its assembly as determined by the cell plate surface area calculations. This lack of cell plate volume reduction compared to the volume reduction of the dumbbells may be due to having only a fraction of the vesicles that contribute to cell plate growth converted into shrunken dumbbell-type intermediates prior to their fusion with the cell plate. Furthermore, because the synthesis of callose by cell plate located enzymes commences before the breakdown of the CPAM (Samuels et al. 1995), any cell plate volume reduction caused by the dynamin motor proteins may be masked by the accumulation of this polysaccharide in the cell plate lumen.

At the end of the tubulo-vesicular cell plate growth phase (Fig. 8B), the cell plate contains more than enough membrane to produce the new cell wall (Seguí-Simarro et al. 2004). This state of the cell plate appears to produce a maturation signal that results in the cessation of new vesicle delivery to the cell plate region by causing a breakdown of the CPAM (Figs. 2D, 4C) and the disassembly the solid phragmoplast MTs that guide the vesicles to that region (Figs. 2D, 3D; Samuels et al. 1995; Ueda et al. 2003; Seguí-Simarro et al. 2004; Austin et al. 2005; Szechynska-Hebda et al. 2006). As this breakdown occurs (transitional stage of cell plate assembly) the CPAM components are moved to the periphery of the tubulo-vesicular cell plate (Figs. 2D, 4C), and a new set of MTs become stabilized in a ring around the cell plate margins (Figs. 2E, 3E). In this manner, the new vesicles are directed to the now centrifugally growing cell plate regions. Mechanistically, centrifugal cell plate growth resembles growth of the initial cell plate in the solid phragmoplast, and this growth ceases when the cell plate tubules start fusing with the plasma membrane (Fig. 4D; Samuels et al. 1995; Seguí-Simarro et al. 2004).

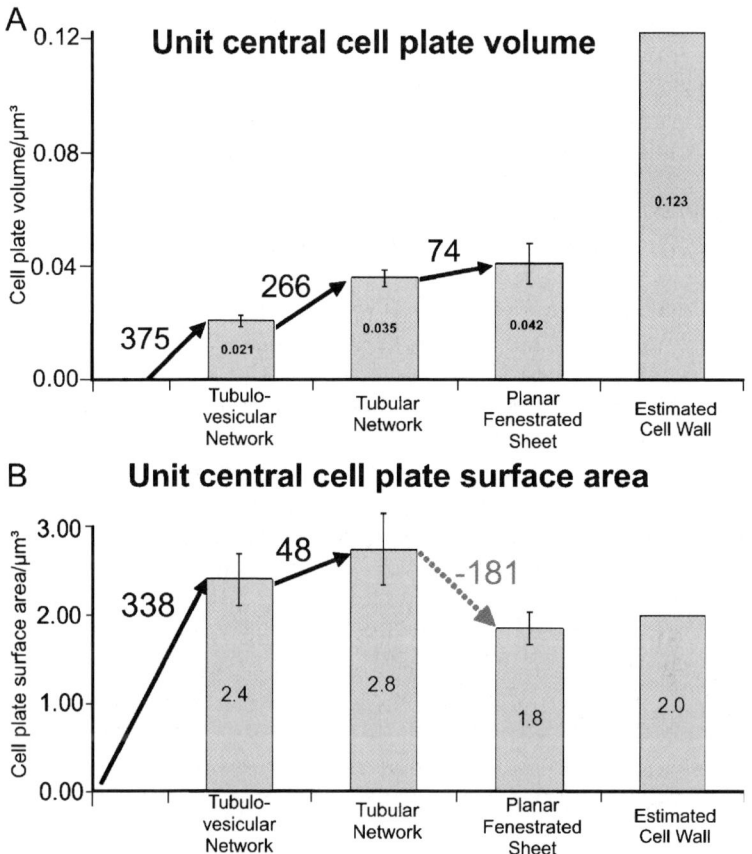

Fig. 8 Changes in cell plate volume and surface area during the different stages of somatic-type cytokinesis. The numbers in the bars indicate the mean cell plate volume (**A**) and surface area (**B**) per 1 µm^3 ± SD, and the numbers next to the *black arrows* reflect the net growth unit cell plate volume or surface area in terms of Golgi-derived vesicle equivalents. The *gray dotted arrow* indicates net loss changes in cell plate surface area in terms of clathrin-coated vesicle equivalents. As a reference, the estimated final cell wall volume and surface area values also are displayed

5.4
Maturation of a Tubulo-Vesicular-Type Cell Plate into a Tubular Network and a Sheet-like Cell Plate Requires Callose Synthesis and Membrane Recycling, but not a CPAM

Breakdown of the solid phragmoplast CPAM and of the associated MTs signals the end of tubulo-vesicular cell plate growth and the onset of cell plate maturation. The transformation of the growing tubulo-vesicular cell plate into a maturing tubular network cell plate manifests itself by the deposition

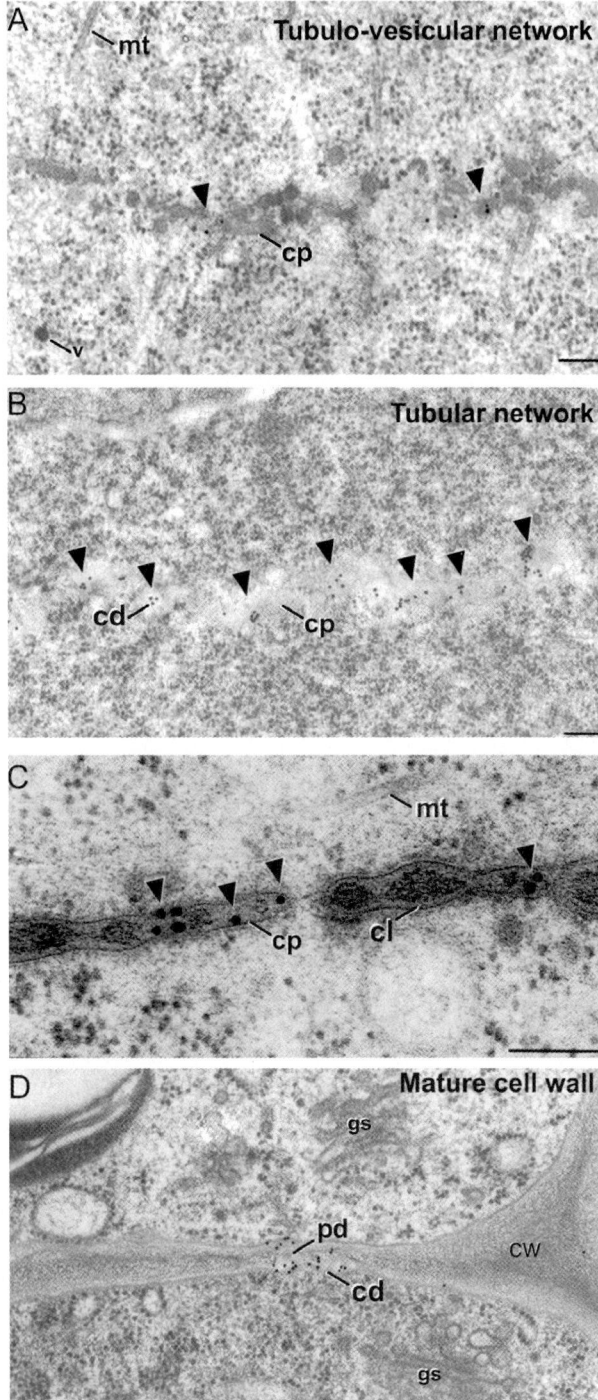

◀ **Fig. 9** Immunogold labeling of callose in the cell plate lumen at different stages of somatic-type cytokinesis. **A** tubulo-vesicular network cell plate. The few gold particles (*arrowheads*) are located mostly over discrete, lighter cell plate (cp) regions. **B** Tubular network. Immunolabeling is much more abundant over the callose deposits (cd) homogeneously distributed throughout the cell plate lumen. **C** Tubular network. Newly synthesized callose is deposited as a layer (cl) at the luminal side of the cell plate, surrounding the previous cell plate material (*darker contents*). **D** Mature cell wall. Massive deposition of cellulose fibrils replaces callose as the main constituent of mature cell walls (cw). Callose deposits (cd) are limited to the spaces around the plasmodesmata (pd). gs: Golgi stack; mt: microtubule; v: vesicle. The *scale bars* represents 100 nm

of copious amounts of callose in the cell plate lumen (compare Fig. 9A and B), the disappearance of the vesicular (spherical) cell plate domains (Fig. 4C), and the formation of increasing numbers of wide tubules that develop into fenestrated sheet-type cell plate domains (Fig. 4D).

The widening of the cell plate tubules into fenestrated sheets appears to be driven by the deposition of newly synthesized callose, a linear b-1,3-linked glucose polymer, into the cell plate lumen (Fig. 9B; Samuels et al. 1995). Callose synthases are large, membrane-spanning protein complexes that excrete their product directly into the cell plate lumen (Verma and Hong 2001), and as this happens, the callose is seen to spread over the membrane surface in a $\sim$15 nm thick layer that has been postulated to provide the force that drives tubule widening and fenestrated sheet formation (Fig. 9C; Samuels et al. 1995). At the end of this spreading phase, the cell plate resembles a planar sheet with numerous openings (fenestrae). Small, secondary CPAMs and associated clusters of MTs assemble de novo over and around these openings, guiding new vesicles to these regions to bring about closure of the gaps. However, where tubular strands of endoplasmic reticulum (ER) become trapped within closing fenestrae, those fenestrae never completely close (Fig. 2F). Instead, they are transformed into the first symplastic connections between the daughter cells, the primary plasmodesmata (Fengshan and Peterson 2001).

While callose synthesis is transforming the general architecture of the cell plate, the concomitant removal of cell plate membrane by clathrin-coated vesicles produces parallel changes in cell plate membrane composition (Samuels et al. 1995; Rensing et al. 2002; Seguí-Simarro et al. 2004; Seguí-Simarro and Staehelin 2006a). For example, upon completion of the rapid, vesicle-mediated growth phase, the need for vesicle tethering and vesicle fusion membrane components is greatly reduced, and thus these membrane proteins have to be removed to allow for maturation of the cell plate membrane to occur. Considering that the recycling of plasma membrane proteins is typically carried out by clathrin-coated pits and vesicles (Alberts et al. 2002), and that the cell plate membrane is a precursor form of the plasma membrane, the clathrin-coated vesicles seen budding from maturing cell plates are assumed to be involved in membrane recycling. The parallel increase of clathrin-coated vesicles and multivesicular bodies during cell plate maturation has been interpreted to suggest that these vesicles are recycled via multivesicular bodies (Seguí-Simarro and Staehelin 2006a).

The final transformation of the planar fenestrated sheet-type cell plate into a new primary cell wall (Fig. 9D) involves the progressive replacement of the callose deposits by cellulose fibrils (Kakimoto and Shibaoka 1992; Samuels et al. 1995; Otegui and Staehelin 2000b). Because the first cellulose molecules can be detected via cellobiohydrolase I binding during the tubular network phase of cell plate development (Samuels et al. 1995), and the large cellulose synthase complexes would not be able to translocate through such a network while spinning out cellulose microfibrils (Brown et al. 1996), it is possible that the first cell plate cellulose molecules are produced by a different isoform of the cellulose synthase enzyme. Upon deposition in the cell plate lumen/cell wall, the cellulose molecules and microfibrils are assumed to trigger the assembly of the primary cell wall by the incorporation of the already secreted hemicellulose molecules into a cellulose-hemicellulose network adjacent to each daughter plasma membrane. In turn, the formation of this cellulose-hemicellulose network could be the driving force behind the general reorganization of the pectic polysaccharides and the formation of the pectin-rich middle lamella (Carpita and McCann 2000).

5.5
Increasing Amounts of Endoplasmic Reticulum Progressively Associate with the Cell Plate During the Maturation Stages of Cytokinesis

Cell plate-associated ER membranes were already described and discussed in some of the earliest electron microscope studies of cytokinesis in plant cells (Porter and Caulfield 1958; Hepler and Newcomb 1967; Bajer 1968). However, the functional importance of these ER membranes in cytokinesis remains an enigma. After the large lytic vacuole, the ER is one of the principal stores of Ca^{2+} in plant cells (Sanders et al. 2002). Calcium, in turn, is known to regulate a number of cellular activities involved in cell plate formation, including vesicle fusion (Hepler et al. 1990; Hepler 1994), the dynamics of cytoskeletal elements (Hepler et al. 1990), callose synthase activity (Kakimoto and Shibaoka 1992; Verma and Hong 2001), and dynamin dynamics (Verma 2001), Thus, the cell plate-associated ER cisternae are generally assumed to function in the local regulation of free Ca^{2+} within the phragmoplast and to thereby control cell plate assembly activities.

By defining in more precise terms the organization of the ER membranes during cytokinesis, electron tomography studies of cryofixed cells have provided a means for narrowing the potential functions of the ER during different stages of cell plate formation. For example, it was suggested that the ER membranes regulate cell plate formation via control of the positioning and fusion of the cell plate forming vesicles (Hepler 1982). A prediction of this hypothesis is that ER plate-associated ER membranes should be very prominent during early stages of cell plate assembly when most vesicles are delivered to the forming cell plate. This is not the case. Instead, during the early stages of cytokinesis, ER membranes in the cell plate forming region are very sparse and the thin tubules are oriented at nearly right angles to the plane of the cell plate (Seguí-Simarro et al. 2004). Thus, it is unlikely that the ER is either involved in the positioning or the control of vesicle fusion during the initial stages of cell plate assembly.

Increasing numbers of ER cisternae begin to appear at about the same time that the deposition of callose in the cell plate lumen is observed, suggesting that ER cisternae could be involved in the control of the Ca^{2+}-activated callose synthases in the cell plate membranes (Samuels et al. 1995; Verma and Hong 2001). In addition, the later appearing ER cisternae most likely serve as templates for the formation of primary plasmodesmata (Lopez-Saez et al. 1966; Robards and Lucas 1990; Fengshan and Peterson 2001). Finally, the appearance of ER cisternal domains that are attached through linkers to late stage cell plate membranes is consistent with observations in other systems where membrane lipids are recycled by means of a lipid hopping mechanism (reviewed in Staehelin 1997; Seguí-Simarro et al. 2004). This hypothesis, however, has yet to be tested.

A recent study of living tobacco NT-1 cells stably expressing ER targeted GFP has provided experimental support for the tomography results showing that the cell plate associated ER increases during cytokinesis (Gupton et al. 2006). In particular, during late anaphase, when the chromosomes approach the spindle poles and the phragmoplast initials are being formed (Fig. 2B), the amount of ER membranes in the equatorial plane is very low, and only during mid telophase (transitional phragmoplast stage, Fig. 2D) is a dramatic increase in the cell plate-associated ER signal seen.

5.6
Syncytial- and Somatic-Type Cell Plate Assembly Involves Different Structural Intermediates but Similar Molecular Mechanisms

Cells such as nuclear endosperms (Fig. 10) and meiocytes (Fig. 11) that exhibit syncytial-type cytokinesis, do not couple nuclear and cytoplasmic division as somatic cells do. Prior to cytokinesis, these cells undergo several cycles of nuclear division, repositioning of the nuclei, and then the simultaneous development of multiple cell walls between sister and non-sister nuclei (Figs. 10A, 11A; Otegui and Staehelin 2000b; Brown and Lemmon 2001). In addition, no preprophase bands are formed in syncytial cells. Thus, although the syncytial- and somatic-type cytokinesis mechanisms utilize many of the same enzymatic and regulatory molecules and the process of cell plate assembly exhibits many commonalities (Jürgens 2005b), there are also important differences and adaptations of the cell plate assembly process that can only be understood in the spatial and temporal context in which the cell plates are assembled.

A. Nuclear cytoplasmic domains

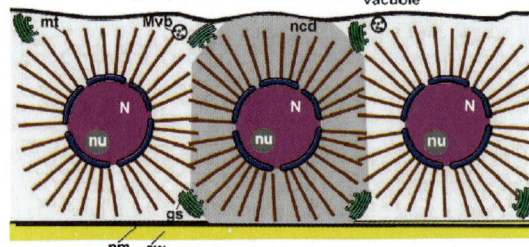

B. Mini-phragmoplast assembly-vesicle fusion

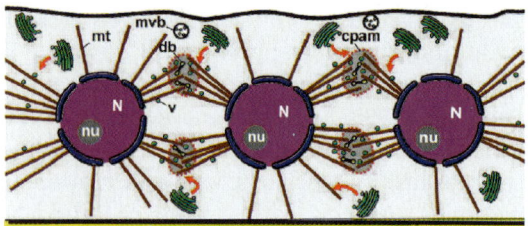

C. Wide tubular netwroks

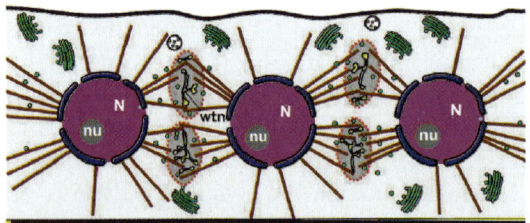

D. Convoluted sheets

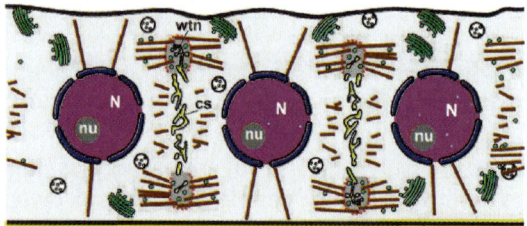

E. Fenestrated sheet

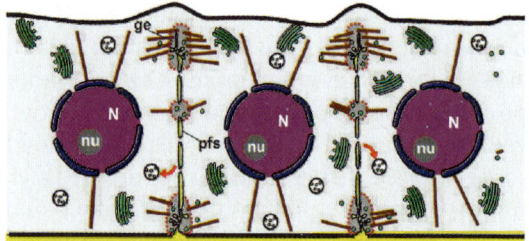

◄ **Fig. 10** Developmental stages of syncytial-type cytokinesis in the Arabidopsis endosperm. **A** Organization of the syncytial endosperm into nuclear cytoplasmic domains (ncd) defined by radial systems of microtubules (mt) prior to cytokinesis. **B** Mini-phragmoplast assembly at the boundaries of adjacent nuclear cytoplasmic domains and vesicle fusion at the division plane. **C** Formation of wide tubular networks (wtn) in multiple cell plate assembly sites. **D** Formation of a coherent cell plate by fusion of multiple assembly sites and maturation of the central cell plate domains into convoluted sheets (cs). **E** Fusion of the cell plate with the parental plasma membrane and conversion of convoluted sheets into planar fenestrated sheets (pfs). cpam: cell plate assembly matrix; cw: cell wall; db: dumbbell-shaped intermediate; ge: cell plate growing edge; gs: golgi stack; mvb: multivesicular body; n: nucleus; ne: nuclear envelope; nu: nucleolus; pm: plasma membrane; v: golgi-derived vesicle

Before syncytial-type cytokinesis occurs, microtubules assemble on MT-organizing centers associated with the outer membrane of the nuclear envelope (Canaday et al. 2000) and radiate outwards, intersecting the MTs extending from adjacent nuclei (Figs. 10A, 11A). The cytoplasmic region defined by a radial array of microtubules is called nuclear-cytoplasmic domain (Brown et al. 1994; Pickett-Heaps et al. 1999; Brown and Lemmon 2001). In male meiocytes (microsporocytes), four nuclear cytoplasmic domains corresponding to the four haploid nuclei are established, whereas up to several hundreds cytoplasmic domains are established in nuclear endosperms. Syncytial-type cell plates originate in CPAMs associated with mini-phragmoplasts (Figs. 10B, 11B; Otegui and Staehelin 2000b, 2004; Otegui et al. 2001). In structural terms, mini-phragmoplasts resemble phragmoplast initials. However, unlike the phragmoplast initials that arise from clusters of anaphase spindle MTs, mini-phragmoplast, MT clusters form at the boundaries of adjacent nuclear cytoplasmic domains (Brown and Lemmon, 1992; Pickett-Heaps et al. 1999). Each mini-phragmoplast contains on average 2×10 MTs, and the (+)-ends of these MTs overlap in the region where the CPAM develops (Otegui and Staehelin 2000b). In contrast, the number of MTs associated with the phragmoplast initials tends to be much higher and more variable, and MT overlap is only seen during the earliest stages of the spindle to phragmoplast transition period (Fig. 3A).

Another difference pertains to the architecture of the cell plate assembly intermediates and their mechanism of maturation. Thus, whereas somatic-type cytokinesis first involves the assembly of a single, coherent, tubulovesicular cell plate, which is rapidly followed by the centrifugal expansion along the cell plate periphery (Figs. 2, 4), syncytial type cell plate formation occurs simultaneously in all of the many mini-phragmoplasts that are aligned along the edges of each nuclear-cytoplasmic domain (Figs. 10B, 11B), and only at a later stage do the multiple cell plate assemblies merge into a coherent and continuous cell plate membrane system (Figs. 10D, 11D).

In the endosperm, just before cellularization begins (Fig. 10A), the nuclei are positioned in a peripheral cytoplasmic layer lining the plasma membrane

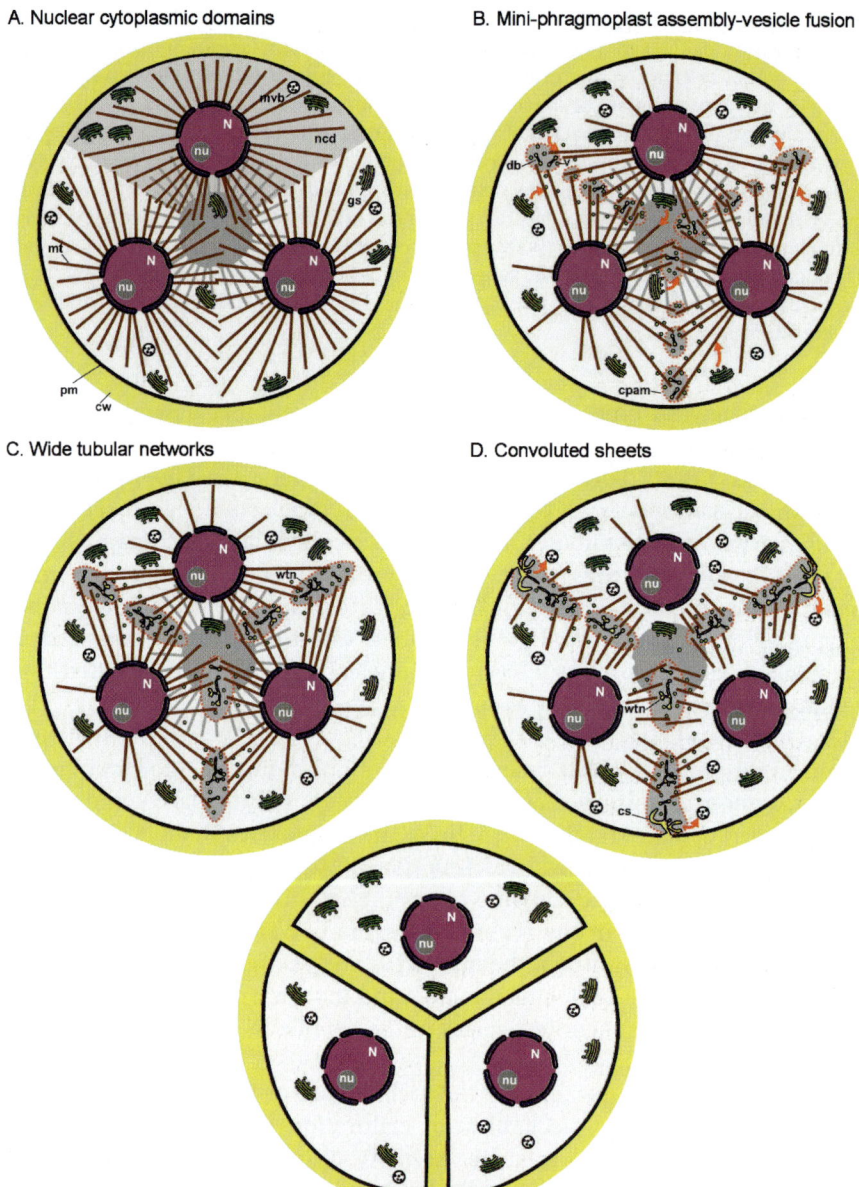

E. Fusion of cell plate with parental plasma membrane

and the central vacuole (or the developing embryo in the endosperm micropylar domain). Cellularization starts at the micropylar domain, where the embryo is located, with the formation of the mini-phragmoplasts (Fig. 10B). The multiple cell plate assemblies that arise between endosperm nuclei possess a wide tubular network type of membrane architecture (Fig. 10C) that

◀ **Fig. 11** Developmental stages of syncytial-type cytokinesis in Arabidopsis microsporocytes. **A** Organization of four nuclear cytoplasmic domains (ncd) after meiosis. Six division planes between the four nuclei are established. **B** Formation of multiple miniphragmoplasts and associated cell plate assembly sites across the entire division plane. **C** Assembly of multiple wide tubular netwoks (wtn). **D** Early fusion of the most peripheral cell plate assembly sites with the parental plasma membrane. These peripheral domains mature into convoluted sheets (cs) and stub-like projections by the accumulation of callose. **E** Upon completion of cytokinesis, six callose-rich cell walls with no plasmodesmata are formed. cpam: cell plate assembly matrix; cw: cell wall; db: dumbbell-shaped intermediate; gs: golgi stack; mt: microtubule; mvb: multivesicular body; n: nucleus; ne: nuclear envelope; nu: nucleolus; pm: plasma membrane

persists until all of the assemblies have merged into a coherent network system (Fig. 10D). The maintenance of the wide tubular network tubules in the endosperm of *Arabidopsis*, is presumably achieved by the transient association of dynamin-like rings and spirals with the tubules (Otegui et al. 2001). Only after all of the wide tubular networks have merged are the tubules converted into sheet-like cell plates and cell walls (Fig. 10E). The advantage for syncytial cells of having cell plate assembly intermediates with a wide tubular membrane network geometry instead of a tubulo-vesicular network/fenestrated sheet geometry as in meristematic cells, might be that the tubular network configuration provides more mechanical flexibility. Tubular membrane networks can readily stretch, twist and be compressed without undergoing major structural changes. As in the somatic-type cytokinesis system, the transformation of the syncytial-type wide tubular networks into convoluted and fenestrated sheets also seems to involve exploitation of the spreading forces associated with callose deposition along the cell plate membranes (Otegui and Staehelin 2000b). Cell plates forming in the thin layer of cytoplasm located between the plasma membrane and the tonoplast of the central vacuole fuse first with the parental plasma membrane and with the adjacent cell plates, whereas the cell plate edges facing the central vacuole keep on growing. This particular pattern of cellularization leads to the formation of open alveoli containing one nucleus each. Once this first set of anticlinal cell walls has formed, any additional cycles of nuclear division are coupled to cytokinesis and thus follow the somatic-type cytokinesis paradigm (Sorensen et al. 2002; Olsen 2004).

In cellularizing microsporocytes (Fig. 11A), the cell plate assembly sites form across the entire division plane (Fig. 11B). However, those wide tubular networks located in the cell periphery begin to fuse with the parental plasma membrane before the more central cell plate assembly sites become organized into a coherent cell plate (Fig. 11C). Fusion of the peripheral cell plate domains with the parental plasma membrane triggers a rapid accumulation of callose at these fusion sites, and the concomitant transformation of the wide tubular networks into stub-like projections that grow towards the center of the syncytium (Fig. 11D). As this occurs, the stub-like projections expand by fusing with the remaining cell plate assembly sites in the division plane until the four new callosic cell walls are completed (Fig. 11E). Interestingly, no dynamin rings have been identified in the wide tubular networks of the cellularizing microsporocytes.

Transformation of the syncytial-type wide tubular networks into fenestrated sheets also seems to involve exploitation of the spreading forces associated with callose deposition along the cell plate membranes as in the somatic-type cytokinesis systems (Otegui and Staehelin 2000b). However, unlike in somatic-type cytokinesis, the callose deposits produced in the syncytial walls persist after completion of the cell walls. There are several reasons why endosperm callose-rich walls might be advantageous for these cell types. For example, in contrast to cellulose, callose can be readily broken down into its glucose building blocks by specific cell wall enzymes (Verma and Hong 2001). This property makes the callose-rich endosperm cell walls suitable for use as carbohydrate reserves for developing embryos, as is the case in *Arabidopsis*, where most of the endosperm tissue is consumed by the embryo during seed development. In addition, callose has gel-like properties that can confer high plasticity to the endosperm cell walls, allowing for the spatial remodeling of the endosperm around the expanding embryo. In the case of the post-meiotic tetrad, the callose walls that break the symplastic connections and isolate the microspores from each other seems to be essential for the expression of the gametophytic genome without interference either from the parental sporophyte or from neighboring microspores (Shivanna et al. 1997). In addition, callose walls can be locally degraded to make room for the primexine layer, the blueprint of the exine, which is deposited on the outer surface of the microspores (Shivanna et al. 1997). Massive callose degradation by callose-specific hydrolytic enzymes (callases) secreted from the surrounding tapetal cells is the mechanism that permits the release of the microspores from the tetrad during maturation.

References

Alberts B, Johnson A, Lewis J, Raff M, Roberts K, Walter P (2002) Intracellular vesicular traffic. In: Alberts B, Johnson A, Lewis J, Raff M, Roberts K, Walter P (eds) Molecular Biology of the Cell. Garland Science, New York, pp 711–766

Assaad FF, Huet Y, Mayer U, Jürgens G (2001) The cytokinesis gene KEULE encodes a Sec1 protein that binds the syntaxin KNOLLE. J Cell Biol 152:531–543

Austin JR, Seguí-Simarro JM, Staehelin LA (2005) Quantitative analysis of changes in spatial distribution and plus-end geometry of microtubules involved in plant-cell cytokinesis. J Cell Sci 118:3895–3903

Bajer AS (1968) Fine structure studies on phragmoplast and cell plate formation. Chromosoma 24:383–417

Baskin TI (2000) The cytoskeleton. In Buchanan BB, Gruissem W, Jones RL (eds) Biochemistry and molecular biology of plants. American Society of Plant Physiologists, Rockville, pp 202–258

Bednarek SY, Falbel TG (2002) Membrane trafficking during plant cytokinesis. Traffic 3:621–629
Brown RC, Lemmon BE (1992) Pollen development in Orchids. 4. Cytoskeleton and ultrastructure of the unequal pollen mitosis in Phalenopsis. Protoplasma 167:183–192
Brown RC, Lemmon BE (2001) The cytoskeleton and spatial control of cytokinesis in the plant life cycle. Protoplasma 215:35–49
Brown RC, Lemmon BE, Olsen OA (1994) Endosperm development in barley: Microtubule involvement in the morphogenetic pathway. Plant Cell 6:1241–1252
Brown RM, Saxena IM, Kudlicka K (1996) Cellulose biosynthesis in higher plants. Trends Plant Sci 1:149–156
Canaday J, Stoppin-Mellet V, Mutterer J, Lambert AM, Schmit AC (2000) Higher plant cells: Gamma-tubulin and microtubule nucleation in the absence of centrosomes. Microsc Res Tech 49:487–495
Carpita NC, McCann MC (2000) The cell wall. In Buchanan BB, Gruissem W, Jones RL (eds) Biochemistry and molecular biology of plants. American Society of Plant Physiologists, Rockville, pp 52–108
Carvalho P, Tirnauer JS, Pellman D (2003) Surfing on microtubule ends. Trends Cell Biol 13:229–237
Chretien D, Fuller SD, Karsenti E (1995) Structure of growing microtubule ends – 2-dimensional sheets close into tubes at variable rates. J Cell Biol 129:1311–1328
Cole RA, Synek L, Zarsky V, Fowler JE (2005) SEC8, a subunit of the putative *Arabidopsis* exocyst complex, facilitates pollen germination and competitive pollen tube growth. Plant Physiol 138:2005–2018
Cvrckova F, Elias M, Hala M, Obermeyer G, Zarsky V (2001) Small GTPases and conserved signalling pathways in plant cell morphogenesis: From exocytosis to the exocyst. In: Geitmann A, Cresti M, Heath IB (eds) Cell biology of plant and fungal tip growth. IOS press, Amsterdam, pp 105–122
Elias M, Drdova E, Ziak D, Bavlnka B, Hala M, Cvrckova F, Soukupova H, Zarsky V (2003) The exocyst complex in plants. Cell Biol Int 27:199–201
Endle MC, Stoppin V, Lambert AM, Schmit AC (1998) The growing cell plate of higher plants is a site of both actin assembly and vinculin-like antigen recruitment. Eur J Cell Biol 77:10–18
Euteneuer U, Jackson WT, McIntosh JR (1982) Polarity of spindle microtubules in Haemanthus endosperm. J Cell Biol 94:644–653
Falbel TG, Koch LM, Nadeau JA, Segui-Simarro JM, Sack FD, Bednarek SY (2003) SCD1 is required for cell cytokinesis and polarized cell expansion in *Arabidopsis* thaliana. Development 130:4011–4024
Fengshan MA, Peterson CA (2001) Plasmodesmata: dynamic channels for symplastic transport. Acta Bot Sin 43:441–460
Finger FP, Hughes TE, Novick P (1998) Sec3p is a spatial landmark for polarized secretion in budding yeast. Cell 92:559–571
Gilkey JC, Staehelin LA (1986) Advances in ultra-rapid freezing for the preservation of cellular ultrastructure. J Electron Microsc Tech 3:177–210
Gillingham AK, Munro S (2003) Long coiled-coil proteins and membrane traffic. Biochim Biophys Acta 1641:71–85
Gilson PR, Vergara CE, Kjer-Nielsen L, Teasdale RD, Bacic A, Gleeson PA (2004) Identification of a Golgi-localised GRIP domain protein from *Arabidopsis* thaliana. Planta 219:1050–1056
Gu X, Verma DPS (1996) Phragmoplastin, a dynamin-like protein associated with cell plate formation in plants. EMBO J 15:695–704

Gu X, Verma DPS (1997) Dynamics of phragmoplastin in living cells during cell plate formation and uncoupling of cell elongation from the plane of cell division. Plant Cell 9:157–169

Gunning BES (1982) The cytokinetic apparatus: Its developmental and spatial regulation. In: Lloyd C (ed) The cytoskeleton in plant growth and development. Academic Press, London, pp 229–292

Guo W, Tamanoi F, Novick P (2001) Spatial regulation of the exocyst complex by Rho1 GTPase. Nat Cell Biol 3:353–360

Gupton SL, Collings DA, Allen NS (2006) Endoplasmic reticulum targeted GFP reveals ER organization in tobacco NT-1 cells during cell division. Plant Physiol Biochem 44:95–105

Hepler PK (1982) Endoplasmic reticulum in the formation of the cell plate and plasmodesmata. Protoplasma 111:121–133

Hepler PK (1994) The role of calcium in cell division. Cell Calcium 16:322–330

Hepler PK, Newcomb EH (1967) Fine structure of cell plate formation in the apical meristem of Phaseolus roots. J Ultrastruct Res 19:499–513

Hepler PK, Palevitz BA, Lancelle SA, McCauley MM, Lichtscheidl I (1990) Cortical Endoplasmic-Reticulum in Plants. J Cell Sci 96:355–373

Hong Z, Zhang Z, Olson JM, Verma DPS (2001a) A novel UDP-glucose transferase is part of the callose synthase complex and interacts with phragmoplastin at the forming cell plate. Plant Cell 13:769–779

Hong Z, Delauney AJ, Verma DPS (2001b) A cell plate specific callose synthase and its interaction with phragmoplastin. Plant Cell 13:755–768

Hong Z, Geisler-Lee CJ, Zhang Z, Verma DPS (2003) Phragmoplastin dynamics: multiple forms, microtubule association and their roles in cell plate formation in plants. Plant Mol Biol 53:297–312

Hsu SC, Hazuka CD, Foletti DL, Scheller RH (1999) Targeting vesicles to specific sites on the plasma membrane: the role of the sec6/8 complex. Trends Cell Biol 9:150–153

Hsu SC, Hazuka CD, Roth R, Foletti DL, Heuser J, Scheller RH (1998) Subunit composition, protein interactions, and structures of the mammalian brain sec6/8 complex and septin filaments. Neuron 20:1111–1122

Hush JM, Wadsworth P, Callaham DA, Hepler PK (1994) Quantification of microtubule dynamics in living plant cells using fluorescence redistribution after photobleaching. J Cell Sci 107:775–784

Inoué S (1964) Organization and function of the mitotic spindle. In: Allen RD, Kamiya N (eds) Primitive motile systems in cell biology. Academic Press, New York, pp 549–598

Jürgens G (2005a) Plant cytokinesis: fission by fusion. Trends Cell Biol 15:277–283

Jürgens G (2005b) Cytokinesis in higher plants. Annu Rev Plant Biol 56:281–299

Jurgens G, Geldner N (2002) Protein secretion in plants: From the trans-Golgi network to the outer space. Traffic 3:605–613

Kakimoto T, Shibaoka H (1992) Synthesis of polysaccharides in phragmoplasts isolated from tobacco BY-2 cells. Plant Cell Physiol 33:353–361

Kang B-H, Busse JS, Bednarek SY (2003) Members of the *Arabidopsis* dynamin-like gene family, ADL1, are essential for plant cytokinesis and polarized cell growth. Plant Cell 15:899–913

Kang BH, Busse JS, Dickey C, Rancour DM, Bednarek SY (2001) The *Arabidopsis* cell plate-associated dynamin-like protein, ADL1Ap, is required for multiple stages of plant growth and development. Plant Physiol 126:47–68

Konopka CA, Schleede JB, Skop AR, Bednarek SY (2006) Dynamin and cytokinesis. Traffic 7:239–247

Latijnhouwers M, Hawes C, Carvalho C (2005) Holding it all together? Candidate proteins for the plant Golgi matrix. Curr Opin Plant Biol 8:632–639

Lauber MH, Waizenegger I, Steinmann T, Schwarz H, Mayer U, Hwang I, Lukowitz W, Jürgens G (1997) The *Arabidopsis* KNOLLE protein is a cytokinesis-specific syntaxin. J Cell Biol 139:1485–1493

Lipschutz JH, Mostov KE (2002) Exocytosis: The many masters of the exocyst. Curr Biol 12:R212–R214

Lopez-Saez JF, Gimenez-Martin G, Risueño MC (1966) Fine structure of the plasmodesmata. Protoplasma 61:81–84

Mandelkow EM, Mandelkow E, Milligan RA (1991) Microtubule dynamics and microtubule caps – a time-resolved cryoelectron microscopy study. J Cell Biol 114:977–991

Mathur J, Mathur N, Kernebeck B, Srinivas BP, Hulskamp M (2003) A novel localization pattern for an EB1-like protein links microtubule dynamics to endomembrane organization. Curr Biol 13:1991–1997

McIntosh R, Nicastro D, Mastronarde D (2005) New views of cells in 3D: An introduction to electron tomography. Trends Cell Biol 15:43–51

Molchan TM, Valster AH, Hepler PK (2002) Actomyosin promotes cell plate alignment and late lateral expansion in Tradescantia stamen hair cells. Planta 214:683–693

Molendijk AJ, Bischoff F, Rajendrakumar CS, Friml J, Braun M, Gilroy S, Palme K (2001) *Arabidopsis* thaliana Rop GTPases are localized to tips of root hairs and control polar growth. EMBO J 20:2779–2788

Muller S, Smertenko A, Wagner V, Heinrich M, Hussey PJ, Hauser MT (2004) The plant microtubule-associated protein AtMAP65-3/PLE is essential for cytokinetic phragmoplast function. Curr Biol 14:412–417

Munro S (2004) Organelle identity and the organization of membrane traffic. Nat Cell Biol 6:469–472

Olsen OA (2004) Nuclear endosperm development in cereals and *Arabidopsis* thaliana. Plant Cell 16:S214–S227

Otegui MS, Austin JR (2006) Visualization of membrane-cytoskeletal interactions during plant cytokinesis. In: McIntosh JR (ed) Cellular electron microscopy. Elsevier Academic Press, San Diego CA, pp 222–241

Otegui MS, Staehelin LA (2000a) Cytokinesis in flowering plants: more than one way to divide a cell. Curr Opin Plant Biol 3:493–502

Otegui MS, Staehelin LA (2000b) Syncytial-type cell plates: a novel kind of cell plate involved in endosperm cellularization of *Arabidopsis*. Plant Cell 12:933–947

Otegui MS, Staehelin LA (2004) Electron tomographic analysis of post-meiotic cytokinesis during pollen development in *Arabidopsis* thaliana. Planta 218:501–515

Otegui MS, Mastronarde DN, Kang BH, Bednarek SY, Staehelin LA (2001) Three-dimensional analysis of syncytial-type cell plates during endosperm cellularization visualized by high resolution electron tomography. Plant Cell 13:2033–2051

Pickett-Heaps JD, Gunning BES, Brown RC, Lemmon BE, Cleary AL (1999) The cytoplast concept in dividing plant cells: Cytoplasmic domains and the evolution of spatially organized cell division. Am J Bot 86:153–172

Porter KR, Caulfield JB (1958) The formation of the cell plate during cytokinesis in Allium cepa L. In: Proceedings of the fourth International Congress on Electron Microscopy. Berlin, pp 503–507

Renna L, Hanton SL, Stefano G, Bortolotti L, Misra V, Brandizzi F (2005) Identification and characterization of AtCASP, a plant transmembrane Golgi matrix protein. Plant Mol Biol 58:109–122

Rensing KH, Samuels AL, Savidge RA (2002) Ultrastructure of vascular cambial cell cytokinesis in pine seedlings preserved by cryofixation and substitution. Protoplasma 220:39-49

Robards AW, Lucas WJ (1990) Plasmodesmata. Annu Rev Plant Physiol Plant Mol Biol 41:369-419

Rose A, Manikantan S, Schraegle SJ, Maloy MA, Stahlberg EA, Meier I (2004) Genome-wide identification of *Arabidopsis* coiled-coil proteins and establishment of the ARABI-COIL database. Plant Physiol 134:927-939

Samuels AL, Giddings TH Jr, Staehelin LA (1995) Cytokinesis in tobacco BY-2 and root tip cells: a new model of cell plate formation in higher plants. J Cell Biol 130:1345-1357

Sanders D, Pelloux J, Brownlee C, Harper JF (2002) Calcium at the Crossroads of Signaling. Plant Cell 14:S401-417

Seguí-Simarro JM, Staehelin LA (2006a) Cell cycle-dependent changes in Golgi stacks, vacuoles, clathrin-coated vesicles and multivesicular bodies in meristematic cells of *Arabidopsis* thaliana: a quantitative and spatial analysis. Planta 223:223-236

Seguí-Simarro JM, Staehelin LA (2006b) Mechanisms of cytokinesis in flowering plants: new pieces for an old puzzle. In: Teixeira da Silva JA (ed) Floriculture, ornamental and plant biotechnology: Advances and topical issues. Global Science Books, London, pp 185-196

Seguí-Simarro JM, Austin JR, White EA, Staehelin LA (2004) Electron tomographic analysis of somatic cell plate formation in meristematic cells of *Arabidopsis* preserved by high-pressure freezing. Plant Cell 16:836-856

Shivanna KR, Cresti M, Ciampolini F (1997) Pollen development and pollen-pistil interaction. In: VK Sawhney (ed) Pollen biotechnology for crop production and improvement. Cambridge University Press, Cambridge UK, pp 15-39

Sorensen MB, Mayer U, Lukowitz W, Robert H, Chambrier P, Jurgens G, Somerville C, Lepiniec L, Berger F (2002) Cellularisation in the endosperm of *Arabidopsis* thaliana is coupled to mitosis and shares multiple components with cytokinesis. Development 129:5567-5576

Sorensen SO, Pauly M, Bush M, Skjot M, McCann MC, Borkhardt B, Ulvskov P (2000) Pectin engineering: Modification of potato pectin by in vivo expression of an endo-1,4-beta-D-galactanase. Proc Natl Acad Sci USA 97:7639-7644

Staehelin LA (1997) The plant ER: a dynamic organelle composed of a large number of discrete functional domains. Plant J 11:1151-1165

Staehelin LA, Moore I (1995) The plant Golgi apparatus – Structure, functional organization and trafficking mechanisms. Annu Rev Plant Physiol Plant Mol Biol 46:261-288

Stowell MHB, Marks B, Wigge P, McMahon HT (1999) Nucleotide-dependent conformational changes in dynamin: evidence for a mechanochemical molecular spring. Nat Cell Biol 1:27-32

Szechynska-Hebda M, Wedzony M, Dubas E, Kieft H, van Lammeren A (2006) Visualisation of microtubules and actin filaments in fixed BY-2 suspension cells using an optimised whole mount immunolabelling protocol. Plant Cell Rep 25:758-766

Terbush DR, Guo W, Dunkelbarger S, Novick P (2001) Purification and characterization of yeast exocyst complex. Methods Enzymol 329:100-110

TerBush DR, Maurice T, Roth D, Novick P (1996) The Exocyst is a multiprotein complex required for exocytosis in Saccharomyces cerevisiae. EMBO J 15:6483-6494

Thakur BR, Singh RK, Handa AK (1997) Chemistry and uses of pectin – A review. Crit Rev Food Sci Nutr 37:47-73

Tirnauer JS, Grego S, Salmon ED, Mitchison TJ (2002) EB1-microtubule interactions in Xenopus egg extracts: Role of EB1 in microtubule stabilization and mechanisms of targeting to microtubules. Mol Biol Cell 13:3614–3626

Ueda K, Sakaguchi S, Kumagai F, Hasezawa S, Quader H, Kristen U (2003) Development and disintegration of phragmoplasts in living cultured cells of a GFP::TUA6 transgenic *Arabidopsis* thaliana plant. Protoplasma 220:111–118

van Damme D, Bouget FY, van Poucke K, Inze D, Geelen D (2004) Molecular dissection of plant cytokinesis and phragmoplast structure: a survey of GFP-tagged proteins. Plant J 40:386–398

Vantard M, Levilliers N, Hill AM, Adoutte A, Lambert AM (1990) Incorporation of Paramecium axonemal tubulin into higher plant cells reveals functional sites of microtubule assembly. Proc Natl Acad Sci USA 87:8825–8829

Verma DP, Hong Z (2001) Plant callose synthase complexes. Plant Mol Biol 47:693–701

Verma DPS (2001) Cytokinesis and building of the cell plate in plants. Annu Rev Plant Physiol Plant Mol Biol 52:751–784

Waizenegger I, Lukowitz W, Assaad F, Schwarz H, Jürgens G, Mayer U (2000) The *Arabidopsis* KNOLLE and KEULE genes interact to promote vesicle fusion during cytokinesis. Curr Biol 10:1371–1374

Wang HY, Tang X, Balasubramanian MK (2003) Rho3p regulates cell separation by modulating exocyst function in Schizosaccharomyces pombe. Genetics 164:1323–1331

Wen TJ, Hochholdinger F, Sauer M, Bruce W, Schnable PS (2005) The roothairless1 gene of maize encodes a homolog of sec3, which is involved in polar exocytosis. Plant Physiol 138:1637–1643

Whalen SL, Valster AH, Hepler PK (2001) A role for myosin in the alignment of the cell plate during cytokinesis in Tradescantia stamen hair cells. Mol Biol Cell 12:294a–294a

Whaley WG, Mollenhauer HH (1963) The Golgi apparatus and cell plate formation. A postulate. J Cell Biol 17:216–221

Zhang GF, Staehelin LA (1992) Functional compartmentation of the Golgi apparatus of plant cells: Immunocytochemical analysis of high-pressure frozen- and freeze-substituted sycamore maple suspension culture cells. Plant Physiol 99:1070–1083

Zhang Z, Hong Z, Verma DP (2000) Phragmoplastin polymerizes into spiral coiled structures via intermolecular interaction of two self-assembly domains. J Biol Chem 275:8779–8784

Zheng HY, Bednarek SY, Sanderfoot AA, Alonso J, Ecker JR, Raikhel NV (2002) NPSN11 is a cell plate-associated SNARE protein that interacts with the syntaxin KNOLLE. Plant Physiol 129:530–539

Vesicle Traffic at Cytokinesis

Anton Sanderfoot

Department of Plant Biology, University of Minnesota-Twin Cities,
250 Biological Sciences Center, 1445 Gortner Ave, Saint Paul, MN 55108, USA
sande099@umn.edu

Abstract Cytokinesis, the mechanical process by which one cell becomes two, is essential to all life, be it single-celled bacteria or complex multicellular organisms like land plants or animals. It is now clear that new membrane must be added to the mother cell to accomplish the complete enclosure of the two daughters. In land plants, this is especially apparent because they accomplish cytokinesis with the cell plate, a structure newly assembled by addition of new membrane vesicles. This review will focus on what we know about the source of the vesicles that make up the cell plate membrane, the molecular machinery involved in assembling those vesicles, and just what those vesicles are carrying to the cell plate.

1
Making Two Cells from One

Cytokinesis is essential to the propagation of all life, be it single-celled bacteria or complex multicellular organisms like land plants or animals. Fundamentally, the binary fission of bacteria is analogous to the morphologically complex cell-plate mediated division of the land plants, since each accomplishes the division of the cytosol by addition of new membrane between the two daughters. The only difference being that the membrane is derived from the "outside-in" movement of the plasma membrane (PM) during binary fission (e.g. the "cleavage furrow"), while the cell plate is considered to function in an "insideout" movement of the expanding nascent cell plate towards the membrane (though see Cutler and Ehrhardt 2003). As more is learned about these processes in each kind of organism, the "analogy" becomes more replaced by "homology", at least in terms of the role of the vesicle trafficking machinery in the eukaryotic processes of cytokinesis.

It is in the land plants that one can most easily observe the role of vesicle trafficking in cell division, since the cell plate is obviously constructed by fusion of vesicles derived from the secretory pathway (Jürgens 2005). However, based on both mathematics and biochemistry, new membrane must be added in all forms of cytokinesis. For example, volume (i.e. the cell contents) scales with the cube of the radius, while surface area (i.e. the membrane) scales with the square. Basically, assuming that the mother cell gives each daughter half the cell volume, half of the surface area of the mother cell is not sufficient to cover the half of the volume given to each daughter cell. The apparent "stretching" that accompanies cells which use a cleavage furrow may appear to temporarily compensate, yet a "stretched" membrane is a leaky membrane. Since a cell cannot be allowed to leak, cells must also add new membrane at the point of cytokinesis (Xu et al. 2002). Recent investigations into the vesicle trafficking machinery has shed light upon the mechanisms of cytokinesis, and this review will attempt to incorporate this research into a better understanding of the components and functions of the secretory organelles and their associated proteins in the assembly of the cell plate in land plants. First, I will look into the source of the vesicles that make up the cell plate. Second I will try to identify what is known about the molecular machinery that turns these vesicles into the cell plate membrane. Finally, we will discuss what is known about the contents of the vesicles, and how the vesicle contents work with the new membrane to produce the organelle that separates the daughter cells.

2
The Secretory Organelles and the Source of the Cell Plate Membrane

The cell plate is essentially a new secretory organelle that is formed during telophase from membrane derived from other organelles of the secretory pathway. The major source of this membrane has always been thought to be vesicles that bud from the Golgi, though certainly other organelles contribute membrane as well (see below). It is generally thought that the cell plate was invented in the algal ancestor to the land plants, and organelles with similar morphologies are observed to be involved in the cytokinetic process of several green algae, including those considered to be the sisters to the land plant lineage (see López-Bautista et al. 2003). On the other hand, many other green algae have a cytokinetic process superficially similar to the cleavage furrow-mediated mechanisms of animal cells (e.g. *Chlamydomonas* and related chlorophytes). In turn, a structure similar to the cell plate is found in some brown algae from eukaryotic lineages that are unrelated to the green plant lineage (e.g. Bisgrove and Kropf 2004), suggesting that there may be other morphological constraints that encourage cytokinetic mechanisms outside of an actin-myosin cleavage furrow. Finally, some have even suggested that some aspect of a cell plate-like structure is involved in completion of the midbody cleavage in animal cells (Otegui et al. 2005), suggesting that all these differences may simply be variations on a singular theme. Overall, that theme seems to involve addition of membrane, specifically in the form of vesicles derived from the organelles of the secretory system.

Traditionally, it has been believed that vesicles derived from the Golgi are the source of the membrane that forms the cell plate. Many cell biological studies have shown that the Golgi apparati form specialized arrays surrounding the nascent cell plate in a cell cycle-specific manner (Nebenführ et al. 2000; and see Nebenführ, this volume), or form morphologically distinct stacks that seem to specialize in cell plate formation in some cell types (Otegui et al. 2001; and see Segui-Simarro et al., this volume). Genetic and molecular biological studies have implicated specialized components of the traditional secretory pathway (see Sect. 3 below, Segui-Simarro et al., this volume, and Hong and Verma, this volume), leading to the suggestion that cell plate formation is simply a modified form of secretion (Bednarek and Falbel 2002). Considering the essential role of the Golgi in the secretory pathway, this also implicates the Golgi as the source of the cell plate membrane. Moreover, the Golgi would also be the natural source of the new cell wall material, and newly synthesized proteins that are delivered to the cell plate. Thus, the models tend to indicate a specialized form of vesicle carrying cargo bearing "cell plate" signals is formed from the trans-Golgi network (TGN). These vesicles are then borne to the nascent cell plate on cytoskeletal elements using one of the many motor proteins implicated in cell plate formation (see Hasezawa, this volume, and Lee and Liu, this volume). Upon arrival at the cell plate, these vesicles fuse with each other (homotypic fusion) or with already-assembled fragments of the nascent cell plate (Fig. 1). Eventually, the cell plate

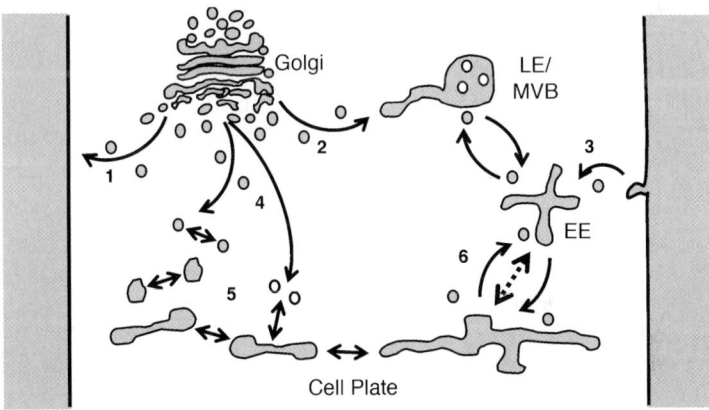

Fig. 1 Likely sources of membrane for the cell plate. During both interphase and mitosis/meiosis, regular anterograde traffic delivers membrane and cargo from the Golgi to the PM (*1*) or from the Golgi to the late endosome (*2*). Similarly, retrograde traffic carries material from the PM to early and late endosomes (*3*). During the late stages of mitosis, a new kind of Golgi vesicle is observed that presumably carried membrane and cargo for the nascent cell plate (*4*). These cell-plate targeted vesicles are thought to undergo homotypic fusion to assemble larger membrane enclosed organelles which are subsequently reformed into sheets through the action of dynamins until they form a sheet like membrane that expands across the point of division (*5*). Endocytosis and clathrin coated vesicles have long been observed budding from the cell plate, and recent evidence has suggested that exocytosis of material back from the endosomes may be a major source of membrane for the cell plate (*6*). See text for more information and references

fuses with the PM, either at the completion of the centrifugal expansion from the center or closing like a curtain across the cytosol from one side to the other (Cutler and Ehrhardt 2003).

Recently, the role of other organelles in the supply of membrane to the cell plate has begun to be examined. For many years, EM-observations have indicated that the cell plate is a major site of coated vesicle formation, likely as part of an endocytic process (see Seguí-Simarro et al. 2004). This makes sense considering that some aspect of the vesicle trafficking machinery must be recycled during the cell plate forming processes, but also suggests the possibility that exocytosis from endosomes may also be an active participant in supplying membrane to the cell plate. In fact, Dhonukshe et al. (2006) have suggested that, rather than the Golgi, the endosomes are the major player in cell plate initiation and expansion. Their results, based on dye tracers and pharmacology, suggest that the major source of membrane and material for the initiating cell plate is derived by flow through the endosomes from the PM (Dhonukshe et al. 2006). Still, decades of results have supported a role for delivery of Golgi-derived vesicles, and associated trafficking machinery to the cell plate. Indeed, such alternate routes for cell plate membrane may support the ability of cell plates (or at least fragments) to form in many cytokinetic mutants, which might otherwise be expected to fail at the first zygotic division. Such results are intriguing, and may allow a better fundamental understanding of the sources of vesicles that create the cell plate. Now that the cell biological tools such as electron tomography, tracer dyes and other well-known endocytic markers can be aimed at the problem, knowledge of the organellar players in cell plate formation can progress along with the known molecular components of the vesicle trafficking machinery.

3
Cytokinetic Mutants Lead to the Vesicle Trafficking Machinery

Several years of forward and reverse genetic examination has provided a glimpse of the molecular machinery involved in assembling the cell plate compartment. Not surprisingly, the parallels between cell plate assembly and regular secretion have become apparent; just as the parallels between all aspects of vesicle trafficking among the endomembrane organelles have indicated a deep conservation of the core machinery involved in moving membrane from one part of the cell to another. Much of the mechanics of operating the endomembrane system are conserved throughout eukaryotic evolution (SNAREs, Rab-GTPases, SM-proteins, coatomers, etc.), though many duplications and specializations have occurred to coordinate different aspects of trafficking (Sanderfoot and Raikhel 2003). Certainly, one expects that some novel components will be necessary, and that future work will identify these factors. Yet, to this day, much of the known components of the cell-plate-fusion machinery represent homologues of proteins used for vesicle trafficking elsewhere.

3.1
KNOLLE, the First Piece of the Puzzle

The first component of the cell-plate-specific vesicle trafficking machinery to be identified was the Qa-SNARE (syntaxin) KNOLLE (Lukowitz et al. 1996). First identified in a screen for embryo/seedling lethal mutants, *knolle* mutants die as young seedlings, with many cell wall stubs, and incomplete cytokinesis (Lukowitz et al. 1996). *KNOLLE* is only expressed in dividing cells, and the KNOLLE protein accumulates quickly on the cell plate, as well as the Golgi and endosome-like structures (Lauber et al. 1997). Control is not only mediated at the message level, since the KNOLLE protein quickly disappears at the end of cytokinesis (Lauber et al. 1997), indicating that the protein must also contain signals (unknown at this point) for cell-cycle-specific destruction. As a Qa-SNARE, KNOLLE was well suited for a role in recognizing and driving fusion of incoming vesicles to the cell plate (Lauber et al. 1997), just as related syntaxins drive fusion of vesicles at the neuronal synapse (for example, Li and Chin 2003). Thus, discovery of KNOLLE drove research towards discovering other aspects of the SNARE-machinery that may be involved.

SNAREs are a well-conserved family of proteins that contain a diagnostic coiled-coil motif (a SNARE-helix) that is the mechanism used to assemble specific heteromeric SNARE complexes. On the basis of the sequences of the SNARE helices from many eukaryotes, Bock et al. (2000) classified SNARE helices into four groups: Qa, Qb, Qc, and R. Appropriately, all known SNARE complexes are made up from four helices, with one from each of the four classes. For example, the well-known synaptic SNARE complex is made from Syntaxin 1 (Qa), SNAP25 (a peptide containing two SNARE helices, an N-terminal Qb, and a C-terminal Qc), and synaptobrevin (R). As shown in Fig. 2A, the presynaptic membrane (a specialized PM) serves as the location where syntaxin and SNAP25 coil together creating a Qa + Qb + Qc heterotrimer of SNARE helices (called a "target-(t)-SNARE complex"). This complex contains a surface groove into which the R-helix of synaptobrevin can specifically fit. Synaptobrevin is found on the synaptic vesicles, such that when the vesicle docks, its R-helix begins to coil into the groove. The coiling motion of the SNARE helices acts like a "molecular twist tie", bringing the vesicle membrane and the target membrane into close proximity (Hong et al. 2005). The coiling also releases large amounts of free energy (Fasshauer et al. 2002), enough to drive vesicle fusion in vitro (for example, McNew et al. 2000). Similar heterotetrameric complexes, each having their own unique set of Qa + Qb + Qc + R-SNAREs, have been identified in vesicle trafficking among endomembrane compartments, suggesting that this model may be the basis for all vesicle trafficking in the cell (e.g. Mc New et al. 2000; Parlati et al.

2002). Whether in vivo SNAREs are the specificity determinant, the fusion machinery, or both is a topic of great debate, though their importance in some aspect of the process is not really questioned. Regardless, many different levels of regulation of the SNARE recognition and assembly process have been identified which would seem to assist in controlling the action of the SNARE proteins. For example, members of the SM (Sec1p/Munc18)-family of proteins seem to regulate the assembly state of the Qa-SNAREs (Toonen and Verhage 2003), members of the Rab-GTPase family seem to regulate vesicle migration and choice of target membrane (Stenmark and Olkkonen 2001), and a long list of "docking factors" seem to assist in the specificity of vesicle trafficking (Whyte and Munro 2002).

3.2
Almost All the Other SNAREs Soon Follow

Qa-SNAREs need three other SNAREs to make a complex, so the search for other partners for KNOLLE proceeded. Heese et al. (2001) showed that the Qb+Qc SNARE SNAP33 was one SNARE partner with KNOLLE (Fig. 2B). However, while SNAP33 is found on the cell plate in dividing cells, it does not have a cell cycle-specific expression pattern, and is found on the PM in similar abundance to the cell plate in dividing cells, indicating that it is not cell plate specific (Heese et al. 2001). Moreover, in non-dividing cells, SNAP33 has been found to interact with other PM Qa-SNAREs as part of both regular (Kargul et al. 2001) and a form of defense-related polarized secretion (Collins et al. 2003; Assaad et al. 2004). In addition, *snp33* mutants do not have a seedling lethal phenotype; instead, they die at the plant equivalent of "teenagers" from the spread of necrotic lesions across the surface of the plant (Heese et al. 2001). These lesions do contain cell wall stubs, and signs of cytokinetic defects (Heese et al. 2001), but it does not directly phenocopy *knolle* mutants. Again, redundancy may be the cause, since SNAP33 is one of a three-member family of SNAREs, each of which also interacts with KNOLLE to a similar extent as SNAP33 in yeast two-hybrid experiments (Heese et al. 2001). Nonetheless, a genetic interaction of *snp33* with *knolle* and *keule* mutants (Heese et al. 2001) showed that KNOLLE and SNAP33 likely represents a Qa + Qb + Qc-SNARE complex involved in cell plate formation.

Interestingly, another SNARE was also shown to interact with KNOLLE and be found on the cell plate. Zheng et al. (2003) identified a Qb-SNARE called NPSN11 which, like SNAP33, was found localized to the cell plate in dividing cells and interacted with KNOLLE by immunoprecipitation, yet was not expressed in a cell specific manner, and was found on other organelles in non-dividing cells (Zheng et al. 2003). Again, NPSN11 was one of three NPSN-like Qb-SNAREs in Arabidopsis (Sanderfoot et al. 2000). In this case the redundancy seems more significant since there was no observed phenotype in *npsn11* mutants (Zheng et al. 2003); in fact, single mutants in each of the other *NPSN1* families are all phenotypically normal (L. Kong and AAS, unpublished data), suggesting a strong redundancy among the members of this gene family. With a Qb-SNARE in hand, a Qc-SNARE was missing, and my group has recently shown that a Qc-SNARE called SYP71 can also interact with KNOLLE and is found on the cell plate in dividing cells (L. Conner and AAS, unpublished data). Again, SYP71 is also expressed in non-dividing cells, and is found on the PM and endosomes in these cells (L. Conner and

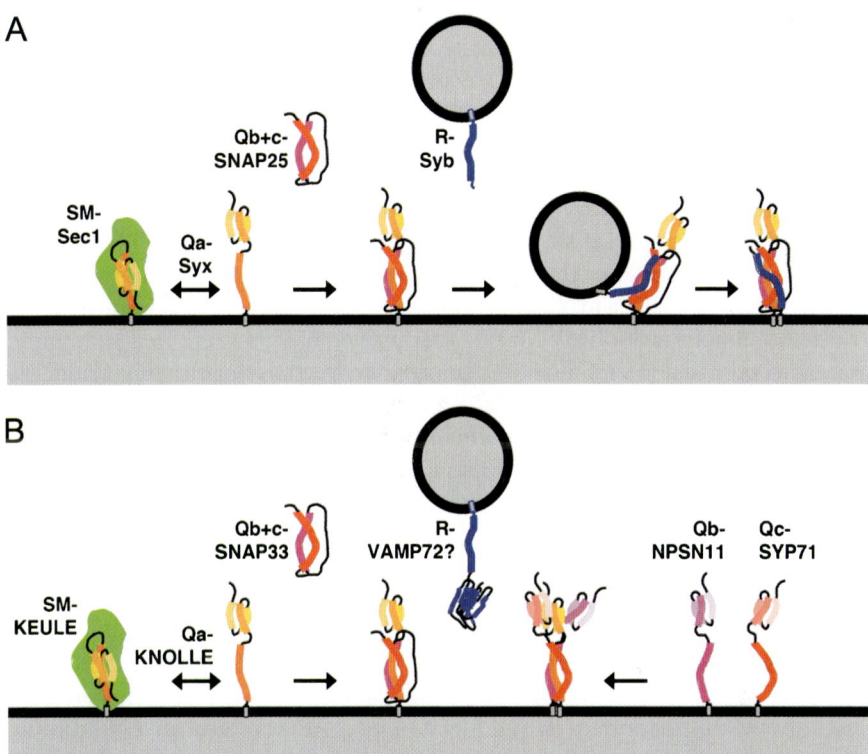

Fig. 2 The SNARE hypothesis and relevance to cell plate formation. **A** In the well-studied fusion of synaptic vesicles, the Qa-SNARE syntaxin 1 is regulated by association with the SM-family protein Sec1. Once free from Sec1, Syntaxin binds to the Qb + c-SNARE SNAP25, forming the pre-synaptic membrane t-SNARE complex. The R-SNARE Synaptobrevin is found on the synaptic vesicle, and when triggered for fusion, coils into the t-SNARE complex, pulling the vesicle membrane close to the synaptic membrane, and likely driving fusion. Many other factors are involved (not shown), see text for details. **B** The equivalent SNARE complexes for the cell plate include the Qa-SNARE KNOLLE, the SM-protein KEULE, the Qb + c-SNARE SNAP33; there is also a second complex made from Qa-KNOLLE, Qb-NPSN11 and the Qc-SYP71. The R-SNARE for either of these two complexes is unknown, yet is expected to be one of the many members of the R-VAMP72 clade. See text for details and references

AAS, unpublished data). Interestingly, in cells of the expanding root, SYP71 is found polarly localized to the acropetal PM (i.e. towards the root meristem; L. Conner and AAS, unpublished data), suggesting a role for this protein in non-dividing cells, outside of cell plate formation. Though SYP71 is one of three SYP7-like Qc-SNAREs in Arabidopsis, there appears to be little redundancy in this gene family, since *syp71* mutants are embryo lethal (L. Conner and AAS, unpublished data). Both the NPSN1- and the SYP7-group of SNAREs are not found in animals or fungi (Sanderfoot et al. 2000), suggesting that such a group may be uniquely placed for a novel function such as cell plate formation. However, it should be noted that NPSN1- and SYP7-like SNAREs are found in non-animal/fungal, i.e. ophistokont eukaryotes that do not form cell plates during cytokinesis (*Chlamydomonas*, *Dictyostelium*, etc; Sanderfoot, 2007), so this is not necessarily a direct relationship. Nonetheless, it is likely that KNOLLE+NPSN11+SYP71 represents a second Qa + Qb + Qc-SNARE complex in dividing cells (Fig. 2B).

Still missing from both of these complexes is an R-SNARE. It is widely believed, though not yet shown experimentally, that the R-SNARE for PM-complexes is a member of the VAMP72 group. The VAMP72 group of SNAREs is represented by seven genes in Arabidopsis, and appears to be unique to green plants (Sanderfoot et al. 2000). Uemura et al. (2004) showed that fluorescent protein fusions to VAMP72-family members were generally found on the PM when transiently expressed in leaf protoplasts, supporting the possibility that these may represent the R-SNARE(s) for a cell-plate complex. Since protoplasts do not divide, their methods cannot help us to understand whether (and which of) the seven members may be involved. Further work in plants is necessary to clearly indicate the proper R-SNAREs for the two Qa+Qb+Qc-SNARE complexes on the cell plate.

3.3
KEULE and the First Signs of Regulation

Since the identification of KNOLLE as a SNARE involved in cytokinesis, other SNARE and SNARE-interacting proteins have also been found to have a role in cell plate formation. KEULE, a member of the SM-family of SNARE-regulators, was found to interact with KNOLLE on the cell plate (Fig. 2B; Assaad et al. 2001). Like *knolle* mutants, *keule* mutants are also seedling lethal (Assaad et al. 1996). Indeed, it is somewhat surprising, considering what should be an essential role, that embryos lacking either *knolle* or *keule* are able to survive many rounds of division prior to their loss of viability at the young seedling stage (Assaad et al. 1996; Lukowitz et al. 1996). Clearly, there must be some level of redundancy to allow the embryos to survive long enough to reach seedling stages. In the case of KEULE, two other PM-type SM-proteins are found in the Arabidopsis genome (Sanderfoot et al. 2000), which also might serve to regulate KNOLLE at some level sufficient to provide some weak cytokinetic function (though SEC1a, one of the KEULE-like SM proteins, did not interact with KNOLLE as KEULE did; Assaad et al. 2001). Similarly, KNOLLE is one of nine PM-type Qa-SNAREs (Sanderfoot et al. 2000), though it appears to be the only one expressed in a cell-cycle specific manner (Volker et al. 2001). Again, some minimal level of cytokinetic function may be supplied by the other PM Qa-SNAREs that allows survival to the seedling stage. Still, the combination of these two proteins is unquestionably important, since *knolle keule* double mutants are "synthetic lethal"—they arrest as a multinucleate zygote that fails to complete even the first division (Waizenegger et al. 2000).

3.4
Other Regulators of Vesicle Fusion

Beyond the SNAREs and SM-proteins, other regulators have been suggested to be involved in vesicle tethering. Tomographic examination of cell plates has suggested that vesicle-docking complexes, of a shape similar to the Sec6/8-complex of mammalian cells, are involved in collecting the vesicles necessary for forming the cell plate (Seguí-Simarro et al. 2004). Actual involvement of these and other exocyst components in vesicle docking and fusion related to cell plate formation has not been shown, but one would expect that something like the exocyst complex (perhaps made from cell-plate-specific isoforms) would be involved. Vesicle docking and fusion also requires the action of a Rab-GTPase, and the identity of this (or these) cell-plate-specific Rab remains unknown. A very large group of Rab11-like GTPases (RABA in Vernoud et al. 2003) is found in land plants, and is likely to be involved in post-Golgi trafficking similar to the mammalian homologues. Could one of these 26 proteins be involved in regulating cell plate formation? On the other hand, there is also a group of Rab8-like GTPases (RABE in Vernoud et al. 2003) that may play an important role in polarized secretion based upon similarity to their yeast and mammalian homologues. Unfortunately, each of these (in fact all the Rabs in Arabidopsis) are present in very large gene families, and the potential for redundancy prevents a simple investigation of the role of these Rabs in cell plate formation (Vernoud et al. 2003). Considering the large gene families typical in most land plants, this potential for redundancy prevents simple characterization of many potential cell-plate-specific factors. Undoubtedly, many other levels of regulation in vesicle docking and fusion will be revealed by future research.

3.5
A Role for Dynamins in Turning Tubes into Sheets

Once the vesicles fuse, work remains before a macroscopic organelle like the cell plate can be formed. The morphology of the small cell plate fragments at early stages suggested a role for a member of the Dynamin family in shaping and preparing the small fragments for subsequent fusion into larger membranes (Otegui et al. 2001; Seguí-Simarro et al. 2004). Green plants have a unique gene family of dynamin-like proteins, the DRP1 family (formerly called ADL1 or Phragmoplastin; Hong et al. 2003), and their essential role in cell plate development has been confirmed by both cell biological and genetic analysis (Gu and Verma 1997; Kang et al. 2001; Kang et al. 2003; Seguí-Simarro et al. 2004). The analogous roles played by dynamins in the cytokinetic processes of animal and plant cells have recently been reviewed (Konopka et al. 2006). That a known membrane deforming/remodeling factor is involved in deforming and remodeling the cell plate membranes should not be a surprise. That green plants have dedicated a specialized family of dynamin-related proteins to this task underlines the importance of this process in cell plate formation. But, who regulates the dynamins? What other factors could be involved in the lipid-aspects of the cell plate? Are specialized kinds of lipases, flippases, scramblases, etc. required for the various membrane gymnastics that occur during the assembly and development of the cell plate? Such things have not been thoroughly investigated, and much is left to be discovered.

4
Just What Do the Vesicles Bring to the Cell Plate?

While it has been clear for a long time that vesicles supply membrane for the cell plate, what those vesicles carry as cargo (besides their constitutive lipids) is still an open question. While many proteins have been shown to associate with the cell plate (see Jürgens 2005), most are soluble/cytosolic proteins, and could presumably just associate directly with the cell plate membrane from the cytosol. Of most relevance would be membrane proteins that would be transported from other endomembrane organelles to the cell plate, though very few of these types of proteins are known: the cell plate SNARE proteins and CalS1.

Components of the vesicle trafficking machinery are found on the cell plate, probably concomitant with cell plate formation (e.g. KNOLLE, SNAP33, NPSN11, SYP71, etc.; see above), there must be some signal that allows them to be redirected away from other endomembrane destinations toward the cell plate vesicles. As one might expect, this signal is protein specific, and appears to be dependent on the cell cycle. For example, expressing *KNOLLE* constitutively does not rescue its cytokinetic defect, and leads to localization of the PM (Völker et al. 2001). This result suggests that the machinery that recognizes the signals in KNOLLE is only expressed (or functional) during the later phases of the cell cycle. However, placing the late endosomal Qa-SNARE SYP21 (i.e. AtPEP12) under the control of the *KNOLLE* promoter does not lead to localization of SYP21 to the cell plate (Müller et al. 2003), indicating that the signal for cell plate localization is a part of the proteins themselves, similar to signals that target proteins to other organelles (see Sanderfoot and Raikhel, 2002). Nonetheless, this putative signal remains unknown.

Aside from the cell plate SNAREs, the only known membrane protein to be targeted as cargo to the cell plate is callose synthase. Because the cell plate lumen fills with the carbohydrate-polymer callose during the tubulovesicular stage (Samuels et al. 1995), such a protein had been assumed to be involved for a long time. Still, it was not until Hong et al. (2001) identified a large family of callose synthase-like enzymes in Arabidopsis that the identity of this enzyme was known. Through their work, they showed that the Arabidopsis CalS1 isoform is found on the cell plate and interacts with DRP1a (i.e. phragmoplastin) and an associated UDP-Glucose transferase (Hong et al. 2001). Being such a large protein, it would be difficult to assign significance to small regions of homology between CalS1 and KNOLLE are (i.e. potential "cell plate signals"), and no such signals have been tested. Until more known cargo proteins are identified, such searches for "cell plate signals" are unlikely to be fruitful.

Other membrane proteins perhaps do not require "cell plate" signals, and can simply arrive at the cell plate by diffusion in the plane of the membrane from the regular PM or through endo-/exocytosis (Dhonukshe et al. 2006) from endosomal organelles. Even these proteins are few and poorly known. Until some mechanism for isolating an enriched source of cell plates for proteomic investigation, cell plate proteins will likely come one at a time through genetic or candidate gene approaches.

5
Summary

The cell plate of the land plants is the most striking example of how vesicle trafficking and the addition of new membrane is required for eukaryotic cytokinesis. As discussed above, much recent work has begun to identify the sources of the membrane that makes up the cell plate, though some questions remain as to how much is derived from secretory or endocytic processes. Considerable effort has also identified many of the molecular machines that are involved in targeting and fusing these membranes at the cell plate, but many components remain elusive, and many more are still completely unknown. Finally, we still know little about what proteins are actually being transported to the cell plate. It may turn out that the cell plate does not have many unique cargo proteins, but more likely, we just have not looked hard enough. Nonetheless, much progress has been made in our understanding of cell plate assembly, and this progress may already be helping to illuminate the study of cytokinesis in other organisms that have less obvious roles for vesicle trafficking. We just might find that studying this esoteric structure of land plants will tell us many things about how our own cells divide.

References

Assaad FF, Mayer U, Wanner G, Jürgens G (1996) The KEULE gene is involved in cytokinesis in Arabidopsis. Mol Gen Genet 253:267–77

Assaad FF, Huet Y, Mayer U, Jürgens G (2001) The cytokinesis gene KEULE encodes a Sec1 protein that binds the syntaxin KNOLLE. J Cell Biol 152:531–43

Assaad FF, Qiu JL, Youngs H, Ehrhardt D, Zimmerli L, Kalde M, Wanner G, Peck SC, Edwards H, Ramonell K, Somerville CR, Thordal-Christensen H (2004) The PEN1 syntaxin defines a novel cellular compartment upon fungal attack and is required for the timely assembly of papillae. Mol Biol Cell 15:5118–29

Bednarek SY, Falbel TG (2002) Membrane trafficking during plant cytokinesis. Traffic 3:621–9

Bisgrove SR, Kropf DL (2004) Cytokinesis in brown algae: studies of asymmetric division in fucoid zygotes. Protoplasma 223:163–73

Bock JB, Matern HT, Peden AA, Scheller RH (2001) A genomic perspective on membrane compartment organization. Nature 409:839–841

Collins NC, Thordal-Christensen H, Lipka V, Bau S, Kombrink E, Qiu JL, Huckelhoven R, Stein M, Freialdenhoven A, Somerville SC, Schulze-Lefert P (2003) SNARE-protein-mediated disease resistance at the plant cell wall. Nature 425:973–977

Cutler SR, Ehrhardt DW, Griffitts JS, Somerville CR (2000) Random GFP::cDNA fusions enable visualization of subcellular structures in cells of Arabidopsis at a high frequency. Proc Natl Acad Sci USA 97:3718–3723

Dhonukshe P, Baluska F, Schlicht M, Hlavacka A, Samaj J, Friml J, Gadella TW Jr (2006) Endocytosis of cell surface material mediates cell plate formation during plant cytokinesis. Dev Cell 10:137–150

Fasshauer D, Antonin W, Subramaniam V, Jahn R (2002) SNARE assembly and disassembly exhibit a pronounced hysteresis. Nat Struct Biol 9:144–151

Heese M, Gansel X, Sticher L, Wick P, Grebe M, Granier F, Jürgens G (2001) Functional characterization of the KNOLLE-interacting t-SNARE AtSNAP33 and its role in plant cytokinesis. J Cell Biol 155:239–249

Hong Z, Delauney AJ, Verma DP (2001) A cell plate-specific callose synthase and its interaction with Phragmoplastin. Plant Cell 12:755–768

Hong Z, Bednarek SY, Blumwald E, Hwang I, Jurgens G, Menzel D, Osteryoung KW, Raikhel NV, Shinozaki K, Tsutsumi N, Verma DP (2003) A unified nomenclature for Arabidopsis dynamin-related large GTPases based on homology and possible functions. Plant Mol Biol 53:261–265

Hong W (2005) SNAREs and traffic. Biochim Biophys Acta 1744:120–144

Jürgens G (2005) Plant cytokinesis: fission by fusion. Trends Cell Biol 15:277–283

Kargul J, Gansel X, Tyrrell M, Sticher L, Blatt MR (2001) Protein-binding partners of the tobacco syntaxin NtSyr1. FEBS Lett 508:253–258

Konopka CA, Schleede JB, Skop AR, Bednarek SY (2006) Abstract Dynamin and cytokinesis. Traffic 7:239–247

Lopez-Bautista JM, Waters DA, Chapman RL (2003) Phragmoplastin, green algae and the evolution of cytokinesis. Int J Syst Evol Microbiol 53:1715–1718

Lauber MH, Waizenegger I, Steinmann T, Schwarz H, Mayer U, Hwang I, Lukowitz W, Jürgens G (1997) The Arabidopsis KNOLLE protein is a cytokinesis-specific syntaxin. J Cell Biol 139:1485–1493

Li L, Chin LS (2003) The molecular machinery of synaptic vesicle exocytosis. Cell Mol Life Sci 60:942–960

Lukowitz W, Mayer U, Jürgens G (1996) Cytokinesis in the Arabidopsis embryo involves the syntaxin-related KNOLLE gene product. Cell 84:61–71

McNew JA, Parlati F, Fukuda R, Johnston RJ, Paz K, Paumet F, Sollner TH, Rothman JE (2000) Compartmental specificity of cellular membrane fusion encoded in SNARE proteins. Nature 407:153–159

Müller I, Wagner W, Volker A, Schellmann S, Nacry P, Küttner F, Schwarz-Sommer Z, Mayer U, Jürgens G (2003) Syntaxin specificity of cytokinesis in Arabidopsis. Nat Cell Biol 5:531–534

Nebenführ A, Frohlick JA, Staehelin LA (2000) Redistribution of Golgi stacks and other organelles during mitosis and cytokinesis in plant cells. Plant Physiol 124:135–152

Otegui MS, Mastronarde DN, Kang BH, Bednarek SY, Staehelin LA (2001) Three dimensional analysis of syncytial-type cell plates during endosperm cellularization visualized by high resolution electron tomography. Plant Cell 13:2033–2051

Otegui MS, Verbrugghe KJ, Skop AR (2005) Midbodies and phragmoplasts: analogous structures involved in cytokinesis. Trends Cell Biol 15:404–413

Parlati F, Varlamov O, Paz K, McNew JA, Hurtado D, Sollner TH, Rothman JE (2002) Proc Natl Acad Sci USA 99:5424–5429

Samuels AL, Giddings TH, Staehelin LA (1995) Cytokinesis in tobacco BY-2 and root tip cells: A new model of cell plate formation in higher plants. J Cell Biol 130:1345–1357

Sanderfoot AA, Assaad FF and Raikhel NV (2000) The Arabidopsis Genome: An abundance of SNAREs. Plant Physiol 124:1558–1569

Sanderfoot AA (2007) The expansion of SNARE gene families parallels the rise of multicellularity among the green plants. Plant Physiol 144:6–17

Sanderfoot AA, Raikhel NV (2003) Vesicle Trafficking. In: Somerville CR, Meyerowitz EM (eds) The Arabidopsis Book. American Society of Plant Biologists, Rockville MD (DOI 10.1199/tab.0098)

Seguí-Simarro JM, Austin JR 2nd, White EA, Staehelin LA (2004) Electron tomographic analysis of somatic cell plate formation in meristematic cells of Arabidopsis preserved by high-pressure freezing. Plant Cell 16:836–856

Stenmark H, Olkkonen VM (2001) The Rab GTPase family. Genome Biol 2:Reviews 3007.1–3007.7

Toonen RF, Verhage M (2003) Vesicle trafficking: pleasure and pain from SM genes. Trends Cell Biol 13:177–186

Uemura T, Ueda T, Ohniwa RL, Nakano A, Takeyasu K, Sato MH (2004) Systematic analysis of SNARE molecules in Arabidopsis: dissection of the post-Golgi network in plant cells. Cell Struct Funct 29:49–65

Vernoud V, Horton AC, Yang Z, Nielsen E (2003) Analysis of the small GTPase gene superfamily of Arabidopsis. Plant Physiol 13:1191–1208

Völker A, Stierhof YD, Jürgens G (2001) Cell cycle-independent expression of the Arabidopsis cytokinesis-specific syntaxin KNOLLE results in mistargeting to the plasma membrane and is not sufficient for cytokinesis. J Cell Sci 114:3001–3012

Waizenegger I, Lukowitz W, Assaad F, Schwarz H, Jurgens G, Mayer U (2000) The Arabidopsis KNOLLE and KEULE genes interact to promote vesicle fusion during cytokinesis. Curr Biol 10:1371–1374

Whyte JR, Munro S (2002) Vesicle tethering complexes in membrane traffic. J Cell Sci 115:2627–2637
Xu H, Boulianne GL, Trimble WS (2002) Membrane trafficking in cytokinesis. Semin Cell Dev Biol 13:77–82
Zheng H, Bednarek SY, Sanderfoot AA, Alonso J, Ecker JR, and Raikhel NV (2003) NPSN11 is a cell plate associated SNARE protein that interacts with the syntaxin KNOLLE. Plant Physiol 129:530–539

Molecular Analysis of the Cell Plate Forming Machinery

Zonglie Hong[1] (✉) · Desh Pal S. Verma[2]

[1]Department of Microbiology, Molecular Biology and Biochemistry, University of Idaho, Moscow, Idaho 83844, USA
zhong@uidaho.edu

[2]Department of Molecular Genetics and Plant Biotechnology Center, The Ohio State University, Columbus, Ohio 43210, USA

Abstract The cell plate is built using Golgi-derived vesicles carrying various proteins, sugars and lipids, which are required for the *de novo* synthesis of the cell wall during cytokinesis. The processes of vesicle fusion, cell plate expansion and maturation are initiated and controlled by a large number of proteins that serve as structural components, transporters, enzymes, and regulatory elements. Since the identification of phragmoplastin, the first protein marker of the cytokinetic organelle phragmoplast, a number of cell plate-associated proteins have been identified and characterized. Some of these proteins appear to be unique to the process of cell plate formation, while others have functions in different subcellular locations and are recruited to the forming cell plate transiently during cytokinesis. A temporal and spatial orchestration of basic exocytotic and endocytotic processes culminate in to formation of this unique subcellular compartment. Completion of this process in a defined time is essential for a proper cell division.

1
Phragmoplastin, the First Protein Marker of the Cell Plate

Phragmoplastin was first cloned from soybean in an attempt to identify proteins required for vacuole biogenesis (Gu and Verma 1996). The cloning was based on its homology with yeast VPS1, a dynamin-like GTPase required for the formation of vacuoles (Rothman et al. 1990; Ekena and Stevens 1995). This protein was found to be associated with the forming cell plate and phragmoplast-microtubules using its polyclonal antibodies (Fig. 1) and the GFP tag (Gu and Verma 1996, 1997; Hong et al. 2003b), and was accordingly named as phragmoplastin. Although the phragmoplast structure was known for a long time, no specific protein marker for this "organelle" was available at the time when phragmoplastin was identified. Earlier studies on cell plate formation were based primarily on anatomical observations and histochemical staining such as aniline blue that stains callose deposited on the forming cell plate (Gunning and Wick 1985; Mineyuki and Gunning 1990). The event of callose deposition, however, occurs much later then the initiation of cell plate by fusion of the Golgi-derived vesicles.

Phragmoplastin is a 68 kD GTPase related to the dynamin family of proteins, but does not contain either the pleckstrin homology (PH) or proline-

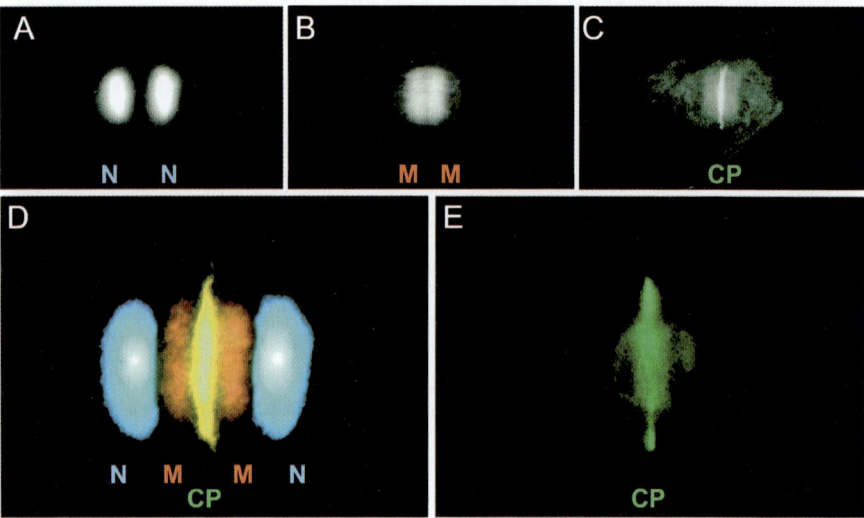

Fig. 1 Subcellular localization of phragmoplastin, the first protein marker of the cell plate. **A–D** A cytokinetic tobacco BY-2 cell was stained with DAPI for nuclear DNA (**A**), anti-tubulin monoclonal antibody for phragmoplast microtubules (**B**) or anti-soybean phragmoplastin polyclonal antibodies (**C**). The fluorescence signals were recorded separately in black-and-white photographs. Artificial colors are assigned to DAPI staining (in *blue*), microtubules (in *red*) and cell plate (in *green*). The three photographs are superimposed to indicate the relative subcellular localization of phragmoplastin (**D**; Gu and Verma 1996). N, nuclear DNA. M, phragmoplast microtubules. CP, cell plate. **E** Green fluorescent protein (GFP)-tagged phragmoplastin expressed in a tobacco BY-2 cell (Gu and Verma 1997)

rich (PR) domain, which both are characteristic of dynamins (Fig. 2; see below). It is present largely in the membrane fractions (presumably the Golgi vesicles) and can also be detected in the cytosol (Gu and Verma 1995). It contains two separable self-assembly (SA1 and SA2) domains that are responsible for the formation of phragmoplastin polymers (Zhang et al. 2000). Intermolecular interaction between SA1 and SA2 leads to the formation of staggered helical polymer structures that wrap around vesicles. Polymerization of phragmoplastin is dependent on GTP binding (Zhang et al. 2000) and may be regulated by other protein components (Verma 2001; Verma and Hong 2005). Phragmoplastin has five closely related homologs (DRP1A-E) in *Arabidopsis* (see below; Hong et al. 2003a), which appear to act as tubulase in creating dumbbell-shaped tubular structures at the cell plate (Samuels et al. 1995; Verma 2001; Segui-Simarro et al. 2004; Verma and Hong 2005). Formation of such tubuler structures is essential to prevent ballooning of the fused vesicles and to give rise to a plate, instead of large vacuole (Verma 2001).

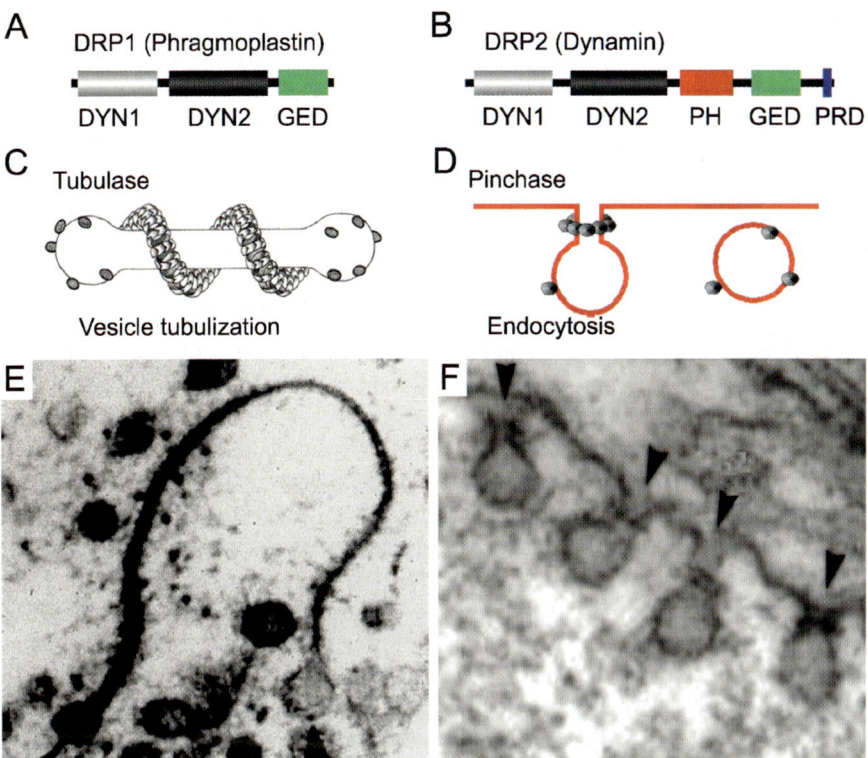

Fig. 2 DRP1 and DRP2 act as tubulase and pinchase that control membrane tubulization and vesiculation, respectively. **A, C** DRP1 (phragmoplastin) acts as a tubulase and forms highly ordered spiral polymers on the surface of vesicles. Upon GTP hydrolysis, DRP1 provides mechanical force to convert vesicles in to tubular structures. **B, D** DRP2 (plant dynamin) acts as a pinchase and forms a collar ring surrounding the junction region between a budding endocytic vesicle and plasma membrane. **E** A long dumbbell-shaped tubule observed at the nascent cell plate of tobacco BY-2 cells (Samuels et al. 1995). Note that the tubule is coated with DRP1 tubulase. **F** The pinching off of endocytic vesicles from the plasma membrane is blocked in fruit flies with a temperature-sensitive mutation at the *shibire* locus that encodes dynamin (Koenig and Ikeda 1989). DYN1, dynamin GTPase domain includes a GTP-binding motif (GXXXSGKS/T) and a dynamin signature. DYN2, dynamin central region. PH, pleckstrin homology domain which binds to membrane phospholipids such as PI-3-P and PI-4-P. GED, dynamin GTPase effector domain. PRD, proline-rich domain which interacts with SH3-domain proteins

2
Tubulase and Pinchase at the Cell Plate

Since the characterization of soybean phragmoplastin (Gu and Verma 1995) and *Arabidopsis* ADL1 (Dombrowski and Raikhel 1995), many other dynamin-related proteins (DRPs) have been studied in plants. They participate in a variety of membrane fusion and fission processes. Cell plate formation involves at least two subfamilies of DRPs: tubulase (DRP1) and pinchase (DRP2). Each of these have isoforms which may differentially interact with other proteins in a cell-type-specific manner.

2.1
DRPs, A Superfamily of Dynamin-Related Proteins

Arabidopsis contains 16 genes that may potentially encode proteins with the dynamin-signature (L-P-[PK]-G-[STN]-[GN]-[LIVM]-V-T-R) (Prosite PDOC00362). These proteins have been grouped into six functional subfamilies (DRP1-6) (Hong et al. 2003a). The DRP1 subfamily contains five phragmoplastin-like members (DRP1A-E), and are localized at the forming cell plate (Lauber et al. 1997; Kang et al. 2001, 2003a,b; Hong et al. 2003b). They may also play specific roles in the plasma membrane (Kang et al. 2003a,b) and cytoskeleton (Hong et al. 2003b). The DRP2 subfamily contains the *bona fide* plant dynamins that have a characteristic PH domain in the middle of the molecule and a PR motif (RXPXXP) near the C-terminus (Fig. 2). DRP2 proteins may be involved in clathrin-coated vesicle trafficking between Golgi, plasma membrane including cell plate, and in vacuole membrane biogenesis (Hong et al. 2003b; Jin et al. 2001; Lam et al. 2002). The DRP3 subfamily members do not contain PH or PR motifs, and have been implicated in the division of mitochondria and peroxisomes (Arimura and Tsutsumi 2002; Nishida et al. 2003; Mano et al. 2004; Kuravi et al. 2006). The DRP4 subfamily is related to the animal Mx proteins that have antiviral activity (Haller and Kochs 2002). *Arabidopsis* DRP5 subfamily and its ortholog from *Cyanidioschyzon merolae* (CmDnm2) are localized to the chloroplast division-ring and required for chloroplast division (Gao et al. 2003; Miyagishima et al. 2003, 2006). The function of DRP6 remains to be determined.

2.2
DRP1 Tubulase is Required for the Formation of Tubular Structures at the Cell Plate

The growing edges of the cell plate are filled with the "hourglass"- or "dumbbell-shaped" tubular structures (Samuels et al. 1995; Otegui et al. 2001; Segui-Simarro et al. 2004; Otegui and Staehelin 2004). These tubular structures are membrane-based transient and fragile structures that are very sensitive to chemical fixation treatments. They can be preserved in tissue samples that are cryofixed (Segui-Simarro et al. 2004; Samuels et al. 1995; Otegui et al. 2001; Otegui and Staehelin 2004). These tubular structures serve as basic building blocks for the formation of cell plate. The volume of one tubular structure is approximately twice the size of a cell plate vesicle, suggesting that the tubular structures are generated from fusion of two cell plate vesicles (Otegui and Staehelin 2004). Dumbbell-shaped tubular structures have a layer of protein coat on the surface of the tubular structures (Samuels et al. 1995), while the hourglass-shaped structures have two collars at the neck regions (Segui-Simarro et al. 2004). These protein coats and collars have been shown to be DRP1 proteins using antibodies against DRP1a (Otegui et al. 2001). DRP1 proteins have been proposed to act as a tubulase that mediates vesicle tubulation and thus prevent ballooning of the fused structures(Verma, 2001; Verma and Hong 2005). Animal dynamins are pinchases (see below), but can also act as a tubulase under certain circumstances such as in the presence of GTP-γ-S (Takei et al. 1995), overexpression of a clathrin antisense RNA (Iversen et al. 2003) or in vitro polymerization on tubular liposomes (Takei et al. 1998; Sweitzer and Hinshaw 1998; Stowell 1999). Thus, plant DRP1 family is a dedicated group of proteins that are responsible for the formation of tubular structures from fused vesicles at the forming cell plate. This novel mechanism serves as a space-filling model from the architectural view and is ideal for building a cell plate structure de novo.

2.3
DRP2 Pinchase is Required for Membrane Recycling at the Cell Plate

Arabidopsis DRP2 proteins are the *bona fide* plant dynamins which appear to function as pinchases (Fig. 2; Verma and Hong 2005). The dynamin pinchase hydrolyzes GTP and provides mechanical force to pinch the vesicles off the plasma membrane, an essential step in endocytosis and membrane recycling (Sweitzer and Hinshaw 1998; McNiven 1998; Verma et al. 2005). Cell plate formation involves the de novo construction of two "flat" layers of the plasma membrane using phospholipids and membrane proteins that are delivered via vesicles. Coalescing of the tubular structure described above leaves behind extensive membrane which needs to be recycled. It has been estimated that 10-14-fold more membranes are required to build one unit of the plasma membrane, suggesting that most membrane constituents that have been delivered to the cell plate through the exocytic pathway need to be recycled back through endocytosis and membrane recycling pathways. DRP2A has been shown to be associated with the cell plate and may be responsible for pinching-off of the vesicles from the forming cell plate to recycle the extra membrane compartment (Hong et al. 2003b).

3
Homotypic and Heterotypic Membrane Fusion at the Cell Plate

Cell plate formation involves intensive activities of membrane fusion events that can be grouped into two categories: homotypic and heterotypic membrane fusions. Whereas the heterotypic fusion at the cell plate may employ redistributed fusion machinery from the plasma membrane, homotypic fusion may require the expression of a new set of proteins because this process is highly amplified during cell plate formation. The entire exocytosis machinery appears to be engaged in generating a population of homotypic vesicles during the early cell plate formation while both homotypic and heterotypic vesicles are produced as the cell plate matures.

3.1
Cell Plate Vesicles

Cell plate vesicles originating from the Golgi apparatus fuse in a unique manner at the center of the phragmoplast, giving rise to the new plasma membranes of the two daughter cells (Fig. 3). They carry cargo for the formation of the cell plate matrix (Gunning and Wick 1985; Hepler and Bonsignore 1990; Staehelin and Moore 1995). Disassembly of the Golgi apparatus by treatment with brefeldin-A (BFA) blocks the completion of cytokinesis by shutting off the supply of cell plate vesicles (Yasuhara et al. 1995; Staehelin and Moore 1995; Satiat-Jeunemaitre et al. 1996). However, initiation of cell

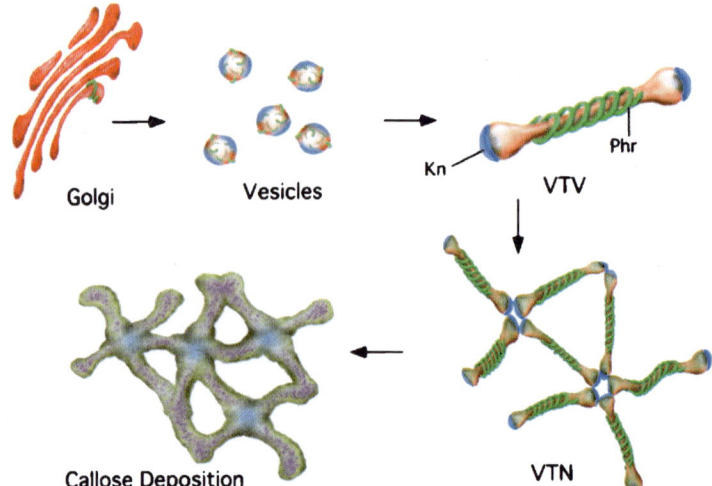

Fig. 3 A proposed model for cell plate formation. The Golgi-derived vesicles carrying callose synthase, phragmoplastin (Phr) and other associated proteins are transported along the microtubules. Once they reach the equatorial plane of the future cell plate, they are squeezed by phragmoplastin into tubular structures, which fuse with each other in an end-to-end fashion, creating a tubular fence (vesiculo-tubular network or VTN). The conversion of the round-shaped vesicles into tubular structures is a fundamental and essential step in de novo membrane formation, and avoids ballooning of the fused membrane structures. The fusion brought about by KNOLLE syntaxin (Kn) is followed by activation of callose synthase. The callose synthesized dilates these tubules and closes the gaps, forming a flat sheet at the forming cell plate. Phragmoplastin may be removed once the callose synthase is activated

plate formation is not blocked, suggesting the presence of a pool of vesicles that participate in the initial cell plate formation. The disruption of microfilaments causes oblique cell plates, but does not prevent the formation of the cell plate *per se* (Schopfer and Hepler 1991). Thus, a coordinated delivery of the Golgi vesicles is essential for the formation of the cell plate. Little is known about the vesicle fusion machinery at the forming cell plate except the identification of the phragmoplastin GTPase (Gu and Verma 1996, 1997; Zhang et al. 2000) and a pair of syntaxin/Sec1-like proteins (KNOLLE/KEULE; Lauber et al. 1997; Assaad et al. 2001). Other proteins must be required for the various events in vesicle fusion.

3.2
Homotypic Membrane Fusion

Homotypic fusion refers to the merging of two membranes of the same type. Cell plate formation involves two types of homotypic membrane fusion, i.e., creation of tubular structures from vesicle-vesicle fusion, and formation of honeycomb-like network via end-to-end fusion of tubular structures. Phragmoplastin and its associated proteins are known to play pivotal roles in creating the dumbbell-shaped tubular structures, whilst Knolle syntaxin is required for the formation of honeycomb-like tubular network (Lauber et al. 1997; Jurgens 2004). In cell plate formation, the end-to-end fusion of tubular structures is unique in the aspect that the fusion involves the alignment of several tubular structures. Under electron microscope, the junction of these tubular structures look like a "star" (Hepler 1982). Mutant cells defective at Knolle cannot form tubular networks at the cell plate (Lauber et al. 1997). Thus, homotypic fusion of vesicles is an essential step in building the cell plate.

Proteins required for homotypic fusion has been studied in other systems such as the yeast mitotic vacuole fusion. After mitosis in yeast, small vacuoles fuse with each other, giving rise to the formation of bigger, round vacuoles (Wickner 2002; Jun et al. 2006). In contrast, homotypic fusion at the cell plate results in the formation of the "dumbbell-shaped" structures due to the presence of phragmoplastin-like DRP1 proteins. These structures are then expended and connected with each other in a unique end-to-end fusion manner. The termini of several dumbbell-shaped structures align together to form "star"-like junctions, which were observed by electron microscopy three decades ago (Hepler 1982). Interconnection of tubular structures gives rise to a tubulo-vesicular network (TVN) (Staehelin and Hepler 1996). Merging of these tubular termini is also thought to occur via homotypic membrane fusion (Verma 2001; Verma and Hong 2005). Thus homotypic fusion of vesicles is an essential step in building the cell plate. Moreover, the homotypic fusion machinery must be highly efficient because the completion of the cell plate takes about an hour. During this time, not only plate must be flattened but callose must be formed to fill the growing cell plate.

Homotypic fusion is a complex and highly ordered process consisting of three stages, priming, docking and membrane fusion. Fusion of mitotic vacuoles in yeast is a well studied case of homotypic fusion. Both types of v- and t-SNARE proteins are present on the same vesicle and can form stable *cis*-SNARE complexes. At the priming step, the *cis*-SNARE complex is disassembled into activated t- and v-SNAREs. The activated t-SNARE on one vesicle can interact with a v-SNARE on another vesicle, forming a *trans*-SNARE complex. The formation of *trans*-SNARE renders irreversible docking and fusion of two vesicles. Nearly two dozen proteins are required for the homotypic vesicle fusion in a cell-free assays in yeast (Wickner and Haas 2000; Wickner 2002). Understanding the precise role of each of these proteins in plants is essential in order to know how different phases of cell plate building are orchestrated and completed in a time-bound manner.

3.3
Heterotypic Membrane Fusion

Heterotypic membrane fusion, i.e., the incorporation of two different types of membranes such as vesicles and the plasma membrane, is mediated by a **v**esicle-associated SNARE (v-SNARE) and a **t**arget membrane-associated SNARE (t-SNARE) (Hawes et al. 1999; McNiven et al. 2000). The growing cell plate is actually a tubular network (Staehelin and Hepler 1996; Verma 2001), and a large quantity of vesicles carrying proteins and cell wall materials fuse directly with this network, eventually leading to the formation of the mature and "flat" plasma membrane, a step referred to as the fenestration of tubular network (Staehelin and Hepler 1996). The protein complexes required for docking of exocytic vesicles to the plasma membrane are believed to be employed for the fenestration of tubular network during cell plate formation.

3.4
Membrane Fusion Machinery

A growing number of SNAREs and other proteins involved in vesicle docking/fusion events localize to the cell plate and are critical for normal cytokinesis. SNAREs comprise a large diverse group of proteins including syntaxins, membrins, VAMPS and others (Sanderfoot et al. 2000). In plants, the cytokinesis-specific syntaxin Knolle (Q-SNARE) interacts with AtSNAP33 (SNAP25 homolog), AtNPSN11 (plant-specific SNARE), AtCDC48, and Keule protein (Sec1 homolog) (Lukowitz et al. 1996; Lauber et al. 1997; Waizenegger et al. 2000; Assaad et al. 2001; Heese et al. 2001; Rancour et al. 2002; Zheng et al. 2002; Jurgens 2004).

Yeast Cdc48 has been implicated in the ER-to-ER homotypic membrane fusion (Latterich et al. 1995), and its *Arabidopsis* homolog (AtCdc48) colocalizes at the cell plate with Knolle and another plate-associated SNARE, SYP31, and may play a similar role in cell plate formation (Feiler et al. 1995; Rancour et al. 2002). CDC48 is one of the AAA proteins (ATPases associated with various cellular activities) that regulate vesicle fusion events by disassembling the SNARE complexes to allow for subsequent rounds of fusion (May et al. 2001). *Arabidopsis* Patellin 1 (PATL1), a Sec-14 related protein implicated in membrane trafficking in yeast, is recruited from the cytoplasm to the expanding cell plate and may play a role in membrane-trafficking events associated with cell-plate maturation (Peterman et al. 2004). RanGAP, a nuclear membrane-associated protein is also observed at the early stages of cell plate (Jeong et al. 2005), but its precise role in the cell plate formation process is not yet clear.

4
Other Cell Plate-Associated Proteins

Several hundred proteins may participate in the processes of cytokinesis in plants. Some of them have been identified by genetic studies (Assaad 2001, Sollner et al. 2002). In addition to the two major categories of proteins (i.e., dynamin-related proteins and proteins involved in membrane fusion) mentioned above, several other groups of proteins have been implicated in cell plate formation. They include proteins associated with the callose synthase complex, microtubule-associated proteins, small GTP-binding proteins, and cytokinesis signaling proteins. Since the building of cell plate also involves formation of plasmodesmata channels, many other proteins involved in the formation of these structures remain to be identified.

4.1
Cell Plate-Specific Callose Synthase Complex

Unlike many other cell wall polysaccharides that are synthesized in the Golgi and delivered to the cell wall, callose and cellulose are synthesized by plasma membrane enzymes and deposited directly to the cell plate. Callose is the first polysaccharide synthesized at the growing cell plate (Northcote et al. 1989; Samuels et al. 1995; Bowser and Reddy 1997). Synthesis of callose begins as the tubular structure is formed and continues until the plate has touched the parental cell wall. Callose appears to apply a spreading force and helps stabilize the structure of the membranous tubular network. In *Arabidopsis*, this callose is synthesized by a cell plate-specific callose synthase (CalS1) that is a member of the large callose synthase family (Hong et al. 2001a). A T-DNA knockout mutant of *cals1* did not display any observable cytokinesis phenotype and contains callose at the cell plate, suggesting that more than one callose synthase is involved in cell plate formation. Callose is a multimeric enzyme which may include homomeric or heteromeric subunits in a cell specific manner. Of the 12 callose synthase genes in *Arabidopsis*, CalS5, CalS11 and CalS12 are known to be involved in callose wall formation during pollen development (Dong et al. 2005; Nishikawa et al. 2005; Enns et al. 2005). CalS12 is responsible for pathogenesis-related callose deposition (Østergaard et al. 2002; Jacobs et al. 2003; Nishimura et al. 2003).

When purified in a sucrose gradient, CalS1 exists in a complex with large molecular mass (Hong et al. 2001b). The CalS complex may include a UDP-glucose transferase (UGT1), a Rho-like small GTPase (Rop1), phragmoplastin, sucrose synthase (SuSy), and annexin (Verma and Hong 2001). SuSy generates UDP-glucose which can be transferred via UGT1 to the active site of CalS forming a substrate channel (Fig. 4) (see also, Amor et al. 1995; Verma 2001). CalS has also been shown to interact with a membrane protein hav-

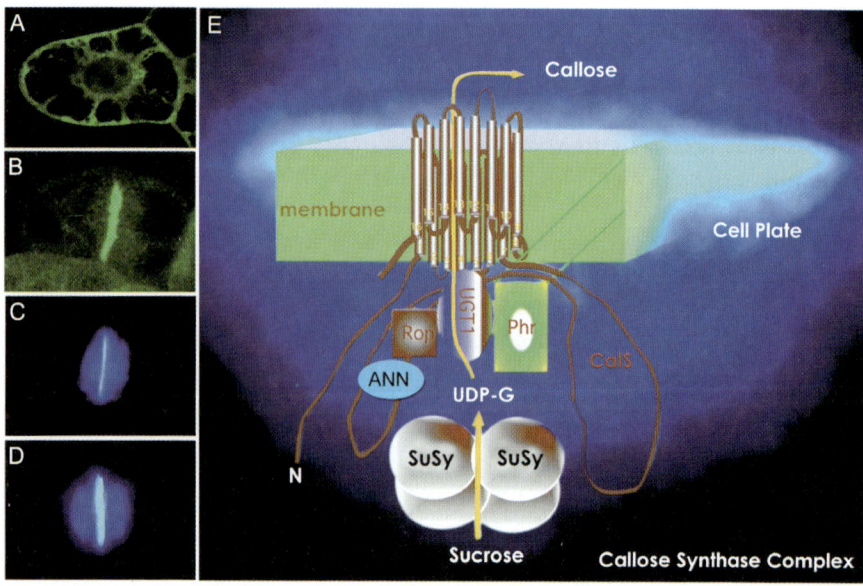

Fig. 4 Cell plate-specific callose synthase complex. **A, B** Tobacco BY-2 cells expressing GFP-tagged *Arabidopsis* callose synthase 1 (CalS1) under the control of the 35S promoter. In non-dividing cells (**A**), CalS1 is present in punctates in the cytoplasm. In cytokinetic cells (**B**), CalS1 is targeted to the cell plate. **C, D** More callose is deposited in the cell plate of the transgenic tobacco BY-2 cells overexpressing CalS1 (**D**) than in control cells (**C**). These cells were stained with DAPI for nuclear DNA and aniline blue for callose, and photographed in a fluorescent microscope with a UV filter set. **E** Proposed model of the callose synthase complex at the cell plate. *Arabidopsis* CalS1 contains 16 predicted transmembrane domains and utilizes UDP-glucose transferase (UGT1) for substrate binding. UGT1 interacts with phragmoplastin (Phr) and Rho-like GTPase (Rop). Sucrose synthase (SuSy) and annexin (ANN) may also be part of the CalS complex (Andrawis et al. 1993; Shin and Brown 1999)

ing similarity with annexin and protein kinase C (Andrawis et al. 1993). This annexin-like protein has GTPase activity which is inhibited by Ca^{2+} and stimulated by Mg^{2+} (Shin and Brown 1999). Rop1 may act as a spatial regulator of CalS as Rho has been shown to be a regulatory subunit of β-1,3-glucan synthase in yeast (Qadota et al. 1996). Phragmoplastin may activate this enzyme by squeezing the vesicles into tubules (Gu and Verma 1996, 1997; Zhang et al. 2000) at the forming cell plate (Verma 2001) or direct interaction via UGT1 as has been demonstrated (Hong et al. 2001b).

4.2
Microtubule-Associated Proteins

The γ-tubulin has been localized towards the minus ends of the phragmoplast microtubules (Joshi and Palevitz 1996). AIR9, a 187 kDa microtubule-associated protein, decorates the preprophase band (PPB) and appears in the junction between the outwardly growing cell plate and the mother cell wall (Buschmann et al. 2006). Kinesin-related proteins with either the plus-end, or the minus-end directed microtubule motors have been found to be associated with the phragmoplast. TKRP125, a tobacco kinesin-related protein and a plus-end directed motor, requires another protein of 120 kD to display microtubule-translocating activity (Asada and Shibaoka 1994; Asada et al. 1997). KatA, an *Arabidopsis* kinesin-like protein and a minus-end-directed motor, is localized in the midzone of an anaphase cell and is associated with phragmoplast during cytokinesis (Liu et al. 1996). Another *Arabidopsis* kinesin-like protein, KCBP, binds to calmodulin and is confined to the mature phragmoplast (Bowser and Reddy 1997). AtPAKRP2, which localizes to Brefeldin A-sensitive puncta during early plate development, is a likely candidate for the motor that drives vesicle movement along the phragmoplast microtubules (Lee et al. 2001; Smith 2002), and the kinesin-related protein, product of the HINKEL/NACK1 genes, functions in microtubule dynamics during plate expansion in collaboration with a mitogen-activated protein kinase-signaling pathway (Nishihama et al. 2002; Strompen et al. 2002). The dynamics of these various motor proteins may facilitate not only the mobilization of tubulin cytoskeleton but also vesicle traffic.

4.3
Cytokinesis-Related Signaling Proteins

Vesicle trafficking and fusion events during cell plate formation may also be regulated by MAP kinases through phosphorylation of rab and dynamin-like GTPases as shown in animals (Earnest et al. 1996, Cormont et al. 1994). A tobacco MAP kinase $p43^{Ntf6}$ is activated in late anaphase and early telophase and is localized to the cell plate in anaphase cells (Calderini et al. 1998). The enzymatic activity of MMK3, an alfalfa MAP kinase associated with cell plate during cytokinesis, was found to be transient in mitosis and correlated with the timing of phragmoplast formation (Bogre et al. 1999). In contrast to the distribution of phragmoplastin, MMK3 is not redistributed from the center to the periphery as the cell plate expands (Bogre et al. 1999).

Centrin, a protein involved in nucleation of centromeres in animal cells and binds Ca^{2+}, has been detected at the forming cell plate (Stoppin-Mellet et al. 1999). This protein appears to be a component of the Ca^{2+}-sensitive contractile nanofilaments in the neck region of plasmodesmata as shown by EM immunolocalization (Blackman et al. 1999). Using antibodies, a vinculin-like protein has also been localized at the cell plate but its plant homolog has not yet been identified (Endlé et al. 1998).

4.4
Cytokinesis-Related Small GTP-Binding Proteins

Rho, Rab and Ran, three families of small GTP-binding proteins that function in membrane trafficking and mRNA transport, have also been shown to participate in cell-plate biogenesis. Rho-related GTPases localize to the cell plate (Molendijk et al. 2001; Brembu et al. 2006). *Arabidopsis* SCD1 encodes a protein that may interact with Rab proteins and regulate intracellular protein transport (Falbel et al. 2003). The *scd1* (*stomatal cytokinesis-defective 1*) mutant displays severe cytokinesis defects in leaf guard mother cells, and affects polar cell expansion in trichomes and root hairs. *Arabidopsis* Ran proteins and their GTPase-activating proteins (RanGAP) are localized to the forming cell plate (our unpublished data; Jeong et al. 2005).

In addition, certain mRNAs species may become localized near the cell plate as has been shown for the actin mRNA in *Fucus* (Bouget et al. 1996). In this regard, a phragmoplastin-interacting protein (PhrIP1) has been identified which has properties in common with RNA-binding proteins and is shown to bind Ran RNA (our unpublished data). A LIM domain protein (WLIM1) characterized by the presence of one or several double zinc finger motifs has been isolated from sunflower (Mundel et al. 2000). This protein accumulates in the phragmoplast in early telophase and may be involved in anchoring the actin cytoskeleton to site of cell plate and plasma membrane adhesion. Antibodies against MTOC (microtubule organizing center) and phosphoprotein MPM-2 have been shown to be associated with the phragmoplast (Smirnova et al. 1995).

5
Future Perspectives

The presence of DRPs in plants with diverse functions in membrane tubulation and pinching, as well as fission of organelles, suggests that the basic property of this group of proteins, i.e., the ability to form helical structures regulated by their GTPase activity, is central to their functions. Despite the high similarity in the structure of various members of the DRP family, it is apparent that they interact with distinct proteins to perform specific roles in each of these events. A detailed dissection of various domains of these proteins and their interaction with specific partners may reveal their precise function in various cellular processes that involve membrane tubulation, pinching as well as organelle divisions. Development of *in vitro* membrane vesiculation, tubulation and fusion systems will further aid in dissection of the details of these mechanisms and the role of various DRPs.

The cell plate is one of a few organelles that have yet to be characterized using modern proteomic technology. The reason for the lack of proteomic characterization of the cell plate is mainly due to the technical challenge in preparing a highly purified cell plate fraction from a plant species with completed genome sequences, such as *Arabidopsis*. This technical hurdle may disappear soon as the tobacco genome sequencing project progresses. Tobacco BY-2 cells are easy to synchronize and have been used successfully for cell plate purification. Proteomic studies on tobacco BY-2 cells will promise the discovery of many expected as well as many unknown proteins that are targeted to this essential cytokinetic organelle. Besides the components of the cell plate itself, other proteins that are involved in the orientation of cell plate, preprophase band formation and the decision of the cell in which direction to lay down the cell plate, remain to be determined.

Acknowledgements This work was supported by NSF grants to DPSV (IBN-0095112) and ZH (MCB-0548525 and IOB 0543923).

References

Amor Y, Haigler CH, Johnson S, Wainscott M, Delmer DP (1995) A membrane-associated form of sucrose synthase and its potential role in synthesis of cellulose and callose in plants. Proc Natl Acad Sci USA 92:9353–9357

Andrawis A, Solomon M, Delmer DP (1993) Cotton fiber annexins: a potential role in the regulation of callose synthase. Plant J 3:763–772

Arimura S, Tsutsumi N (2002) A dynamin-like protein (ADL2b), rather than FtsZ, is involved in *Arabidopsis* mitochondrial division. Proc Natl Acad Sci USA 99:5727–5731

Asada T, Kuriyama R, Shibaoka H (1997) TKRP125, a kinesin-related protein involved in the centrosome-independent organization of the cytokinetic apparatus in tobacco BY-2 cells. J Cell Sci 110:179–189

Asada T, Shibaoka H (1994) Isolation of polypeptide with microtubule-translocating activity from phragmoplasts of tobacco BY-2 cells. J Cell Sci 107:2249–2257

Assaad FF (2001) Plant cytokinesis. Exploring the links. Plant Physiol 126:509–516

Assaad FF, Huet Y, Mayer U, Jurgens G (2001) The cytokinesis gene *KEULE* encodes a Sec1 protein which binds the syntaxin KNOLLE. J Cell Biol 152:531–543

Blackman LM, Harper JD, Overall RL (1999) Localization of a centrin-like protein to higher plant plasmodesmata. Eur J Cell Biol 78:297–304

Bogre L, Caldrini O, Binarova P, Mattauch M, Till S, Kiegerl S, Jonak C, Pollaschek C, Barker P, Huskisson NS, Hirt H, Heberle-Bors E (1999) A MAP kinase is activated late in plant mitosis and becomes localized to the plane of cell division. Plant Cell 11:101–103

Bowser J, Reddy ASN (1997) Localization of a kinesin-like calmodulin-binding protein in dividing cells of *Arabidopsis* and tobacco. Plant J 12:1429–1437

Brembu T, Winge P, Bones A, Yang Z (2006) A RHOse by any other name: a comparative analysis of animal and plant Rho GTPases. Cell Res 16:435–445

Buschmann H, Chan J, Sanchez-Pulido L, Andrade-Navarro MA, Doonan JH, Lloyd CW (2006) Microtubule-associated AIR9 recognizes the cortical division site at pre-prophase and cell-plate insertion. Curr Biol 16:1938–1943

Calderini O, Bögre L, Vicente O, Binarova P, Heberle-Bors E, Wilson C (1998) A cell cycle regulated MAP kinase with a possible role in cytokinesis in tobacco cells. J Cell Sci 111:3091–3100

Cormont M, Tanti JF, Zahraoui A, Van Obberghen E, Le Marchand-Brustel Y (1994) Rab4 is phosphorylated by the insulin-activated extracellular-signal-regulated kinase ERK1. Eur J Biochem 219:1081–1085

Cui X, Shin H, Song C, Laosinchai W, Amano Y, Brown RM Jr (2001) A putative plant homolog of the yeast β-1,3-glucan synthase subunit FKS1 from cotton (*Gossypium hirsutum* L.) fibers. Planta 213:223–230

Dombrowski JE, Raikhel NV (1995) Isolation of cDNA encoding a novel GTP-binding protein of *Arabidopsis thaliana*. Plant Mol Biol 28:1121–1126

Dong X, Hong Z, Sivaramakrishnan M, Mahfouz M, Verma DPS (2005) Callose synthase (CalS5) is required for exine formation during microgametogenesis and for pollen viability in *Arabidopsis*. Plant J 42:315–328

Earnest S, Khokhlatchev A, Albanesi JP, Barylko B (1996) Phosphorylation of dynamin by ERK2 inhibits the dynamin microtubule interaction. FEBS Lett 396:62–66

Endlé M-C, Stoppin V, Lambert A-M, Schmit A-C (1998) The growing cell plate of higher plants is a site of both actin assembly and vinculin-like antigen recruitment. Eur J Cell Biol 77:10–18

Falbel TG, Koch LM, Nadeau JA, Segui-Simarro JM, Sack FD, Bednarek SY (2003) SCD1 is required for cytokinesis and polarized cell expansion in *Arabidopsis thaliana*. Development 130:4011–4024

Feiler HS, Desprez T, Santoni V, Kronenberger J, Caboche M, Traas J (1995) The higher plant *Arabidopsis thaliana* encodes a functional CDC48 homologue which is highly expressed in dividing and expanding cells. EMBO J 14:5626–5637

Gao H, Kadirjan-Kalbach D, Froehlich J, Osteryoung K (2003) ARC5, a cytosolic dynamin-like protein from plants, is part of the chloroplast division machinery. Proc Natl Acad Sci USA 100:4328–4333

Gu X, Verma DPS (1996) Phragmoplastin, a dynamin-like protein associated with cell plate formation in plants. EMBO J 15:695–704

Gu X, Verma DPS (1997) Dynamics of phragmoplastin in living cells during cell plate formation and uncoupling of cell elongation from the plane of cell division. Plant Cell 9:157–169

Gunning BES, Wick SM (1985) Preprophase bands, phragmoplasts, and spatial control of cytokinesis. J Cell Sci Suppl 2:157–179

Haller O, Kochs G (2002) Interferon-induced Mx proteins: dynamin-like GTPases with antiviral activity. Traffic 3:710–717

Hawes CR, Brandizzi F, Andreeva AV (1999) Endomembranes and vesicle trafficking. Curr Opin Plant Biol 2:454–461

Heese M, Gansel X, Sticher L, Wick P, Grebe M, Granier F, Jurgens G (2001) Functional characterization of the KNOLLE-interacting t-SNARE AtSNAP33 and its role in plant cytokinesis. J Cell Biol 155:239–249

Hepler PK (1982) Endoplasmic reticulum in the formation of the cell plate and plasmodesmata. Protoplasma 111:121–133

Hepler PK, Bonsignore C (1990) Caffeine inhibition of cytokinesis: ultrastructure of cell plate formation/degradation. Protoplasma 157:182–192

Hong Z, Bednarek S, Blumwald E, Hwang I, Jurgens G, Menzel D, Osteryoung K, Raikhel N, Shinozaki K, Tsutsumi N, Verma DPS (2003a) A unified nomenclature for *Arabidopsis* dynamin-related large GTPases based on homology and possible functions. Plant Mol Biol 53:261–265

Hong Z, Delauney A, Verma DPS (2001a) A cell plate-specific callose synthase and its interaction with phragmoplastin and UDP-glucose transferase. Plant Cell 13:755–768

Hong Z, Geisler-Lee J, Zhang Z, Verma DPS (2003b) Phragmoplastin dynamics: multiple forms, microtubule association and their roles in cell plate formation in plants. Plant Mol Biol 53:297–312

Hong Z, Zhang Z, Olson J, Verma DPS (2001b) A novel UDP glucose transferase interacts with callose synthase and phragmoplastin at the forming cell plate. Plant Cell 13:769–780

Iversen TG, Skretting G, van Deurs B, Sandvig K (2003) Clathrin-coated pits with long, dynamin-wrapped necks upon expression of a clathrin antisense RNA. Proc Natl Acad Sci USA 100:5175–5180

Jacobs A, Lipka V, Burton R, Panstruga R, Strazhov N, Schulze-Lefert P, Fincher G (2003) An *Arabidopsis* callose synthase, GSL5, is required for wound and papillary callose formation. Plant Cell 15:2503–2513

Jeong SY, Rose A, Joseph J, Dasso M, Meier I (2005) Plant-specific mitotic targeting of RanGAP requires a functional WPP domain. Plant J 42:270–282

Jin J, Kim Y, Kim S, Lee S, Kim D, Cheong G, Hwang I (2001) A new dynamin-like protein, ADL6, is involved in trafficking from the trans-Golgi network to the central vacuole in *Arabidopsis*. Plant Cell 13:1511–1526

Joshi H, Palevitz B (1996) γ-Tubulin and microtubule organization in plants. Trends Cell Biol 6:41–44

Jun Y, Thorngren N, Starai VJ, Fratti RA, Collins K, Wickner W (2006) Reversible, cooperative reactions of yeast vacuole docking. EMBO J 25:5260–5269

Jurgens G (2004) Membrane trafficking in plants. Annu Rev Cell Dev Biol 20:481–504

Kang B, Busse J, Bednarek S (2003a) Members of the *Arabidopsis* dynamin-like gene family, ADL1, are essential for plant cytokinesis and polarized cell growth. Plant Cell 15:899–913

Kang B, Busse J, Dickey C, Rancour D, Bednarek S (2001) The *Arabidopsis* cell plate-associated dynamin-like protein, ADL1Ap, is required for multiple stages of plant growth and development. Plant Physiol 126:47–68

Kang B, Rancour D, Bednarek S (2003b) The dynamin-like protein ADL1C is essential for plasma membrane maintenance during pollen maturation. Plant J 35:1–15

Kang S, Jing J, Hai P, Kyeong P, Hyun J, Jeong L, Hwang I (1998) Molecular cloning of an *Arabidopsis* cDNA encoding a dynamin-like protein that is localized to plastids. Plant Mol Biol 38:437–447

Koenig JH, Ikeda K (1989) Disappearance and reformation of synaptic vesicle membrane upon transmitter release observed under reversible blockage of membrane retrieval. J Neurosci 9:3844–3860

Kuravi K, Nagotu S, Krikken A, Sjollema K, Deckers M, Erdmann R, Veenhuis M, van der Klei I (2006) Dynamin-related proteins Vps1p and Dnm1p control peroxisome abundance in Saccharomyces cerevisiae. J Cell Sci 119:3994–4001

Lam BC, Sage TL, Bianchi F, Blumwald E (2002) Regulation of ADL6 activity by its associated molecular network. Plant J 31:565–576

Latterich M, Frohlich KU, Schekman R (1995) Membrane fusion and the cell cycle: Cdc48p participates in the fusion of ER membranes. Cell 82:885–893

Lauber MH, Waizenegger I, Steinmann T, Schwarz H, Mayer U, Hwang I, Lukowitz W, Jurgens G (1997) The *Arabidopsis* KNOLLE protein is a cytokinesis-specific syntaxin. J Cell Biol 139:1485–1493

Lee Y-R, Giang HM, Liu B (2001) A novel plant kinesin-related protein specifically associated with the phragmoplast organelles. Plant Cell 13:2427–2439

Liu B, Cyr RJ, Palevitz BA (1996) A kinesin-like protein, KatAp, in the cells of *Arabidopsis* and other plants. Plant Cell 8:119–132

Lukowitz W, Mayer U, Jurgens G (1996) Cytokinesis in the *Arabidopsis* embryo involves the syntaxin-related KNOLLE gene product. Cell 84:61–71

Mano S, Nakamori C, Kondo M, Hayashi M, Nishimura M (2004) An *Arabidopsis* dynamin-related protein, DRP3A, controls both peroxisomal and mitochondrial division. Plant J 38:487–498

May A, Whiteheart S, and Weis W (2001) Unraveling the Mechanism of the Vesicle Transport ATPase NSF, the N-Ethylmaleimide-sensitive Factor. J Biol Chem 276:21991–21994

McNiven MA (1998) Dynamin: a molecular motor with pinchase action. Cell 94:151–154

Mineyuki Y, Gunning B (1990) A role for preprophase bands of microtubules in maturation of new cell walls, and a general proposal on the function of preprophase band sites in cell division in higher plants. J Cell Sci 97:527–537

Miyagishima S, Froehlich J, Osteryoung K (2006) PDV1 and PDV2 mediate recruitment of the dynamin-related protein ARC5 to the plastid division site. Plant Cell 18:2517–2530

Miyagishima S, Nishida K, Mori T, Matsuzaki M, Higashiyama T, Kuroiwa H, Kuroiwa T (2003) A plant-specific DRP forms a ring at the chloroplast division site. Plant Cell 15:655–665

Molendijk A, Bischoff F, Rajendrakumar C, Friml J, Braun M, Gilroy S, Palme K (2001) *Arabidopsis thaliana* Rop GTPases are localized to tips of root hairs and control polar growth. EMBO J 20:2779–2788

Mundel C, Baltz R, Eliasson Å, Bronner R, Grass N, Kräuter R, Evrard J-L, Steinmetz A (2000) A LIM-domain protein from sunflower is localized to the cytoplasm and/or nucleus in a wide variety of tissues and is associated with the phragmoplast in dividing cells. Plant Mol Biol 42:291–302

Nishida K, Takahara M, Miyagishima S, Kuroiwa H, Matsuzaki M, Kuroiwa T (2003) Dynamic recruitment of dynamin for final mitochondrial severance in a primitive red alga. Proc Natl Acad Sci USA 100:2146–2151

Nishihama R, Ishikawa M, Araki S, Soyano T, Asada T, Machida Y (2001) The NPK1 mitogen activated protein kinase kinase kinase is a regulator of cell plate formation in plant cytokinesis. Gen Develop 15:352–363

Nishihama R, Machida Y (2001) Expension of the phragmoplast during plant cytokinesis: a MAP kinase pathway may MAP it out. Curr Open Plant Biol 4:507–512

Nishihama R, Soyano T, Ishikawa M, Araki S, Tanaka H, Asada T, Irie K, Ito M, Terada M, Banno H, Yamazaki Y, Machida Y (2002) Expansion of cell plate in plant cytokinesis requires a kinesin-like protein/MAPKKK complex. Cell 109:87–99

Nishikawa S, Zinkl G, Swanson R, Maruyama D, Preuss D (2005) Callose (β-1,3 glucan) is essential for *Arabidopsis* pollen wall patterning, but not tube growth. BMC Plant Biol 5:22

Nishimura M, Stein M, Hou B, Vogel J, Edwards H, Somerville S (2003) Loss of a callose synthase results in salicylic acid-dependent disease resistance. Science 301:969–972

Northcote DH, Davet R, Lay J (1989) Use of antisera to localize callose, xylan and arabinogalactan in the cell plate, primary and secondary walls of plant cells. Planta 178:353–366

Østergaad L, Petersen M, Mattson O, Mundy J (2002) An *Arabidopsis* callose synthase. Plant Mol Biol 49:559–566

Otegui M, Staehelin L (2004) Electron tomographic analysis of post-meiotic cytokinesis during pollen development in *Arabidopsis thaliana*. Planta 218:501–515

Otegui M, Mastronarde D, Kang B, Bednarek S, Staehelin L (2001) Three-dimensional analysis of cellularization visualized by high resolution electron tomography. Plant Cell 13:2033–2051

Qadota H, Python CP Inoue SB, Arisawa M, Anraku Y, Zheng Y, Watanabe T, Levin DE, Ohya Y (1996) Identification of yeast Rho1p GTPase as a regulatory subunit of 1,3-β-glucan synthase. Science 272:279–281

Rancour D, Dickey C, Park S, Bednarek S (2002) Characterization of AtCDC48. Evidence for multiple membrane fusion mechanisms at the plane of cell division in plants. Plant Physiol 130:1241–1253

Samuels A, Giddings T, Staehelin L (1995) Cytokinesis in tobacco BY-2 and root tip cells: A new model of cell plate formation in higher plants. J Cell Biol 130:1345–1357

Sanderfoot A, Assaad F, Raikhel N (2000) The *Arabidopsis* Genome. An abundance of soluble N-ethylmaleimide-sensitive factor adaptor protein receptors. Plant Physiol 124:1558–1569

Satiat-Jeunemaitre B, Cole L, Bourett T, Howard R, Hawes C (1996) Brefeldin A effects in plant and fungal cells: something new about vesicle trafficking? J Microscopy 181:162–177

Schopfer CR, Hepler PK (1991) Distribution of membrane and the cytoskeleton during cell plate formation in pollen mother cells of *Tradescantia*. J Cell Sci 100:717–728

Shin H, Brown RM (1999) GTPase activity and biochemical characterization of a recombinant cotton fiber annexin. Plant Physiol 119:925–934

Smirnova E, Cox D, Bajer A (1995) Antibody against phosphorylated proteins (MPM-2) recognizes mitotic microtubules in endosperm cells of higher plant *Haemanthus*. Cell Motil Cytoskel 31:34–44

Smith L (2002) Plant cytokinesis: motoring to the finish. Curr Biol 12:R206–208

Sollner R, Glasser G, Wanner G, Somerville C, Jurgens G, Assaad F (2002) Cytokinesis-defective mutants of *Arabidopsis*. Plant Physiol 129:678–690

Staehelin LA, Hepler PK (1996) Cytokinesis in higher plants. Cell 84:821–824

Staehelin LA, Moore I (1995) The plant Golgi apparatus: structure, functional organization and trafficking mechanisms. Annu Rev Plant Physiol Plant Mol Biol 46:261–288

Stoppin-Mellet V, Canaday J and Lambert AM (1999) Characterization of microsome associated tobacco BY-2 centrins. Eur J Cell Biol 78:842–848

Stowell MH (1999) Nucleotide-dependent conformational changes in dynamin: evidence for a mechanochemical molecular spring. Nature Cell Biol 1:27–32

Strompen G, El Kasmi F, Richter S, Lukowitz W, Assaad FF, Jurgens G, Mayer U (2002) An *Arabidopsis* HINKEL gene encodes a kinesin-related protein involved in cytokinesis and and is expressed in a cell cycle-depependent manner. Curr Biol 12:153–158

Sweitzer SM, Hinshaw JE (1998) Dynamin undergoes a GTP-dependent conformational change causing vesiculation. Cell 93:1021–1029

Takei K, Haucke V, Slepnev V, Farsad K, Salazar M, Chen H, De Camilli P (1998) Generation of coated intermediates of clathrin-mediated endocytosis on protein-free liposomes. Cell 94:131–141

Takei K, Mcpherson PS, Schmid S, Camilli PD (1995) Tubular membrane invaginations coated by dynamin rings are induced by GTP-γ-S in nerve terminals. Nature 374:186–190

Takei K, Mundigl O, Daniell L, Camilli PDT (1996) The synaptic vesicle cycle: a single vesicle budding step involving clathrin and dynamin. J Cell Biol 133:1237–1250

Verma DPS (2001) Cytokinesis and building of the cell plate in Plants. Annu Rev Plant Physiol Plant Mol Biol 52:751–784

Verma DPS, Hong Z (2001) Plant callose synthase complexes. Plant Mol Biol 47:693–701

Waizenegger I, Lukowitz W, Assaad F, Schwarz H, Jurgens G, Mayer U (2000) The *Arabidopsis KNOLLE* and *KEULE* genes interact to promote vesicle fusion during cytokinesis. Curr Biol 10:1371–1374

Wickner W (2002) Yeast vacuoles and membrane fusion pathways. EMBO J 21:1241–1247

Wickner W, Haas A (2000) Yeast homotypic vacuole fusion: a window on organelle trafficking mechanisms. Annu Rev Biochem 69:247–275

Yasahura H, Sonobe S, Shiboaka H (1995) Effects of brefeldin A on the formation of the cell plate in tobacco BY-2 cells. Eur J Cell Biol 66:2749–281

Zhang Z, Hong Z, Verma DPS (2000) Phragmoplastin polymerizes into spiral coiled structures via intermolecular interaction of two self-assembly domains. J Biol Chem 275:8779–8784

Zheng H, Bednarek SY, Sanderfoot AA, Alonso J, Ecker JR, Raikhel NV (2002) NPSN11 is a cell plate-associated SNARE protein that interacts with the syntaxin KNOLLE. Plant Physiol 129:530–539

Part D
Cell Division and Differentiation

Plant Cell Monogr (9)
D.P.S. Verma and Z. Hong: Cell Division Control in Plants
DOI 10.1007/7089_2007_134/Published online: 21 August 2007
© Springer-Verlag Berlin Heidelberg 2007

Asymmetric Cell Divisions: Zygotes of Fucoid Algae as a Model System

Sherryl R. Bisgrove[1] (✉) · Darryl L. Kropf[2]

[1]Department of Biological Sciences, Simon Fraser University,
8888 University Drive, Burnaby, BC V5A 1S6, Canada
sbisgrov@sfu.ca

[2]Department of Biology, University of Utah, 257 South 1400 East,
Salt Lake City, UT 84112, USA

Abstract Asymmetric cell divisions are commonly used across diverse phyla to generate different kinds of cells during development. Although asymmetric divisions play important roles during development in plants, algae, fungi, and animals, emerging data indicate that there is some variability amongst the mechanisms that are at play in these different organisms. Zygotes of fucoid algae have long served as models for understanding early developmental processes including cell polarization and asymmetric cell division. In addition, brown algae are phylogenetically distant from other organisms, including plant models, a feature that makes them interesting from a comparative perspective (Andersen 2004; Peters et al. 2004). This monograph focuses on advances made toward understanding how asymmetric divisions are regulated in fucoid algae and, where appropriate, comparisons are made to higher plant zygotes.

1
Introduction

How does a single cell, the zygote, give rise to a complex organism with many different cell and tissue types? The answer to this question lies in the ability of cells in a growing embryo to acquire separate identities, a feat that is often accomplished by asymmetric cell divisions. By definition, asymmetric cell divisions produce nonidentical daughter cells and can thereby initiate the process of cell differentiation. Asymmetric cell divisions are known to play important roles in development across diverse plant and algal phyla. Examples include the first cell division in many zygotes (Brownlee 2004; Gallagher and Smith 1997; Okamoto et al. 2005; Zernicka-Goetz 2004), as well as the production of gonidial and somatic cells in *Volvox carteri* (Kirk 2004; Schmitt 2003), reproductive initial cells from caulonema filaments in moss (Cove et al. 2006; Schumaker and Dietrich 1998), rhizoids from prothalli cells in ferns (Murata and Sugai 2000), stomata on the epidermal surfaces of leaves (Lucas et al. 2006; Nadeau and Sack 2002, 2003), and microspores during pollen development (Park et al. 2004; Twell et al. 1998). Because of the importance of asymmetric divisions in development, the mechanisms that regulate the process are under investigation in several model organisms. In this monograph, we focus on advances made toward understanding how asymmetric divisions are regulated in zygotes of fucoid brown algae.

1.1
Asymmetric Divisions and Cell Fate Decisions

Generally, there are three ways by which the products of an asymmetric division acquire separate identities (Fig. 1):

1. Developmental determinants can be differentially partitioned between cells during division. In this case, each cell inherits a different set of cytoplasmic instructions that lead it down a unique developmental pathway. Because cell fate is controlled by determinants located within the cytoplasm, this type of development is often referred to as intrinsic or cell-autonomous (Fig. 1a). Both the first division of the *Caenorhabditis elegans* zygote and the divisions of neuroblasts in *Drosophila melanogaster* embryos represent examples of asymmetric divisions in which intrinsic factors control daughter cell fates (Betschinger and Knoblich 2004; Cowan and Hyman 2004).
2. In some cases, the cytokinetic plane is positioned such that the daughter cells are placed in different locations within the developing organism. Each cell then receives a unique set of positional cues from neighboring cells or the environment that dictate its fate (Fig. 1b). Since cell identities are determined by signals received from external sources, this type of development is known as extrinsic or non-cell-autonomous. In the *Arabidopsis thaliana* root, for example, the decision to become either an endodermal or a cortical cell depends on an asymmetric cell division that places daughter cells in different cell files in the root. Signals from neighboring cells then direct the daughters down different developmental pathways (Heidstra et al. 2004; Scheres et al. 2002).
3. An asymmetric division can produce daughters of different sizes and/or shapes, and these morphological differences determine the developmental pathway that each cell will follow (Fig. 1c). In *V. carteri*, asymmetric divisions generate small and large daughter cell pairs, and the size of the cell then activates either a somatic or a germline developmental program (Cheng et al. 2005; Kirk et al. 1993; Schmitt 2003).

Asymmetric divisions are commonly regulated in a three-step process. In the first step, cells polarize (Fig. 2). Sometimes there are obvious cytological or morphological changes associated with cell polarization while in other cases the polarity is more subtle, and may be manifested simply by the fact that the ends of the cell lie in different positions in the developing organism. After cell polarization, the mitotic apparatus (step 2) and the site of cytokinesis (step 3) must be positioned appropriately with respect to the axis defined by

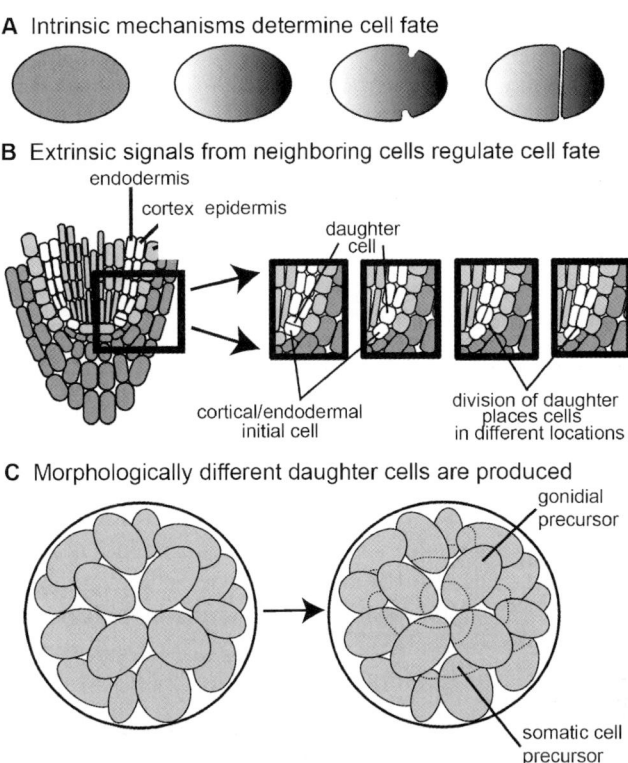

Fig. 1 Mechanisms by which asymmetric cell divisions generate diverse cell types during development. **a** In *C. elegans* zygotes, developmental determinants (*gray shading*) are segregated to one end of the zygote. When cytokinesis occurs, the daughter cells each inherit cytoplasm that is qualitatively different. **b** In *A. thaliana* roots, cortical and endodermal cells are produced through cell divisions that ultimately place cells in different files within the developing root. A cortical/endodermal initial cell divides to produce a daughter cell. When the daughter divides, the cell plate is laid down parallel with the longitudinal axis of the root, placing the two new cells in different cell files. Signals from neighboring cells then direct the adoption of either a cortical or an endodermal fate (redrawn from Scheres et al. 2002). **c** Asymmetric cell divisions in *V. carteri* embryos generate larger cells that will become reproductive gonidia and smaller, somatic cell precursors (Green and Kirk 1981). *Dashed lines* indicate sites of cytokinesis

the polarity of the cell. Alignment of the mitotic apparatus with the cellular axis ensures that each daughter will inherit both the appropriate cellular domains and a full chromosomal complement after division. During cytokinesis, the cell plate is positioned to bisect both the mitotic apparatus and the cellular axis correctly. Although cells usually polarize first, the order in which the latter two steps occur can vary. In zygotes of fucoid algae, for example, the mitotic apparatus is positioned first and then the site of cytokinesis is specified by the position of the mitotic apparatus (Fig. 2a; Bisgrove et al. 2003).

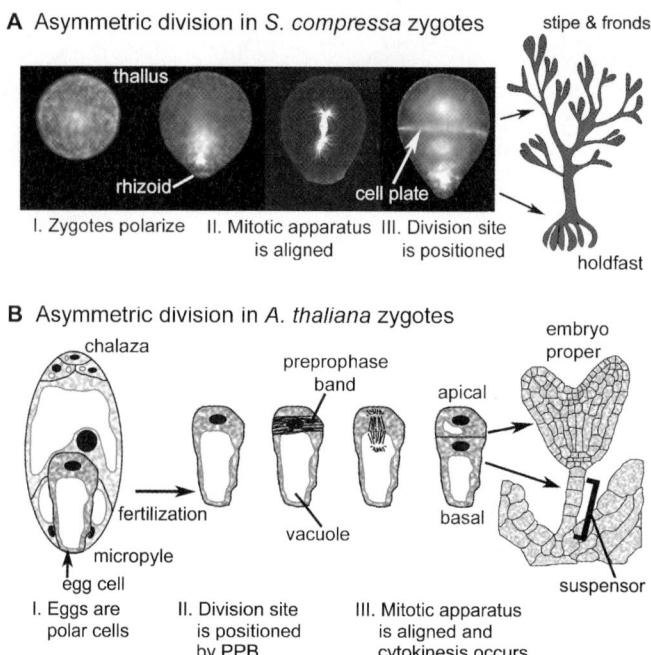

Fig. 2 Asymmetric cell divisions are commonly regulated in three steps. **a** *Silvetia compressa* eggs are spherical in shape with no obvious asymmetries, and polarization (I) is first manifested morphologically several hours after fertilization when increased secretion on one hemisphere produces a bulge, the rhizoid. The opposite end of the zygote is termed the thallus, and the axis defined by the two poles is the rhizoid/thallus axis. Next, the mitotic apparatus aligns parallel with the rhizoid/thallus axis (II). Finally, cytokinesis occurs and the cell plate is positioned perpendicular to the rhizoid/thallus axis (III). The three zygotes shown in the panels corresponding to I and III were stained with fluorescein diacetate which labels the cell plate, perinuclear regions, and cytoplasm. The zygote in II is in metaphase and was labeled with anti-alpha tubulin antibodies (image kindly provided by Nick T. Peters). **b** Asymmetric divisions in *A. thaliana* zygotes are also regulated in a three-step process, but the order in which the steps occur is different than it is in fucoid algae. In plants, polarity is acquired by the egg cell during development of the embryo sac (I). After fertilization, a preprophase band of microtubules marks the position of the first zygotic division (II) and then the mitotic apparatus is positioned with respect to both the cellular axis and the predetermined division site (Webb and Gunning 1991). Drawn using Drews and Yadegari (2002), Mayer et al. (1993), and Webb and Gunning (1991) as guides

In this case, proper placement of the spindle is required for correct specification of the cytokinetic site. Alternatively, the site of cytokinesis can be specified prior to mitosis in accordance with cues located in the cortex of the cell (Fig. 2b). Because the site of cytokinesis is determined before mitosis, this mechanism does not require precise positioning of the spindle. Instead, the mitotic apparatus needs only to align well enough to ensure that each daughter cell inherits a nucleus after telophase. This method is commonly employed by plant somatic cells, including zygotes. In these cells, a preprophase band of microtubules transiently forms in the cell cortex and marks the upcoming division site (Brown and Lemmon 2001; Marcus et al. 2005; Webb and Gunning 1991).

2
Zygotes of Fucoid Algae as a Model System

Zygotes of fucoid algae have, for many years, been a fruitful system in which to study the mechanisms by which cells acquire polarity and regulate asymmetric cell divisions, mainly because they are easy to manipulate and analyze in the laboratory (for recent reviews, see Brownlee 2004; Katsaros et al. 2006). Fucoid algae are marine brown algae, belonging to the Phaeophyceae class of stramenopiles (Andersen 2004). In nature they grow attached to rocks in the intertidal zone where they reproduce by releasing large, spherical eggs and biflagellated, motile sperm into the surrounding seawater. Gamete release can be induced from reproductive fronds in the lab and thousands of synchronously developing zygotes are easy to obtain for experimental analyses. The zygotes are relatively large, up to 100 μm in diameter, a size that renders them amenable to micromanipulation and analyses that require spatial measurements of subcellular features. Soon after fertilization zygotes settle onto the substratum, a rock in the intertidal zone, or a coverslip in the lab, and a sticky adhesive is secreted that firmly anchors them in place. Eggs are spherically shaped cells with no detectable asymmetries. However, within the first few hours following fertilization there are extensive cytoplasmic and morphological changes that result in asymmetric cells with rhizoid and thallus poles (Fig. 2a). To establish polarity, zygotes sense a wide array of environmental cues, although light is probably the dominant signal in nature. Zygotes developing in unidirectional light form rhizoids on their shaded hemispheres. An early sign of polarity is the preferential localization of secretion to the rhizoid pole, and increased secretion at this pole eventually produces a bulge, the tip-growing rhizoid (Fig. 2). When the first division occurs, about 24 h after fertilization (AF), it is oriented transverse to the rhizoid/thallus axis and bisects the zygote into two morphologically distinct cells with different developmental fates. The thallus cell gives rise to most of the photosynthetic and reproductive organs of the mature alga, while the rhizoid cell eventually becomes the holdfast that anchors the alga to its rock on the beach.

The first zygotic division in higher plants is also an asymmetric one that produces two morphologically distinct daughter cells with different developmental fates (Fig. 2b). The smaller apical cell is cytoplasmically dense and its progeny give rise to most of the developing embryo, while the larger, vacuolate basal cell divides only a few more times to form a single file of cells. The uppermost cell in this file becomes part of the root meristem and the remaining cells form the suspensor, a structure that attaches the embryo to the ovule (Laux et al. 2004; Souter and Lindsey 2000; Torres-Ruiz 2004). Although the developmental pattern that is set up by the first zygotic cell division is similar in plants and fucoid algae, there are key mechanistic differences between the two. In many plants, for example, polarity arises in the egg prior to fertilization rather than in the zygote. Plant eggs and zygotes are also buried within the ovule where their development can be influenced by surrounding maternal tissues. Zygotes of fucoid algae, on the other hand, are free-living and they develop in response to vectorial information in the environment such as sunlight from above (Brownlee 2004). Because plant eggs and zygotes are relatively inaccessible, approaches that involve manipulating individual cells are difficult. Instead, molecular/genetic analyses of mutants are being used to address questions of cell polarity and asymmetric divisions. This research is yielding interesting data, but our understanding of how asymmetric divisions are regulated in plant zygotes is still rudimentary. In contrast, the free-living zygotes of fucoid algae are easy to access and are amenable to physical manipulations. Over the years research on fucoid algae has provided a wealth of mechanistic data and, although many questions still remain, we are beginning to understand how asymmetric cell divisions are regulated in these zygotes.

3
Polarization and Germination in Zygotes

Fucoid zygotes have long served as a paradigm for investigating the mechanisms by which polarity is established following fertilization. In 1920, Hurd reported that monochromatic blue light polarizes zygotes (Hurd 1920), and since that time many other vectorial cues, including electrical, ionic, and osmotic gradients, have been shown to induce a growth axis (for a review, see Jaffe 1969). These diverse stimuli likely activate distinct signal transduction pathways that converge at a common response, formation of a growth axis (Kropf et al. 1999). The presumed goal is to maximize the chance that the rhizoid will grow into a crevice on the rocky surface and thereby permanently anchor the developing embryo in the turbulent intertidal environment.

But when is polarity first set up? Is the fertilized egg apolar until it senses its environment? Recent work has shown that in fact polarity is first set up at fertilization (Hable and Kropf 2000). Sperm entry induces a rhizoid pole to form at that site and a branching actin network rapidly assembles in the cell cortex there (Fig. 3a). A zygote has a greater density than seawater and settles rapidly onto the rocky substratum with its sperm-induced rhizoid pole randomly oriented with respect to the surface. Over the next 2 h the sperm

A Polarization

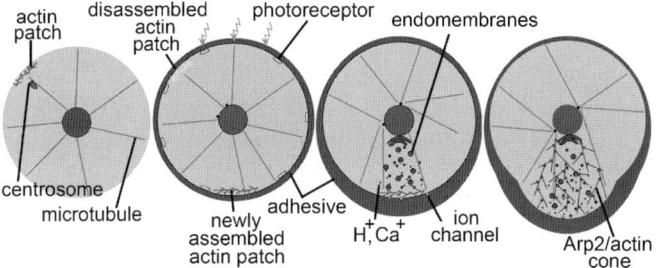

B Centrosomal alignment

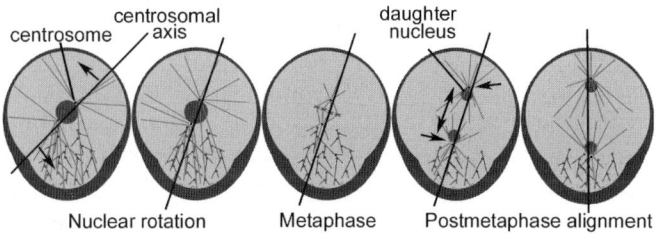

C Cytokinesis

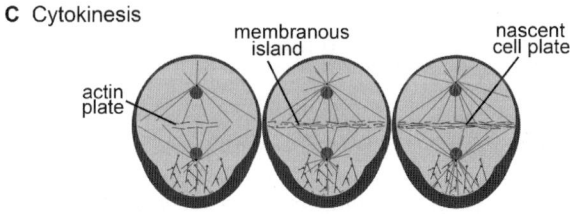

Fig. 3 Mechanism of asymmetric cell division in zygotes of fucoid algae. **a** Fertilization induces formation of a cortical actin patch that marks the rhizoid pole of a default axis. Photopolarization causes disassembly of the sperm-induced patch and assembly of a new patch at the shaded pole. Endomembrane cycling then becomes focused to the rhizoid pole as the nascent axis is amplified, and cytosolic ion gradients are generated. At germination, the actin array is remodeled into a cone nucleated by the Arp2/3 complex. During early development the paternally inherited centrosomes migrate to opposite sides of the nuclear envelope and acquire microtubule nucleation activity, but microtubules play only an indirect role in polarization. See text for details. **b** Centrosomal alignment begins with a premitotic rotation of the nucleus that partially aligns the centrosomal axis (defined by a line drawn through the two centrosomes) with the rhizoid/thallus axis. When the metaphase spindle forms it is partially aligned with the rhizoid/thallus axis. Postmetaphase alignment brings the telophase nuclei into almost perfect register with the rhizoid/thallus axis. *Arrows* indicate directions of nuclear movements. **c** Cytokinesis is positioned between the two daughter nuclei. A plate of actin assembles in the midzone between the nuclei, then membranous islands are deposited in the cytokinetic plane. The islands consolidate and cell plate materials are deposited in the division plane. All of these structures mature centrifugally, beginning in the middle of the zygote and progressing outward to the cell cortex

pronucleus migrates to the egg pronucleus utilizing microtubules (Swope and Kropf 1993), and the zygote secretes a cell wall (Quatrano 1982) and an adhesive that attaches it firmly to the rock (Hable and Kropf 1998). Once attached, the young zygote monitors its environment for positional information. Perceived environmental cues are integrated and used to specify a new growth axis that is appropriate for the environmental context. Under normal growth conditions the sperm-induced axis is usually overridden by environmental cues, and it can therefore be considered a default axis to be used only if the zygote fails to perceive positional information.

Unidirectional light is probably the most relevant vector in the intertidal environment, and is easy to apply in a laboratory setting. Photopolarization induces a new rhizoid pole on the shaded hemisphere (Fig. 3a), toward the rocky substratum. Although zygotes can perceive different light qualities, blue light is most effective. The photoreceptor is thought to reside at or near the plasma membrane (Jaffe 1958), and may be a rhodopsin-like protein (Gualtieri and Robinson 2002; Robinson et al. 1998). How light perception on one hemisphere of the zygote is transduced into a rhizoid pole on the opposite hemisphere is not well understood, but may involve formation of cGMP gradients resulting from differential photoreceptor activation (Robinson and Miller 1997) and/or activation of a plasma membrane redox chain on the shaded hemisphere (Berger and Brownlee 1994). Pharmacological studies indicate that photopolarization also requires signaling through a tyrosine kinase-like protein (Corellou et al. 2000a). At the downstream end, signal transduction results in depolymerization of the cortical actin at the sperm-entry site and polymerization of a new branching actin network nucleated by the Arp2/3 complex at the new rhizoid pole (Alessa and Kropf 1999; Hable et al. 2003; Hable and Kropf 2005). Thus, cortical actin localization is a faithful marker of the existing developmental axis.

Beginning about 4 h AF, the existing axis becomes steadily reinforced, or amplified. The essence of axis amplification is targeting of the endomembrane system and generation of cytosolic ion gradients (Fig. 3a). Both endocytotic and exocytotic limbs of membrane cycling are dispersed throughout the cytoplasm in young zygotes, but gradually become focused to the rhizoid pole (Hadley et al. 2006). This results in preferential secretion of adhesive at the rhizoid and may also establish a cortical domain with unique molecules in the rhizoid membrane and/or cell wall (Belanger and Quatrano 2000b; Fowler and Quatrano 1997). Simultaneously, cytosolic gradients of H^+ and Ca^{2+} are generated with highest activity at the rhizoid pole (Berger and Brownlee 1993; Kropf et al. 1995; Pu and Robinson 2003). Cytosolic H^+ and Ca^{2+} gradients and endomembrane cycling may comprise a positive feedback loop in which local elevation of H^+ and Ca^{2+} activity stimulate secretion and insertion of ion transporters at the rhizoid pole, thereby strengthening the ion gradients and promoting further secretion. However, it should be noted that to date there is no direct evidence for transporter accumulation at the rhizoid pole. Surprisingly, the axis remains labile throughout the amplification period; when the direction of the light vector is changed actin, endomembranes, and ion gradients reposition to the new rhizoid pole.

Just prior to germination, the developmental axis becomes fixed in space and insensitive to subsequent environmental cues. Axis fixation is thought to involve formation of axis-stabilizing complexes at the rhizoid pole comprised of transmembrane bridges from the cortical actin to sulfated polysaccharides in the cell wall (Fowler and Quatrano 1997). Total mRNA accumulates at the thallus pole during axis fixation (Bouget et al. 1996), and some localized mRNAs may serve as developmental determinants that are asymmetrically partitioned when the zygote divides.

Rhizoid outgrowth denotes germination and is driven by an increase in targeted secretion. The branching Arp2/actin network expands dramatically at germination forming a continuum that extends from the rhizoid face of the nuclear envelope to the cortical domain in the rhizoid tip (Fig. 3a; Hable and Kropf 2005). The very apex is relatively devoid of cytoskeleton and is filled with secretory vesicles, as has been observed in other tip growing cells including pollen tubes (Lovy-Wheeler et al. 2005). Germinated zygotes exhibit negative phototropism, which is preceded by a shift in the actin array and the vesicle accumulation zone toward the shaded side of rhizoid where new growth becomes focused (Hable and Kropf 2005). These and other findings (Brawley and Quatrano 1979) suggest that the extensive actin array transports secretory vesicles from Golgi to the apical growth site. Microtubules are not required for polarization or germination, but may help organize the actin/endomembrane system. Microtubule depolymerization or stabilization results in a more dispersed endomembrane system (Hadley et al. 2006) and fat rhizoids (Kropf et al. 1990).

4
Microtubules and Asymmetric Cell Division

Although microtubules are not required for polarization or germination, they are essential for cell division. They are the major structural component of the mitotic spindle, and their organization within the cell determines both the position of the mitotic apparatus and the placement of the cell plate during division. How, then, are microtubules organized in developing zygotes? Like animals, fucoid algae have discrete microtubule organizing centers called centrosomes that regulate the distribution and organization of microtubules in the cell (Fig. 3b; Bisgrove et al. 1997; Motomura and Nagasato 2004; Nagasato et al. 1999). Hence, the location of the centrosomes during cell division determines both the position of the mitotic apparatus and the subsequent site of cell plate deposition. Because of their importance, the centrosomes have been monitored in zygotes during polarization and cell division.

4.1
Microtubule Organization During Polarization

Unfertilized eggs do not have centrosomes and microtubules emanate from the nucleus in an array that is evenly dispersed around the nuclear periphery (Bisgrove et al. 1997; Motomura 1994; Nagasato et al. 1999). The centriolar components of the centrosomes are acquired from the flagellar basal bodies of the sperm at fertilization (Fig. 3a). Since sperm are biflagellated, the egg receives two centrioles; they migrate with the sperm pronucleus through the cytoplasm and are deposited on the nuclear envelope at karyogamy (Bisgrove et al. 1997; Motomura 1994; Motomura and Nagasato 2004; Nagasato et al. 1999; Nagasato 2005; Swope and Kropf 1993). As development proceeds, the centrosomes slowly separate from each other by migrating around the nucleus until they reach positions on opposite sides of the nuclear envelope. At the same time, there is a gradual reorganization of the microtubules into an array in which microtubules emanate mainly from the two perinuclear centrosomes outward into the cortex of the cell. These steps occur over several hours and are not completed until shortly before zygotes enter mitosis, about 16 h AF. Although centrosomal separation does occur concurrently with polarization of the zygote, the two processes appear to proceed independently of each other since treatments that inhibit polarization or tip growth do not affect centrosomal separation and vice versa (Bisgrove and Kropf 1998).

Just prior to mitosis, the centrosomes come to rest on opposite sides of the nucleus and microtubules extend from them out into the cortex of the cell. The rhizoid appears to provide a favorable environment for microtubules, since they are more abundant in this part of the zygote. In addition to the microtubules that emanate from the centrosomes, recent studies in living zygotes microinjected with fluorescently labeled tubulin have revealed a cortical array that is not seen in fixed preparations (Corellou et al. 2005). In young zygotes the cortical microtubules are randomly arranged and distributed evenly around the cell. However, as zygotes develop, the cortical microtubules localize preferentially to the presumptive rhizoid where they become denser as zygotes germinate and the rhizoid elongates. Although the function of these cortical microtubules is unknown, it has been postulated that they might be involved in shaping the rhizoid as it grows. The microtubules appear to originate in the cell cortex where they form an array that is not contiguous with the centrosomes. It is, therefore, unlikely that the cortical microtubules are involved in positioning the mitotic apparatus or the division site (Corellou et al. 2005). Nonetheless, the abundance of both centrosomal and cortical microtubules suggests that the rhizoid provides an environment conducive to microtubule assembly and/or stabilization.

4.2
Positioning the Mitotic Apparatus

When zygotes enter mitosis the centrosomes form the poles of the metaphase spindle, and their position determines the placement of the spindle. Initially, the centrosomal axis, defined by a line drawn through the two centrosomes, is not well aligned with the rhizoid/thallus growth axis (Fig. 3b). However, before zygotes enter mitosis there is a nuclear rotation that partially aligns the centrosomal axis with the growth axis and results in crudely aligned metaphase spindles (Allen and Kropf 1992; Bisgrove and Kropf 1998, 2001; Corellou et al. 2000b). Alignment of the centrosomes continues as zygotes progress through mitosis, and by the end of telophase the centrosomal axis is parallel with the growth axis (Bisgrove and Kropf 2001).

Treating zygotes with a battery of inhibitors at different times during centrosomal alignment disrupts the premetaphase rotation of the nucleus but does not affect alignment during telophase, suggesting that the pre- and postmetaphase alignments are mechanistically different. The existing evidence supports a model in which premetaphase nuclear rotation is effected by microtubules that extend from the centrosomes out toward the cortex of the zygote (Allen and Kropf 1992; Bisgrove and Kropf 1998). These microtubules are most likely dynamic, growing out from the centrosomes and disassembling back toward them. Microtubules that reach the cell cortex appear to be captured in the actin-containing bridges that link the plasma membrane to the cell wall, since treatments that affect actin or the cell wall also disrupt nuclear rotation (Alessa and Kropf 1999; Bisgrove and Kropf 2001; Henry et al. 1996). Actin–cell wall bridges are concentrated in the rhizoid apex (Henry et al. 1996) and so microtubules are preferentially captured there. Motors located either at the centrosome or the cortex are postulated to exert a pulling force on the captured microtubules. By chance, one centrosome usually resides closer to the rhizoid apex; this centrosome will have more microtubules in contact with the cortex and will be pulled toward the rhizoid apex. The other centrosome will move toward the thallus pole, resulting in a rotation that partially aligns the centrosomal axis. A similar microtubule-based "search and capture" mechanism is thought to align the mitotic apparatus in budding yeast and animal cells (reviewed by McCarthy and Goldstein 2006). When the metaphase spindle forms, it is crudely aligned with the rhizoid/thallus axis. Spindle formation requires the activities of Kinesin-5 motors to maintain spindle bipolarity. In addition, Kinesin-5 motors also appear to be involved in maintaining the integrity of spindle poles in fucoid algae, an activity that has not yet been reported for these motors in other cell types (Peters and Kropf 2006).

As zygotes exit metaphase, the centrosomal axis continues to align, albeit by a mechanism that appears to be different from the nuclear rotation that occurs before mitosis. Although this phase of alignment is not well understood, it is temporally associated with an elongation of the mitotic apparatus that occurs during anaphase and telophase (Fig. 3b). One possibility is that microtubule-based centering mechanisms acting on the centrosomes during spindle elongation could contribute to this phase of alignment (Bisgrove and Kropf 2001). Centrosomal centering involves interactions of microtubule ends with stationary objects such as the periphery of the cell. Polymerizing microtubules that impact the cell boundary can exert a force that pushes the centrosome toward the middle of the cell or, alternatively, cytoplasmic motors acting on shortening microtubules can pull the centrosome toward the cortex (Howard 2006). In theory, in a cell that is longer than it is wide, centrosomal centering forces could align the anaphase/telophase mitotic apparatus if the centrosomes move as a unit. Similar microtubule-based forces appear to be involved in centering the nucleus in fission yeast cells (for example see Daga et al. 2006).

4.3
Cytokinesis

By the end of telophase, the centrosomal axis is aligned parallel with the rhizoid/thallus axis. Microtubules radiating from the centrosomes on the daughter nuclei meet and interdigitate in the midzone of the remnant spindle. The zone of microtubule overlap extends outward to the cell cortex and predicts the position of the future division site (Bisgrove et al. 2003). During cytokinesis, a plate of actin first appears in the zone where microtubules meet, and then membrane is deposited in islands throughout the cytokinetic plane (Fig. 3c). The membranous islands fuse into a continuous compartment into which cell wall materials are deposited. All of these structures mature in a centrifugal fashion, from the center of the cell outward (Belanger and Quatrano 2000a; Bisgrove and Kropf 2004). Similar cytoskeletal arrays have been observed in other brown algal cells during cytokinesis (Karyophyllis et al. 2000; Katsaros et al. 1983, 2006; Katsaros and Galatis 1992; Nagasato and Motomura 2002a,b; Varvarigos et al. 2005). Plant cells also divide centrifugally, but they utilize a unique microtubule-based structure, the phragmoplast, during cytokinesis (see Jurgens 2005 for a recent review).

How is the division site chosen? In general, there are two ways by which cells determine a site for cytokinesis:

1. In metazoan, protist, and some plant cells the position of the mitotic apparatus during metaphase/anaphase or telophase determines the site of cytokinesis. In animal cells cytokinesis occurs by furrowing, and spindle microtubules appear to deliver signals to the cell cortex that determine the site of furrow formation (reviewed by Wadsworth 2005). Similarly, during cellularization in endosperm and female gametophytes, radial microtubules define cellular spaces around nuclei and cell plate deposition

occurs at the boundaries (Brown and Lemmon 2001; Otegui and Staehelin 2000; Pickett-Heaps et al. 1999).
2. Alternatively, in somatic plant cells, fission yeast, and budding yeast, cell polarity specifies the site of cytokinesis in accordance with localized cortical cues. In these cells the site of cytokinesis is determined before mitosis rather than by the mitotic apparatus during or after the nuclear division (Arkowitz 2001; Hoshino et al. 2003; Marcus et al. 2005; Wasteneys 2002; Wu et al. 2003).

In fucoid algae, the position of the two daughter nuclei at the end of telophase determines the division site. This conclusion is based on experiments in which the colinearity between the telophase nuclei and the rhizoid/thallus axis was uncoupled. Cytokinesis always occurred between telophase nuclei rather than perpendicular to the rhizoid/thallus axis, indicating that it is nuclear position and not cell polarity that defines the site of cytokinesis (Bisgrove et al. 2003). At the time of cytokinesis, microtubules radiating from the centrosomes define domains around the nuclei. Cytokinesis occurs in the zone of microtubule overlap between telophase nuclei, in a manner similar to the cellularization that occurs in endosperm and female gametophytes.

5
Zygotic Cell Division and Cell Fate Decisions

Is proper placement of the zygotic division developmentally important in fucoid algae? If so, why? Generally, there are three ways by which asymmetric cell divisions influence cell fate decisions: via the intrinsic, extrinsic, and morphological pathways discussed above. In fucoid algae, there is evidence indicating that all three pathways may be operational. Zygotic polarity develops in response to positional cues from the environment (extrinsic signals). Sperm entry and environmental vectors, light or ion gradients for instance, determine where the rhizoid will form. When the zygote divides, rhizoid and thallus cells of different shapes are produced, and there is evidence to support the idea that these morphological differences are developmentally important. Pulse-treating zygotes with pharmacological agents that perturb either the cytoskeleton or secretion disrupts placement of the division and can affect subsequent embryogenesis. In particular, severely misaligned divisions in which the cell plate bisects the rhizoid tip disrupt the ability of embryos to elongate their rhizoids normally. Rhizoid extension is either blocked or two rhizoids are initiated, depending on the pharmacological agent used (Bisgrove and Kropf 1998; Shaw and Quatrano 1996). Finally, there is also evidence that suggests developmental determinants may be asymmetrically partitioned between rhizoid and thallus cells when the zygote divides (intrinsic signals). Poly(A)+ RNA is preferentially segregated to the thallus in germinated zygotes and two-celled embryos, and this asymmetric distribution of mRNA could play a role in determining cell fates (Bouget et al. 1995). Also, in an elegant set of laser microsurgery experiments, Berger et al. (1994) found that thallus cells quickly redifferentiated into rhizoid cells once they contacted residual cell wall from an ablated rhizoid, suggesting that developmental determinants might be localized in the rhizoid cell wall. Curiously, Bisgrove and Kropf (1998) found that moderate misalignments of the zygotic division had little effect on subsequent development. This observation suggests that if the division does segregate determinants, they are either tightly localized to the apical wall or daughter cell fates do not depend on precisely partitioning them.

In plants, assessing how the first zygotic division influences development is difficult because the relevant cells are buried deep within the maternal tissues of the ovule. Nonetheless, there is reason to believe that intrinsic, extrinsic, and morphological pathways may also have roles in plant zygotes and young embryos. Plant cells commonly make cell fate decisions in response to positional information (extrinsic cues), and genetic studies indicate that gametophytic and sporophytic tissues surrounding the zygote contribute to its development (reviewed by Laux et al. 2004). In addition, analyses of expression patterns have identified transcripts that are expressed in the zygote and differentially localized to either the apical or the basal cell of the two-celled embryo, suggesting that the first zygotic division differentially partitions determinants (Haecker et al. 2004; Laux et al. 2004; Lukowitz et al. 2004; Okamoto et al. 2005; Weterings et al. 2001). Finally, embryos of mutants with mispositioned division planes, such as *fass* and *gnom*, have morphological defects, suggesting that division plane alignment is important for morphogenesis (Busch et al. 1996; Geldner et al. 2003; Mayer et al. 1993; McClinton and Sung 1997; Shevell et al. 1994; Torres-Ruiz and Jurgens 1994).

6
Conclusions and Future Directions

Analyses conducted over the last several years have provided us with a basic understanding of how asymmetric cell divisions are regulated in zygotes of fucoid algae. The emerging evidence indicates that there are mechanistic differences between asymmetric divisions in brown algal and plant zygotes, a fact that is not surprising given the large phylogenetic distances that separate the two groups. In plants, the availability of genomic resources and molecular/genetic techniques are facilitating the identification of molecules that may play roles in asymmetric divisions and cell fate decisions. The lack of these resources for any species in the phaeophyte lineage has been perhaps the largest technical hurdle hindering molecular analyses in the brown algae. Recently, a project to sequence the genome of the marine brown alga *Ectocarpus siliculosis* was initiated by the French sequencing center GENOSCOPE (Peters et al. 2004). This project will move brown algae forward into the molecular/genomics era and enhance the feasibility of comparative analyses between phaeophytes and other eukaryotic lineages at the molecular level.

References

Alessa L, Kropf DL (1999) F-actin marks the rhizoid pole in living *Pelvetia compressa* zygotes. Development 126:201–209

Allen VW, Kropf DL (1992) Nuclear rotation and lineage specification in *Pelvetia* embryos. Development 115:873–883

Andersen RA (2004) Biology and systematics of heterokont and haptophyte algae. Am J Bot 91:1508–1522

Arkowitz RA (2001) Cell polarity: connecting to the cortex. Curr Biol 11:R610–R612

Belanger KD, Quatrano RS (2000a) Membrane recycling occurs during asymmetric tip growth and cell plate formation in *Fucus distichus* zygotes. Protoplasma 212:24–37

Belanger KD, Quatrano RS (2000b) Polarity: the role of localized secretion. Curr Opin Plant Biol 3:67–72

Berger F, Brownlee C (1993) Ratio confocal imaging of free cytoplasmic calcium gradients in polarising and polarised *Fucus* zygotes. Zygote 1:9–15

Berger F, Brownlee C (1994) Photopolarization of the *Fucus* sp zygote by blue light involves a plasma membrane redox chain. Plant Physiol 105:519–527

Berger F, Taylor A, Brownlee C (1994) Cell fate determination by the cell wall in early *Fucus* development. Science 263:1421–1423

Betschinger J, Knoblich JA (2004) Dare to be different: asymmetric cell division in *Drosophila*, *C. elegans*, and vertebrates. Curr Biol 14:R674–R685

Bisgrove SR, Kropf DL (1998) Alignment of centrosomal and growth axes is a late event during polarization of *Pelvetia compressa* zygotes. Dev Biol 194:246–256

Bisgrove SR, Kropf DL (2001) Asymmetric cell division in fucoid algae: a role for cortical adhesions in alignment of the mitotic apparatus. J Cell Sci 114:4319–4328

Bisgrove SR, Kropf DL (2004) Cytokinesis in brown algae: studies of asymmetric division in fucoid zygotes. Protoplasma 223:163–173

Bisgrove SR, Nagasato C, Motomura T, Kropf DL (1997) Immunolocalization of centrin in *Fucus distichus* and *Pelvetia compressa* (Fucales, Phaeophyceae). J Phycol 33:823–829

Bisgrove SR, Henderson DC, Kropf DL (2003) Asymmetric division in fucoid zygotes is positioned by telophase nuclei. Plant Cell 15:854–862

Bouget F-Y, Gerttula S, Quatrano RS (1995) Spatial redistribution of poly(A)+ RNA during polarization of the *Fucus* zygote is dependent upon microfilaments. Dev Biol 171:258–261

Bouget F-Y, Gerttula S, Shaw SL, Quatrano RS (1996) Localization of actin mRNA during the establishment of cell polarity and early cell divisions in *Fucus* embryos. Plant Cell 8:189–201

Brawley SH, Quatrano RS (1979) Sulfation of fucoidin in *Fucus* embryos. IV. Autoradiographic investigations of fucoidin sulfation and secretion during differentiation and the effect of cytochalasin treatment. Dev Biol 73:193–205

Brown RC, Lemmon BE (2001) The cytoskeleton and spatial control of cytokinesis in the plant life cycle. Protoplasma 215:35–49

Brownlee C (2004) From polarity to pattern: early development in fucoid algae. In: Lindsey K (ed) Annual plant reviews, vol. 12. Blackwell, Oxford, pp 139–155

Busch M, Mayer U, Jurgens G (1996) Molecular analysis of the *Arabidopsis* pattern formation gene *GNOM*: gene structure and intragenic complementation. Mol Gen Genet 250:681–691

Cheng Q, Pappas V, Hallmann A, Miller SM (2005) Hsp70A and GlsA interact as partner chaperones to regulate asymmetric division in *Volvox*. Dev Biol 286:537–548

Corellou F, Potin P, Brownlee C, Kloareg B, Bouget F-Y (2000a) Inhibition of the establishment of zygotic polarity by protein tyrosine kinase inhibitors leads to an alteration of embryo pattern in *Fucus*. Dev Biol 219:165–182

Corellou FC, Bisgrove SR, Kropf DL, Meijer L, Kloareg B, Bouget F-Y (2000b) A S/M DNA replication checkpoint prevents nuclear and cytoplasmic events of cell division including centrosomal axis alignment and inhibits activation of cyclin dependent kinase-like proteins in fucoid zygotes. Development 127:1651–1660

Corellou F, Coelho SMB, Bouget F-Y, Brownlee C (2005) Spatial re-organisation of cortical microtubules in vivo during polarisation and asymmetric division of *Fucus* zygotes. J Cell Sci 118:2723–2734

Cove D, Benzanilla M, Harries P, Quatrano R (2006) Mosses as model systems for the study of metabolism and development. Annu Rev Plant Biol 57:497–520

Cowan CR, Hyman AA (2004) Asymmetric cell division in *C. elegans*: cortical polarity and spindle positioning. Annu Rev Cell Dev Biol 20:427–453

Daga RR, Yonetani A, Chang F (2006) Asymmetric microtubule pushing forces in nuclear centering. Curr Biol 16:1544–1550

Drews GN, Yadegari R (2002) Development and function of the angiosperm female gametophyte. Annu Rev Genet 36:99–124

Fowler JE, Quatrano RS (1997) Plant cell morphogenesis: plasma membrane interactions with the cytoskeleton and cell wall. Annu Rev Cell Dev Biol 13:697–743

Gallagher K, Smith LG (1997) Asymmetric cell division and cell fate in plants. Curr Opin Cell Biol 9:842–848

Geldner N, Anders N, Wolters H, Keicher J, Kornberger W, Muller P, Delbarre A, Ueda T, Nakano A, Jurgens G (2003) The *Arabidopsis* GNOM ARF-GEF mediates endosomal recycling, auxin transport, and auxin-dependent plant growth. Cell 112:219–230

Green KJ, Kirk DL (1981) Cleavage patterns, cell lineages, and development of a cytoplasmic bridge system in *Volvox* embryos. J Cell Biol 91:743–755

Gualtieri P, Robinson KR (2002) A rhodopsin-like protein in the plasma membrane of *Silvetia compressa* eggs. Photochem Photobiol 75:76–78

Hable WE, Kropf DL (1998) Roles of secretion and the cytoskeleton in cell adhesion and polarity establishment in *Pelvetia compressa* zygotes. Dev Biol 198:45–56

Hable WE, Kropf DL (2000) Sperm entry induces polarity in fucoid zygotes. Development 127:493–501

Hable WE, Kropf DL (2005) The Arp2/3 complex nucleates actin arrays during zygote polarity establishment and growth. Cell Motil Cytoskel 61:9–20

Hable WE, Miller NR, Kropf DL (2003) Polarity establishment requires dynamic actin in fucoid zygotes. Protoplasma 221:193–204

Hadley R, Hable WE, Kropf DL (2006) Polarization of the endomembrane system is an early event in fucoid zygote development. BMC Plant Biol 6:5

Haecker A, Gross-Hardt R, Geiges B, Sarkar A, Breuninger H, Herrmann M, Laux T (2004) Expression dynamics of *WOX* genes mark cell fate decisions during early embryonic patterning in *Arabidopsis thaliana*. Development 131:657–668

Heidstra R, Welch D, Scheres B (2004) Mosaic analyses using marked activation and deletion clones dissect *Arabidopsis SCARECROW* action in asymmetric cell division. Genes Dev 18:1964–1969

Henry CA, Jordan JR, Kropf DL (1996) Localized membrane–wall adhesions in *Pelvetia* zygotes. Protoplasma 190:39–52

Hoshino H, Yoneda A, Kumagai F, Hasezawa S (2003) Roles of actin-depleted zone and preprophase band in determining the division site of higher-plant cells, a tobacco BY-2 cell line expressing GFP-tubulin. Protoplasma 222:157–165

Howard J (2006) Elastic and damping forces generated by confined arrays of dynamic microtubules. Phys Biol 3:54–66

Hurd AM (1920) Effect of unilateral monochromatic light and group orientation of germinating *Fucus* spores. Bot Gaz 70:25–50

Jaffe LF (1958) Tropistic responses of zygotes of the Fucaceae to polarized light. Exp Cell Res 15:282–299

Jaffe LF (1969) On the centripetal course of development, the *Fucus* egg, and self-electrophoresis. Dev Biol Suppl 3:83–111

Jurgens G (2005) Cytokinesis in higher plants. Annu Rev Plant Biol 56:281–299

Karyophyllis D, Katsaros C, Dimitriadis I, Galatis B (2000) F-actin organization during the cell cycle of *Sphacelaria rigidula* (Phaeophyceae). Eur J Phycol 35:25–33

Katsaros C, Galatis B (1992) Immunofluorescence and electron microscopic studies of microtubule organization during the cell cycle of *Dictyota dichotoma* (Phaeophyta, Dictyotales). Protoplasma 169:75–84

Katsaros C, Galatis B, Mitrakos K (1983) Fine structural studies on the interphase and dividing apical cells of *Sphacelaria tribuloides* (Phaeophyta). J Phycol 19:16–30

Katsaros C, Karyophyllis D, Galatis B (2006) Cytoskeleton and morphogenesis in brown algae. Ann Bot 97:679–693

Kirk DL (2004) Volvox. Curr Biol 14:R599–R600

Kirk MM, Ransick A, McRae SE, Kirk DL (1993) The relationship between cell size and cell fate in *Volvox carteri*. J Cell Biol 123:191–208

Kropf DL, Maddock A, Gard DL (1990) Microtubule distribution and function in early *Pelvetia* development. J Cell Sci 97:545–552

Kropf DL, Money NP, Gibbon BC (1995) Role of cytosolic pH in axis establishment and tip growth. Can J Bot 73:S126–S130

Kropf DL, Bisgrove SR, Hable WE (1999) Establishing a growth axis in fucoid algae. Trends Plant Sci 4:490–494

Laux T, Wurschum T, Breuninger H (2004) Genetic regulation of embryonic pattern formation. Plant Cell 16:S190–S202

Lovy-Wheeler A, Wilsen KL, Baskin TI, Hepler PK (2005) Enhanced fixation reveals the apical cortical fringe of actin filaments as a consistent feature of the pollen tube. Planta 221:95–104

Lucas JR, Nadeau JA, Sack FD (2006) Microtubule arrays and *Arabidopsis* stomatal development. J Exp Bot 57:71–79

Lukowitz W, Roeder A, Parmenter D, Somerville C (2004) A MAPKK kinase gene regulates extra-embryonic cell fate in *Arabidopsis*. Cell 116:109–119

Marcus AI, Dixit R, Cyr RJ (2005) Narrowing of the preprophase microtubule band is not required for cell division plane determination in cultured plant cells. Protoplasma 226:169–174

Mayer U, Buttner G, Jurgens G (1993) Apical–basal pattern formation in the *Arabidopsis* embryo: studies on the role of the *gnom* gene. Development 117:149–162

McCarthy EK, Goldstein B (2006) Asymmetric spindle positioning. Curr Opin Cell Biol 18:79–85

McClinton RS, Sung R (1997) Organization of cortical microtubules at the plasma membrane in *Arabidopsis*. Planta 201:252–260

Motomura T (1994) Electron and immunofluorescence microscopy on the fertilization of *Fucus distichus* (Fucales, Phaeophyceae). Protoplasma 178:97–110

Motomura T, Nagasato C (2004) The first spindle formation in brown algal zygotes. Hydrobiologia 512:171–176

Murata T, Sugai M (2000) Photoregulation of asymmetric cell division followed by rhizoid development in the fern *Ceratopteris* prothalli. Plant Cell Physiol 41:1313–1320

Nadeau JA, Sack FD (2002) Stomatal development in *Arabidopsis*. In: Somerville CR, Meyerowitz EM (eds) The *Arabidopsis* book. American Society of Plant Biologists, Rockville, MD, pp 1–28

Nadeau JA, Sack FD (2003) Stomatal development: cross talk puts mouths in place. Trends Plant Sci 8:294–299

Nagasato C (2005) Behavior and function of paternally inherited centrioles in brown algal zygotes. J Plant Res 118:361–369

Nagasato C, Motomura T (2002a) Influence of the centrosome in cytokinesis of brown algae: polyspermic zygotes of *Scytosiphon lomentaria* (Scytosiphonales, Phaeophyceae). J Cell Sci 115:2541–2548

Nagasato C, Motomura T (2002b) Ultrastructural study on mitosis and cytokinesis in *Scytosiphon lomentaria* zygotes (Scytosiphonales, Phaeophyceae) by freeze-substitution. Protoplasma 219:140–149

Nagasato C, Motomura T, Ichimura T (1999) Influence of centriole behavior on the first spindle formation in zygotes of the brown alga *Fucus distichus* (Fucales, Phaeophyceae). Dev Biol 208:200–209

Okamoto T, Scholten S, Lorz H, Kranz E (2005) Identification of genes that are up- or down-regulated in the apical or basal cell of maize two-celled embryos and monitoring their expression during zygote development by a cell manipulation- and PCR-based approach. Plant Cell Physiol 46:332–338

Otegui M, Staehelin LA (2000) Syncytial-type cell plates: a novel kind of cell plate involved in endosperm cellularization of *Arabidopsis*. Plant Cell 12:933–947

Park SK, Rahman D, Oh SA, Twell D (2004) *Gemini pollen 2*, a male and female gametophytic cytokinesis defective mutation. Sex Plant Rep 17:63–70

Peters AF, Marie D, Scornet D, Kloareg B, Cock JM (2004) Proposal of *Ectocarpus siliculosus* (Ectocarpales, Phaeophyceae) as a model organism for brown algal genetics and genomics. J Phycol 40:1079–1088

Peters NT, Kropf DL (2006) Kinesin-5 motors are required for organization of spindle microtubules in *Silvetia compressa* zygotes. BMC Plant Biol 6:19

Pickett-Heaps J, Gunning BE, Brown R, Lemmon B, Cleary A (1999) The cytoplast concept in dividing plant cells: cytoplasmic domains and the evolution of spatially organized cell. Am J Bot 86:153–172

Pu R, Robinson KR (2003) The involvement of Ca^{2+} gradients, Ca^{2+} fluxes, and CaM kinase II in polarization and germination of *Silvetia compressa* zygotes. Planta 217:407–416

Quatrano RS (1982) Cell-wall formation in *Fucus* zygotes: a model system to study the assembly and localization of wall polymers. In: Brown RM (ed) Cellulose and other natural polymer systems: biogenesis, structure, and degradation. Plenum, New York, pp 45–59

Robinson KR, Miller BJ (1997) The coupling of cyclic GMP and photopolarization of *Pelvetia* zygotes. Dev Biol 187:125–130

Robinson KR, Lorenzi R, Ceccarelli N, Gualtieri P (1998) Retinal identification in *Pelvetia fastigiata*. Biochem Biophys Res Commun 243:776–778

Scheres B, Benfey P, Dolan L (2002) Root development. In: Somerville CR, Meyerowitz EM (eds) The *Arabidopsis* book, DOI:10.1199/tab.0101. American Society of Plant Biologists, Rockville, MD, pp 1–18

Schmitt R (2003) Differentiation of germinal and somatic cells in *Volvox carteri*. Curr Opin Microbiol 6:608–613

Schumaker KS, Dietrich MA (1998) Hormone-induced signaling during moss development. Annu Rev Plant Phys Plant Mol Biol 49:501–523

Shaw SL, Quatrano RS (1996) The role of targeted secretion in the establishment of cell polarity and the orientation of the division plane in *Fucus* zygotes. Development 122:2623–2630

Shevell DE, Leu W-M, Gillmor CS, Xia G, Feldmann KA, Chua N-H (1994) EMB30 is essential for normal cell division, cell expansion, and cell adhesion in *Arabidopsis* and encodes a protein that has similarity to Sec7. Cell 77:1051–1062

Souter M, Lindsey K (2000) Polarity and signalling in plant embryogenesis. J Exp Bot 51:971–983

Swope RE, Kropf DL (1993) Pronuclear positioning and migration during fertilization in *Pelvetia*. Dev Biol 157:269–276

Torres-Ruiz RA (2004) Polarity in *Arabidopsis* embryogenesis. In: Lindsey K (ed) Annual plant reviews, vol 12. Blackwell, Oxford, pp 157–191

Torres-Ruiz R, Jurgens G (1994) Mutations in the *FASS* gene uncouple pattern formation and morphogenesis in *Arabidopsis* development. Development 120:2967–2978

Twell D, Park SK, Lalanne E (1998) Asymmetric division and cell-fate determination in developing pollen. Trends Plant Sci 3:305–310

Varvarigos V, Galatis B, Katsaros C (2005) A unique pattern of F-actin organization supports cytokinesis in vacuolated cells of *Macrocystis pyrifera* (Phaeophyceae) gametophytes. Protoplasma 226:241–245

Wadsworth P (2005) Cytokinesis: Rho marks the spot. Curr Biol 15:R871–R874

Wasteneys GO (2002) Microtubule organization in the green kingdom: chaos or self-order? J Cell Sci 115:1345–1354

Webb MC, Gunning BE (1991) The microtubular cytoskeleton during development of the zygote, proembryo and free-nuclear endosperm in *Arabidopsis thaliana* (L.) Heynh. Planta 184:187–195

Weterings K, Apuya NR, Bi Y, Fischer R, Harada JJ, Goldberg RB (2001) Regional localization of suspensor mRNAs during early embryo development. Plant Cell 13:2409–2425

Wu J-Q, Kuhn JR, Kovar DR, Pollard TD (2003) Spatial and temporal pathway for assembly and constriction of the contractile ring in fission yeast cytokinesis. Dev Cell 5:723–734

Zernicka-Goetz M (2004) First cell fate decisions and spatial patterning in the early mouse embryo. Semin Cell Dev Biol 15:563–572

Stomatal Patterning and Guard Cell Differentiation

Keiko U. Torii

Department of Biology, University of Washington, Seattle, WA 98195, USA
ktorii@u.washington.edu

Abstract Gas exchange between plants and the atmosphere takes place through stomata (*singular*, stoma), which are microscopic valves on the plant epidermis composed of paired guard cells. Stomatal differentiation involves a series of asymmetric divisions of precursor cells followed by a single symmetric cell division that produces terminally differentiated guard cell pairs. Stomatal development emerged as a model system to study how environmental- and cell-cell signals translate into site/orientation of asymmetric cell division and cell-type differentiation. This chapter focuses on cell-state transition events leading to guard cell differentiation in the model plant Arabidopsis, and cell-cell signaling mechanisms controlling stomatal patterning. Understanding how cell-cycle regulators influence stomatal patterning and differentiation will advance our knowledge of cell division control in plant development.

1
Introduction

The evolution of land plants relied on the acquisition of mechanisms that protected themselves from the dry atmosphere and harmful UV rays, while allowing gas exchange for photosynthesis; and transpiration for stimulating water movement from the soil to aboveground tissues. The innovation of two distinct cell types on the plant epidermis was critical for solving this challenge. Epidermal pavement cells are tightly-sealed interlocking cells with thick cuticle layers. Stomata act as turgor-driven valves that allow gas exchange and transpiration. It is therefore not surprising that evolutionary biologists believe that the emergence of stomata predates the evolution of leaves, flowers, or even vasculature (Edwards et al. 1998). A stoma consists of a microscopic pore surrounded by a pair of guard cells, which open and close upon sensing environmental signals, such as drought, light, and CO_2 concentrations. Given the importance of stomatal function for plant growth and survival, significant research has been done on physiological and molecular bases of stomatal opening/closure as well as their eco-physiological and environmental consequences (Assmann and Shimazaki 1999; Schroeder et al. 2001, 2001; Hetherington and Woodward 2003).

For developmental biology, stomata serve as a superb system to understand cell-cell signaling, cell division, stem cell differentiation, cell polarity, and cellular morphogenesis in plants. The steps leading to the differentiation of guard cells are uniquely coupled with specific types of cell divisions, the reiterative asymmetric division of precursor stem cells and a single symmetric division that generates a pair of guard cells. The simplicity and tractable nature of the leaf epidermis makes the study of stomatal development technically amenable (Nadeau and Sack 2002; Bergmann et al. 2004). Recent advances in model plant molecular genetics have begun to unravel how genetic and environmental signals act in controlling stomatal patterning. In this chapter, I will introduce the cellular processes of stomatal development with emphasis on the model plant Arabidopsis, and provide the latest updates on emerging cell-cell signaling mechanisms specifying the correct spacing and differentiation of stomata. Potential interactions of cell-cell signaling with intrinsic developmental regulators as well as environmental cues will be explored. Finally, future prospects on integrating cell cycle regulators in the context of stomatal patterning will be presented.

2
Stomatal Development in Arabidopsis

Arabidopsis stomata are typically found in complexes with three subsidiary cells, one being distinctly smaller than the others, surrounding a pair of guard cells (Esau 1977; Zhao and Sack 1999; Serna and Fenoll 2000; Nadeau and Sack 2002). These are characteristic "anisocytic" stomatal complexes and are generated through stereotypical cell division patterns (Esau 1977). Stomatal development initiates post-embryonically when populations of protodermal cells, termed meristemoid mother cells (MMC), enter into asymmetric division (Nadeau and Sack 2002). This initial asymmetric division generates two daughter cells with distinct fates. The larger daughter cell differentiates into an epidermal pavement cell. In contrast, the smaller daughter cell, termed a meristemoid, possesses stem-cell like characteristics, as it continues to divide asymmetrically to renew itself over several rounds of divisions (Nadeau and Sack 2002). Typically, meristemoids reiterate 3 rounds of asymmetric division. The repeated asymmetric division of meristemoids will be hereafter referred to as amplifying asymmetric division, as each division increases the number of cells, which we termed stomatal-lineage ground cells (SLGC), larger daughter cells that function as subsidiary cells (Shpak et al. 2005). SLGCs are also referred to as "subsidiary cells" or "pavement cells" in literature. The meristemoid then differentiates into a round guard mother cell (GMC), which divides symmetrically once to generate a pair of guard cells (Nadeau and Sack 2002). This results in an anisocytic complex with three clonally-related subsidiary cells (Berger and Altmann 2000; Serna et al. 2002). However, the number of asymmetric divisions as well as the clonal relationship among cells constituting the stomatal complex is plastic and variable. For example, a detailed clonal analysis of stomatal complexes in the adaxial epidermis of Arabidopsis L*er* accession by Serna et al. (2002) revealed that, while the vast majority (87%) of anisocytic stomatal complexes derived from single precursor cells, the rest were of polyclonal origins. Geisler et al. (2000) reported that the number of asymmetric divisions in the Columbia accession varies from zero to three. This plastic nature of stomatal ontogeny reflects a dynamic intrinsic developmental program that integrates external cues for adaptation and survival. In fact, environmental factors, such as humidity and CO_2 concentrations, are known to affect stomatal density and patterning (Gray et al. 2000; Lake et al. 2002).

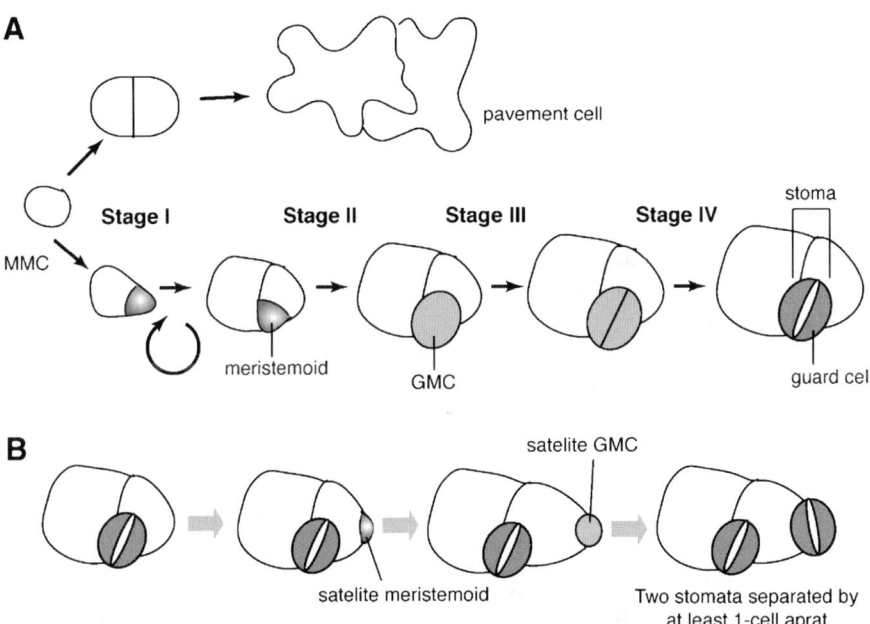

Fig. 1 Stomatal development in Arabidopsis. **A** Cartoon showing the key steps of stomatal differentiation. Undifferentiated cells in the protoderm can undergo either proliferative division to form pavement cells or asymmetric division to initiate stomatal development. *Stage I*: a subset of protodermal cells, a meristemoid mother cell (MMC) divides asymmetrically and forms a self-renewing meristemoid that reiterate a few rounds of asymmetric division. *Stage II*: the meristemoid then differentiates into a round, guard mother cell (GMC). *Stage III*: the GMC undergoes a single symmetric division. *Stage IV*: a pair of immature guard cells achieves final morphogenesis to form a functional stoma. The amplifying asymmetric division of meristemoids generates surrounding stomatal-lineage ground cells (SLGCs) that provide water and ions for stomatal opening and closure. **B** A polarity of asymmetric division during satellite meristemoid formation. The secondary asymmetric division occurs away from the existing stoma, thereby assuring that two stomata are separated by at least one cell apart (1-cell spacing rule). Modified from Torii (2006)

The entire process of stomatal patterning and differentiation in Arabidopsis can be divided into the following four critical stages (Fig. 1A). Stage I initiates the entry into the stomatal-lineage via emergence of MMCs (Stage I-a) and commitment to reiterative asymmetric division (Stage I-b). Stage II represents the differentiation of meristemoids into GMCs associated with the loss of potential for asymmetric division. Stage III includes the acquisition of symmetric division potential in GMCs, and finally Stage IV, or guard cell morphogenesis, concludes the process of stomatal development (Fig. 1A).

Occasionally, SLGCs initiate asymmetric division and produce satellite meristemoids (Fig. 1B). This secondary asymmetric division occurs in a non-random fashon away from the existing stoma (Yang and Sack 1995; Geisler et al. 2000, 2003; Nadeau and Sack 2002). As a consequence, stomata are sep-

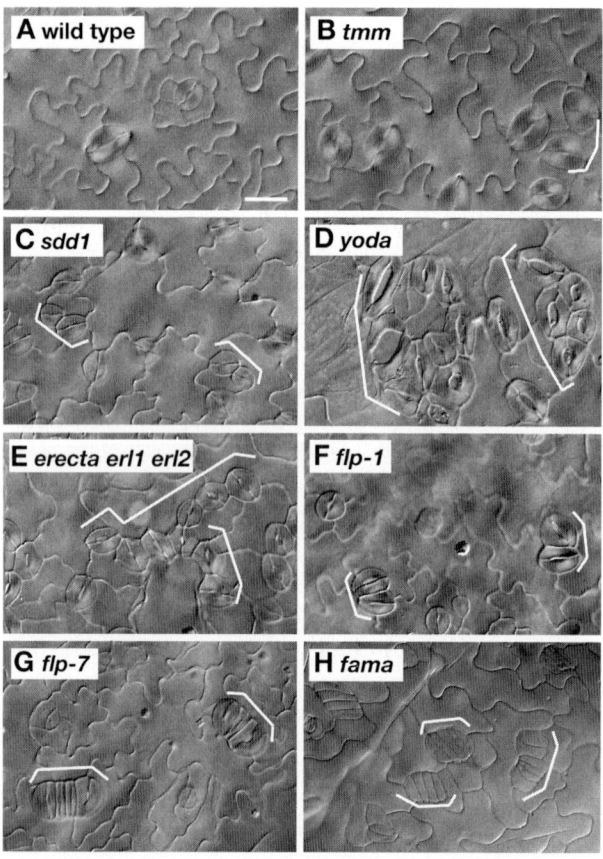

Fig. 2 Stomatal patterning mutants. Shown are the DIC (differential interference contrast) microscopy images of the abaxial rosette leaf epidermis of: **A** wild type; **B** *tmm*; **C** *sdd1*; **D** *yoda*; **E** *erecta erl1 erl2*; **F** *flp-1*; **G** *flp-7*; and **H** *fama*. Images are taken under the same magnification. A *scale bar* = 20 μm

arated by at least one cell (known as the "one-cell spacing rule") (Fig. 1B, Fig. 2) (Nadeau and Sack 2002). Proper spacing is critical for physiological functions of stomata, because guard cells must exchange water and ions (e.g. K^+ and Cl^-) with surrounding subsidiary cells in order to open and close (Assmann and Shimazaki 1999; Schroeder et al. 2001, 2001; Hetherington and Woodward 2003). The observed, "one-cell spacing rule" indicates that the newly forming meristemoid "knows" the location of pre-existing stoma and avoids stomatal cluster formation by orienting the site of secondary asymmetric division. This suggests the presence of cell–cell communication.

In the following sections, I will describe the cytological events during stomatal differentiation and emerging roles of key regulatory genes of stomatal patterning.

3
Stage I-a: Entry Into Asymmetric Division

3.1
Regulation of Orientation and Frequency of Asymmetric Division

Thus far, no molecular markers have been reported for MMC identity. The earliest cytological event that clearly distinguishes the MMC is the polarization of the cytoskeleton, which predicts the site of asymmetric division (Lucas et al. 2005).

The orientation and frequency of the initial asymmetric divisions and cell-cell interaction among the daughter cells determine the proper density and spacing of stomata. Genes implicated in signal transduction play important roles in stomatal patterning. They include *TOO MANY MOUTHS* (*TMM*), *STOMATAL DENSITY AND DISTRIBUTION1* (*SDD1*), *YODA* (*YDA*), and three *ERECTA*-family genes, *ERECTA* (*ER*), *ERECTA-LIKE1* (*ERL1*), and *ERL2* (Table 1) (Berger and Altmann 2000; Nadeau and Sack 2002; Bergmann et al. 2004; Shpak et al. 2005). Loss-of-function mutations in these genes confer clustered stomata, thus violating the "1-cell spacing rule" (Fig. 2). However, phenotypes of these mutants are not identical, suggesting that their relationships are not simply linear.

TMM encodes a receptor-like protein with an extracellular leucine-rich repeat (LRR-RLP), which likely acts as a receptor for a positional cue that specifies the site of asymmetric division (Nadeau and Sack 2002). The phenotypes of *tmm* mutant plants are organ-dependent and complex: the cotyledons and leaves produce clustered stomata (Fig. 2B); the stems produce no stomata; and pedicels exhibit a gradient of no stomata to stomatal clusters (Yang and Sack 1995; Geisler et al. 1998). This complex phenotype implies that TMM triggers contrasting developmental events in a dosage-dependent manner and that each organ requires a different dosage of TMM. Perhaps, TMM potentiates

Table 1 Arabidopsis genes regulating stomatal development

Gene name	AGI number	Gene Product (putative)	Refs.
1: Genes regulating asymmetric division and stomatal patterning			
ERECTA (ER)[a,b]	At2g26330	LRR-receptor-like kinase	(Shpak et al. 2005)
ERECTA-LIKE1 (ERL1)[a,b]	At5g62230	LRR-receptor-like kinase	(Shpak et al. 2005)
ERECTA-LIKE2 (ERL2)[a]	At5g07180	LRR-receptor-like kinase	(Shpak et al. 2005)
STOMATAL DENSITY AND DISTRIBUTION1 (SDD1)[a]	At1g04110	subtilicin-like proteinase	(Berger & Altmann 2000; von Groll et al. 2002)
TOO MANY MOUTHS (TMM)[a]	At1g80080	LRR-receptor protein	(Yang and Sack 1995; Nadeau and Sack 2002)
YODA (YDA)[a]	At1g63700	MAPkinase kinase kinase	(Bergmann et al. 2004)
2: Genes regulating guard cell differentiation			
FAMA[a]	A3g24140	bHLH protein	(Bergmann et al. 2004)
FOUR LIPS (FLP)[a]	At1g14350	R2R3 Myb protein	(Lei et al. 2005)
MYB88[b]	At2g02820	R2R3 Myb protein	(Lei et al. 2005)
3: Genes regulating guard cell cytokinesis/morphogenesis			
CYCLIN-DEPENDENT KINASE B1;1 (CDKB1;1)	At3g54180	cyclin-dependent kinase	(Boudolf et al. 2004)
CYTOKINESIS DEFECTIVE1 (CYD1)	NA	NA (not cloned)	(Yang et al. 1999)
KEULE[c]	At1g12360	Sec1 protein	(Sollner et al. 2002)
STOMATAL CYTOKINESIS DEFECTIVE1 (SCD1)[c]	At1g49040	DENN-WD40 protein	(Fabel et al. 2003)
4: Genes mediating environmental control of stomatal density			
HIGH CARBON DIOXIDE (HIC)	At2g46720	Long-chain fatty acid biosynthesis	(Gray et al. 2000)

Notes: Genes are in alphabetical orders
[a] Photographic images provided in Fig. 2
[b] Phenotypes largely redundant. Combination of double (or triple) mutations among closely related paralogs revealed synergistic interactions
[c] Weak- or temperature sensitive alleles show guard cell cytokinesis defects. The defects in severe alleles are pleiotropic

the entry into the stomatal pathway at lower concentrations (e.g. in MMC), but at high concentration (e.g. in meristemoids) TMM ensures guard cell differentiation, while inhibiting its neighboring cells from further asymmetric division.

SDD1 encodes a subtilicin-like putative extracytoplasmic protease (Berger and Altmann 2000). In animals, this protease family is known to process peptidic ligands to a mature form (Cui et al. 1998). The *sdd1* mutant exhibits high stomatal density but with few stomatal clusters (Fig. 2C). This implies that *SDD1* may primarily regulate the frequency of initial asymmetric division with lesser effects on specifying the orientation in relation to existing stoma.

YDA encodes a putative mitogen activated protein kinase kinase kinase (MAPKKK), a cytoplasmic protein kinase acting at the entry point of MAPK cascades (Bergmann et al. 2004). In plants, animal, and fungi, MAPK cascades integrate and amplify signals transmitted from the upstream cell-surface receptors and activate downstream gene expression in the nucleus to regulate cellular processes. (Serger and Krebs 1995). Unlike *tmm* and *sdd1*, the *yda* mutation is highly pleiotropic. *yda* was first reported as a mutation defective in the initial asymmetric division of zygote with disrupted embryo patterning (Lukowitz et al. 2004). In addition, *yda* plants show severe dwarfism, disrupted floral patterning, and male and female sterility (Lukowitz et al. 2004). The epidermis of *yda* leaves produces high-density stomatal clusters (Bergmann et al. 2004). This phenotype is much more severe than in *tmm* and *sdd1* (Fig. 2D). YDA may act as an on-off switch to repress initial entry into stomatal development when a cell receives a signal from its neighbors, while TMM and SDD1 translate the signal gradient. Consistently, the overly-active form of YDA (*YDA△NB*) severely inhibits the onset of initial asymmetric divisions, resulting in the epidermis consisting solely of pavement cells (Bergmann et al. 2004).

Unexpectedly, the *ERECTA*-family genes were recently shown to play a role in stomatal patterning. *ERECTA* is a well-known gene regulating plant architecture, in which loss-of-function mutation confers a characteristic compact inflorescence with short pedicels and blunt fruits (Torii et al. 1996). *ERECTA* encodes an LRR receptor-like kinase (LRR-RLK), a prevalent family of RLKs that play important roles in developmental, steroid-hormone signal transduction, and defense against pathogens (Torii et al. 1996, 2004; Becraft 2002). *ERECTA* and its two paralogous genes, *ERL1* and *ERL2*, interact in a synergistic manner in regulating stomatal patterning (Shpak et al. 2005). The *erecta erl1 erl2* triple loss-of-function mutations confer severe dwarfism, disrupted floral patterning, male- and female sterility, and a high-density stomatal clustering phenotype (Fig. 2E) (Shpak et al. 2004, 2005). Overall, these phenotypes highly resemble those of the *yda* single mutant, suggesting that the YDA MAPK cascade may function downstream of ERECTA-family RLKs in multiple developmental processes.

3.2
Genetic Interactions and Hierarchy of Signal Transduction

Studies of genetic interactions are now illuminating possible cell-cell signal transduction mechanisms regulating stomatal patterning. The overexpression of SDD1 inhibited the initial asymmetric division and reduced stomatal density (von Groll et al. 2002). However, the effect was reversed by the *tmm* mutation, thus placing *TMM* downstream of *SDD1* (von Groll et al. 2002). The simplest interpretation of these results is that the excessive production of mature ligands by overexpression of SDD1 overly inhibited entry into stomatal development via the TMM receptor.

YDA most likely acts downstream of the *SDD1-TMM* pathway. A single copy of *YDA△NB*, which by itself does not completely inhibit the entry into stomatal development, was able to suppress the stomatal cluster phenotype of *sdd1* and *tmm* (Bergmann et al. 2004). This indicates that a slight increase in YDA activity was able to resume a normal level of signal transduction in the absence of *SDD1* or *TMM*. Perhaps, signals mediated via TMM are transmitted to the YDA-MAPK cascades to suppress neighboring cells to enter the stomatal-lineage.

TMM does not possess any cytoplasmic stretch (Nadeau and Sack 2002). In fact, the TMM protein molecule ends with a 23 amino-acid-long membrane-spanning region, suggesting that the C-terminal end of TMM does not extend from the plasma membrane to the cytoplasm. In addition, the C-terminus of TMM possesses a GPI (glycosylphosphatidylinositol)-anchor motif, suggesting that TMM may be anchored to the membrane surface (Nadeau and Sack 2002). Such structural features make TMM unlikely to transmit signals by itself, and it is reasonable to speculate that TMM forms a receptor complex with a partner molecule that possesses a cytoplasmic effector domain. Scientists predicted that the partner of TMM would be an LRR-RLK, in light of other systems such as CLAVATA (CLV) in shoot apical meristem development, whereby CLV1 LRR-RLK is thought to form a receptor dimer with CLV2 LRR-RLP (Clark et al. 1997; Jeong et al. 1999).

The three ERECTA-family LRR-RLKs are attractive candidates of the TMM partner. The genetic interaction of *TMM* and *ERECTA*-family by Shpak et al. (2005) suggest that TMM may restrict the inhibitory action of ERECTA-family RLKs on the initial entry into asymmetric division. This was evident from their interactions in the stems. The *tmm* stems give rise to no stomata, while the *erecta erl1 erl2* stems form high-density stomatal clusters (Nadeau and Sack 2002; Shpak et al. 2005). The epistatic relationship of *TMM* and *ERECTA*-family genes in the stem epidermis exhibited stoiochiometric dynamics: While *TMM* is epistatic to each one of the *ERECTA*-family genes, three *ERECTA*-family genes together are epistatic to *TMM*. Removing *TMM* and two out of three *ERECTA*-family genes resumed nearly wild-type stomatal phenotype (Shpak et al. 2005). It is exciting to speculate that TMM modulates the activity of ERECTA-family RLKs via direct association. However, in cotyledons and leaves, both *tmm* and *erecta*-family triple mutants display stomatal clusters. Do they act cooperatively in these organs while acting antagonistically in the stems? Establishing the molecular bases of receptor interactions and identifying their ligands is critical for elucidating the complex action of TMM and ERECTA-family receptors. By analogy to known LRR-RLPs and LRR-RLKs, the ligands for TMM/ERECTA-family receptors are most likely small peptides.

4
Stage I-b and II: Amplifying Asymmetric Division and Differentiation of Meristemoids

The amplifying asymmetric division of a meristemoid occurs in an inward spiral. Serna et al. (2002) documented that the division angle of meristemoids in the Ler adaxial leaf epidermis is exactly 60 degrees with little deviation. Therefore, a newly formed, triangular-shaped meristemoid likely re-establishes its polarity away from the polar end of the previous asymmetric division. This implies the presence of a chemical gradient and a cellular system to translate gradient to determine the site of cytokinesis. Lucas et al. (2006) reported that it is common to have two adjacent meristemoids in wild-type epidermis, given that the initial entry asymmetric division occurs randomly. However, two adjacent meristemoids always divide away from each other to avoid further contact. Such polarity is disrupted in the *tmm* mutant, indicating that *TMM* functions in perceiving the positional cues during amplifying asymmetric division.

Flexible numbers of amplifying asymmetric division allows developmental plasticity to adjust stomatal density in response to environmental changes. Furthermore, it provides a way to "correct" erroneous division to avoid clustered stomata. The number of amplifying asymmetric divisions may be regulated by *ERECTA-LIKE* (*ERL*) genes. The pedicel epidermis of *erl1 erl2* double mutants gave rise to stomatal complexes with reduced number of SLGC (stomatal lineage ground cells), implying that the meristemoids completed reduced rounds of asymmetric division and precociously differentiated into GMCs.

ERECTA-family regulates the binary decision of daughter cells of an asymmetric division to adopt stomata vs. SLGC fates. Consistently, the epidermis of *erecta* single mutant produced occasional patches of 2–3 cells that appear to have undergone asymmetric division but failed to differentiate into guard cells. Expression of stomatal-lineage markers, *TMM::GUS* and *ERL1::GUS* in these groups of cells supports the hypothesis that both daughter cells became SLGCs. The *tmm* mutation greatly enhanced this "patches of cells with no stomata" phenotype. (Shpak et al. 2005). Indeed, the *tmm erecta* double mutations completely eliminated stomata from cauline-leaf and carpel epidermis, leaving numerous small cells that likely underwent asymmetric division and then became SLGC without accompanying guard cells. (Shpak et al. 2005). In both cases, termination of stomatal fate is likely due to misregulation of *ERL1*, which becomes overly inhibitory in repressing GMC differentiation. Consistent with this hypothesis, the additional *erl1* mutation reversed the no-stomata phenotype of *tmm erecta*. Whether TMM regulates ERL1 via direct association awaits further biochemical analysis.

A positive regulator of differentiation of a meristemoid has not yet been identified. If such a gene exists, then the loss-of-function mutation may extend the lifespan of meristemoids' stem cell-like activity, and consequently mutant plants should exhibit excessive rounds of amplifying asymmetric division. If a default pathway of a meristemoid is to adopt pavement cell fate, then the mutant plants lacking the positive regulator may form two pavement cells after entry into asymmetric division, just like those observed in the *tmm erecta* double mutant background (Shpak et al. 2005). Further isolation of such mutants and molecular cloning of causal genes will help us elucidate the molecular mechanisms of stem cell differentiation in the plant epidermis.

5
Stage VI: GMC Division – Intrinsic Regulation by Transcriptional Factors

Once the meristemoid commits to becoming a GMC, it loses its potential for asymmetric division. GMC differentiation is evident from its changes in cellular morphology. The triangular meristemoid cell expands and become oval in shape with characteristic, end wall thickening (Lucas et al. 2006). The GMC achieves, precisely, a single symmetric division. This reflects the fundamental importance of having a pair of guard cells for proper stomatal function. Therefore, mechanisms must be present to ensure exactly one symmetric division occurs. Two genes, *FOUR LIPS* (*FLP*) and *FAMA*, prevent excessive (more than one) symmetric division of the GMC (Bergmann et al. 2004; Lai et al. 2005)

The *flp-1* mutant was initially isolated from the phenotype of paired stomata, which gives four aligned guard cells (Fig. 2F) (Lai et al. 2005). While adjacent stomata are arranged randomly in stomatal patterning mutants (*tmm*, *sdd1*, *erecta*-family, and *yoda*), paired stomata in *flp* are in parallel due to their origin from a single GMC. In the severe allele *flp-7*, the GMC undergoes reiterative symmetric divisions, which result in formation of a row of "caterpillar-like" guard cell clusters (Fig. 2G). *FLP* encodes an atypical R2R3-type Myb protein, which most likely functions as DNA-binding transcription factor (Lai et al. 2005). Therefore, an attractive hypothesis is that FLP suppresses expression of positive regulators of cell cycle progression promoting GMC division. Interestingly, both weak- (*flp-1*, *flp-2*) and severe alleles (*flp-7*) are predicted to produce truncated proteins with incomplete Myb domains, with weaker alleles producing shorter fragments than the severe one (Lai et al. 2005). This apparent discrepancy between the severity of phenotypes and impacts on protein structure implies a dominant-negative activity of *flp-7* gene products, which may interfere with redundant components. Consistent with this idea, the T-DNA inserted knockout alleles of *flp* display weak phenotypes (Lai et al. 2005).

The paralogous Myb gene, *MYB88*, is most likely the redundant factor. *MYB88* shares high sequence identity (91% amino-acid identity in the MYB domain, 71% overall) and exhibits similar expression pattern with *FLP* (Lai et al. 2005). Interestingly, while complete loss-of-function mutations in *MYB88* failed to confer any visible phenotype, they dramatically enhanced the size of "caterpillar-like" guard cell stacks in the *flp* mutant (Lai et al. 2005). Moreover, introduction of an extra genomic copy of *MYB88* with its own promoter rescued the *flp* phenotype. The combined dosage of *FLP/MYB88* may be critical for guard cell differentiation.

FAMA was identified from transcriptional profiling as a gene upregulated in *yoda* (in which the epidermis is predominantly stomata) compared to *YDAΔNB* (in which the epidermis is predominantly pavement cells) (Bergmann et al. 2004). The loss-of-function *fama* phenotype highly resembles that of *flp*, suggesting that *FAMA* and *FLP* function in the same step of GMC differentiation. However, unlike *flp*, the abnormal "caterpillar-like" clusters in *fama* never form mature guard cells, suggesting that *FAMA* suppresses GMC division but in addition promotes the guard cell differentiation program (Fig. 2H). *FAMA* encodes a bHLH (basic-Helix Loop Helix) protein and most likely acts as a DNA- binding transcription factor.

The molecular identity of FLP and FAMA as Myb and bHLH transcription factors highlights an intriguing link to underlying mechanisms of epidermal cell-type differentiation in leaves and roots. Differentiation of trichomes and root hairs requires orchestrated actions of Myb transcriptional activators, which associate with bHLH proteins (Schiefelbein 2000, 2003). It would be therefore of special interest to address whether FLP and FAMA physically associate with each other and constitute a transcriptional regulatory complex.

6
Stage IV: Guard Cell Morphogenesis

The final stage of stomatal differentiation involves guard cell morphogenesis, a step leading to the formation of paired guard cells. After a symmetric division of the GMC, the new cell wall forms along the side of division, which develops a pore. The guard cells adopt a characteristic microtubule and microfibril organization (Hepler and Palevitz 1974). Defective cytokinesis leads to abnormal guard cell morphology. The *cytokinesis defective1* (*cyd1*) mutant forms abnormal guard cells with various degrees of cytokinesis defects: $\sim$ 20% form a single, large round cell lacking any ventral wall or pore, $\sim$ 10% have incomplete wall with pore, and $\sim$ 5% form a single round guard cell with partial cell wall protrusions (Yang et al. 1999). These abnormal cells are either single- or bi-nucleated, each correlating with the extent of cytokinetic defects (Yang et al. 1999). The molecular identity of *CYD1* is not known. The temperature-sensitive *stomatal cytokinesis-defective1-1* (*scd1-1*) mutant exhibits abnormal guard cells similar to the *cyd1* mutant (Falbel et al. 2003). *SCD1* encodes a protein with two domains (DENN domain and WD-40 repeats), and these structural features imply a possible role for SCD1 in vesicle trafficking during cell-plate formation. Consistently, the null alleles of *scd1* exhibit pleiotropic defects in cytokinesis and polar cell expansion. Therefore, abnormal guard cell morphology may be a sensitive indicator of general cytokinesis defects, which can be exploited to recover weak alleles of key regulatory genes for cytokinesis. Consistently, the weak alleles of *KNOLLE* and *KEULE*, two genes initially isolated as regulators of embryogenesis, display abnormal stomatal morphology (Sollner et al. 2002). *KNOLLE* and *KEULE* encode syntaxin and Sec1, respectively, two physically-interacting proteins required for vesicle fusions at the nascent cell plate (Lukowitz et al. 1996; Waizenegger et al. 2000). In addition to cytokinesis, the cell cycle defects may confer abnormal guard cells (see below).

7
Cell Cycle Regulation in Stomatal Patterning

Stomatal patterning and differentiation is tightly coupled with specific types of cell division: the initial asymmetric division of MMCs, the amplifying asymmetric division of meristemoids, and a single symmetric division of GMCs. In addition, genome replication is strictly controlled during stomatal development, as guard cells remain at 2C (diploid) unlike the rest of epidermal cells that undergo endoreduplication (Melaragno et al. 1993). The obvious questions are whether specific cell cycle regulators control distinct cell division types during stomatal development and, if so, whether forcing cell cycle switches can invoke/suppress stomatal development. Studies suggest that cell cycle regulators may influence stomatal patterning, but they do not impinge on stomatal differentiation.

The promoter of the Arabidopsis *CTD1* gene, which encodes a subunit of the DNA-replication licensing complex together with *AtCDC6*, is highly active in stomatal-lineage cells (Castellano Mdel et al. 2004). The *AtCTD1::GUS* promoter activity resembles that of *TMM* and *ERL1*, with highest activity in meristemoids and GMCs, and moderate activity in SLGCs (Castellano Mdel et al. 2004). Overexpression of *AtCDT1* and *AtCDC6* slightly increased the numbers of stomata, but it did not lead to formation of adjacent stomata. These results suggest that the DNA-replication licensing complex may promote stomatal asymmetric division and that forcing G1-to-S phase transition may slightly increase the MMC specification. However, this is not sufficient to overcome negative regulation by cell–cell signaling components encoded by *SDD1*, *TMM*, *YODA*, and *ERECTA*-family genes.

Arabidopsis B-type cyclin-dependent kinase gene *CDKB1,1* is also expressed in stomatal-lineage cells with high expression in meristemoids, GMCs, and in guard cells (Boudolf et al. 2004). Overexpression of a dominant-negative form of *CDKB1,1* led to a significant reduction in SLGCs due to reduced amplifying asymmetric division. Intriguingly, the mature stomata in the dominant-negative transgenic plants exhibited aberrant morphology, with unicellular round or kidney-shaped single guard cells without a pore (Boudolf et al. 2004). These unicellular stomata have a nuclear content of 4C, indicating that they are arrested in the G2 phase. Therefore, inhibition of *CDKB1,1* prevents division of both meristemoids and GMCs without interfering with the guard cell differentiation program.

How stomatal developmental regulatory genes influence cell cycle machinery is an open question. At least four members of stomatal cell-cell signal transduction, YDA and three ERECTA-family RLKs, are required for cell proliferation during normal plant growth, as both *yda* and *erecta erl1 erl2* triple mutant plants are severely dwarfed with reduced cell numbers (Lukowitz et al. 2004; Shpak et al. 2004). Conversely, the overly-active *YDAΔNB* plants show excessive stem elongation due to increased cell numbers (McAbee and Torii, unpublished). How do YDA and three ERECTA-family RLKs promote cell proliferation while suppressing entry into the stomatal lineage? RT-PCR analysis of *erecta erl1 erl2* triple mutant plants by Shpak et al. (2004) did not reveal any increase in mRNA levels of G1-cyclins that are known to promote auxin-mediated organ growth (Mizukami and Fischer 2000; Hu et al. 2003). It is possible that cell proliferation is modulated by a mechanism other than G1-cyclin expression. Better understanding of the exact cell cycle defects in these mutants may link cell cycle regulation and stomatal patterning.

8
Environmental Control of Stomatal Patterning

Plants sense environmental changes and adjust stomatal density accordingly. Numerous environmental factors, including light, humidity, drought, ozone, and atmospheric CO_2 concentrations affect stomatal density and/or stomatal index (Holroyd et al. 2002). Among these factors, CO_2 concentrations and stomatal density show an inverse correlation in a wide variety of plant species (Holroyd et al. 2002). How do plants integrate environmental signals to modulate intrinsic stomatal developmental programs? Identification of the *HIGH CARBON DIOXIDE* (*HIC*) gene by Gray et al. (2000) brought new insight into this important question. The Arabidopsis *hic* mutant has no apparent phenotype in ambient conditions. However, the *hic* mutant is greatly increased in stomatal density (approx. 40% increase) under the elevated CO_2 concentration (Gray et al. 2000). *HIC* encodes a putative 3-keto acyl Co-A synthase, an enzyme regulating synthesis of very-long-chain fatty acids (VLCFA), which constitute epicuticular wax. *HIC* is expressed specifically in developing guard cells (but not in the meristemoid or GMC).

Consistent with the role of *HIC* in epicuticular wax biosynthesis, mutations in two additional epicuticular wax biosynthesis genes, *CER1* and *CER6*, confer significant increases in stomatal density even in the ambient CO_2 levels (Gray et al. 2000; Holroyd et al. 2002). Unlike *HIC*, *CER1* and *CER6* affect wax composition in the entire epidermis, including pavement cells (Aarts et al. 1995; Fiebig et al. 2000). One scenario is that the altered composition in the guard cell extracellular matrix changes the concentration gradient of a diffusible inhibitor of stomatal development to neighboring cells. Under high CO_2 concentrations, altering wax composition only in the guard cells (but not the entire epidermis) is sufficient to trigger excess stomatal formation. Obviously, identifying the elusive diffusible signal is the key for understanding environmental control of stomatal patterning.

9
Future Perspectives

Recent years have seen a dramatic advancement in our understanding of molecular mechanisms of stomatal patterning and differentiation. The identification of *SDD1*, *TMM*, *YODA*, and *ERECTA*-family genes now allows us to investigate the biochemical basis of stomatal cell-cell signaling. Establishing molecular interactions among these signaling molecules is the obvious next step. It would be particularly interesting to see whether TMM and ERECTA-family RLKs form receptor heterodimers. However, several key regulatory molecules are still missing. For example, we do not know the identity of ligands or downstream MAPK components for stomatal patterning. Likewise, nothing is known about the positive regulators, which specify meristemoid identity as well as differentiation of meristemoids to GMCs. In animals, control of asymmetric division and cell-type differentiation is controlled by the orchestrated actions of cell-cell signaling, cell division programs, and transcription factors that drive the fate decision. Some key transcription factors for stomatal differentiation may have yet to be discovered. An integrated approach, taking advantage of modern "omics" as well as classical forward- and reverse genetics, may lead to a breakthrough in filling the gap in our knowledge of the molecular bases of stomatal development.

Updates: Since the original book chapter was submitted, four significant publications have appeared. Wang et al. (2007) identified two MAPKs (AtMPK3 and AtMPK6) and two upstream MAPKKs (AtMKK4/AtMKK5) as redundant negative regulators of stomatal differentiation. Both biochemical and genetic data indicate that these kinases act downstream of YODA. Interestingly, AtMPK3/6 and AtMKK4/5 are known to regulate environmental-

and biotic (pathogen) stress. Findings by Wang et al. (2007) provide tantalizing evidence that both developmental and stress-induced signaling pathways converge at the downstream MAPK cascades.

Second, a trio of genes directing three key steps of stomatal differentiation was identified (MacAlister et al. 2007; Ohashi-Ito and Bergmann, 2006; Pillitteri et al. 2007). Loss-of-function mutations in the gene *SPEECHLESS* (*SPCH*) confer an epidermis solely made of pavement cells, thus lacking any stomatal lineage cells. Loss-of-function mutations in *MUTE* lead to excessive asymmetric division of the meristemoids that fail to differentiate into GMC. Both *SPCH* and *MUTE* encode basic helix-loop-helix (bHLH) proteins closely-related with each other as well as with *FAMA*. Therefore, stomatal differentiation is directed by sequential actions of the three "key switch" bHLH genes: *SPCH* at initiation (from MMCs to meristemoids), *MUTE* at precursor differentiation (from meristemoids to GMCs), and *FAMA* at terminal differentiation of guard cells (from GMCs to guard cells). The findings highlight an intriguing parallel between stomatal cell-type differentiation and muscle- and neuron cell-type differentiation in animals.

Acknowledgements I thank members of my laboratory, especially Lynn Pillitteri, Jessica McAbee, and Naomi Bogenschutz for providing photographic images and helpful comments. My research programs are supported by the US-Department of Energy (DE-FG02-03ER15448), Japan Science & Technology Agency (CREST award), and by the US-National Science Foundation (IOB-0520548).

References

Aarts M, Keijzer C, Stiekema W, Pereira A (1995) Molecular characterization of the *CER1* gene of Arabidopsis involved in epicuticular wax biosynthesis and pollen fertility. Plant Cell 7:2115–2127

Assmann S, Shimazaki K (1999) The multisensory guard cell. Stomatal responses to blue light and abscisic acid. Plant Physiol 119:809–816

Becraft PW (2002) Receptor kinase signaling in plant development. Annu Rev Cell Develop Biol 18:163–192

Berger D, Altmann T (2000) A subtilisin-like serine protease involved in the regulation of stomatal density and distribution in *Arabidopsis thaliana*. Genes Dev 14:1119–1131

Bergmann DC, Lukowitz W, Somerville CR (2004) Stomatal development and pattern controlled by a MAPKK kinase. Science 304:1494–1497

Castellano Mdel M, Boniotti MB, Caro E, Schnittger A, Gutierrez C (2004) DNA replication licensing affects cell proliferation or endoreplication in a cell type-specific manner. Plant Cell 16:2380–2393

Clark SE, Williams RW, Meyerowitz EM (1997) The *CLAVATA1* gene encodes a putative receptor kinase that controls shoot and floral meristem size in Arabidopsis. Cell 89:575–585

Cui Y, Jean F, Thomas G, Christian JL (1998) BMP-4 is proteolytically activated by furin and/or PC6 during vertebrate embryonic development. EMBO J 17:4735–4743

Edwards D, Kerp H, Haas H (1998) J Exp Bot 49:255-278
Esau K (1977) Stomata. In: Anatomy of seed plants. Wiley, New York, pp 88-99
Falbel TG, Koch LM, Nadeau JA, Segui-Simarro JM, Sack FD, Bednarek SY (2003) SCD1 is required for cytokinesis and polarized cell expansion in Arabidopsis thaliana [corrected]. Development 130:4011-4024
Fiebig A, Mayfield J, Miley N, Chau S, Fischer R, Preuss D (2000) Alteration in CER6, a gene identical to CUT1, differentially affect long-chain lipid contects on the surface of pollen and stems. Plant Cell 12:2001-2008
Geisler M, Nadeau J, Sack FD (2000) Oriented asymmetric divisions that generate the stomatal spacing pattern in Arabidopsis are disrupted by the too many mouths mutation. Plant Cell 12:2075-2086
Geisler M, Yang M, Sack FD (1998) Divergent regulation of stomatal initiation and patterning in organ and suborgan regions of the Arabidopsis mutants too many mouths and four lips. Planta 205:522-530
Geisler MJ, Deppong DO, Nadeau JA, Sack FD (2003) Stomatal neighbor cell polarity and division in Arabidopsis. Planta 216:571-579
Gray JE, Holroyd GH, van der Lee FM, Bahrami AR, Sijmons PC, Woodward FI, Schuch W, Hetherington AM (2000) The HIC signalling pathway links CO_2 perception to stomatal development. Nature 408:713-716
Hepler P, Palevitz B (1974) Microtibules and microfilaments. Annu Rev Plant Physiol 25:309-362
Hetherington AM, Woodward FI (2003) The role of stomata in sensing and driving environmental change. Nature 424:901-908
Holroyd GH, Hetherington AM, Gray JE (2002) A role for the cuticular waxes in the environmental control of stomatal development. New Phytol 153:433-439
Hu Y, Xie Q, Chua NH (2003) The Arabidopsis auxin-inducible gene ARGOS controls lateral organ size. Plant Cell 15:1951-1961
Jeong S, Trotochaud AE, Clark SE (1999) The Arabidopsis CLAVATA2 gene encodes a receptor-like protein required for the stability of the CLAVATA1 receptor-like kinase. Plant Cell 11:1925-1933
Lai LB, Nadeau JA, Lucas J, Lee EK, Nakagawa T, Zhao L, Geisler M, Sack FD (2005) The Arabidopsis R2R3 MYB proteins FOUR LIPS and MYB88 restrict divisions late in the stomatal cell lineage. Plant Cell 17:2754-2767
Lake JA, Woodward FI, Quick WP (2002) Long-distance CO_2 signalling in plants. J Exp Bot 53:183-193
Lucas JR, Nadeau JA, Sack FD (2006) Microtubule arrays and Arabidopsis stomatal development. J Exp Bot 57:71-79
Lukowitz W, Mayer U, Jurgens G (1996) Cytokinesis in the Arabidopsis embryo involves the syntaxin-related KNOLLE gene product. Cell 84:61-71
Lukowitz W, Roeder A, Parmenter D, Somerville C (2004) A MAPKK kinase gene regulates extra-embryonic cell fate in Arabidopsis. Cell 116:109-119
MacAlister CA, Ohashi-Ito K, Bergmann DC (2007) Transcription factor control of asymmetric cell divisions that establish the stomatal lineage. Nature 445:537-540
Mizukami Y, Fischer RL (2000) Plant organ size control: AINTEGUMENTA regulates growth and cell numbers during organogenesis. Proc Natl Acad Sci USA 97:942-947
Nadeau JA, Sack FD (2002) Control of stomatal distribution on the Arabidopsis leaf surface. Science 296:1697-1700
Nadeau JA, Sack FD (2002) Stomatal Development in Arabidopsis. In: Somervile CR, Meyerowitz EM (eds) Arabidopsis Book. ASPB, Rockville, MD

Ohashi-Ito K, Bergmann DC (2006) Arabidopsis FAMA controls the final proliferation/differentiation switch during stomatal development. Plant Cell 18:2493–2505

Pillitteri LJ, Sloan DB, Bogenschutz NL, Torii KU (2007) Termination of asymmetric cell division and differentiation of stomata. Nature 445:501–505

Schiefelbein J (2003) Cell-fate specification in the epidermis: a common patterning mechanism in the root and shoot. Curr Opin Plant Biol 6:74–78

Schiefelbein JW (2000) Constructing a plant cell. The genetic control of root hair development. Plant Physiol 124:1525–1531

Schroeder JI, Allen G, Hugouvieux V, Kwak JM, Wagner D (2001) Guard cell signal transduction. Annu Rev Plant Physiol Plant Mol Biol 52:627–658

Schroeder JI, Kwak JM, Allen GJ (2001) Guard cell abscisic acid signalling and engineering drought hardiness in plants. Nature 410:327–330

Serger R, Krebs E (1995) The MAPK signaling cascade. FASEB J 9:726–735

Serna L, Fenoll C (2000) Stomatal development and patterning in *Arabidopsis* leaves. Physiol Plant 109:351–358

Serna L, Torres-Contreras J, Fenoll C (2002) Clonal analysis of stomatal development and patterning in Arabidopsis leaves. Dev Biol 241:24–33

Shpak ED, Berthiaume CT, Hill EJ, Torii KU (2004) Synergistic interaction of three ERECTA-family receptor-like kinases controls Arabidopsis organ growth and flower development by promoting cell proliferation. Development 131:1491–1501

Shpak ED, McAbee JM, Pillitteri LJ, Torii KU (2005) Stmatal patterning and differentiation by synergistic interactions of receptor kinases. Science 309:290–293

Sollner R, Glasser G, Wanner G, Somerville CR, Jurgens G, Assaad FF (2002) Cytokinesis-defective mutants of Arabidopsis. Plant Physiol 129:678–690

Torii KU (2004) Leucine-rich repeat receptor kinases: structure, function, and signal transduction pathways. Int Rev Cytol 234:1–46

Torii KU (2006) Stomatal patterning and differentiation: emerging role of cell-cell signaling (in Japanese). Tanpakushitsu Kakusan Koso 51:145–154

Torii KU, Mitsukawa N, Oosumi T, Matsuura Y, Yokoyama R, Whittier RF, Komeda Y (1996) The Arabidopsis *ERECTA* gene encodes a putative receptor protein kinase with extracellular leucine-rich repeats. Plant Cell 8:735–746

von Groll U, Berger D, Altmann T (2002) The subtilisin-like serine protease SDD1 mediates cell-to-cell signaling during Arabidopsis stomatal development. Plant Cell 14:1527–1539

Wang H, Nugwenyama N, Liu Y, Walker JC, Zhang S (2007) Stomatal development and patterning are regulated by environmentally responseive mitogen activated protein kinases in Arabidopsis. Plant Cell doi10.1105/tpc106.048298

Waizenegger I, Lukowitz W, Assaad F, Schwarz H, Jurgens G, Mayer U (2000) The Arabidopsis *KNOLLE* and *KEULE* genes interact to promote vesicle fusion during cytokinesis. Curr Biol 10:1371–1374

Yang M, Nadeau JA, Zhao L, Sack FD (1999) Characterization of a cytokinesis defective (cyd1) mutant of Arabidopsis. J Exp Bot 50:1437–1446

Yang M, Sack FD (1995) *The too many mouths* and *four lips* mutations affect stomatal production in Arabidopsis. Plant Cell 7:2227–2239

Zhao L, Sack FD (1999) Ultrastructure of stomatal development in *Arabidopsis* (Brassicaceae) leaves. Am J Bot 86:929–939

Genetic Control of Anther Cell Division and Differentiation

Carey L. H. Hord[1,2,3] · Hong Ma[1,2] (✉)

[1]Department of Biology, The Pennsylvania State University,
University Park, PA 16802, USA
hxm16@psu.edu

[2]The Intercollege Graduate Program in Plant Biology, The Huck Institutes of Life Sciences, The Pennsylvania State University, University Park, PA 16802, USA

[3]*Present address:*
3015 Quinby Dr. Columbus, OH 43232, USA

Abstract Anther development requires the coordinated differentiation of several cell types. Recent molecular genetic analyses have led to exciting advances in our understanding of anther cell division and differentiation. The *SPOROCYTELESS/NOZZLE* (*SPL/NZZ*) gene is a putative transcription factor critical for early anther cell division and/or differentiation. Several genes that regulate tapetum formation and differentiation have been isolated, including *EXCESS MICROSPOROCYTES1/EXTRA SPOROGENOUS CELLS* (*EMS1/EXS*), *SOMATIC EMBRYOGENESIS1*(*SERK1*), *SERK2*, and *TAPETUM DETERMINANT1* (*TPD1*). Also, genes important for normal tapetum differentiation and function have been uncovered, including two closely related MYB genes, *MYB33* and *MYB65*, which are post-transcriptionally regulated by microRNAs, and the rice *Udt1* gene. Finally, genes encoding putative transcription regulators, *ABORTED MICROSPORES* (*AMS*) and *MALE STERILITY1* (*MS1*), and a lipid metabolism enzyme, *AtGPAT1* are important for pollen development. These discoveries have ushered in a new era for understanding the control of cell division and differentiation during plant male reproductive development.

1
Introduction

Flowering plants rely on microsporogenesis and microgametogenesis for male reproduction and the propagation of the species (Li and Ma 2002; Ma 2005). Microsporogenesis is the formation of the microspores that precedes the development of the male gametophytes (pollen). Microgametogenesis occurs within the developing pollen grain and leads to the generation of sperm cells. These complex processes require coordinated development of sporophytic and gametophytic cell types within the anther. In *Arabidopsis*, the anther cross section is butterfly-shaped, with two adaxial and two abaxial lobes. At early stages, the development of the adaxial lobes lags behind slightly that of the abaxial lobes. Anther development in *Arabidopsis* has been divided into 14 stages on the basis of the anther morphology (Sanders et al. 1999) and additional features can be detected using transmission electron microscopy (TEM) (Owen and Makaroff 1995) (Table 1).

Table 1 Stages of *Arabidopsis* anther development

Anther stage[a]	Major events and morphological markers	Tissues present
1	Stamen is round with three major cell layers	L1, L2, L3
2	Stamen becomes oval shaped; archesporial cells are in the four corners of the L2 layer	Ep, Ar, L3
3	Archesporial cells have divided parallel to the epidermis to form the primary parietal and primary sporogenous cells in the lobes	Ep, PP, PSp, L3
4	Primary parietal cells have divided parallel to the epidermis to form the inner secondary parietal and outer secondary parietal cell layers; they begin to form concentric rings around the sporgenous cells. The Vascular and connective become distinguishable	Ep, ISP, OSP, Sp, V
5	One of the secondary parietal layers has divided periclinally at the outer surface, three L2-derived cell layers surround the pollen mother cells. At late stage five, callose begins to form around PMCs	Ep, En, ML, T, PMC, C, V
6	A thick callose layer is apparent surround the pollen mother cells, which enter and go through meiosis. The middle layer becomes compressed and appears as a thin line. Tapetum becomes vacuolated	Ep, En, ML, T, Mc, PMC, V
7	Meiosis has completed with cytokinesis. Tetrads of microspores appear in the locules.	Ep, En, ML, T, Td, C, V
8	Microspores are free within in the locules after the callose wall is degenerated. Middle layer is no longer visible.	Ep, En, T, MSp, C, V
9	Microspores generate an exine wall and become vacuolated. Septum is visible using TEM	Ep, En, T, Msp, C, V, Sm
10	Tapetum degeneration is initiated	Ep, En, T, Msp, C, V, Sm
11	Microspores undergo three rounds of mitotic divisions. Tapetum continues to degenerate. Endothecium and epidermis appear highly vacuolated. Septum cells begin to degenerate	Ep, En, T, Msp, C, V, Sm, St
12	Tricellular pollen grains are visible. Septum degeneration and breakage cause the anthers to become bilocular. Differentiated stonium visible using TEM.	Ep, En, PG, C, V, St
13	Breakage along stonium allows for pollen release-dehiscence	Ep, En, PG, C, V, St
14	The stamen senesces as the cells shrink	Ep, En, C, V

[a] Anther stages were taken from Sanders et al. (1999).
Ar, archesporial cells; Ep, epidermis; PP, primary parietal cells; PSp, primary sporogenous cells; ISP, inner secondary parietal cells; OSP, outer secondary parietal cells; Sp, sporogenous cells; V, vascular tissue; En, endothecium; ML, middle layer; T, tapetum; PMC, pollen mother cells; C, connective tissue; Mc, meiocyte; Td, tetrads; Msp, microspores; Sm, septum; St, stonium; PG, pollen grain

The establishment of anther cell layers, histogenesis, occurs within the first five stages. Although the relative roles that position and cell lineage play in anther cell differentiation have yet to be elucidated (Goldberg et al. 1993; Scott et al. 2004), by early anther stage five, the cell layers are fully established with the endothecium, middle layer and tapetum forming rings around the pollen mother cells (PMCs). At this stage, the PMCs and tapetal cells are connected to each other and to their own cell type by plasmodesmata (Owen and Makaroff 1995). Post-histogenesis growth also requires the coordinated development and function of sporophytic and gametophytic cells, particularly the tapetum, which supports the growth of the developing pollen at later stages. In addition, tapetum degeneration and its timing are very important for proper microspore development (Wu and Cheun 2000). Following tapetum degeneration, the anther becomes bilocular as the septum separating the locules on each side of the anther breaks. Subsequently, anther dehiscence occurs to allow the release of pollen. The morphological description of *Arabidopsis* anther development provides a basis for the characterization of mutants defective in anther development (Owen and Makaroff 1995; Sanders et al. 1999). This chapter presents recent advances in our understanding of the control of anther cell division and differentiation from genetic and molecular analyses.

2
Control of Early Anther Development by SPL/NZZ

An early acting anther gene is *SPOROCYTELESS/NOZZLE* (*SPL/NZZ*), which is required for the formation of the PMCs and also has a role in ovule development (Balasubramanian and Schneitz 2000; Schiefthaler et al. 1999; Yang et al. 1999). In the stage 5 *spl* anthers, there are no normal somatic or sporogenous cells; however, the nature of the earliest defect is not clear. Yang et al. (1999) indicated that the archesporial cells of *spl* anthers underwent a normal periclinal cell division to form the primary parietal cells (PPCs) and primary sporogenous cells (PSCs) at stage 3. On the other hand, Schiefthaler et al. (1999) reported that the *nzz* mutant anther failed to show differentiation of the archesporial cells. It is possible that the difference in phenotypic descriptions reflect allelic variations and/or environmental effects. However, the lack of molecular markers for early anther cell types makes it difficult to identify them. Nevertheless, it is clear that *SPL/NZZ* is required for normal early anther development.

SPL/NZZ encodes a putative transcription factor (Schiefthaler et al. 1999; Yang et al. 1999). It was recently shown that the floral MADS-box protein AGAMOUS (AG) can directly bind to the 3' end of *SPL/NZZ* and induce its transcription (Ito et al. 2004). Using a fusion of *SPL* to the rat glucocorticoid receptor gene (*35S::SPL-GR*), it was shown that SPL/NZZ is sufficient to induce microsporogenesis in a strong *ag* mutant background, indicating that

SPL/NZZ is a central mediator for the promotion of microsporogenesis by AG (Ito et al. 2004). These *35S::SPL-GR* flowers produced whorl 3 petals with locules, instead of stamens; therefore, *SPL/NZZ* alone was not sufficient to specify stamens and other whorl-specific factors are involved in this process (Ito et al. 2004).

There is evidence that *SPL/NZZ* is expressed at anther stage 2, consistent with the idea that *SPL/NZZ* is necessary for promoting cellular division and differentiation of another cell type (Schiefthaler et al. 1999). It is possible that *SPL/NZZ* promotes the formation or differentiation of the PPCs and PSCs. Alternatively, *SPL/NZZ* might not be necessary for the differentiation of PPCs and PSCs, but is required for promoting the differentiation of the sporogenous cells beginning at late stage 3. Due to the lack of specific molecular markers for these early anther cell types, it is difficult to know whether PPCs and PSCs are correctly formed at stage 3 and these hypotheses cannot yet be distinguished.

Although the parietal cell types are affected in the *spl/nzz* mutants, it is not known how *SPL/NZZ* regulates the development of the parietal cell types. A current theory in anther development is that the sporogenous cells promote the differentiation and development of the parietal cell layers possibly through cell-cell communication (Albrecht et al. 2005; Scott et al. 2004; Yang et al. 1999, 2003). This model would support an indirect role for *SPL/NZZ* in sporophytic development, by regulating genes necessary for the formation of sporogenous cells, which then promote the differentiation of adjacent cells into the parietal cell types. Alternatively, it is possible that the *spl/nzz* mutants are not able to properly specify the archesporial cell type. Thus, the observation that the anther was filled with vacuolated parenchyma cells (Yang et al. 1999) may be the result of a defect in this specification. In this case, *SPL/NZZ* would play an earlier and more direct role in specifying the parietal cells.

3
Tapetum Specification Requires Cell-Cell Signaling

The tapetum cell layer supports pollen development by producing and releasing essential proteins (Izhar and Frankel 1971; Stieglitz 1977). When the tapetum cells are selectively destroyed the plant fails to produce pollen (Koltunow et al. 1990; Mariani et al. 1990, 1991). Recently several genes have been shown to be critical for tapetum formation and may be components of the same signaling pathway. Two groups independently isolated mutant alleles of the same gene called *EXCESS MICROSPOROCYTES1* (*EMS1*)/*EXTRA SPOROGENOUS CELLS* (*EXS*) (Canales et al. 2002; Zhao et al. 2002). Both the *ems1* and *exs* mutant alleles appear to lack the tapetal cell layer and produce an increased number of PMCs (Fig. 1). This observation was further supported by the results that the expression domain of the meiosis-specific marker *SDS*

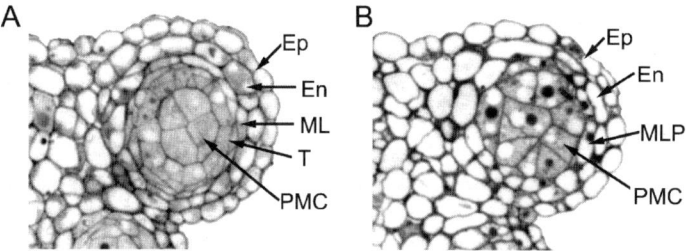

Fig. 1 Cross-sections of wild-type and *ems1* anthers at late stage 5. **A** The wild anther lobe has the five distinct cell layers. **B** The *ems1* anther lobe has a higher than normal number of PMCs in the center of the lobe, and the tapetum layer is missing. Ep, epidermis; En, endothecium; ML, middle layer; T, tapetum; PMC, pollen mother cells; MLP, middle layer-positioned cells

was expanded to the extra PMCs in the *ems1* mutant, whereas no signal was detected for the tapetum-specific probe *ATA7* (Zhao et al. 2002).

In a normal stage 6 anther, PMCs are detached from each other and the tapetum; in contrast, the *ems1/exs* mutant PMCs were abnormally enlarged and adhered to adjacent cells (Zhao et al. 2002). This may be due to abnormal callose accumulation/deposition on the mutant meiocytes (Canales et al. 2002), supporting the hypothesis that tapetum is necessary for normal callose deposition and the physical separation of the PMCs. Nevertheless, the *ems1/exs* mutants undergo meiotic nuclear events from prophase I to telophase II (Canales et al. 2002; Zhao et al. 2002). However, the *ems1* meiocytes fail to undergo cytokinesis and degenerate without forming microspores (Zhao et al. 2002).

The *EMS1/EXS* transcript was detected in archesporial cells at stage 2 and subsequently in the L2-derived cells at stages 3 and 4 (Canales et al. 2002; Zhao et al. 2002). At stage 5, the *EMS1/EXS* transcript was detected strongly in the tapetum and less so in PMCs. The *EMS1/EXS* expression is greatly reduced starting at stage 6. Canales et al. (2002) also detected *EMS1/EXS* expression in the floral meristem and ovule primordia. The *EMS1/EXS* gene encodes a leucine-rich repeat receptor-like kinase (LRR-RLKs) (Canales et al. 2002; Zhao et al. 2002), a member of the largest family of RLKs in plants (Shiu and Bleecker 2001). Transient expression in onion epidermal cells of an EMS1-GFP fusion suggests that EMS1 is localized to the cell surface; furthermore, an in vitro assay showed that EMS1/EXS has autophosphorylation activity (Zhao et al. 2002). These results support the hypothesis that it functions as a receptor-like protein kinase to mediate an important developmental signal.

In addition to *EMS1/EXS*, two other LRR-RLKs, *SERK1* and *SERK2*, play critical roles in tapetum formation (Albrecht et al. 2005; Colcombet et al. 2005). Although the *serk1* and *serk2* single mutants seem normal, the *serk1 serk2* double mutant has the same male fertility and anther development phenotypes as those of *ems1/exs* mutants. Therefore, *SERK1* and *SERK2* act redundantly in a way similar to *EMS1/EXS*. The expression patterns of *SERK1* and *SERK2* in the anthers largely coincide with the expression pattern of *EMS1/EXS*. Fluorescence Resonance Energy Transfer (FRET) analysis showed that SERK1 and SERK2 can form homo- or hetero-dimers in the plasma membrane, suggesting that SERK1 and SERK2 act interchangeably in the same complex (Albrecht et al. 2005). *SERK1* and *SERK2* transcripts were detected in the L2-derived cells at stages 4 and early 5, including the PMCs, strongly in the tapetum at late stage 5, and at reduced levels after stage 8 or 9.

The similarity of *ems1* and *serk1 serk2* anther phenotypes suggests that these genes might mediate the same signaling pathway. EMS1/EXS is closely related to the BRI1 brassinosteroid receptor, except that EMS1 lacks the hormone binding domain (Zhao et al. 2002). In addition, the SERK1/SERK2 close homolog BAK1 is known to interact with BRI1 (Albrecht et al. 2005; Colcombet et al. 2005; Nam and Li 2002). Therefore, it is possible that EMS1/EXS forms a receptor complex with SERK1 or SERK2 that binds to one or more ligands (Albrecht et al. 2005; Colcombet et al. 2005). Currently, the most likely candidate for the ligand is TAPETUM DETERMINANT1 (TPD1), a small 176 amino acid protein (Yang et al. 2003). The *tpd1* mutant exhibits an identical phenotype (Yang et al. 2003) to those of the *ems1/exs* single mutant and *serk1 serk2* double mutant, with no tapetum and extra PMCs. A double mutant between *tpd1* and another *ems1/exs* allele has nearly identical anther phenotypes to those of the *ems1* and *tpd1* single mutants (Yang et al. 2003), suggesting that these genes act in the same signaling pathway.

It was shown that *TPD1* expression in the anther appears to be in all of the L2-derived cells at stages 2–4 (Yang et al. 2003). At late stage 4 and early stage 5, *TPD1* and *EMS1/EXS* are expressed in the PMCs and in tapetal precursors, with *TPD1* being strongest in the PMCs and *EMS1/EXS* being strongest in tapetal precursors, suggesting that they might mediate a signaling event between these two cell types. The expression of *TPD1* was not reduced in the *ems1/exs1-2* mutant background, nor was *EMS1/EXS* expression reduced in the *tpd1* mutant background, indicating that they do not regulate one another at the mRNA level (Yang et al. 2005). Expression of *TPD1* under the 35S promoter in transgenic lines caused a range of abnormal phenotypes, including abnormal tapetal development and degeneration and wide carpels (Yang et al. 2005). In addition, the wide carpel phenotype was dependent on the presence of a wild type copy of *EMS1/EXS*. Together these results indicate that *TPD1* requires *EMS1/EXS* in order to regulate reproductive development (Yang et al. 2005).

In rice, the *MULTIPLE SPOROCYTE1* (*MSP1*) gene encodes an LRR-RLK that shares 63.8% amino acid sequence identity with EMS1/EXS (Nonomura et al. 2003). The *msp1* mutant is phenotypically similar to *ems1/exs*, suggesting that *EMS/EXS* and *MSP1* are functional homologs (Nonomura et al. 2003). In addition, the *msp1* mutant ovules also produce multiple sporogenous cells. Thus, perhaps analogous to *EMS1/EXS* in *Arabidopsis*, *MSP1* is important for restricting sporogenous cell number and promoting normal anther wall formation in rice (Nonomura et al. 2003). A similar mutant, *multiple archesporial cells1* (*mac1*), has been described in maize, where the cell lineage in anther development is well defined (Sheridan et al. 1999) (Fig. 2). In the

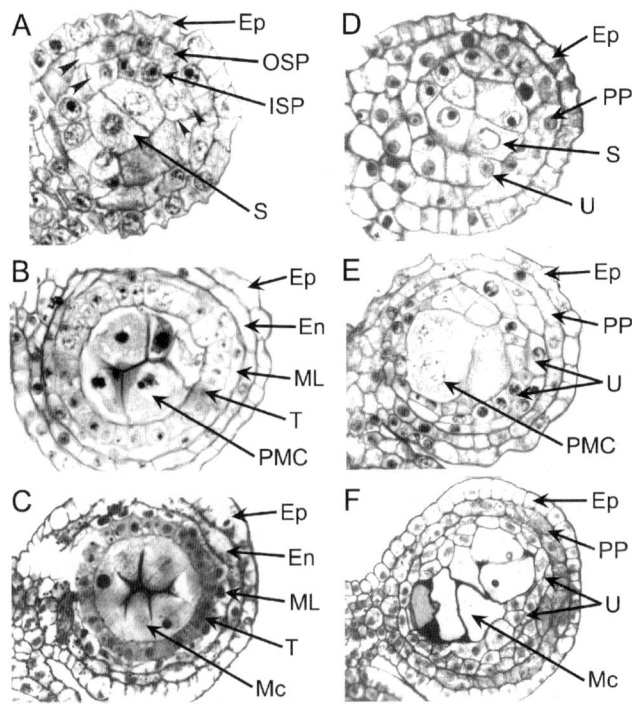

Fig. 2 Cross-sections of maize wild-type and *mac1* mutant anthers. **A–C** Wild-type; **D–F** *mac1*. **A** A stage comparable to *Arabidopsis* late stage 4, the anther has the four cell layers, and the inner secondary parietal layer is dividing to form the middle layer and tapetum (*small arrowheads*). **B** An anther with five distinct cell layers. **C** A stage comparable to *Arabidopsis* early stage 6, the PMCs are in prophase I of meiosis. **D** A stage similar to that in **A**. Cell patterning is disrupted at this stage. The primary parietal layer has not divided periclinally and the inner sporogenous cells appear to have proliferated abnormally. **E** A stage similar to that in **B**, the *mac1* mutant primary parietal layer continues to persist, while the sporogenous cells have formed additional layers, the innermost of which has the appearance of PMCs. **F** A stage comparable to that in **C** in the *mac1* mutant anther, the inner cell layers are disrupted; the primary parietal cells persist in the endothecial position, no tapetum is formed, and while the inner-most PMC-like cells have entered prophase I of meiosis, they are irregularly shaped and are surrounded by a largely disorganized array of unknown cell types. Ep, epidermis; PP, primary parietal; OSP, outer secondary parietal; ISP, inner secondary parietal; S, sporogenous; En, endothecium; ML, middle layer; T, tapetum; PMC, pollen mother cells; Mc, meiocyte; U, unknown cell type. This figure was modified from Figs. 1 and 2 of Sheridan et al. (1999) with permission from the Genetics Society of America

mac1 mutant, the primary parietal cells do not seem to undergo any periclinal division, and the sporogenous cells proliferate abnormally (Sheridan et al. 1999) (Fig. 2). The innermost sporogenous cells enter meiosis, but they are abnormal in shape, organization, and callose-deposition, and fail to proceed to beyond prophase I. Like *msp1* in rice, the *mac1* mutation in maize also results in the over-proliferation of archesporial cells in the ovule (Nonomura et al. 2003; Sheridan et al. 1999). This implies that these genes act to negatively regulate sporogenous cell division and to positively regulate parietal cell differentiation, similar to *EMS1/EXS*, *SERK1*, *SERK2*, and *TPD1* in *Arabidopsis*.

The results from *Arabidopsis*, rice, and maize suggest that there is a conserved cell-cell signaling pathway that promotes the formation of the tapetum layer and limits the number of PMCs. The mutant phenotypes also support the idea that the default developmental program favors the formation of PMCs. Furthermore, molecular studies described here suggest that these genes represent signaling pathway(s) that respond to signal(s) from the central region of the PMCs and direct differentiation of the tapetum, which then support meiosis and pollen development (Ma 2005).

4
Regulation of Normal Tapetum Development and Function

The coordinated differentiation and function of tapetal cells and the male gametophyte is essential for normal pollen development. Several genes have been identified that are important for tapetum differentiation and/or function, including the closely related *MYB33* and *MYB65* genes (Millar and Gubler 2005). Disruption of either gene has no apparent effect on plant development, whereas the *myb33 myb65* double mutant plants are conditionally male sterile and fail to produce pollen, indicating that they function redundantly (Millar and Gubler 2005). Cross sections of the mutant anthers revealed that the male sterile phenotype was due to hypertrophy (abnormal enlargement) of the tapetal cells from stages 5 to 6 (Fig. 3A and B). In addition to having an increased size, the tapetal cells do not separate from one another, as they do in the wild type. Furthermore, the tapetum appears unable to perform its normal functions. For example, callose deposits remain in the locules where the tapetum has expanded, indicating that the tapetum is not releasing callase (Millar and Gubler 2005).

Interestingly, male sterility in the *myb33 myb65* double mutant is affected by environmental conditions. Occasionally normal locules were seen adjacent to hypertrophied locules (Fig. 3C), suggesting that the *MYB33/MYB65* function might not be absolutely essential, but is important under certain conditions. Fertility of the double mutant is significantly increased under relatively high light ($\sim 300\,\mu mol\,\mu m^{-2}\,s^{-1}$ vs. $95\,\mu mol\,\mu m^{-2}\,s^{-1}$) and at low tempera-

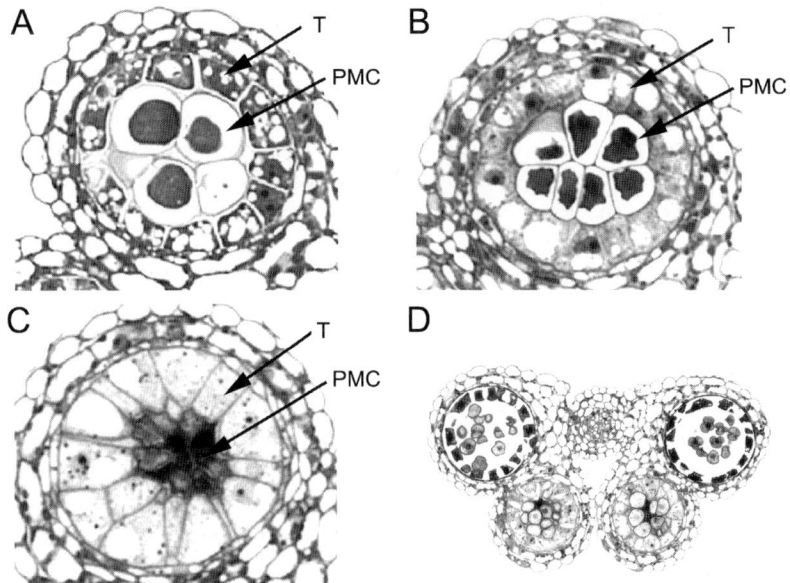

Fig. 3 Cross-sections of *myb33 myb65* double mutant anthers. **A** Around anther stage 6, pre-meiotic PMCs have a thick callose layer, but have become irregularly shaped under an abnormally expanding tapetum layer. **B** A later stage anther showing tapetal cells that are highly vacuolated and have expanded radially inward around degrading PMCs. **C** An anther with adaxial lobes undergoing normal development and abaxial lobes with hypertrophied tapetal cells. T, tapetum; PMC, pollen mother cells. This figure was modified from Figs. 3 and 4 from Millar and Gubler (2005) with permission from the American Society of Plant Biologists

tures (16 °C vs. 22 °C), and were almost as fertile as the wild type (75% vs. 90–96% filled siliques). Both high light and low temperatures increase the level of soluble carbohydrates in plants (Hurry et al. 1995). Carbohydrate reserves are mobilized in the tapetum just before meiosis; this timing coincides with the mutant defect (Clément et al. 1994; Millar and Gubler 2005). Thus *MYB33* and *MYB65* may play a role in starch mobilization similar to the function of a known *MYB* gene from barley (*HvGAMYB*) (Millar and Gubler 2005). Indeed, *MYB33* and *MYB65* were shown be capable of activating the α-amylase promoter in the place of *HvGAMYB* (Gocal et al. 2001). Furthermore, the *GAMYB* mutants in rice and barley have similar anther defects, exhibiting tapetum hypertrophy prior to meiosis (Kaneko et al. 2004; Murray et al. 2003). Therefore, it is possible that the role of *MYB* genes in anther development may be conserved between monocots and eudicots (Millar and Gubler 2005).

MYB33 is expressed at low levels in anthers at stages 5 and 6, and at much higher levels at stage 7 in all four cell layers of the anther wall, as well as in the tetrads. *MYB33* expression in the tapetum just before meiosis is consistent with the tapetum hypertrophy in the *myb33 myb65* double mutant at this stage. The *MYB33* and *MYB65* genes are also post-transcriptionally regulated by microRNAs (miRNAs) (Millar and Gubler 2005). miRNAs can guide cleavage of the cognate mRNAs, and cleavage products of both *MYB33* and *MYB65* have been isolated (Palatnik et al. 2003). In addition, it was shown that overexpression of the miRNA miR159a causes cleavage of *MYB33* and male sterility (Achard et al. 2004).

A rice mutant named *undeveloped tapetum1* (*udt1*) was recently isolated from a T-DNA insertional screen (Jung et al. 2005). In the *udt1* mutant, anther development appears normal through the pre-meiotic stages (stage 5) (Jung et al. 2005). However, at the time of meiosis the mutant tapetal cells become highly vacuolated and continue to increase in size through stage 7; at the same time, the middle layer cells do not flatten or degenerate. From meiosis onward all of the cell layers of the anther wall cell layers appear irregularly shaped and abnormally stained. The *udt1* PMCs undergo normal nuclear division, forming dyads at the end of meiosis I. However, at the stage where tetrads are normally formed the meiocytes are degraded. Thus, *Udt1* plays an important role in tapetum differentiation, beginning around meiosis, and may be involved in the development of the other anther cell wall layers (Jung et al. 2005).

Semi-quantitative RT-PCR showed that *Udt1* is strongly expressed within the anther, from meiosis to pollen release. Using a GUS reporter gene, researchers first observed *Udt1* expression after the initiation of tapetum development. *Udt1* encodes a protein with a predicted basic helix-loop-helix (bHLH) domain (Jung et al. 2005), which define a large family of putative transcription factors, including the *Arabisdopsis* ABORTED MICROSPORES (AMS, see below) and mammalian Myc proteins. Transient expression of a *35S:Udt1-GFP* fusion construct revealed that GFP signal was localized to the nucleus, indicating that UDT1 is likely a nuclear protein (Jung et al. 2005). Therefore, *Udt1* appears to be a key regulator of tapetum differentiation and/or function at the time of meiosis and soon afterwards.

Because *Udt1* encodes a putative transcription factor, the transcript level of three rice tapetum-specific markers was tested: *Osc4* and *Osc6*, which encode protease inhibitors, and *Cys protease1* (Jung et al. 2005; Lee et al. 2004; Tsuchiya et al. 1994). Each of these genes is expressed in the wild type from meiosis (stage 6) through the vacuolated pollen stage (about stage 10). In the *udt1* mutant anthers, however, transcripts for these genes were undetectable (Jung et al. 2005). This suggests that they may be downstream of *Udt1* (Jung et al. 2005). In addition, a potential rice homolog (Os02g02820) of *AMS*, also important for tapetum development (see below), is strongly expressed in the wild-type anther but reduced in the *udt1* mutant (Jung et al. 2005). Further investigation using microarray experiments revealed a large number of genes that show reduced expression in the *udt1* mutant, including genes coding for bHLH, MYB, WRKY, and APETELA2 transcription factors (Jung et al. 2005).

5
Genes Affecting Tapetal Function and Pollen Development

Following the completion of meiosis, the tapetum is critical for normal pollen development. The *ABORTED MICROSPORES* (*AMS*) gene plays a critical role in the post-meiotic tapetum that supports microspore development (Sorensen et al. 2002). The *ams* mutant is male sterile and cannot produce pollen. Although *ams* is normal in anther histogenesis, meiosis and microspore formation, the microspores soon degenerate. In addition, the *ams* tapetum appears to begin premature degeneration. Eventually both microspores and tapetum cells completely disintegrate. *AMS* expression was strong in post-meiotic floral buds, but not in open flowers, siliques, leaves or roots (Sorensen et al. 2002). The highest levels of GUS activity from an *AMS-GUS* fusion were seen in the tapetum and the nuclei of microspores, consistent with a role for *AMS* in promoting tapetum function and pollen mitosis I. The regulation of pollen mitosis I may be direct, via *AMS* function in the microspores, or indirect, through regulation of tapetum function, or both. The *AMS* gene codes for a Myc-like bHLH transcription factor (Sorensen et al. 2002) and is homologous to the rice gene Os02g02820, which is positively regulated by *Udt1* (Jung et al. 2005). It is possible that AMS is part of a regulatory network that controls tapetum differentiation and function (Sorensen et al. 2002).

Part of tapetal function includes biosynthesis of macromolecules that support pollen wall formation, such as lipids and proteins. One of the major pathways for Lipid synthesis in plants generally follows one of two major pathways: incorporation into triacylglycerol (TAG) or into membrane glycerolipids. Glycerolipids are essential components of the cell membrane and are likely necessary for maintaining membrane integrity as well as membrane biogenesis (Zheng et al. 2003). Glycerol-3-phosphate acyltransferases (GPATs) are enzymes that mediate the initial step of glycerolipid biosynthesis. Tapetum and microspores are both known to be quite active in lipid metabolism (Ferreira et al. 1997; Piffanelli et al. 1998; Platt et al. 1998), and the *Arabidopsis* mutant *atgpat1* is defective in the development of both of these cell types (Zheng et al. 2003). Although the *atgpat1-1* mutant appears normal up to the time of microspore formation, subsequently the tapetal cells begin to enlarge abnormally (Zheng et al. 2003). This defect underscores the importance of lipid biosynthesis in tapetum function. In vitro assays showed that ATGPAT1 is an acyltransferases that can specifically acetylate glycerol-3-phosphate.

Another gene important for tapetum function is the *MALE STERILE1* (*MS1*) gene, which appears to function later than *AtGPAT1* and in a different pathway. The *ms1* mutant appears to develop normally through stage 8, but the microspores fail to form a complete exine wall (stage 9) and both the tapetum and microspores become vacuolated and eventually degenerate (Fig. 4) (Ito and Shinozaki 2002; Wilson et al. 2001). Also, the cytoplasm

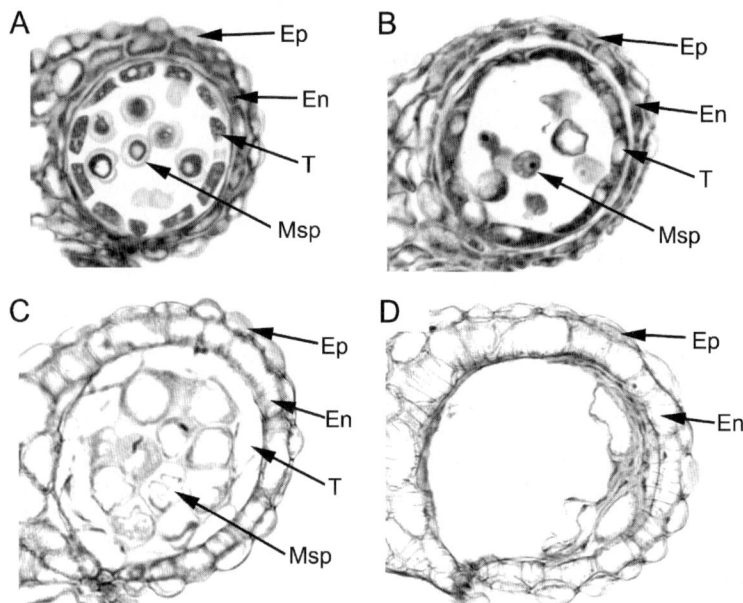

Fig. 4 Cross-sections of wild-type No-0 and *ms1* anther lobes. **A** Wild-type at anther stage 9 has microspores with an exine layer. **B** The *ms1* mutant microspores at stage 9 have no exine layer. **C** In *ms1*, subsequent to stage 9, the tapetum and microspores become vacuolated and eventually degrade **D** leaving the locules empty. Ep, epidermis; En, endothecium; T, tapetum; Msp, microspores (Courtesy of Takuya Ito)

of the microspores appears granular and vacuolated prior to degeneration (Wilson et al. 2001). *MS1* expression was only found in the inflorescence (Ito and Shinozaki 2002; Wilson et al. 2001). In addition, in situ hybridization showed that in anthers, *MS1* transcript could only be detected in the tapetal cells at the tetrad stage (stage 7) (Ito and Shinozaki 2002). Locules of the same anther at a slightly different developmental stage did not show any *MS1* expression, indicating a very short period of expression. Taken together, the mutant phenotype and expression pattern imply that *MS1* expression at stage 7 is required for the tapetum to produce materials essential for subsequent exine development on microspores at stage 9 (Ito and Shinozaki 2002). Therefore, *MS1* function is necessary for normal tapetum function after stage 8.

MS1 codes for a protein with three predicted domains: a nuclear localization signal, a PHD-finger motif and a leucine zipper-like sequence with sequence similarity to an *Arabidopsis* mitochondrial ORF. PHD-finger motifs are thought to be involved in protein-protein interactions and proteins containing these motifs may participate in transcriptional regulation (Aasland et al. 1995; Ito and Shinozaki 2002). Using a GFP fusion construct and particle bombardment, it was shown that the N-terminal region of MS1 is sufficient to direct localization to the nucleus (Ito and Shinozaki 2002). Thus, MS1 appears to be a nuclear protein involved in regulating transcription either through altering chromatin structure or as a transcription factor (Aasland et al. 1995; Ito and Shinozaki 2002; Wilson et al. 2001).

6
Summary and Perspectives

Many exciting discoveries have been made in the realm of anther development within the last ten years. The early acting *SPL/NZZ* gene is a key regulator of sporogenous cell differentiation and encodes a putative transcription factor. Its function is likely coordinated with other factors important for differentiation of the early anther cell types. Between stages 2 and 5, cells within the anther lobes undergo highly oriented cell divisions and the factors governing this process are not yet known. Further studies in the near future will likely uncover additional genes that control anther differentiation, including genes that are regulated by SPL/NZZ.

EMS/EXS, *SERK1*, *SERK2*, and *TPD1* appear to function in the same signaling pathway to promote tapetum differentiation. It is possible that the proteins coded for by these genes physically interact. EMS/EXS may form a functional receptor complex with SERK1 and/or SERK2, and TPD1 may be the ligand for this complex. In addition, other LRR-RLKs may also play a role in other aspects of anther development. The currently ongoing 2010 project on the functional analysis of 30 LRR-RLKs (http://www.mcb.arizona.edu/tax/2010/index.htm) may yield further insight into the regulation of anther development by members of this important gene family.

Finally, several genes have been described that are important for both tapetum and pollen development. Interestingly, *Udt1* appears to regulate the expression of a potential rice homolog for *AMS* (Os02g02820). Thus, the *Arabidopsis* homolog of *Udt1* may regulate the expression of *AMS*. If this is the case, it would indicate that this transcriptional pathway for tapetum development and/or function is highly conserved between monocots and dicots. Tapetum and pollen development are complex and important processes, future studies using multiple approaches offer the promise of new discoveries about the molecular control of plant development.

Acknowledgements We greatly appreciate the contributions of figures by Drs. Takuya Ito, Anthony Millar, and William Sheridan. We would also like to thank Bridget Leyland and Gavilange Nestor for comments on the manuscript. The work in our laboratory has been supported by a grant from the US Department of Energy to H.M. (DE-FG02-02ER15332). C.L.H.H. was partially supported by the Integrative Bioscience Graduate Degree Program at the Pennsylvania State University.

References

Aasland R, Gibson TJ, Stewart AF (1995) The PHD finger: implications for chromatin-mediated transcriptional regulation. Trends Biochem Sci 20:56–59

Achard P, Herr A, Baulcombe DC, Harberd NP (2004) Modulation of floral development by a gibberellin-regulated microRNA. Development 131:3357–3365

Albrecht C, Russinova E, Hechtm V, Baaijens E, de Vries S (2005) The *Arabidopsis thaliana* SOMATIC EMBRYOGENESIS RECEPTOR-LIKE KINASES1 and 2 control male sporogenesis. Plant Cell 17:3337–3349

Balasubramanian S, Schneitz K (2000) *NOZZLE* regulates proximal-distal pattern formation, cell proliferation and early sporogenesis during ovule development in *Arabidopsis thaliana*. Development 127:4227–4238

Canales C, Bhatt AM, Scott R, Dickinson H (2002) *EXS*, a putative LRR receptor kinase, regulates male germline cell number and tapetal identity and promotes seed development in *Arabidopsis*. Curr Biol 12:1718–1727

Clément C, Chavant L, Burrus M, Audran JC (1994) Anther starch variations in *Lilium* during pollen development. Sex Plant Reprod 7:347–356

Colcombet J, Boisson-Dernier A, Ros-Palau R, Vera CE, Schroeder JI (2005) *Arabidopsis* SOMATIC EMBRYOGENESIS RECEPTOR KINASES1 and 2 are essential for tapetum development and microspore maturation. Plant Cell 17:3350–3361

Ferreira MA, de Almeida Engler J, Miguens FC, van Montagu M, Engler G, de Oliveira DE (1997) Oleosin gene expression in *Arabidopsis thaliana* tapetum coincides with accumulation of lipids in plastids and cytoplasmic bodies. Plant Physiol Biochem 35:729–739

Gocal GF, Sheldon CC, Gubler F, Moritz T, Bagnall DJ, MacMillan CP, Li SF, Parish RW, Dennis ES, Weigel D, King RW (2001) *GAMYB-like* genes, flowering, and gibberellin signaling in *Arabidopsis*. Plant Physiol 127:1682–1693

Goldberg RB, Beals TP, Sanders PM (1993) Anther development: basic principles and practical applications. Plant Cell 5:1217–1229

Hurry VM, Strand A, Tobiaeson M, Gardestrom P, Oquist G (1995) Cold hardening of spring and winter wheat and rape results in differential effects on growth, carbon metabolism, and carbohydrate content. Plant Physiol 109:697–706

Ito T, Shinozaki K (2002) The *MALE STERILITY1* gene of *Arabidopsis*, encoding a nuclear protein with a PHD-finger motif, is expressed in tapetal cells and is required for pollen maturation. Plant Cell Physiol 43:1285–1292

Ito T, Wellmer F, Yu H, Das P, Ito N, Alves-Ferreira M, Riechmann JL, Meyerowitz EM (2004) The homeotic protein AGAMOUS controls microsporogenesis by regulation of *SPOROCYTELESS*. Nature 430:356–360

Izhar S, Frankel R (1971) Mechanism of male sterility in *Petunia*: The relationship between pH, callase activity in the anthers, and the breakdown of the microsporogenesis. Theo Appl Genet 44:104–108

Jung KH, Han MJ, Lee YS, Kim YW, Hwang I, Kim MJ, Kim YK, Nahm BH, An G (2005) Rice *Undeveloped Tapetum1* is a major regulator of early tapetum development. Plant Cell 17:2705–2722

Kaneko M, Inukai Y, Ueguchi-Tanaka M, Itoh H, Izawa T, Kobayashi Y, Hattori T, Miyao A, Hirochika H, Ashikari M, Matsuoka M (2004) Loss-of-function mutations of the rice *GAMYB* gene impair alpha-amylase expression in aleurone and flower development. Plant Cell 16:33–44

Koltunow AM, Truettner J, Cox KH, Wallroth M, Goldberg RB (1990) Different temporal and spatial gene expression patterns occur during anther development. Plant Cell 2:1201–1224

Lee S, Jung KH, An G, Chung YY (2004) Isolation and characterization of a rice cysteine protease gene, *OsCP1*, using T-DNA gene-trap system. Plant Mol Biol 54:755–765

Li W, Ma H (2002) Gametophyte development. Curr Biol 12:R718–721

Ma H (2005) Molecular genetic analyses of microsporogenesis and microgametogenesis in flowering plants. Annu Rev Plant Biol 56:393–434

Mariani C, Goldberg RB, Leemans J (1991) Engineered male sterility in plants. Symp Soc Exp Biol 45:271–279

Mariani C, De Beuckeleer M, Truettner J, Leemans J, Goldberg RB (1990) Induction of male sterility in plants by a chimaeric ribonuclease gene. Nature 347:737–741

Millar AA, Gubler F (2005) The *Arabidopsis GAMYB-like* genes, *MYB33* and *MYB65*, are microRNA-regulated genes that redundantly facilitate anther development. Plant Cell 17:705–721

Murray F, Kalla R, Jacobsen J, Gubler F (2003) A role for HvGAMYB in anther development. Plant J 33:481–491

Nam KH, Li J (2002) BRI1/BAK1, a receptor kinase pair mediating brassinosteroid signaling. Cell 110:203–212

Nonomura K, Miyoshi K, Eiguchi M, Suzuki T, Miyao A, Hirochika H, Kurata N (2003) The *MSP1* gene is necessary to restrict the number of cells entering into male and female sporogenesis and to initiate anther wall formation in rice. Plant Cell 15:1728–1739

Owen HA, Makaroff CA (1995) Ultrastructure of microsporogenesis and microgametogenesis in *Arabidopsis thaliana* (L.) Heynh. ecotype Wassilewskija (Brassicaciae). Protoplasma 185:7–21

Palatnik JF, Allen E, Wu X, Schommer C, Schwab R, Carrington JC, Weigel D (2003) Control of leaf morphogenesis by microRNAs. Nature 425:257–263

Piffanelli P, Ross JHE, Murphy DJ (1998) Biogenesis and function of the lipidic structures of pollen grains. Sex Plant Reprod 11:65–80

Platt KA, Huang AHC, Thomson WW (1998) Ultrastructural study of lipid accumulation in tapetal cells of *Brassica napus* L. cv. Westar during microsporogenesis. Int J Plant Sci 159:724–737

Sanders PM, Bui AQ, Weterings K, McIntire KN, Hsu Y, Lee PY, Truong MT, Beals TP, Goldberg RB (1999) Anther development defects in *Arabidopsis thaliana* male-sterile mutants. Sex Plant Reprod 11:297–322

Schiefthaler U, Balasubramanian S, Sieber P, Chevalier D, Wisman E, Schneitz K (1999) Molecular analysis of *NOZZLE*, a gene involved in pattern formation and early sporogenesis during sex organ development in *Arabidopsis thaliana*. Proc Natl Acad Sci USA 96:11664–11669

Scott RJ, Spielman M, Dickinson HG (2004) Stamen structure and function. Plant Cell 16:S46–60

Sheridan WF, Golubeva EA, Abrhamova LI, Golubovskaya IN (1999) The *mac1* mutation alters the developmental fate of the hypodermal cells and their cellular progeny in the maize anther. Genetics 153:933–941

Shiu SH, Bleecker AB (2001) Receptor-like kinases from *Arabidopsis* form a monophyletic gene family related to animal receptor kinases. Proc Natl Acad Sci USA 98:10763–10768

Sorensen A, Guerineau F, Canales-Holzeis C, Dickinson HG, Scott RJ (2002) A novel extinction screen in *Arabidopsis thaliana* identifies mutant plants defective in early microsporangial development. Plant J 29:581–594

Stieglitz H (1977) Role of beta-1,3-glucanase in postmeiotic microspore release. Dev Biol 57:87–97

Tsuchiya T, Toriyama K, Ejiri S, Hinata K (1994) Molecular characterization of rice genes specifically expressed in the anther tapetum. Plant Mol Biol 26:1737–1746

Wilson ZA, Morroll SM, Dawson J, Swarup R, Tighe PJ (2001) The *Arabidopsis* MALE STERILITY1 (*MS1*) gene is a transcriptional regulator of male gametogenesis, with homology to the PHD-finger family of transcription factors. Plant J 28:27–39

Wu HM, Cheun AY (2000) Programmed cell death in plant reproduction. Plant Mol Biol 44:267–281

Yang SL, Xie LF, Mao HZ, Puah CS, Yang WC, Jiang L, Sundaresan V, Ye D (2003) *TAPETUM DETERMINANT1* is required for cell specialization in the *Arabidopsis* anther. Plant Cell 15:2792–2804

Yang SL, Jiang LX, Puah CS, Xie LF, Zhang XQ, Chen LQ, Yang WC, Ye D (2005) Overexpression of *TAPETUM DETERMINANT1* alters the cell fates in the *Arabidopsis* carpel and tapetum via genetic interaction with *EXCESS MICROSPOROCYTES1/EXTRA SPOROGENOUS CELLS*. Plant Physiol 139:186–191

Yang WC, Ye D, Xu J, Sundaresan V (1999) The *SPOROCYTELESS* gene of *Arabidopsis* is required for initiation of sporogenesis and encodes a novel nuclear protein. Genes Dev 13:2108–2117

Zhao DZ, Wang GF, Speal B, Ma H (2002) The *EXCESS MICROSPOROCYTES1* gene encodes a putative leucine-rich repeat receptor protein kinase that controls somatic and reproductive cell fates in the *Arabidopsis* anther. Genes Dev 16:2021–2031

Zheng Z, Xia Q, Dauk M, Shen W, Selvaraj G, Zou J (2003) *Arabidopsis AtGPAT1*, a member of the membrane-bound glycerol-3-phosphate acyltransferase gene family, is essential for tapetum differentiation and male fertility. Plant Cell 15:1872–1887

Coordination of Cell Division and Differentiation

Crisanto Gutierrez

Centro de Biologia Molecular "Severo Ochoa", Consejo Superior de Investigaciones, Científicas and Universidad Autónoma de Madrid, Cantoblanco, 28049 Madrid, Spain
cgutierrez@cbm.uam.es

Abstract Cell division is a highly regulated process in individual cells. Multicellularity has introduced extra layers of regulatory complexity since maintenance of a strict cellular homeostasis is crucial for proper development. In the case of plants, where organogenesis is a post-embryonic and continuous process, the coordination between cell proliferation and cell differentiation is of primary importance. The last 10 years have witnessed an unprecedented advance in our understanding of cell division and how it integrates with differentiation and development. These studies have benefited enormously from the availability of the full genome sequence of *Arabidopsis thaliana* and the genomic tools generated. We now face the challenge to integrate the functional relationships of cell cycle regulators into common pathways and to define the complex transcriptional networks that coordinate cell proliferation and cell differentiation during plant development.

1
Introduction

Progression through the cell division cycle requires duplication of the genetic material and the delivery of the newly duplicated genomes to the two daughter cells during mitosis. However, an important consideration is that the cell cycle, as we normally understand it, is indeed the functional integration of multiple cycles (Fig. 1). Thus, cells develop a "growth cycle" during which they increase in total mass, a process that largely occurs in an almost continuous manner. They also develop a "DNA replication cycle", which is a discrete process whose duration defines the S-phase. In other words, in a population of asynchronously proliferating cells, only a proportion of them is, at a given time, engaged in duplicating their genome. During the "chromosome segregation and cytokinesis cycle" the duplicated genome is transferred to the daughter cells and it defines the mitotic phase, which includes cytokinesis. More recent molecular and biochemical studies are serving to establish the occurrence of other cycles, e.g., the "CDK cycle", which is defined by a succession of periods with high (from just before the S-phase until the end of metaphase) and low (from anaphase to late G1) CDK activity. The temporal superimposition of these functional cycles in a particular proliferating cell originates what we define as a cell cycle with the typical G1, S, G2 and M phases, although the rest of functional activities should also be taken into account (Fig. 1).

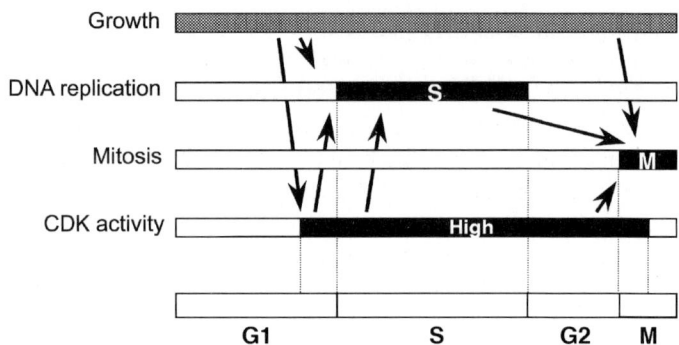

Fig. 1 Each proliferating cell develops different processes during its life. Some of them are depicted in the figure. These cycles frequently communicate and their superimposition gives rise to different functional stages that we define as the cell cycle

However, as expected in a system under a strict homeostasis, these different cycles do not run independently. Rather, they are coupled in various ways, and complex crosstalks occur among them. In fact, the coordination between the cycles is crucial for proper cell cycle progression. For example, cell growth signals accumulate to favor the accumulation of components necessary to increase CDK activity late in G1. This allows the inactivation of the RBR protein and the release of E2F activity that triggers the G1/S transition (Fig. 1). Therefore, understanding the mechanisms that regulate cell cycle progression requires the interplay between different processes. In fact, the coordination among all of them is at the basis of the large variety of genes that either directly or indirectly have an impact on cell proliferation.

The coordinated activities that occur during cell cycle progression also contribute to its unidirectionality. Thus, the regulated transition from one functional stage to the next occurs in a manner where way back is not possible. The molecular basis for this is that the activity of specific cell cycle drivers very frequently modifies their targets in such a way that once they execute their function, they are irreversibly converted into an "inactive" form. This may occur, for example, by changing their cellular localization, by targeting them for selective proteolysis, or by modifying the cellular transcriptional program.

Plants, like animals, as multicellular organisms have evolved regulatory mechanisms that allow the integration of processes at the cellular level with those related to organogenesis. These coordination mechanisms regulate, in the context of a developing or growing organism, the cell division potential, both in proliferating cells and in stem cell populations, cell cycle arrest and its withdrawal back to cell cycle, cell differentiation and cell death. Furthermore, plants and animals have evolved very different developmental strategies. While organogenesis in animals occurs during embryogenesis, organ initiation and growth in plants is a post-embryonic and continuous process that occurs over the entire lifespan of the organism.

In some cases, after completing one full genome replication cycle during the S-phase, cells take the decision of exiting the normal cell cycle to enter the endocycle, a different cycle in which repeated rounds of genome replication occur in the absence of cytokinesis. Endoreplication, by which cells increase exponentially their nuclear DNA content, occur in all eukaryotic cells (Edgar and Orr-Weaver 2001), but in plants is a much more frequent event in many cell types (Kondorosi et al. 2000; Larkins et al. 2001). Interestingly, a variety of specific differentiation programs depend on the occurrence of endocycle rounds.

Another important aspect in this context is the ability of plants to regenerate. By this process certain cells can dedifferentiate and revert to a totipotent or pluripotent state, proliferate and take various cell fates to originate an organ de novo. It is obvious that cell cycle control should have a crucial impact in coordinated regeneration although the molecular details have not been studied yet with sufficient detail.

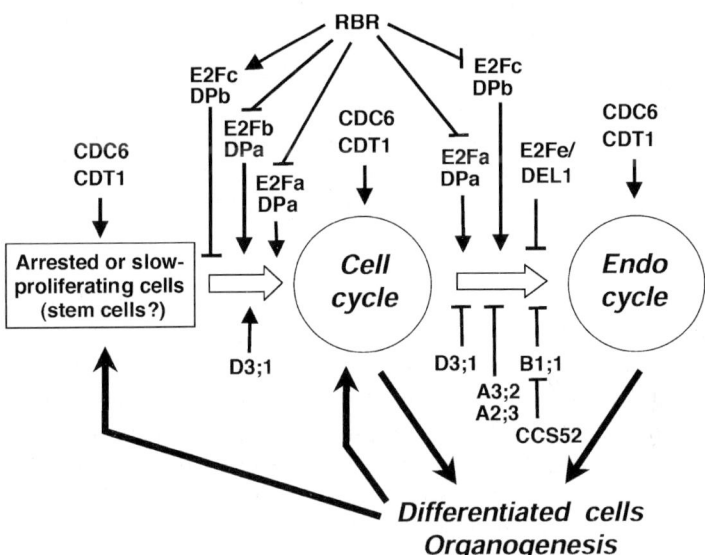

Fig. 2 Different cellular pools can be distinguished in a developing plant. Cells that are arrested or slow proliferating (some of these could be stem cells) respond to hormonal or developmental signals to initiate an active proliferative phase. Later in development, some cells exit the cell cycle and are recruited to enter the endocycle program before they differentiate within particular organs. Some cell cycle regulatory components affect these transitions. Cyclins (the acronym CYC has been omitted for simplicity) has been placed separately, although they most likely exert their function together by activating specific CDK which phosphorylate a variety of factors. The individual combinations of cyclins, CDK and their targets in vivo are not known

Therefore, in the plant organs, cellular pools including slow proliferating and stem cells, proliferating cells and endoreplicating cells have to maintain a strict balance for a proper homeostasis (Fig. 2). The control of cell proliferation and differentiation during development depends, in most cases, on the concerted action of plant hormones. Quite interestingly, hormones directly impinge on the availability of several cell cycle regulators either at the level of transcriptional regulation or, later, on postranslational modifications, typically, selective proteolysis (del Pozo et al. 2005).

2
Cell Cycle Control in a Developmental Context

Studies in single-cell model systems (yeast and mammalian cells in culture) have provided during the last two decades most of our current knowledge of molecular mechanisms regulating cell cycle transitions. Phylogenetic analysis of cell cycle regulators, together with molecular, cellular and genetic approaches, has revealed that the basic cell cycle machinery is also highly conserved in all eukaryotes. The identification and further cloning of the first plant cell cycle regulators (John et al. 1989, 1990; Feiler and Jacobs 1990; Hata et al. 1990; Ferreira et al. 1991) showed a striking similarity with those identified in yeast and human cells, just one or two years earlier (Lee and Nurse 1987; Nurse, 1990). However, one possibility, based on the different complexities and properties of animals and plants, was that cell cycle regulation could be relatively more similar to that occurring in yeast. Intense efforts in the following years, highly helped by genomic information from *A. thaliana*, revealed that this was not the case. In addition, plant-specific genes and circuits have been identified and they are reviewed in other chapters in this volume.

Detailed information of each group of cell cycle regulators has been provided in previous chapters. Studies in various model systems are revealing unanticipated roles of cell cycle regulators in the context of a developing or growing organism. It seems that cell cycle regulators are also the targets to integrate cell proliferation with other cellular activities as well as with organogenesis (Gutierrez 2005; de Jager et al. 2005). Here, I will present an overview of cell proliferation control under a developmental perspective with the aim of providing current evidence in support of the need to reassess the role of cell cycle regulators in the context of a multicellular organism.

2.1
CDKs, Cyclins and CDK Inhibitors

Different types of CDKs have been identified in *Arabidopsis*. Type A is constituted by CDKs containing the typical amino acid motif PSTAIRE. A variation of this motif (PP/STA/TLRE) is present in type B CDKs. Others contain the PITAIRE or the SPTAIRE motifs (Vandepoele et al. 2002).

Overexpression of *Arabidopsis* CDKA;1 does not have any detectable macroscopic consequences (Hemerly et al. 1995). Likewise, overexpression of CDKA in maize endosperm does not influence the endoreplication process associated with endosperm development (Leiva-Neto et al. 2004). However, overexpression of a dominant negative version of CDKA inhibits cell division (Hemerly et al. 1995) and ploidy level (Leiva-Neto et al. 2004). Loss-of-function of CDKA;1 produces a male gametophytic phenotype due to failure of the generative cell during male gametogenesis (Iwakawa et al. 2006). Thus, double fertilization does not take place and the embryo arrests early at the globular stage. Auxin is sufficient to up regulate CDKA;1 gene expression but cell cycle progression also requires cytokinin (Zhang et al. 1996), consistent with the need of a concerted action, which probably impinges on the availability of other components that modulate CDK activity. In fact, cytokinin seems to mediate CDK activation by stimulating tyrosine dephosphorylation (Zhang et al. 1996).

Altering the levels of the plant-specific CDKB also has developmental consequences. Thus, CDKB1;1 regulates the balance between dividing and endoreplicating cells, by negatively regulating endocycle occurrence (Boudolf et al. 2004a). A decrease in CDBK1;1 activity produces a short hypocotyl phenotype which is partially rescued by increasing brassinosteroid level (Yoshizumi et al. 1999; Hu et al. 2000). Reduction of CDKB1;1 levels also negatively affects stomatal development through inhibition of cell division of the stomatal lineage (Boudolf et al. 2004b). Interestingly, these cell cycle arrested cells acquire a rather normal stomatal cell identity, providing one example (among others discussed below) of uncoupling cell division and cell differentiation.

Mutations in the *PROPORZ1* (*PRZ1*) gene were identified and produce a high tendency to form calli only in presence of either auxin or cytokinin (Sieberer et al. 2003). PRZ1 is a transcriptional adapter protein that regulates gene expression, including that of several cell cycle genes. Thus, *CDKB1;1* (as well as of *E2Fc*) is up regulated in the *prz1* mutant grown in hormone-containing medium, leading to the formation of undifferentiated callus-like structures (Sieberer et al. 2003). In hormone free medium, it is the expression of B- and D-types cyclins that is reduced in the *prz1* mutant.

Arabidopsis (Vandepoele et al. 2002; Wang et al. 2004) and rice (La et al. 2006) contain a large family of cyclins, suggesting that this may likely be the case for other plant species. Expression of *CYCA3* genes peaks in S-phase (Menges et al. 2005), suggesting that they are major elements of the kinases involved in S-phase-related events. Consistent with this idea, constitutive overexpression of tobacco *CYCA3;2* in *Arabidopsis* leads to a suppression of the endoreplication program associated with normal leaf development concomitantly with an induction of hyperplasia (Yu et al. 2003). *CYCA3;2* gene expression is up regulated at the G1/S transition and excess expression prevents cell differentiation and regeneration from leaf disks (Yu et al. 2003). CYCA2;3 acts also as a major repressor of endoreplication. However, as revealed by the phenotype of null mutants, endocycle occurrence is promoted and ploidy level is increased, but the number of cells undergoing endocycle does not change (Imai et al. 2006).

Adequate levels of CYCB are also important for correct development. CYCB1 degradation depends on CCS52, a Fizzy-related (Fzr) activator of the anaphase-promoting complex (APC; Capron et al. 2003), originally identified in alfalfa (Cebolla et al. 1999). The relevance of CYCB1 is further supported by studies where ectopic expression of *CYCB1;2* (but not of *CYCB1;1*) in *Arabidopsis* trichomes results in the appearance of multicellular structures, containing 2C nuclei, instead of the normal unicellular, polyploid branched trichomes (Schnittger et al. 2002). Interestingly, mutations in other APC components also lead to CYCB1 accumulation and to mitosis-related developmental defects (Blilou et al. 2002; Capron et al. 2003; Kwee and Sundaresan 2003).

Overexpression of *Arabidopsis CYCD3;1* (Riou-Khamlichi et al. 1999; Dewitte et al. 2003), but not of *CYCD2* (Cockcroft et al. 2000), allows cytokinin-independent growth and induces ectopic divisions producing leaves with more but smaller cells. Consistent with this, *CYCD3;1* is up regulated in the *Arabidopsis siamese* mutants, which has multicellular trichomes (Walker et al. 2000). Furthermore, overexpression of *AINTEGUMENTA* (*ANT*; Mizukami and Fischer 2000) or the auxin-inducible *ARGOS* (Hu et al. 2003) genes induce cell proliferation, leading to an increase in organ size, an effect also mediated by upregulating *CYCD3* expression. It should be pointed out that the transcriptional activation of *CYCD3* by cytokinins is the main evidence for the involvement of cytokinin in G1/S regulation (Soni et al. 1995). A recent study implicates the microRNA *JAW-D* as a direct regulator of *TCP* genes (Palatnik et al. 2003), a repressor of *CYCD3* (Gaudin et al. 2000).

Seed germination is another process where D-type cyclins play crucial roles. Before root emergence, expression of six *CYCD* (and two *CYCA*) genes is activated (Masubelele et al. 2005). Current transcriptomic data support a model in which the cumulative action (rather than functional redundancy), of several CYCD determine cell cycle activation. A role in germination has been also shown for maize CYCD2, based on its cytokinin-dependent transcriptional activation (Gutierrez et al. 2005). Furthermore, the importance of hormones (benzyladenine, abscisic acid) during maize germination, most likely controlling the activity of different CDK-cyclin complexes, has been also demonstrated (Sanchez et al. 2005). CYCD4;1, a distinct type of cyclin that lacks the RBR-binding motif, forms active kinase complexes with CDKA;1 and its function is required to promote callus proliferation in hypocotyl explants (Kono et al. 2006).

Arabidopsis KRP1 and *KRP2* are expressed mainly in endoreplicating cells (Wang et al. 2000; De Veylder et al. 2001). Overexpression of *KRP1* or *KRP2* inhibits cell division in leaves, which contain less but larger cells (Wang et al. 2000; De Veylder et al. 2001). This effect is partially reversed by overexpression of *CYCD3* (Zhou et al. 2003), an inducer of cell division. Similarly, the altered leaf phenotype of plants expressing *CYCD3* ectopically is reversed by overexpression of tobacco CDK inhibitor (KIS1) (Jasinski et al. 2002). Interestingly, *KRP1* overexpression does not have an appreciable effect on cell fate specification that occurs independently of changes in endoreplication level (Weinl et al. 2005). The function of KRP2 seems to be exerted through CDKA;1 and, when overexpressed in mitotically dividing cells, it also has an effect on endocycle progression (Verkest et al. 2005). Maize KRP1 and KRP2 play a role in endocycle control through inhibition of the endosperm-associated CDK activity, and consequently during maize endosperm formation (Coelho et al. 2005).

CDK activity, modulated by specific cyclins and CDK inhibitors, seems to be one of the limiting components, most likely not the only one, that regulate the arrested state of pericycle cells before triggering the lateral root initiation program (Casimiro et al. 2003; Vanneste et al. 2005). Pericycle cells contain high levels of *CDKA* and *KRP2* transcripts, but lack *CYCB1* (Himanen et al. 2002), likely contributing to maintenance of cell cycle arrest. Pericycle cells can be synchronized by treatment with the auxin transport inhibitor NPA and the released again upon auxin application (Himanen et al. 2002). With this system, over 900 genes have been identified to change specifically during lateral root initiation (Himanen et al. 2004). Thus, auxin addition triggers the expression of *CYCD3;1*, *E2Fa* and *histone H4* genes, which are quickly induced, then *CDKB* was induced, although later than the S-phase genes. In addition, the expression of CDK inhibitors *KRP1* and *KRP2* was rapidly and strongly reduced (Himanen et al. 2002).

2.2
Retinoblastoma, E2F-DP Transcription Factors and their Targets

CDK activity drives cell cycle transitions by phosphorylating specific targets. Changes in their phosphorylation state allows the transition from G1 to S, through S and from G2 to M. In G1/S, the retinoblastoma-related (RBR) protein is the main target of G1 CDKs (Nakagami et al. 1999; Boniotti and Gutierrez 2001; Nakagami et al. 2002). RBR is a negative regulator of cell proliferation, as revealed by loss-of-function mutations in the *RBR* gene, which results in an impairment to restricts mitosis in the haploid nuclei of the female gametophyte (Ebel et al. 2004). Local inactivation of *RBR* in *Arabidopsis* roots stimulates stem cell renewal, as it occurs in plants overexpressing *CYCD3;1* or *E2Fa-DPa* (Wildwater et al. 2005).

One of the major roles of RBR is to repress the expression of genes regulated by the E2F/DP family of transcription factors, as discussed below. Recent evidence derived from RBR protein inactivation strategies is consistent with this role. Virus-induced transcriptional silencing of the tobacco *RBR* gene extends cell proliferation activity and induces extra endocycles in leaves (Park et al. 2005). Similar phenotypes are observed after inactivation of *Arabidopsis* RBR by expressing the geminivirus RepA protein (Desvoyes et al. 2006). The consequences of RBR inactivation depends on the developmental stage and the cell type, e.g., hyperplasia in young leaves and extra endocycles in older leaves (Desvoyes et al. 2006). These effects are mediated by an increase in E2Fa and E2Fc activities, suggesting that they act in the same pathway to regulate the transcriptional program during development. Stimulation of cell division in the leaf epidermis is also the outcome of altering other, less well-characterized pathways. Thus, overexpression of the *STRUWWELPETER* (*SWP*) gene (Autran et al. 2002) or silencing of *DEK* gene (a calpain homologue) expression in tobacco also leads to a hyperplastic phenotype (Ahn et al. 2004). Although the mechanism behind these effects is not fully understood, these mutant plants have an altered pattern of expression of several cell cycle genes. The range of control pathways that operate in conjunction with the RBR/E2F complexes seems to increase as new data are obtained. Thus, pull-down experiments have shown that *Arabidopsis* RBR and FIE interact, supporting a role of FIE-containing polycomb complexes in inhibiting premature division of the central cell of the embryo sac (Mosquna et al. 2004).

Compelling evidence indicates that RBR, in a manner analogous to RB in animal cells, modulate the activity of E2F-DP transcription factors (Korenjak and Brehm 2005). E2F transcription factors were originally identified by their ability to interact and activate the human adenovirus E2 promoter (Helin et al. 1992). Its partner, DP, was identified soon afterwards (Helin et al. 1993). E2F transcription factors regulate the expression of a variety of genes required for cell cycle progression and their activity is modulated by the retinoblastoma (RB) protein. About ten years ago, the identification of proteins containing a LxCxE amino acid motif (Soni et al. 1995; Dahl et al. 1995; Xie et al. 1995) strongly pointed to the, somewhat unexpected, existence of a RB-related pathway in plants. This was fully confirmed with the identification of the first plant RBR (Xie et al. 1996; Grafi et al. 1996), E2F (Ramirez-Parra et al. 1999; Sekine et al. 1999) and DP (Ramirez-Parra and Gutierrez 2000; Magyar et al. 2000). These studies were followed by others that have led to the identification of a large family of E2F-DP factors in different plant species, including *Arabidopsis*, rice, maize, tobacco, carrot, and the unicellular algae *Chlamydomonas reinhardtii* and *Ostreococcus tauri* (Ramirez-Parra et al. 2007). Three *Arabidopsis* E2F, named E2Fa, E2Fb and E2Fc, share a domain organization similar to that of human E2F1-5 (Shen 2002; Ramirez-Parra et al. 2007). The other three members, known as E2Fd/DEL2, E2Fe/DEL1 and E2Ff/DEL3, are atypical since they contain a duplicated DNA binding domain and function independently of DP.

Constitutive overexpression of both *E2Fa* and *DPa* induces cell division and endoreplication (De Veylder et al. 2002; Rossignol et al. 2002). The E2Fa-

DPa-mediated hyperplasia is inhibited by co-expressing a dominant-negative mutant of *CDKB1* but not the endoreplication phenotype, which is actually enhanced (Boudolf et al. 2004a). *E2Fa* overexpression up regulates a number of cell cycle genes, such as *RBR*, *KRP3* and *KRP5* in rosette leaves, and this may contribute to the leaf phenotype observed including reduction in cell number and increase in cell size (He et al. 2004). Genome-wide analysis of plants expressing a dominant-negative version of DP (Ramirez-Parra et al. 2003) or E2Fa- DPa (Vandepoele et al. 2005) have identified a few hundred genes that are most likely direct E2F targets. These are, among others, *PCNA*, *RNR*, *CDC6*, *MCM3*, *CDC45*, DNA polymerase and DNA primase genes (DNA replication genes), and *E2Fb*, *E2Fc*, *RBR1*, *E2Ff/DEL3*, and *CYCA3;2* (cell cycle genes), but also others whose relationship with cell proliferation, if any, still needs to be established.

E2Fb, which prefers DPa for heterodimerization (Kosugi and Ohashi 2002; Magyar et al. 2005), plays a role in controlling cell proliferation dependent on auxin signaling. Coexpression of *E2Fb*, but not *E2Fa*, with *DPa* stimulates cell proliferation in the absence of auxin (Magyar et al. 2005). High levels of E2Fb leads to phenotypes in roots, leaves and cotyledons consistent with hyperplasia, that correlates with up-regulation of a variety of E2F target genes required for G1/S and G2/M (Sozzani et al. 2006). Furthermore, E2Fb itself seems to be an E2F target as shown in chromatin immunoprecipitation experiments (Sozzani et al. 2006). E2Fc and DPb interact in vitro (Kosugi and Ohashi 2002) and in vivo (del Pozo et al. 2006). Overexpression of *E2Fc* is highly detrimental for development of leaf primordia (del Pozo et al. 2002), while a strong reduction of *E2Fc* mRNA levels produces leaves with a reduced ploidy level, suggesting that E2Fc-DPb regulate the switch from proliferation to the endocycle program (del Pozo et al. submitted). *E2Fe/DEL1* is expressed in proliferating cells and it has been implicated in restricting endocycle progression (Vlieghe et al. 2005). Interestingly, the endoreplication phenotype of the E2Fa-DPa overexpressing plants, but not the hyperplasia, is reduced by *E2Fe/DEL1* overexpression (Vlieghe et al. 2005). E2Ff/DEL3 is required for cell expansion, a process clearly observed in hypocotyl cells, without apparently affecting the endocycle program (Ramirez-Parra et al. 2004). Its role in differentiated cells is mediated by negatively regulating the expression of genes such *EXP3*, *EXP7*, *EXP9*, and *UGT*, involved in cell wall biosynthesis.

The pre-replication complex (pre-RC), required for initiation of chromosomal DNA replication, is constituted by the association of CDC6 and CDT1 with ORC, the six subunit origin recognition complex, and the MCM complex. Ectopic expression of CDC6 (Castellano et al. 2001) or CDT1 (Castellano et al. 2004) is sufficient to induce extra endocycles or cell division in a cell type-specific manner (Castellano et al. 2004). It is worth mentioning that increasing Cdt1 activity in animals, similar consequences are obtained (Del Bene et al. 2004). Gametophytic development (see McCormick 2004; Yadegari and Drews 2004) and the early embryonic stages seem to be highly sensitive to pre-RC dysfunction. Thus, (1) mutations in the *ORC2* gene, an E2F target (Diaz-Trivino et al. 2005), leads to failure in nuclear division control (Collinge et al. 2004), (2) in the *PROLIFERA* (*PRL*) gene, that encodes MCM7, produces abnormal patterns of division planes (Holding and Springer 2002) and (3) in the *CDC45* gene, which acts downstream pre-RC, to sterility (Stevens et al. 2004).

2.3
Other Genes

An increasing number of genes, some of them cited in the paragraphs above, are reported to affect cell morphogenesis, organogenesis or the balance between proliferation and endoreplication. While the individual mechanism of action is not known, it is worth discussing them in the context of this article.

Defects in the *TONSOKU* gene (Suzuki et al. 2004), expressed in S-phase and involved in meristem maintenance (Suzuki et al. 2005a), delay cell cycle progression, cause G2/M arrest and slightly increases the number of 4C cells (Suzuki et al. 2005b). Whether they are actually endoreplicating cells is not known, but interestingly, they also have increased levels of *CYCB1;1* expression. Disruption of the *RPN1a* gene, encoding a component of the regulatory particle of the 26S proteasome, causes embryo lethality by promoting arrest at the globular stage (Brukhin et al. 2005). Whether and how the high levels of *CYCB1;1* maintained in the *rpn1a* mutant plants contribute to the phenotype is unknown. *CIA1* encodes an amidotransferase (ATase2) involved in the first step of de novo purine biosynthesis. *cia1* mutants have leaves slightly smaller in size than the wt but contains about half the number of cells (Hung et al. 2004).

One of the phenotypes exhibited by the *root hairless2* (*rhl2*) and *hypcotyl6* (*hyp6*) mutants is reduction in the ploidy level (Hartung et al. 2002; Sugimoto–Shirazu et al. 2002). These genes encode the A and B subunits of topoisomerase VI (TOPVI). ROOT HAIRLESS1 (RHL1), also known as HYPOCOTYL7 (HYP7), is an essential component of the topoisomerase VI complex, with similarities to the C-terminus of mammalian topoisomerase IIα. (Sugimoto–Shirazu et al. 2005). The *rhl1* mutant, as well as the *rhl2* and *hyp6* mutants, has a maximum of 8C of DNA content in their cells, revealing the importance of topoisomerase function in proper endocycle progression.

Auxin response factors ARF6 and ARF10, which are targets of *miR160*, control root cap formation. Decrease in ARF10 and ARF6 expression (either by overexpressing *miR160* or in the double mutant *arf10 arf6*), displays root tip defects, uncontrolled cell division and a blockage of cell differentiation in the distal parts. These phenotypes suggest that ARF10 and ARF6 restrict the stem cell niche in the columella (Wang et al. 2005). A thermosensitive mutant, *rpd1* (*root primordium defective 1*) has been identified on the basis of its impairment in producing adventitious roots from the hypocotyl in response to auxin

(Konishi and Sugiyama 2006). Cell proliferation in cultured calli is also affected. RPD1 seems to be a plant-specific protein containing a winged-helix fold similar to that present in the DNA binding domain of the E2F family members, although the molecular basis for its function is not known.

3
Future Prospects

The last 5–10 years have witnessed an unprecedented advance in our understanding of cell proliferation control. The mechanism of action of individual genes controlling different aspects of cell cycle is being elucidated. Their integration with developmental processes is also beginning to be understood. These studies should definitely continue until comprehensive information of each gene impinging on cell proliferation is available. However, we must face, at least, two major challenges. One is to integrate the functional relationships of different cell cycle regulators into common pathways. Another is the identification and understanding of the very complex transcriptional networks that regulate cell proliferation and its coordination with development. The availability of an increasing amount of plant lines with identified alterations of individual genes affecting cell proliferation control in combination with molecular, genetic and genomic tools should promote the next qualitative advance in the years to come.

References

Ahn JW, Kim M, Lim JH, Kim GT, Pai HS (2004) Phytocalpain controls the proliferation and differentiation fates of cells in plant organ development. Plant J 38:969–981

Autran D, Jonak C, Belcram K, Beemster GT, Kronenberger J, Grandjean O, Inze D, Traas J (2002) Cell numbers and leaf development in *Arabidopsis*: a functional analysis of the STRUWWELPETER gene. EMBO J 21:6036–6049

Blilou I, Frugier F, Folmer S, Serralbo O, Willemsen V, Wolkenfelt H, Eloy NB, Ferreira PC, Weisbeek P, Scheres B (2002) The *Arabidopsis* HOBBIT gene encodes a CDC27 homolog that links the plant cell cycle to progression of celldifferentiation. Genes Dev 16:2566–2575

Boniotti MB, Gutierrez C (2001) A cell-cycle-regulated kinase activity phosphorylates plant retinoblastoma protein and contains, in *Arabidopsis*, aCDKA/cyclin D complex. Plant J 28:341–350

Boudolf V, Vlieghe K, Beemster GT, Magyar Z, Torres Acosta JA, Maes S, Van Der Schueren E, Inze D, De Veylder L (2004a) The plant-specific cyclin-dependent kinase CDKB1;1 and transcription factor E2Fa-DPa control the balance of mitotically dividing and endoreduplicating cells in *Arabidopsis*. Plant Cell 16:2683–2692

Boudolf V, Barroco R, Engler Jde A, Verkest A, Beeckman T, Naudts M, Inze D, De Veylder L (2004b) B1-type cyclin-dependent kinases are essential for the formation of stomatal complexes in *Arabidopsis* thaliana. Plant Cell 16:945–955

Brukhin V, Gheyselinck J, Gagliardini V, Genschik P, Grossniklaus U (2005) The RPN1 subunit of the 26S proteasome in *Arabidopsis* is essential for embryogenesis. Plant Cell 17:2723-2737

Capron A, Serralbo O, Fulop K, Frugier F, Parmentier Y, Dong A, Lecureuil A, Guerche P, Kondorosi E, Scheres B, Genschik P (2003) The *Arabidopsis* anaphase-promoting complex or cyclosome: molecular and genetic characterization of the APC2 subunit. Plant Cell 15:2370-2382

Casimiro I, Beeckman T, Graham N, Bhalerao R, Zhang H, Casero P, Sandberg G, Bennett MJ (2003) Dissecting *Arabidopsis* lateral root development. Trends Plant Sci 8:165-171

Castellano MM, del Pozo JC, Ramirez-Parra E, Brown S, Gutierrez C (2001) Expression and stability of *Arabidopsis* CDC6 are associated with endoreplication. Plant Cell 13:2671-2686

Castellano MM, Boniotti MB, Caro E, Schnittger A, Gutierrez C (2004) DNA replication licensing affects cell proliferation or endoreplication in a cell type-specific manner. Plant Cell 16:2380-2393

Cebolla A, Vinardell JM, Kiss E, Olah B, Roudier F, Kondorosi A, Kondorosi E (1999) The Mitotic Inhibitor ccs52 is required for endoreduplication and ploidy-dependent cell enlargement in plants. EMBO J 18:4476-4484

Cockcroft CE, Den Boer BG, Healy JM, Murray JA (2000) Cyclin D control of growth rate in plants. Nature 405:575-579

Coehlo CM, Dante RA, Sabelli PA, Sun Y, Dilkes BP, Gordon-Kamm WJ, Larkins BA (2005) Cyclin-dependent kinase inhibitors in maize endosperm and their potential in endoreduplication. Plant Physiol 138:2323-2336

Collinge MA, Spillane C, Kohler C, Gheyselinck J, Grossniklaus U (2004) Genetic interaction of an origin recognition complex subunit and the Polycomb group gene MEDEA during seed development. Plant Cell 16:1035-1046

Dahl M, Meskiene I, Bogre L, Ha DT, Swoboda I, Hubmann R, Hirt H, Heberle-Bors E (1995) The D-type alfalfa cyclin gene cycMs4 complements G1 cyclin-deficient yeast and is induced in the G1 phase of the cell cycle. Plant Cell 7:1847-1857

de Jager SM, Maughan S, Dewitte W, Scofield S, Murray JA (2005) The developmental context of cell-cycle control in plants. Semin Cell Dev Biol 16:385-396

De Veylder L, Beeckman T, Beemster GT, Krols L, Terras F, Landrieu I, van der Schueren E, Maes S, Naudts M, Inze D (2001) Functional analysis of cyclin-dependent kinase inhibitors of *Arabidopsis*. Plant Cell 13:1653-1668

De Veylder L, Beeckman T, Beemster GT, de Almeida Engler J, Ormenese S, Maes S, Naudts M, Van Der Schueren E, Jacqmard A, Engler G, Inze D (2002) Control of proliferation, endoreduplication and differentiation by the *Arabidopsis* E2Fa-DPa transcription factor. EMBO J 21:1360-1368

Del Bene F, Tessmar-Raible K, Wittbrodt J (2004) Direct interaction of geminin and Six3 in eye development. Nature 427:745-749

del Pozo JC, Boniotti MB, Gutierrez C (2002) *Arabidopsis* E2Fc functions in cell division and is degraded by the ubiquitin-SCF(AtSKP2) pathway in response to light. Plant Cell 14:3057-3071

del Pozo JC, Lopez-Matas MA, Ramirez-Parra E, Gutierrez C (2005) Hormonal control of the cell cycle. Physiol Plant 123:173-183

del Pozo JC, Diaz-Trivino S, Cisneros N, Gutierrez C (2006) The balance between cell division and endoreplication depends on E2Fc-DPB, transcription factors regulated by the ubiquitin-SCF^{SKP2A} pathway in Arabidopsis. Plant Cell 18:2224-2235

Desvoyes B, Ramirez-Parra E, Xie Q, Chua N-H, Gutierrez C (2006) Cell type-specific role of the retinoblastoma/E2F pathway during *Arabidopsis* leaf development. Plant Phys 140:67–80

Dewitte W, Riou-Khamlichi C, Scofield S, Healy JM, Jacqmard A, Kilby NJ, Murray JA (2003) Altered cell cycle distribution, hyperplasia, and inhibited differentiation in *Arabidopsis* caused by the D-type cyclinCYCD3. Plant Cell 15:79–92

Diaz-Trivino S, Castellano MM, Sanchez MP, Ramirez-Parra E, Desvoyes B, Gutierrez C (2005) The genes encoding *Arabidopsis* ORC subunits are E2F targets and the two ORC1 genes are differently expressed in proliferating and endoreplicating cells. Nucleic Acids Res 33:5404–5414

Ebel C, Mariconti L, Gruissem W (2004) Plant retinoblastoma homologues control nuclear proliferation in the female gametophyte. Nature 429:776–780

Edgar BA, Orr-Weaver TL (2002) Endoreplication cell cycles: more or less. Cell 105:297–306

Feiler HS, Jacobs TW (1990) Cell division in higher plants: a cdc2 gene, its 34-kDa product, and histone H1 kinase activity in pea. Proc Natl Acad Sci USA 87:5397–5401

Ferreira PC, Hemerly AS, Villarroel R, Van Montagu M, Inze D (1991) The *Arabidopsis* functional homolog of the p34cdc2 protein kinase. Plant Cell 3:531–540

Gaudin V, Lunness PA, Fobert PR, Towers M, Riou-Khamlichi C, Murray JA, Coen E, Doonan JH (2000) The expression of D-cyclin genes defines distinct developmental zones in snapdragon apical meristems and is locally regulated by the Cycloidea gene. Plant Physiol 122:1137–1148

Grafi G, Burnett RJ, Heleutjaris T, Larkins BA, DeCaprio JA, Sellers WR, Kaelin WG (1996) A maize cDNA encoding a member of the retinoblastoma protein family: involvement in endoreduplication. Proc Natl Acad Sci USA 93:8962–8967

Gutierrez R, Quiroz-Figueroa F, Vazquez-Ramos JM (2005) Mazie cyclin D2 expression, associated kinase activity and effect of phytohormones during germination. Plant Cell Physiol 46:166–173

Gutierrez C (2005) Coupling cell proliferation and development in plants. Nat Cell Biol 7:535–541

Hartung F, Angelis KJ, Meister A, Schubert I, Melzer M, Puchta H (2002) An archaebacterial topoisomerase homolog not present in other eukaryotes is indispensable for cell proliferation of plants. Curr Biol 12:1787–1791

Hata S, Kouchi H, Suzuka I, Ishii T (1991) Isolation and characterization of cDNA clones for plant cyclins. EMBO J 10:2681–2688

He SS, Liu J, Xie Z, O'Neill D, Dotson S (2004) *Arabidopsis* E2Fa plays a bimodal role in regulating cell division and cell growth. Plant Mol Biol 56:171–184

Helin K, Lees JA, Vidal M, Dyson N, Harlow E, Fattaey A (1992) A cDNA encoding a pRBbinding protein with properties of the transcription factor E2F. Cell 70:337–350

Helin K, Wu CL, Fattaey AR, Lees JA, Dynlacht BD, Ngwu C, Harlow E (1993) Heterodimerization of the transcription factors E2F-1 and DP-1 leads to cooperative trans-activation. Genes Dev 7:1850–1861

Hemerly A, Engler Jde A, Bergounioux C, Van Montagu M, Engler G, Inze D, Ferreira P iterative plant development. EMBO J 14:3925–3936

Himanen K, Boucheron E, Vanneste S, de Almeida Engler J, Inze D, Beeckman T (2002) Auxin-mediated cell cycle activation during early lateral root initiation. Plant Cell 14:2339–2351

Himanen K, Vuylsteke M, Vanneste S, Vercruysse S, Boucheron E, Alard P, Chriqui D, Van Montagu M, Inze D, Beeckman T (2004) Transcript profiling of early lateral root initiation. Proc Natl Acad Sci USA 101:5146–51

Holding DR, Springer PS (2002) The *Arabidopsis* gene PROLIFERA is required for proper cytokinesis during seed development. Planta 214:373–382

Hu Y, Bao F, Li J (2000) Promotive effect of brassinosteroids on cell division involves a distinct CycD3-induction pathway in *Arabidopsis*. Plant J 24:693–701

Hu Y, Xie Q, Chua N-H (2003) The *Arabidopsis* auxin-inducible gene ARGOS controls lateral organ size. Plant Cell 15:1951–1961

Hung WF, Chen LJ, Boldt R, Sun CW, Li HM (2004) Characterization of *Arabidopsis* glutamine phosphoribosyl pyrophosphate amidotransferase-deficient mutants. Plant Physiol 135:1314–1323

Imai KK, Ohashi Y, Tsuge T, Yoshizumi T, Matsui M, Oka A, Aoyama T (2006) The A-type cyclin CYCA2;3 is a key regulator of ploidy levels in *Arabidopsis* endoreduplication. Plant Cell 18:382–396

Iwakawa H, Shinmyo A, Sekine M (2006) *Arabidopsis* CDKA;1, a cdc2 homologue, controls proliferation of generative cells in male gametogenesis. Plant J 45:819–831

Jasinski S, Perennes C, Bergounioux C, Glab N (2002) Comparative molecular and functional analyses of the tobacco cyclin-dependent kinase inhibitor NtKIS1a and its spliced variant NtKIS1b. Plant Physiol 130:1871–1882

John PC, Sek FJ, Lee MG (1989) A homolog of the cell cycle control protein p34cdc2 participates in the division cycle of Chlamydomonas, and a similar protein is detectable in higher plants and remote taxa. Plant Cell 1:1185–1193

John PC, Sek FJ, Carmichael JP, McCurdy DW (1990) p34cdc2 homologue level, cell division, phytohormone responsiveness and cell differentiation in wheat leaves. J Cell Sci 97:627–630

Kondorosi E, Roudier F, Gendreau E (2000) Plant cell-size control: growing by ploidy? Curr Opin Plant Biol 3:488–492

Konishi M, Sugiyama M (2006) A novel plant-specific family gene, ROOT PRIMORDIUM DEFECTIVE 1, is required for the maintenance of active cell proliferation. Plant Physiol 140:591–602

Kono A, Ohno R, Umeda-Hara C, Uchimiya H, Umeda M (2006) A distinct type of cyclin D, CYCD4;2, involved in the activation of cell division in *Arabidopsis*. Plant Cell Rep 12:1–6

Korenjak M, Brehm A (2005) E2F-Rb complexes regulating transcription of genes important for differentiation and development. Curr Opin Genet Dev 15:520–527

Kosugi S, Ohashi Y (2002) Interaction of the *Arabidopsis* E2F and DP proteins confers their concomitant nuclear translocation and transactivation. Plant Physiol 128:833–843

Kwee HS, Sundaresan V (2003) The NOMEGA gene required for female gametophyte development encodes the putative APC6/CDC16 component of the Anaphase Promoting Complex in *Arabidopsis*. Plant J 36:853–866

La H, Li J, Ji Z, Cheng Y, Li X, Jiang S, Venkatesh PN, RAMAchandran S (2006) Genomewide analysis of cyclin family in rice (Oryza sativa L.). Mol Genet Genomics 25:1–13

Larkins BA, Dilkes BP, Dante RA, Coelho CM, Woo YM, Liu Y (2001) Investigating the hows and whys of DNA endoreduplication. J Exp Bot 52:183–192

Lee MG, Nurse P (1987) Complementation used to clone a human homologue of the fission yeast cell cycle control gene cdc2. Nature 327:31–35

Leiva-Neto JT, Grafi G, Sabelli PA, Dante RA, Woo YM, Maddock S, Gordon-Kamm WJ, Larkins BA (2004) A dominant negative mutant of cyclin-dependent kinase A reduces endoreduplication but not cell size or gene expression in maize endosperm. Plant Cell 16:1854–1869

Magyar Z, Atanassova A, De Veylder L, Rombauts S, Inze D (2000) Characterization of two distinct DP-related genes from *Arabidopsis* thaliana. FEBS Lett 486:79–87

Magyar Z, De Veylder L, Atanassova A, Bako L, Inze D, Bogre L (2005) The role of the *Arabidopsis* E2FB transcription factor in regulating auxin-dependent cell division. Plant Cell 17:2527–2541

Masubelele NH, Dewitte W, Menges M, Maughan S, Collins C, Huntley R, Nieuwland J, Scofield S, Murray JA (2005) D-type cyclins activate division in the root apex to promote seed germination in *Arabidopsis*. Proc Natl Acad Sci USA 102:15694–15699

Mccormick S (2004) Control of male gametophyte development. Plant Cell 16:S142–S153

Menges M, de Jager SM, Gruissem W, Murray JA (2005) Global analysis of the core cell cycle regulators of *Arabidopsis* identifies novel genes, reveals multiple and highly specific profiles of expression and provides a coherent model for plant cell cycle control. Plant J 41:546–566

Mizukami Y, Fischer RL (2000) Plant organ size control: AINTEGUMENTA regulates growth and cell numbers during organogenesis. Proc Natl Acad Sci USA 97:942–947

Mosquna A, Katz A, Shochat S, Grafi G, Ohad N (2004) Interaction of FIE, a polycomb protein, with pRb: a possible mechanism regulating endosperm development. Mol Genet Genomics 271:651–657

Nakagami H, Sekine M, Murakami H, Shinmyo A (1999) Tobacco retinoblastoma-related protein phosphorylated by a distinct cyclin-dependent kinase complex with Cdc2/cyclin D in vitro. Plant J 18:243–252

Nakagami H, Kawamura K, Sugisaka K, Sekine M, Shinmyo A (2002) Phosphorylation of retinoblastoma-related protein by the cyclin D/cyclin-dependent kinase complex is activated at the G1/S-phase transition in tobacco. Plant Cell 14:1847–1857

Nurse P (1990) Universal control mechanism regulating onset of M-phase. Nature 344:549–552

Palatnik JF, Allen E, Wu X, Schommer C, Schwab R, Carrington JC, Weigel D (2003) Control of leaf morphogenesis by microRNAs. Nature 425:257–263

Park JA, Ahn JW, Kim YK, Kim SJ, Kim JK, Kim WT, Pai HS (2005) Retinoblastoma protein regulates cell proliferation, differentiation, and endoreduplication in plants. Plant J 42:153–163

Ramirez-Parra E, Gutierrez C (2000) Characterization of wheat DP, a heterodimerization partner of the plant E2F transcription factor which stimulates E2F-DNA binding. FEBS Lett 486:73–78

Ramirez-Parra E, Xie Q, Boniotti MB, Gutierrez C (1999) The cloning of plant E2F, a retinoblastoma-binding protein, reveals unique and conserved features with animal G(1)/S regulators. Nucleic Acids Res 27:3527–3533

Ramirez-Parra E, Frundt C, Gutierrez C (2003) A genome-wide identification of E2F regulated genes in *Arabidopsis*. Plant J 33:801–811

Ramirez-Parra E, Lopez-Matas MA, Frundt C, Gutierrez C (2004) Role of an atypical E2F transcription factor in the control of *Arabidopsis* cell growth and differentiation. Plant Cell 16:2350–2563

Ramirez-Parra E, del Pozo JC, Desvoyes B, Sanchez MP, Gutierrez C (2007) E2F-DP transcription factors. In: Inzé D (ed) Cell Cycle Control and Plant Development. Annu Plant Rev 32:138–162

Riou-Khamlichi C, Huntley R, Jacqmard A, Murray JA (1999) Cytokinin activation of *Arabidopsis* cell division through a D-type cyclin. Science 283:1541–1544

Rossignol P, Stevens R, Perennes C, Jasinski S, Cella R, Tremosaygue D, Bergounioux C (2002) AtE2F-a and AtDP-a, members of the E2F family of transcription factors, induce *Arabidopsis* leaf cells to re-enter S-phase. Mol Gen Gen 266:995–1003

Sanchez MP, Gurusinghe SH, Bradford KJ, Vazquez-Ramos JM (2005) Differential response of PCNSand Cdk-A proteins and associated kinase activities to benzyladenine and abscisic acid during maize germination. J Exp Bot 56:515–523

Schnittger A, Schobinger U, Stierhof YD, Hulskamp M (2002) Ectopic B-type cyclin expression induces mitotic cycles in endoreduplicating *Arabidopsis* trichomes. Curr Biol 12:415–420

Sekine M, Ito M, Uemukai K, Maeda Y, Nakagami H, Shinmyo A (1999) Isolation and characterization of the E2F-like gene in plants. FEBS Lett 460:117–122

Shen WH (2002) The plant E2F-Rb pathway and epigenetic control. Trends Plant Sci 7:505–511

Sieberer T, Hauser MT, Seifert GJ, Luschnig C (2003) PROPORZ1, a putative *Arabidopsis* transcriptional adaptor protein, mediates auxin and cytokinin signals in the control of cell proliferation. Curr Biol 13:837–842

Soni R, Carmichael JP, Shah ZH, Murray JA (1995) A family of cyclin D homologs from plants differentially controlled by growth regulators and containing the conserved retinoblastoma protein interaction motif. Plant Cell 7:85–103

Sozzani R, Maggio C, Varotto S, Canova S, Bergounioux C, Albani D, Cella R (2006) Interplay between *Arabidopsis* activating factors E2Fb and E2Fa in cell cycle progression and development. Plant Physiol 140:1355–1366

Stevens R, Mariconti L, Rossignol P, Perennes C, Cella R, Bergounioux C (2002) Two E2F sites in the *Arabidopsis* MCM3 promoter have different roles in cell cycle activation and meristematic expression. J Biol Chem 277:32978–32984

Sugimoto-Shirasu K, Stacey NJ, Corsar J, Roberts K, Mccann MC (2002) DNA topoisomerase VI is essential for endoreduplication in *Arabidopsis*. Curr Biol 12:1782–1786

Sugimoto-Shirazu K, Roberts GR, Stacey NJ, McCann MC, Maxwell A, Roberts K (2005) RHL1 is an essential component of the plant DNA topoisomerase VI complex and is required for ploidy-dependent cell growth. Proc Natl Acad Sci USA 102:18736–18741

Suzuki T, Inagaki S, Nakajima S, Akashi T, Ohto MA, Kobayashi M, Seki M, Shinozaki K, Kato T, Tabata S, Nakamura K, Morikami A (2004) A novel *Arabidopsis* gene TONSOKU is required for proper cell arrangement in root and shoot apical meristems. Plant J 38:673–684

Suzuki T, Nakajima S, Morikami A, Nakamura K (2005a) An *Arabidopsis* protein with a novel calcium-binding repeat sequence interacts with TONSOKU/MGOUN3/BRUSHY1 involved in meristem maintenance. Plant Cell Physiol 46:1452–1461

Suzuki T, Nakajima S, Inagaki S, Hirano-Nakakita M, Matsuoka K, Demura T, Fukuda H, Morikami A, Nakamura K (2005b) TONSOKU is expressed in S phase of the cell cycle and its defect delays cell cycle progression in *Arabidopsis*. Plant Cell Physiol 46:736–742

Vandepoele K, Raes J, De Veylder L, Rouze P, Rombauts S, Inze D (2002) Genome-wide analysis of core cell cycle genes in *Arabidopsis*. Plant Cell 14:903–916

Vandepoele K, Vlieghe K, Florquin K, Hennig L, Beemster GT, Gruissem W, Van de Peer Y, Inze D, De Veylder L (2005) Genome-wide identification of potential plant E2F target genes. Plant Physiol 139:316–328

Vanneste S, Maes L, De Smet I, Himanen K, Naudts M, Inzé D, Beeckman T (2005) Auxin regulation of cell cycle and its role during lateral root initiation. Physiol Plant 123:139–146

Verkest A, Manes CL, Vercruysse S, Maes S, Van Der Schueren E, Beeckman T, Genschik P, Kuiper M, Inze D, De Veylder L (2005) The cyclin-dependent kinase inhibitor KRP2 controls the onset of the endoreduplication cycle during *Arabidopsis* leaf development through inhibition of mitotic CDKA;1 kinase complexes. Plant Cell 17:1723–36

Vlieghe K, Boudolf V, Beemster GT, Maes S, Magyar Z, Atanassova A, De Almeida Engler J, De Groodt R, Inze D, De Veylder L (2005) The DP-E2F-like gene DEL1 controls the endocycle in *Arabidopsis* thaliana. Curr Biol 15:59–63

Walker JD, Oppenheimer DG, Concienne J, Larkin JC (2000) SIAMESE, a gene controlling the endoreduplication cell cycle in *Arabidopsis* thaliana trichomes. Development 127:3931–3940

Wang H, Zhou Y, Gilmer S, Whitwill S, Fowke LC (2000) Expression of the plant cyclin-dependent kinase inhibitor ICK1 affects cell division, plant growth and morphology. Plant J 24:613–623

Wang G, Kong H, Sun Y, Zhang X, Zhang W, Altman N, Depamphilis CW, Ma H (2004) Genome-wide analysis of the cyclin family in *Arabidopsis* and comparative phylogenetic analysis of plant cyclin-like proteins. Plant Physiol 135:1084–1099

Wang JW, Wang LJ, Mao YB, Cai WJ, Xue HW, Chen XY (2005) Control of root cap formation by MicroRNA-targeted auxin response factors in *Arabidopsis*. Plant Cell 17:2204–2216

Weinl C, Marquardt S, Kuijt SJ, Nowack MK, Jakoby MJ, Hulskamp M, Schnittger A (2005) Novel functions of plant cyclin-dependent kinase inhibitors, ICK1/KRP1, can act non-cellautonomously and inhibit entry into mitosis. Plant Cell 17:1704–1722

Wildwater M, Campilho A, Perez-Perez JM, Heidstra R, Blilou I, Korthout H, Chatterjee J, Mariconti L, Gruissem W, Scheres B (2005) The RETINOBLASTOMA-RELATED gene regulates stem cell maintenance in *Arabidopsis* roots. Cell 123:1337–1349

Xie Q, Suárez-López P, Gutierrez C (1995) Identification and analysis of a retinoblastomabinding motif in the replication protein of a plant DNA virus: requirement for efficient viral DNA replication. EMBO J 14:4073–4082

Xie Q, Sanz-Burgos A, Hannon G, Gutierrez C (1996) Plant cells contain a novel member of the retinoblastoma family of growth regulatory proteins. EMBO J 15:4900–4908

Yadegari R, Drews GN (2004) Female gametophyte development. Plant Cell 16(Suppl): S133–S141

Yoshizumi T, Nagata N, Shimada H, Matsui M (1999) An *Arabidopsis* cell cycle-dependent kinase-related gene, CDC2b, plays a role in regulating seedling growth in darkness. Plant Cell 11:1883–1896

Yu Y, Steinmetz A, Meyer D, Brown S, Shen WH (2003) The tobacco A-type cyclin, Nicta;CYCA3;2, at the nexus of cell division and differentiation. Plant Cell 15:2763–2777

Zhang K, Letham DS, John PC (1996) Cytokinin controls the cell cycle at mitosis by stimulating the tyrosine dephosphorylation and activation of p34cdc2-like H1 histone kinase. Planta 200:2–12

Zhou Y, Wang H, Gilmer S, Whitwill S, Fowke LC (2003) Effects of co-expressing the plant CDK inhibitor ICK1 and D-type cyclin genes on plant growth, cell size and ploidy in *Arabidopsis* thaliana. Planta 216:604–613

Subject Index

Abaxial leaf epidermis, 346
Abaxial lob, 361, 369
ABORTED MICROSPORES (AMS), 361, 370, 371, 373
ABSENCE OF FIRST DIVISION1 (AFD1), 108, 110
Ace2, 16
Acentrics, 147
Acropetal PM, 296
Actin, 172, 252, 257, 262, 267, 290, 328–331, 333, 334
– Actin depleted zone (ADZ), 46, 132
– Actin depolymerizing drugs, 35
– Actin filament, 203, 204, 252, 262, 267
– Actin, 9, 33, 35, 36, 41, 44, 46, 47, 49–51, 125, 126, 132, 133, 349, 350, 354
– Actin-binding domain (ABD), 132
Actomyosin, 169, 204
Adaxial leaf epidermis, 345, 351
Adaxial lob, 361, 369
ADL1 (*Arabidopsis* dynamin-like protein), 198, 305
AFLP, 21
AGAMOUS (AG), 363, 364
Agrobacterium-mediated transformation, 126
AINTEGUMENTA (ANT), 26, 382
AIR9, 313
Alga, 59, 60, 196, 215, 221, 328–337, 384
– Green, 3, 4, 10, 27, 169, 211, 213, 290
– Brown, 290, 323, 324, 327, 334, 336, 337
– Fucoid, 323, 325–331, 335, 336
Allele, 42, 105, 363, 364, 366
– Knockout, 353
– Maternal, 91
– Null, 354
– Paternal, 91
– Severe, 110, 348, 352
– Temperature-sensitive, 348

– Weak, 67, 108, 110, 348, 352, 354
AMEIOTIC (AM1), 105, 106
– Am1-praI, 105
– Am1/SWI1/DYAD, 106
– Mutant, 106
Amidotransferase (ATse2), 386
Amino acid
– Motif, 170, 172, 234, 380, 384
– Of proteins, 131, 171, 217, 240, 241, 243, 350, 366
– Sequence homology, 112, 117, 130, 240, 243, 353, 366
Amiprophos methyl, 21
Amphidiploid, 22
Amylase, 369
Anaphase, 130, 132, 145–148, 169, 176–178, 197, 199, 201, 217, 241–245, 279, 377
– Anaphase A, 142, 147
– Anaphase B, 142, 147, 177
– Anaphase-promoting complex (APC), 383
– Early, 222
– In meiosis I, 103, 114, 117
– In meiosis II, 117
– Late, 147, 217, 222, 238, 242, 252, 255, 256–258, 270, 277, 313
– Transition, 17, 19, 21, 80, 245, 334
Anastral spindles, 169
Angiosperm, 75, 175, 251
Aniline blue, 236, 303, 312
Annexin, 312, 313
ANP (*Arabidopsis* homolog of tobacco NPK1), 233, 235, 237, 242
ANQ (*Arabidopsis* homolog of tobacco NQK1; AtMKK6), 239, 241, 242
Anterograde traffic, 291
Anther, 361–373
– Cell division, 361, 363
– Cell layer, 363
– Cell wall layer, 367, 369

– Development, 361–367, 369, 372, 373
– Gene, 363–369
– Hair, 75
– Morphology, 361, 362, 365, 367, 369, 372
– Mutant, 363–369
Antibody, 42, 48, 51, 145, 174, 175, 178, 181, 199, 216, 217, 221, 236, 241, 243, 244, 268, 303, 304, 307, 314, 326
Antipodal cell, 75
APC (Adenomatous polyposis coli), 48, 52, 149
APC/C (Anaphase promoting complex/cyclosome), 18, 19, 77–81, 93, 382
APETELA2 (AP2), 26, 370
Aphidicolin, 21, 22, 27, 44, 127, 237
Arabidopsis, 4, 8, 9, 13, 19–23, 25, 26, 42–45, 48, 51, 61–66, 68, 75, 76, 78–82, 86–91, 105–108, 111–117, 128–132, 148–151, 155, 171, 175, 182, 183, 186, 198, 200–202, 216, 218, 233–235, 237, 239–243, 255, 257–259, 263–268, 279, 281, 282, 294, 296, 297, 299, 304–307, 310–315, 324, 343–346, 348, 354, 355, 361–363, 366–368, 371–373, 377, 380–384
Arabinogalactan, 271
Archesporial cell, 104, 105, 362–365, 367, 368
ARGOS, 382
Arp2/3 complex, 329, 330
Arp2/actin network, 331
Arum maculatum, 76
Ase1 (Anaphase spindle elongation factor), 130, 148, 243, 245
ASK1 (*Arabidopsis* homolog of SKP1), 107, 216
Aspergillus, 178, 214, 222
Astral microtubules (Astral MTs), 52, 157, 158, 174
ASY2 (At4g32200), 107
Asymmetric division, 39–41, 323–329, 331, 335, 336, 343–352, 354–357
ASYNAPTIC1 (ASY1), 107, 108
ATA7, 365
AtCASP, 265
AtCDC6, 354
AtCDC48, 310
AtCDT1, 354
AtExo, 262

AtGPAT1, 361, 371
ATH1, 22
ATK1, 44, 117, 151, 152, 154, 155
ATK5, 151, 152, 154, 174, 179, 181, 183
AtNPSN11, 310
ATP, 131, 150, 170, 171, 174, 183, 217, 240, 311
AtPAKRP2, 185, 313
ATPase, 171, 183, 217, 311
ATP-binding motif, 170, 240
AtPEP12, 298
AtSec, 262, 267
– AtSec3, 262
– AtSec5, 262
– AtSec6, 262
– AtSec8, 267
– AtSec15, 262
AtSNAP33, 310
AtSPO11, 111, 112
– AtSPO11-1, 111, 112, 115
– AtSPO11-2, 111, 112
– AtSPO11-3, 111, 112
AtVam3, 134
A-type cyclin (*see* Cyclin CycA)
Aurora kinase, 132
– AtAur1, 132
– AtAur2, 132
– AtAur3, 132
Autonomous system, 3
Autophosphorylation, 365
Auxin, 19, 40, 63, 90, 355, 381–383, 385, 386
Auxin response factor (ARF), 386
Axial element (AE), 107
Axis of elongation, 40, 41

BAK1, 366
Basal bodies, 212
Basic helix-loop-helix (bHLH), 348, 353, 357, 370, 371
Basic-leucine zipper (bZIP), 24
BDM (2,3-Butanedione monoxime), 51, 203
BimC, 151, 152, 178, 179, 181
Biological clock, 3
Bipolar spindles, 36, 44, 45, 131, 151, 153–158, 178, 181, 196, 220, 333
Bistheonellide A, 133, 203
Bouquet formation of chromosomes, 108–111, 115, 215, 216
Brassinosteroid, 366, 381
BRCA2, 113, 115

Brefeldin A (BFA), 46, 198, 308
BRI1, 366
Bridge MT, 155–158
B-type cyclin (*see* Cyclin CycB)
Budding yeast, 6, 16, 17, 22, 113, 333, 335
BY-2 cells, 19–21, 25, 42, 44–47, 51, 62, 126–134, 147, 148, 150, 152, 179, 198–202, 218, 235–239, 241, 243–245, 259, 304, 305, 312, 315
BY-GT (BY-2 cell line expressing GFP-tagged tubulin), 128, 130, 131

Caenorhabtis elegans, 92, 214, 222, 243, 324, 325
CAF1 (Chromatin assembly factor 1), 67, 89
Caffeine-induced binucleate cell, 155
CAK (CDK-activating kinase), 61
Callase, 282, 368
Callose
– Callose wall, 281, 282, 312, 362, 365, 368, 369
– Polysaccharide, 237, 270
– Staining with aniline blue, 303
– Synthesis, 270, 272, 273, 275, 308, 309
– At the cell plate, 272, 275–277, 281, 282, 311
Callose synthase (CalS), 262, 275–277, 298, 299, 308, 311–313
– CalS1, 298, 299, 311, 312
– CalS5, 312
– CalS11, 312
– CalS12, 312
– CalS complex, 312, 313
Calmodulin, 172, 174, 175, 187, 221, 313
Calponin-homology domain (CH), 171, 186
Carbohydrate, 282, 299, 369
Cargo-binding domain, 171, 187
Carpel, 61, 351, 366
Carrot, 148, 152, 179, 181, 215, 217, 218, 222, 384
Caryokinesis, 251, 256
Caryopsis, 83, 84
Casein kinase II, 8
Catharanthus roseus, 25
Cauliflower mosaic virus 35S promoter, 135
Cauline-leaf, 351
Caulonema filament, 323
CCA1 (Circadian clock associated protein 1), 8–10
CCS52 (mitotic cyclin inhibitor), 82, 93, 382

CDH1, 80
CDK inhibitor (KIS1), 383
CDK kinase inhibitor (CKI), 6, 18, 60, 80, 82, 85, 86, 88
cDNA, 21, 130, 171, 172, 238, 240
CDK (*see* Cyclin-dependent kinase)
CDT1, 61, 66, 67, 77, 78, 87, 354, 385
Cell autonomous, 324
Cell axis, 170
Cell cycle, 13–29, 104–106, 290–298
– Arrest, 19, 27, 131, 378, 383
– Control, 5, 6, 8, 13–22, 27–29, 59, 65, 75, 80, 90, 379, 380
– Gene, 9, 10, 87, 94, 381, 384, 385
– Gene expression, 9, 13–16, 19–21, 23–27
– Machinery, 4, 6, 59–64, 90, 355, 380
– Phase, 6, 13–16, 19, 21–23, 26–28
– Progression, 5–7, 10, 13, 16, 17, 19, 20, 24, 27, 28, 61, 64, 65, 125–134, 352, 378, 381, 384, 386
– Regulation, 4, 10, 14, 16, 17, 20–27, 60, 61, 65, 75, 86, 94, 354, 355, 380
– Regulator, 22, 23, 26, 27, 94, 106, 131, 343, 344, 354, 377, 379, 380, 387
– Synchronization, 13, 21, 23, 27, 44, 127
– Transition, 16–18, 60, 80, 235, 380, 383
Cell differentiation, 323, 343, 345, 347–349, 351, 352, 355, 357, 377, 378, 381, 382, 386
Cell division
– Asymmetric, 324–336
– Circadian, 3–5
– Commitment, 17, 59, 346
– Cycle, 3, 22, 59, 251, 377
– Plane, 33–52, 127, 135, 143, 150, 180, 185, 186, 197, 199, 201–203, 220, 221, 243, 279, 281, 282, 329, 336, 386
– Rate, 63
– Regulation, 6, 7, 66
– Rhythm, 4
– Site, 33, 35, 44, 46–49, 52, 128, 142, 143, 149, 170, 179, 181–186, 198, 203, 326, 327, 332, 334, 335
Cell division cycle protein/genes
– Cdc2, 78, 81, 82, 116, 244
– Cdc6, 66, 67, 77, 78, 87, 354, 385
– Cdc18, 78
– Cdc25, 6
– Cdc27, 387
– Cdc45, 66, 106, 385, 386
– Cdc48, 310, 311

– Cdc53, 63
Cell elongation, 33, 125, 127, 175
Cell equator, 257, 261
Cell files, 39, 149, 324, 325
Cell mass, 17, 251
Cell plate
– Assembly, 169, 180, 182, 184, 195, 199, 252, 258, 265, 267, 270, 272, 277, 279, 292
– Assembly matrix (CPAM), 179, 251, 252, 255, 256, 259, 261, 262, 271, 272, 279, 281
– Attachment, 49, 50, 52, 127, 180, 270
– Expansion, 183, 186, 198, 235, 237, 258, 273, 292, 313
– Lumen, 270, 272, 275–277, 299
– Membrane, 251, 259, 261–264, 270, 271, 275, 277, 279, 281, 282, 289–292, 298
– Orientation, 46, 47, 50, 52, 149, 186
– Position, 34, 36, 45, 46, 49, 127, 133, 182, 186, 198, 265
– Proteins, 185, 296–299, 303–314
– Signal, 299
– Somatic, 251, 252, 255, 256, 258, 260, 261, 267–270, 273, 275, 277, 279, 281, 282
– Syncytial, 251, 252, 259, 261, 264, 267, 269, 277, 279, 281, 282
– Syntaxin, 26
– Vesicles, 147, 170, 200, 219, 252, 257–259, 265, 270, 271, 276, 291, 293, 298, 306, 308–310
Cell polarity, 35, 38, 39, 328, 335, 343
Cell polarization, 323, 324
Cell proliferation, 62, 66, 80, 82, 355, 377–387
Cell wall
– Material, 51, 271, 291, 310, 334
– Mother, 34, 38, 39, 311, 313
– New, 33, 34, 36, 39, 40, 242, 251, 252, 258, 270–272, 291, 353
– Polysaccharide, 200, 311, 331
– Primary, 127, 276
– Staining, 236
– Synthesis, 23, 33, 34, 65, 242, 243, 246, 385
Cellobiohydrolase, 276
Cellular axis, 325, 326
Cellularization, 279–281, 334, 335
Cellulose, 128, 270, 275, 276, 282, 311
– Microfibril, 26, 127, 128, 270, 275, 276, 282
– Synthase, 26, 128, 270, 276, 282
CenA, 78
CENP-E, 151, 175, 182

Central element (CE), 103, 104
Centriole, 212, 213, 215, 332
Centromere, 84, 108, 116, 117, 208, 215, 314
Centrosome, 129, 141, 145, 153, 154, 156, 157, 172, 177, 181, 208, 209, 215–217, 329, 331–335
CER (wax biosynthesis gene), 356
CESA6 (Cellulose synthase), 128
cGMP, 330
Charales, 213
Charophyta, 211, 213
Checkpoint, 3, 5–7, 21, 59, 78
– Cell size, 6
– Circadian, 6, 7
– Kinase, 5
– Mitotic, 79, 208, 213, 214
– Replication, 21
– Spindle assembly, 235
Chiasmata, 103, 107, 111, 113, 114, 117
Chlamydomonas, 4–8, 10, 212, 221, 290, 296, 384
Chloroplast, 196, 203, 306
Chlorotetracycline, 196
Chromatid, 17, 19, 75, 81, 106–108, 113, 142, 145, 169, 172, 177, 196, 219
Chromatin, 59, 65–68, 75, 77, 78, 82, 84, 89, 91–93, 103, 105, 107, 115–117, 145, 146, 153, 154, 158, 207, 209, 215–217, 219, 220, 222, 223, 373, 374
– Assembly, 67, 69
– Condensation, 75, 84, 93, 104
– Dynamic, 65
– Immunoprecipitation, 385
– Loop, 107
– Remodeling, 65, 66, 91
– Replication, 60, 68
– Silencing, 66
– Structure, 105, 107, 373
Chromokinesin, 151, 175
Chromosome
– Axis, 104, 107, 108
– Condensation, 107, 213, 214, 219
– Congression, 151, 176, 177
– Endoreduplication, 84
– Fragmentation 66, 113, 116
– Homologous, 59, 103, 107–109, 112–115, 172–177
– Meiosis, 103
– Mitosis, 127, 141, 210, 215
– Pairing, 105–109, 113–115

Subject Index

- Recombination, 115
- Segregation, 6, 16, 34, 75, 76, 103, 115, 117, 109, 132, 142, 145, 147, 151, 153, 215, 377
- Structure, 106–108
- Synapsis, 107, 113, 213, 215

CIA1, 386
Circadian
- Checkpoint, 6
- Clock, 3–10, 27
- Gating, 5–7
- Oscillator, 5, 8
- Regulation, 4–10
- Rhythm, 3, 4

cis-Regulatory element, 23, 303
cis-SNARE, 310
Clathrin, 200, 259, 262, 267, 269, 273, 275, 291, 306, 307
CLAVATA (CLV), 350
- CLV1, 350
- CLV2, 350
Cleavage furrow, 212, 213, 289, 290
CLF (CURLY LEAF), 66
CLIP170, 42
Cln3, 15, 17, 18
Clock, 3–10
CLOCK-BMALL1 complex, 9
Closed mitosis, 208, 210–213, 222
CLSM (confocal laser scanning microscopy), 128
Coatomers, 292
Coiled-coil
- Domain, 171, 178, 185, 215, 243, 244, 293
- Protein, 150, 215, 217, 218, 264, 265
- Structure, 218, 235, 236, 238
Colchicine, 21, 111
Coleochaetales, 212, 213
Commitment, 13, 17, 18, 25, 60, 346
Compartmentalization, 81, 208
Convoluted sheets, 279, 281
COP (Coat protein)
- COPI, 265
- COPII, 264
Cortex, 33–36, 39, 41, 42, 49–52, 128, 129, 133, 142–144, 157, 175, 185, 326, 327–329, 332–334
Cortical actin, 46, 329–331
Cortical cytoplasmic ring (CCR), 40
Cortical division site, 33, 35, 49, 149, 185, 198, 203
Cortical mark, 33, 46, 49, 50

Cortical MFs (Cortical microfilaments), 36, 132, 133
Cortical MTs (CMTs, cortical microtubules), 34, 36, 41–43, 125–128, 130, 131, 146, 149, 244, 245, 332
Cotyledon, 76, 237, 239, 347, 350, 385
CPAM (Cell plate assembly matrix), 251, 252, 255–281
CRE1 (Cytokinin receptor 1), 19
Crosslinking factor, 155
Crossover (CO), 103, 110, 113, 114
- Class I, 114
- Class II, 114
- Interference, 114
- Formation 110, 114
- Non-crossover (NCO), 113
- Pathway, 113
Cryptochrome, 8
CSS52A, 80, 81
CTD1, 354
C-type cyclin (see Cyclin CycC)
CUE, 89
Cyanidioschyzon merolae, 306
Cyanobacteria, 3, 4
Cyclin (Cyc), 6–8, 10, 13–15, 17, 19, 379–383
- Binding motif, 88, 382
- CycA, 9, 22, 60–66, 78, 81, 82, 85, 89, 131, 381, 382, 385
- CycB, 4, 15, 21, 22, 26, 78, 81, 82, 85, 87, 88, 131, 238, 381–383, 386
- CycC, 61
- CycD, 6, 9, 26, 60–66, 80, 82, 85, 87, 88, 381–383
- CycE, 61, 65, 79, 81, 92
- Cyc T, 61
- G1-type, 18, 355
- Mitotic, 19, 25, 80, 82
- S1-phase type, 18
Cyclin-dependent kinase (CDK), 6, 14, 60, 63, 65, 77, 81, 171, 235, 178, 185
- CDK-A, 9, 60–63, 66, 68, 82, 84, 85, 88, 89, 381–383
- CDK-B, 9, 10, 61, 63, 84, 348, 355, 381, 383, 385, 386
- CDK-C, 61
- CDK-E, 61
- CDK-F, 60, 61
- CDK-dependent phosphorylation, 15–17, 24, 25, 238

- CDK inhibitor (CDKI), 7, 8, 60, 63, 76, 82, 87, 88, 91, 92, 380, 383
- Complex, 14, 15, 17, 18, 25, 239, 244, 245
- G1-phase-specific, 383
- G2/M-specific, 6, 81, 84
- Phosphorylation site, 185, 238
- Regulation, 6, 17, 18, 148, 377–379
- S-phase-specific, 6, 18, 81, 84

CYD1 (CYTOKINESIS DEFECTIVE1), 348, 353, 354

Cys protease1 (Osc6), 370

Cytochalasin D, 50, 155, 203

Cytokinesis, 16, 25, 33, 41–52, 75, 106, 130–134, 147, 157, 158, 169, 179, 182–187, 207, 212, 214, 218, 233–246, 252–263, 267–270, 276, 279, 289–299, 303, 308–314, 324–326, 329, 334, 335, 348, 353, 354, 362, 365, 377, 379
- Defect, 42, 293, 314, 348, 353, 354
- Early, 197, 277
- Late, 36, 197
- Polarized, 36, 49, 50
- Pollen, 267
- Meiotic, 183, 240, 241
- Modes of, 49, 246
- Signaling protein of, 313
- Site of, 324–326, 334, 335, 351
- Somatic-type, 183, 251, 255, 258, 261, 270, 273, 275, 277, 281, 282
- Spatial control, 34
- Specific syntaxin, 26
- Stomatal, 314, 354
- Syncytial-type, 251, 258, 277, 279, 281
- Symmetric, 36

Cytokinetic apparatus, 34

Cytokinetic plane, 324, 329, 334

Cytokinin, 13, 19, 20, 62, 90, 381, 382
- Cytokinin oxidase, 19
- Cytokinin-independent growth, 62
- Perception, 19
- Synthesis, 19

Cytoplasm, 35, 40, 43, 91, 144, 153, 156, 196–199, 203, 215, 218, 236, 281, 289, 292, 298, 304, 311, 312, 324–327, 330, 332, 350, 371

Cytoplasmic centrioles, 212

Cytoplasmic clear zone, 143

Cytoplasmic motor, 334

Cytoplasmic strands, 35, 38, 133, 134, 142, 143, 156

Cytoskeleton, 21, 33, 34, 36, 39, 111, 125, 128, 132, 134, 195, 202, 203, 255, 282, 306, 314, 331, 335, 347
- Cytoskeletal arrangement, 36
- Cytoskeletal arrays, 252, 334
- Cytoskeletal assembly, 33
- Cytoskeletal component, 111, 157, 207
- Cytoskeletal dynamics, 125
- Cytoskeletal elements, 39, 143, 255, 276, 291
- Cytoskeletal motor protein, 169, 187
- Cytoskeletal-disrupting drugs, 143
- Cytoskeletal structure, 34, 52, 125, 209, 243, 246
- Cytoskeletal system, 253

Cytosol, 289, 292, 298

DAPI, 105, 175, 236, 244, 304, 312

Datura stramonium, 91

Daughter cell, 33, 36, 141, 153, 195, 196, 203, 207, 251, 256, 275, 290, 308, 324, 325, 327, 336, 344, 347, 351, 377

De novo, 44, 111, 114, 117, 261, 263, 275, 303, 307, 308, 379, 386

Dedifferentiation, 65, 379

Dehiscence, 362, 363

DEK (DEFECTIVE KERNEL), 89, 384

DEL (DP and E2F-Like), 64
- AtDEL1, 9, 64, 80, 384, 385
- AtDEL2, 64, 384, 385
- AtDEL3, 64, 384, 385

DENN domain, 348, 354

Dephosphorylation, 6, 20, 33, 43, 48, 81, 381

Depolymerization, 111, 130, 133, 145–147, 170, 180, 182, 183, 242, 243, 263, 330, 331

Depolymerizing drug, 35, 44, 47, 50, 51, 111, 149, 177, 183, 203, 245

Deposition, 33, 107, 128, 259, 273, 275–277, 281, 282, 303, 312, 331, 334, 365, 368

Deubiquitination, 19

Diakinesis, 103

Dictyostelium, 296

DIF1, 108

Differentiation, 59, 76, 80, 369
- Anther, 373
- Cell, 24, 62, 64, 68, 82, 91, 93, 323, 343, 348, 355, 363, 368, 377, 378, 382
- Cell-type, 343, 353, 356, 357
- Coordinated, 361, 368

- GMC (guard mother cell), 343, 344, 348–355
- Nodule, 93
- Sporogenous cell, 373
- Stem cell, 343, 352
- Stomatal, 333–348, 352–357
- Tapetum, 361, 368, 370, 371, 373
- Terminal, 93, 357
- Tracheary element, 129
- Trichome, 86–88, 92

Diplotene, 103, 107, 117
Directionality, 13, 20, 171, 178, 378
Dis1/XMAP215/TOG, 149
Division plane (*see* Cell division plane)
DMC1 (DISRUPTION OF MEIOTIC CONTROL 1), 106, 112, 113, 115, 116
DNA, 3, 5, 6, 13, 17, 18, 21–26, 59, 64–68, 75–81, 84, 85, 87–92, 103, 108, 111–115, 127, 130, 153, 154, 171, 172, 175, 176, 182, 209, 238, 240, 304, 311, 312, 352–354, 370, 377, 379, 384–387
- cDNA, 21, 130, 171, 172, 238, 240
- DNA binding domain, 24, 25
- DNA catenation, 88
- DNA damage, 3, 5, 6, 13, 21, 112
- DNA polymerase, 21, 26, 66, 67, 127, 385
- DNA primase, 385
- DNA repair, 64, 65, 103, 110, 112–114, 116
- DNA replication, 6, 17, 18, 21, 22, 59, 64–68, 77, 78, 80, 84, 85, 108, 354, 377, 385
- DNA-binding domain, 64

Docking factor, 294
Dominant negative mutant, 84, 92, 143, 245, 352, 381, 385
Double fertilization, 381
Double-membrane structure, 207
Double-strand break (DSB), 103, 104, 110–113, 116
Downstream, 59, 60, 88, 106, 112, 234, 240–244, 330, 349, 350, 356, 357, 370, 386
DP (Dimerization partner) protein, 23, 24, 60, 64, 65, 68, 79, 80, 383–385
Drosophila, 79–81, 86, 177, 178, 183, 221, 223, 243, 324
D-type cyclin (*see* Cyclin CycD)
Dumbbell-shaped tubular, 304–306, 309
Dynamin, 266–272, 276, 281, 282, 291, 297, 298, 303, 304, 306, 307, 313
- signature, 305, 306

Dynamin-related protein
- DRP1, 262, 268, 298, 299, 304–309
- DRP2, 262, 269, 305–307
- DRP3, 306
- DRP4, 306
- DRP5, 306
- DRP6, 306
Dynein, 152, 154, 170, 174, 175, 177, 216, 220, 221

E2F (E2 promoter-binding factor), 23, 24, 59, 60, 64–68, 76, 79, 80, 86, 378, 381, 383–386
- E2F/DP, 23, 24, 79, 383
- E2Fa, 64, 68, 79, 80, 383–385
- E2Fb, 64, 65, 384, 385
- E2Fc, 64, 79, 381, 384, 385
- E2Fd, 384
- E2Fe, 80, 384, 385
- E2Ff, 64, 384, 385
E3 ubiquitin ligase, 80
EB1 (End binding protein 1), 44, 51, 52, 144, 149, 154, 155, 181, 262, 264
Ectocarpus siliculosis, 336
Ectopic expression, 24, 78, 85, 87, 88, 382, 383, 385
Eg5 (Kinesin-5), 154
Egg, 61, 152, 153, 214, 326–328, 330, 332
Electron microscopy (EM), 46, 112, 129, 145, 184, 196, 217, 221, 252, 253, 255, 259, 263, 276, 309, 361
Electron tomography, 184, 197, 251–253, 255–257, 260, 261, 263, 264, 267, 268, 272, 276, 277, 292, 297
Embryo, 62, 67, 75, 81, 83, 93, 108, 280, 282, 293, 296, 323–328, 335, 336, 349, 381, 384
Embryogenesis, 61, 67, 183, 335, 354, 361, 378
EMI1 (EARLY MITOTIC INHIBITOR 1), 78
Endocycle, 379, 381–386
Endocytic vesicle, 199, 305
Endocytosis, 35, 46, 291, 307
Endodermal cell, 324, 325
Endokaryotic hypothesis, 208
Endomembrane, 200, 201, 292, 293, 298, 329–331
Endomitosis, 75
Endoplasmic reticulum (ER), 35, 40, 125, 144, 146, 170, 209, 217, 219, 220–222, 255, 275, 276, 310, 346–352, 354–356

Endoreduplication, 62, 63, 66, 75–94, 354, 357, 379, 381–383, 385, 387
Endosome, 198, 200, 291–293, 295
Endosperm, 61, 75, 76, 82–86, 89–91, 93, 144–146, 153, 155, 173, 196, 258, 259, 268, 269, 277, 279–282, 334, 335, 381, 383
Endosymbiosis, 208, 209, 211
Endothecium, 362, 363, 365, 367, 372
End-to-end fusion, 308, 309
Enzyme, 43, 46, 66, 77, 85, 114, 208, 265, 270, 272, 276, 282, 299, 303, 311, 313, 355, 361, 371
Epidermal cell, 86, 89, 92, 343, 344, 354, 365
Epidermis, 76, 78, 91, 201, 323, 343–346, 349–353, 356, 357, 362, 365, 367, 372, 384
Equational division (Meiosis II), 103, 106
Equator, 145, 156, 157, 221, 236, 238, 240, 241, 244, 257, 261
– Equatorial organizer, 155, 157, 158
– Equatorial plane, 255, 257, 258, 260, 277, 308
– Equatorial region, 144, 170
– Equatorial ring, 198
– Equatorial zone, 236, 238
ERECTA (ER), 346–356
Ethylene, 90
E-type cyclin (*see* Cyclin CycE)
Euchromatin, 68
Eudicot, 369
Eukaryote, 4, 6, 27, 33, 78, 106, 109, 132, 158, 177, 185, 187, 195, 207, 208, 210, 211, 216, 219, 233, 242, 293, 296, 380
Eukaryotic cell, 59, 77, 129, 147, 207, 209, 379
– Eukaryotic cell cycle, 13, 15, 17, 18, 20
Evolution, 8, 14, 207–211, 292, 343
EXCESS MICROSPOROCYTES 1 (EMS1), 361
Exine wall, 282, 362, 371, 372
Exocyst, 262, 265–267, 269, 297
Exocytosis, 266, 267, 291, 292, 299, 303, 307, 308, 310, 330
EXPANSIN (EXP)
– EXP3, 385
– EXP7, 385
– EXP9, 385
EXTRA SPOROGENOUS CELLS (EXS), 361
Extracellular signal, 5, 40, 233

F-actin, 41, 46, 47, 157, 158, 262
FAMA, 346, 348, 352, 353, 357
FASCIATA (FAS)
– FAS1, 67
– FAS2, 68
Feedback regulation, 5, 8, 16, 25, 330
Feed-forward regulation, 16
Female sterility, 349
Fern, 323
Fertilization, 61, 91, 326–329, 332, 381
FIE (FERTILIZATION-INDEPENDENT ENDOSPERM), 91, 384
Filament, 112, 117, 156, 201, 203, 204, 217, 242, 252, 262, 267, 309, 323
Fimbrin, 47, 132, 133
FIP37 (Immunophilin FKBP12), 88
FIS (FERTILIZATION-INDEPENDENT SEED), 91
Fission yeast, 6, 14, 17, 82, 85, 109, 334, 335
FKBP12 (FIS), 88
FKH1 (FORKHEAD HOMOLOGUE 1), 15
FKH2 (FORKHEAD HOMOLOGUE 2), 15, 16
Flagella, 212, 332
Flippase, 298
Floral meristem, 61, 365
Flow cytometry, 79
FLP (FOUR LIPS), 346, 348, 352, 353
Fluorescein, 326
Fluorescence resonance energy transfer (FRET), 366
Fluorescent protein, 125, 126, 128, 135, 179, 217–219, 221, 240, 296, 304
FM4-64, 133, 200
Foci, 107, 112–117
Fragmentation, 66, 112, 113, 116, 145
Freeze-substitution, 251, 252
FRILL1, 89
Fucoid algae, 323–329, 331, 333, 335, 336
Fungus, 3, 5, 8, 21, 108, 110, 115, 147, 148, 151, 152, 169, 171, 172, 174, 177–179, 186, 210, 214, 220, 222, 296, 323, 349
FUS3 (MAP kinase), 238
FZR (FIZZY-RELATED), 80

G phases
– G1 phase, 6, 7, 9, 15–19, 22, 23, 36, 59, 62, 64, 77, 81, 86, 106, 127–129, 131, 133, 134, 256, 354, 355, 377, 378, 382, 383, 385

- G1/S transition, 6, 7, 9, 15, 17, 18, 23, 59–68, 76–78, 82, 88, 127, 378, 382, 383, 385
- G2 phase, 6, 7, 9, 15–17, 19, 22, 23, 25, 26, 33–36, 43, 61, 62, 77–79, 127, 128, 132–135, 142, 220, 235, 313, 377, 383–386
- G2/M transition, 6, 7, 9, 15–17, 19, 23, 25, 26, 52, 61, 62, 77–79, 256, 385, 386
- M/G1 transition, 16, 19, 22, 128, 129, 131

Gamete, 111, 327
Gametophyte, 75, 82, 334, 335, 361, 368, 383
GAMYB, 369
GED (GTPase effector domain), 305
Gel shift analysis, 24
Geminin, 78, 81
Gene conversion, 113
Gene expression, 3, 4, 9, 13–16, 20–29, 76, 79, 85, 91–94, 349, 381
Genetic analysis, 298, 328, 361, 363
Genetic approach, 14, 68, 380
Genetic control, 90, 361
Genetic interaction, 294, 349, 350
Genetic material, 103, 169, 170, 177, 207, 208, 251, 377
Genetic study, 51, 59, 61, 63, 186, 311, 336
Genetic tools, 10, 387
Genetics, 10, 88, 109, 344, 356, 367
Genome, 9, 10, 13, 20, 22, 27, 59, 61, 65, 77, 90, 92, 93, 107, 111, 113–116, 130, 148, 151, 152, 169, 170, 171, 177, 182, 195, 208, 215, 233, 264, 265, 282, 296, 315, 336, 354, 377, 379, 385
Genomic imprinting, 91
Genomic resources, 336
Genomic tools, 377, 387
Genomics, 22, 209, 337, 380
Germ cell, 62, 104
Germination, 62, 93, 239, 328, 329, 331, 332, 336, 382
Germline, 324
GFP (green fluorescent protein), 34, 42–44, 47, 51, 125, 128–130, 132–134, 144, 147, 148, 150, 154, 155, 185, 196, 199, 200, 218, 236, 240, 244, 267, 277, 303, 304, 312, 365, 370, 372
- CLIP170:GFP, 42
- EB1-GFP, 44, 51, 155
- EMS1-GFP, 365
- GFP-AtRop4, 267
- GFP-AtVam3, 134
- GFP-CalS1, 312
- GFP:EB1, 144, 154
- GFP-Fimbrin, 132, 133
- GFP-KCA1, 128
- GFP:MAP4, 155
- GFP:MAP70, 150
- GFP-MBD, 128
- GFP-NPK1, 236
- GFP-NQK1, 240
- GFP-NtMap65-1a, 148, 244
- GFP-Phr, 304
- GFP-RabF2b, 200
- GFP tag, 34, 267, 303, 304, 372
- GFP-Talin, 132
- GFP-tubulin, 128, 129, 147
- MAP4:GFP, 42, 51
- RanGAP1-GFP, 218
- SPR1:GFP, 150
- TON2:GFP, 43
- UDT1-GFP, 370
Gl3 (GLABRA3), 89, 91
Globular stage, 381, 386
Glucan, 236
Glucan synthase, 313
Glucocorticoid receptor, 363
Glucose, 275, 282, 299, 312
Glutaraldehyde, 252
Glycerol-3-phosphate acyltransferase (GPAT), 361, 371
Glycerolipid, 371
Golgi apparatus, 35, 46, 175, 183, 195, 197–200, 203, 218, 219, 221, 256, 258, 259, 262, 264, 265, 270, 275, 281, 290–293, 297, 303, 306, 308, 311, 331
- Golgi belt, 198
- Golgi-dependent secretion, 46
- Golgi-derived vesicle, 125, 179, 184, 185, 197, 198, 242, 252, 256, 259, 262, 265, 266, 270, 273, 279, 292, 303, 308, 309
- Golgi enzyme, 46
- Golgi vesicle, 199, 259, 291, 304, 309
- Golgin, 264
Gonidial cell, 323
GPI (glycosylphosphatidylinositol)-anchor motif, 350
GRIP, 265
Growth axis, 40, 328, 330, 333
Growth polarity, 40
Growth regulator, 19
GTP-binding motif, 305

GTP-binding protein, 265, 268, 304, 305, 311, 314
GTP-exchange factor RCC1, 153
GTP hydrolysis, 269, 305, 307
GTPase, 153, 209, 210, 218, 262, 267, 292, 294, 297, 303, 305, 309, 312–315
GTP-γ-S, 307
Guard cell, 237, 343–348, 351–357
Guard mother cell (GMC), 41, 50, 314, 344–346, 351–357
GUS, 351, 354, 370, 371
Gymnosperms, 35, 196

H2A/H2B chaperone, 67
H3/H4 chaperone, 67
Haemanthus, 144, 145, 153, 173, 196
Half-spindles, 142, 147, 155, 170
Haploid, 76, 279, 383
HECT domain protein, 89
Heliotrop plant, 3
Hemicellulose, 199, 270, 271, 276
Hepatocyte, 6, 7, 9
Heterochromatin, 66, 68, 82, 116, 215
Heterodimer, 64, 116, 356, 385
Heteromeric complex, 15
Heteromeric protein kinase, 16, 17
Heteromeric SNARE complex, 293
Heterotypic membrane fusion, 307, 308, 310
HEX binding protein 1a (HBP-1a), 24
Hexamer (HEX) motif/element, 23, 24
Hexaploid, 107, 114
HIC (HIGH CARBON DIOXIDE), 348, 355, 356
High pressure freezing (HPF), 251–253
HINKEL (HIK), 239, 313
Histone, 4, 13, 22–24, 62, 65–68, 91, 107, 383
Hog1 MAP kinase, 240
Homeodomain, 16, 26
Homolog, 5, 10, 42, 48, 61, 63, 65, 66, 78, 91, 103–108, 110–116, 131, 149, 178, 179, 182, 183, 185, 215–218, 223, 233, 235, 237, 239, 241, 244, 265, 293, 297, 304, 310, 314, 366, 370, 373, 384
Homology, 60, 64, 104, 109, 114, 115, 130, 148–150, 171, 176, 186, 240, 269, 289, 299, 303, 305
Homotetramer, 178
Homotypic membrane fusion, 291, 307–310
Hop1, 107

HOP2, 113, 115, 116
Hormone, 13, 21, 40, 62, 75, 88–90, 234, 349, 366, 380–382
HUA ENHANCER 3 (HEN3), 61
Hyperplasia, 62, 381, 384, 385
Hypertrophy, 368, 369
Hypocotyl, 8, 76, 89, 90, 381, 382, 385, 386
Hypodermal cell, 105

ILP1-1D (INCREASED LEVEL OF POLYPLOIDY1-1D), 82, 89
Immunocytochemical screen, 185
Immuno-fluorescence microscopy, 125–127, 129, 130, 133
Immuno-gold labeling, 113, 268, 275
Immunolocalization, 108, 117, 119, 181, 261, 314
Immunoprecipitation, 238, 294, 385
Importin, 78, 209, 223
In silico analysis, 9, 10, 26, 28
In situ hybridization, 372
In vitro, 48, 62, 63, 85, 130, 145, 148, 149, 154, 181, 219, 222, 238, 241, 243–245, 263, 269, 293, 307, 315, 365, 371, 385
In vivo, 10, 24, 63, 149, 154, 220, 241, 244, 245, 294, 379, 385
INCW2, 89
Inhibitor
– Actin inhibitor, 50, 133, 155
– Auxin efflux inhibitor, 40, 383
– CDK inhibitor (CKI), 6–8, 18–20, 60, 63, 81, 82, 85, 89, 92, 383
– Cyclin inhibitor, 82
– DNA polymerase inhibitor, 127
– DNA replication inhibitor, 67, 77, 78
– Endoreduplication inhibitor, 79, 82
– Golgi inhibitor, 46
– Inhibitor of CDK (ICK), 60, 63, 76, 82, 85, 87, 88, 91
– Inhibitor of the SBF transcription factor complex, 17
– Microtubule (MT) inhibitor, 47, 127
– Mitotic spindle inhibitor, 21, 27
– Myosin inhibitor, 51, 203
– Protease inhibitor, 370
– Proteasome inhibitor, 131
– Protein kinase inhibitor, 43, 155
– Protein phosphatase inhibitor, 43
– Protein synthesis inhibitor, 155
INK4, 63

Interphase, 36, 41–43, 105–108, 111, 146, 147, 150, 156, 175, 195, 198–202, 212, 214, 216–218, 220, 235, 235, 291
Interpolar microtubules (MTs), 142, 145–147, 157, 158, 177
Intracellular protein transport, 314
Intracellular signal, 233
Intracellular store of Ca^{2+}, 196
Intracellular structure, 125, 135
Isodiametric cell, 41
Isoprenoid, 19

JAGGED (JAG), 26

KAKTUS (KAK), 89
Kar3p, 149, 174, 179, 181
Karymastigont, 208
Karyogamy, 332
Karyogenic hypothesis, 208
Karyokinesis, 208, 210, 218
Karyopherins, 209, 210, 214, 222
Karyoplasmic ratio, 91
KAT (*Arabidopsis* kinesin-like protein)
– KatA/ATK1, 174, 179, 181, 185, 313
– KatB, 179, 181
– KatC, 179, 181
KCA (Kinesin-like protein in *Arabidopsis*)
– KCA1, 47, 185, 186
– KCA2, 185
KCBP (kinesin-like calmodulin binding protein), 151, 174, 175, 221, 313
KEULE, 262, 294–297, 309, 310, 348, 354
Kinase
– Aurora kinase, 132
– Casein kinase, 8
– Checkpoint kinase, 5
– Cyclin-dependent kinase (CDK), 6, 14, 59, 60–65, 82, 85, 171, 235, 348, 355
– Kinase complex, 62, 382
– Kinase defective mutant, 236, 239–241
– Kinase domain, 233, 235, 236
– Kinase inhibitor, 43, 60, 155
– MAP kinase, 132, 169, 182, 183, 233–235, 240–244, 246, 313, 356
– MAP kinase cascade, 169, 182, 183, 233–238, 240–242, 246, 348–350, 356, 357
– Protein kinase, 10, 16, 18, 43, 45, 129, 131, 132, 183, 234, 236, 313, 349
– Receptor-like kinase, 348, 349, 356, 365

– Serine/threonine kinase, 233
– Tyrosine kinase, 330
– WEE1 kinase, 6
Kinesin, 45, 47, 117, 149–152, 154, 169–186, 221, 234, 238, 313
– Kinesin-1 (POK1), 48, 171
– Kinesin-2 (POK2), 48
– Kinesin-10 (Nod/Kid), 151, 175
– Kinesin-12, 169, 182
– Kinesin-13 (KinI/MCAK), 151, 175, 177, 183
– Kinesin-14 (C-terminal motor), 151, 152, 154, 169, 171, 173, 174, 179, 181, 183, 185
– Kinesin-3 (KatC), 179, 181
– Kinesin-4 (Chromokinesin/KIF4), 151, 175, 176
– Kinesin-5 (BimC), 151, 152, 154, 169, 178, 179, 181, 333
– Kinesin-6 (MKLP1), 151
– Kinesin-7 (CENP-E), 151, 175, 177, 182
– Kinesin-8 (Kip3), 151
– Kinesin family, 150, 151
– Kinesin-like/related protein, 129, 131, 221, 234, 238, 313
Kinetochore, 142, 145–147, 151, 153, 158, 169–179, 213, 214
– K-fiber (kinetochore MT fiber), 142, 145–147
– kMTs (kinetochore MTs), 142
Kinetosome, 209
KinI/MCAK, 151
Kip1, 149
Kip3, 151
KIP-related protein (KRP), 8, 60, 63, 76, 82, 85, 86, 88, 91, 313
– KRP1, 76, 82, 85, 86, 88, 91, 382, 383
– KRP2, 9, 63, 82, 85, 86, 88, 91, 382, 383, 313
– KRP3, 9, 385
– KRP5, 385
KLP (Kinesin-like protein), 234, 235, 237, 238, 243
KNAT-type, 26
KNOLLE (KN), 25, 26, 262, 308, 309, 354

Lamin, 207, 210, 216–218, 220, 223
Lamina, 210, 216, 220, 222
Laser ablation, 156, 336
Laser microdissection microscopy, 27
Laser scanning confocal microscopy, 128, 218

Lateral element (LE), 107
Latrunculin B, 203, 204
LBR (lamin B receptor), 217–219, 223
LCA1, 217, 221
Leaf primordia, 62, 385
Leptotene, 103, 107–110, 112, 117
Leucine-rich repeat (LRR), 347–351, 365, 366, 373
Leucine zipper, 24, 64, 372
LHY (LATE ELONGATED HYPOCOTYL), 8–10
LIM domain protein, 314
Lipase, 298
Locule, 362–364, 368, 372
Loss-of-function mutation, 25, 61, 62, 65–67, 174, 240, 241, 245, 347, 349, 352, 353, 357, 383
Lotus japonicus, 80
Lovastatin, 19
LRR receptor-like kinase (LRR-RLK), 347–351, 365, 366, 373
Lupinus albus, 80

Machinery
– Cell cycle, 4, 6, 59–61, 63, 64, 90, 355, 380
– Cytokinesis 199, 202, 246, 252
– Exocytosis, 308
– Membrane fusion, 293, 294, 298, 307, 309, 310
– Microtubule nucleation, 215
– Nucleocytoplasmic transport, 209, 214, 215
– Proteolysis, 14, 20, 131
– Vesicle trafficking, 289, 290, 292, 293, 298
MADS-box protein, 363
MAF1 (MAR-binding filament-like protein 1), 217–219
MALE STERILITY 1 (MS1), 361, 371–373
Manduca sexta, 76
MAP (Microtubule-associated protein), 129, 130, 141, 148–150, 215, 234, 242, 311, 313
– MAP4, 42, 51, 128, 155
– MAP65, 42, 130, 132, 148, 181, 183, 234, 243–246, 262, 264
– Map70, 150
– MAP190, 150
– MAP200, 149
– MAP215, 42, 131, 149

MAP kinase (MAPK), 132, 169, 182, 183, 233–235, 240–244, 246, 313, 356
MAP kinase cascade, 169, 182, 183, 233–238, 240–242, 246, 348–350, 356, 357
MAPK, 233–238, 240–244, 246, 349, 350, 356, 357
MAPKK, 233, 234, 238, 240, 241, 356
MAPKKK, 182, 183, 233–235, 237–241, 349
Maternal control, 90
Maternal effect, 90, 91
MBD (microtubule-binding domain), 128
MBF (MCB-binding factor), 15
MCAK (Mitotic centromere-associated kinesin), 151, 175, 177
MCM (minichromosome maintenance), 66, 67, 78, 385
– MCM1, 15, 16
– MCM2, 77, 78
– MCM3, 66, 385
– MCM7, 66, 77, 78, 386
MEA (MEDEA), 91
Medicago, 80, 82, 93
Meiocyte, 105, 106, 108–110, 116, 117, 277, 279, 362, 365, 367, 370
Meiosis, 63, 66, 103–109, 111–117, 169, 170, 172, 174, 179, 181, 187, 213, 215, 281, 291, 362, 364, 367–371
– Meiotic cell, 104–106, 108, 114, 116, 152, 155
– Meiotic cytokinesis, 183, 240, 241
– Meiotic double strand break, 103
– Meiotic prophase, 104–106, 108, 115, 215
– Meiotic recombination, 103, 109, 111, 113, 115
– Meiotic sister-chromatid cohesion (SCC) defects, 106
MEL-28 (MATERNAL EFFECT LETHAL 28), 214
Meloidogyne incognita, 80, 90
Membrane
– Cell plate Membrane, 251, 259, 261–264, 270, 271, 275, 277, 279, 281, 282, 289–292, 298
– Endomembrane, 200, 201, 292, 293, 298, 329–331
– Membrane compartment, 200, 265, 275, 293, 307
– Membrane fusion, 267, 306, 307, 309–311

- Membrane organelles, 125, 126, 292, 298
- Membrane protein, 216, 217, 220, 223, 275, 298, 299, 307, 312
- Membrane recycling, 202, 270, 273, 275, 307
- Membrane structure, 125, 126, 133, 207, 221, 271, 308
- Membrane trafficking, 46, 172, 199, 223, 269, 311, 314
- Membrane tubulation, 305, 314, 315
- Membrane tubule, 259
- Membranous islands, 329, 334
- Nuclear membrane, 19, 75, 207, 209, 216, 217, 219–221, 311
- Plasma membrane (PM), 33, 35, 36, 46, 47, 180, 198, 200, 208, 221, 256, 259, 261, 270, 272, 275, 276, 279, 281, 289–299, 305–308, 310, 311, 314, 330, 333, 350, 366
- Transmembrane, 217, 312, 331

Meristem, 4, 20, 61–63, 67, 199–201, 255, 257–259, 263, 264, 268, 281, 296, 328, 350, 365, 386
- Meristematic cell, 50, 143, 155, 147, 198, 201, 218, 233, 235, 258
- Meristemoid mother cells (MMC), 344–348, 351, 352, 354–357

Mesocarp, 76
Metabolism, 3, 13, 22, 59, 92, 361, 371
- Metabolic activity, 75, 76, 91
- Metabolite, 27

Metaphase, 17, 19, 21, 33, 34, 36, 45, 80, 81, 103, 108, 127, 128, 131, 133, 135, 142, 145, 146, 148, 151, 152, 154, 157, 170, 172, 173, 175, 176, 197, 199–201, 203, 212, 217, 219, 221, 222, 236, 237, 245, 258, 259, 326, 329, 333, 334, 377
- Metaphase plate, 142, 145, 151, 172, 176
- Metaphase spindle, 145, 146, 154, 157, 199, 200, 203, 329, 333
- Metaphase to anaphase transition, 13, 19, 21, 80, 334
- Postmetaphase, 329, 333
- Prometaphase, 33, 127, 142, 145, 146, 148, 152, 154, 157, 175, 176, 198, 237, 258

Metazoan, 78, 210, 334
Methylation, 66, 84, 107
MFTP (MF twin peaks), 133
MG-132 (Cell cycle-specific proteasome inhibitor), 131
Microfibril, 127, 128, 270, 276, 353

Microfilament (MF), 33, 34, 36, 38, 40, 44, 46, 47, 49, 50, 52, 125, 126, 132, 133, 142, 150, 170, 172, 186
Microgametogenesis, 361
Microinjection, 125, 126, 132
Micromanipulation, 327
MicroRNA (miRNA), 361, 370, 382
- microRNA *JAW-D*, 382
- *miR159*, 370
- *miR160*, 386

Microspore, 282, 323, 361–363, 365, 370–372
Microsporocyte, 279, 281, 282, 361, 364
Microsporogenesis, 361, 363, 364
Microtubule (MT), 109, 126, 129, 130, 141, 146, 170, 256
- Binding site, 128, 171, 174
- Cortical, 33–36, 41, 125, 130, 332
- Depolymerizing drug, 21, 111, 127
- Dynamics, 34, 127, 128, 182
- Fluorescence marker, 125, 173, 175, 176, 178
- Kinetochore, 142, 172, 173, 176–178, 209
- Midzone, 177–181
- Minus end, 129, 142, 144, 147, 151, 152, 171, 173, 183, 313
- Motor activity, 131, 150, 169–172, 176, 178, 182, 184, 313
- Organizing center (MTOC), 129, 130, 144, 145, 156, 157, 169, 173, 181, 261, 209, 212, 215, 220, 314
- Phragmoplast, 51, 127, 131, 169, 170, 177, 179, 180–185, 251, 255, 303, 304, 313
- Plus end, 42, 52, 142, 144, 147, 149, 151, 153, 154, 170, 171, 176, 181–183
- Polymerization, 170, 179, 242, 331
- Preprophase, 34–41, 175, 212, 221, 326, 327
- Spindle, 117, 132, 146, 147, 151, 152, 169, 170, 172–174, 196, 203, 219, 220, 255, 257, 334, 279

Microtubule-associated protein (*see* MAP)
Midbody, 235, 242, 290
Middle layer, 362, 363, 365, 367, 370
Midzone, 142, 147, 148, 152, 154, 177–181, 183, 235, 241, 242, 244, 245, 313, 329, 334
Mini-phragmoplast, 258, 279, 280
Mitochondria, 195, 196, 198, 201–203, 208, 211, 255, 306, 372
Mitosis, 6, 10, 16–19, 33, 34, 36, 41–49, 52, 75–77, 80–83, 85, 88, 89, 92, 106, 108,

127, 128, 131–134, 141, 145, 148–150, 154, 157, 158, 169–174, 176, 179, 181, 183, 195, 196, 201–203, 207–223, 233, 245, 257–259, 291, 309, 314, 326, 332, 333, 335, 371, 377, 382, 383
- Acytokinetic mitosis, 83
- Animal mitosis, 148, 151
- Closed mitosis, 208, 210, 212, 213
- Endomitosis, 75
- Fenestral mitosis, 210
- Late mitosis, 77
- Open mitosis, 207, 208, 210–213, 219, 220, 222
- Plant mitosis, 141–143, 148, 158, 169, 172, 187, 212, 218–220
- Pollen mitosis, 42, 131, 371
Mitotic, 16, 19, 21, 25, 103–106, 108, 111, 112, 362
Mitotic activity, 89, 218
Mitotic apparatus, 141, 201, 208, 324–326, 331–335
Mitotic cell cycle, 75, 76, 79, 82, 83, 85, 86, 89, 92
Mitotic cells, 77, 79, 80, 82, 108, 174, 179, 203
Mitotic checkpoint, 79, 208, 213, 214
Mitotic cyclin, 16, 19, 25, 80, 82, 87
Mitotic kinesin, 150, 171
Mitotic spindle, 21, 34, 52, 75, 125, 127–132, 134, 141–143, 148–153, 155, 157, 196, 207, 214, 223, 245, 246, 255, 331
Mitotic stage/phase, 85, 90, 197, 212, 259, 377
MKLP1 (Kinesin-6), 151
MLH (MutL homolog protein)
- MLH1, 114
- MLH3, 114
MLP2 (Myosin-like protein 2), 214
MMK3 (*Medicago* MAP kinase 3), 313, 314
MND1 (MEIOTIC NUCLEAR DIVISION 1), 113, 115, 116
Monochromatic blue light, 328
Monocot, 86, 369, 373
Mor1/Map200, 42, 43, 130, 131, 149, 246
Mother cell wall, 34, 38, 39, 270, 313
Motor, 41, 48, 51, 129, 146–149, 151–155, 158, 169–178, 181–185, 187, 202, 216, 221, 236, 238, 239, 272, 291, 313, 333, 334
- Motor activity, 48, 147, 174, 175, 177
- Motor dependent sorting, 154

- Motor domain, 151, 170–174, 176, 178, 182, 236, 238, 239
- Motor kinesin, 169, 170, 181, 185
- Motor MAP, 148
- Motor myosin, 176, 216
- Motor protein, 41, 51, 129, 149, 153, 154, 158, 169, 170, 202, 272, 291, 313
M phases
- M phase, 18, 19, 22, 25, 26, 61, 62, 77, 78, 81, 84, 181, 233, 235, 237, 238, 240, 241, 243, 244, 245
- G2/M transition, 6, 7, 9, 15–17, 19, 23, 25, 26, 52, 61, 62, 77–79, 256, 385, 386
- M/G1 transition, 16, 19, 22, 128, 129, 131
M phase-specific activator (MSA), 23, 25, 26
MPM-2 (Mitotic protein-specific monoclonal antibody 2), 314
MRE11 (MEIOTIC RECOMBINATION 11), 112, 113, 115
MRN (MRE11/RAD50/NBS1) complex, 112
MSH (MutS homolog protein)
- MSH4, 114
- MSH5, 114
MTCC (MT-converging center), 144, 146
MTOC (MT-organizing center), 129, 130, 144, 145, 156, 157, 169, 173, 181, 209, 261, 314
Multicellular organism, 3, 7, 210, 289, 378, 380
Multicellular phenotype, 87, 88
Multinucleated cell, 25, 236, 237, 239, 241, 297
MULTIPLE ARCHESPORIAL CELLS 1 (MAC1), 367, 368
MULTIPLE SPOROCYTE 1 (MSP1), 366, 368
Multivesicular body (MVB), 200, 204, 256, 275, 276, 279, 281
Mutant, 9, 10, 25, 34, 42, 43, 48, 61, 63, 67, 82, 84, 87–89, 91, 94, 105–117, 130, 131, 149, 150, 152, 155, 181, 216, 236, 238, 240, 241, 243, 245, 266–268, 292–294, 296, 297, 309, 311, 314, 328, 336, 346, 347, 349–355, 363–372, 381, 382, 384–386
MutL, 114
MutS, 114
MYA2 (Myosin XI), 186
MYB transcription factor, 8, 23, 25, 26, 348, 352, 353, 361, 368–370
- MYB3R1, 25

– MYB3R4, 25
– MYB88, 348, 353
– 3RMYB, 23, 25, 26
Myc, 370, 371
Myosin, 51, 169, 170, 172, 186, 202–204, 290

NACK, 234, 238–240
– NACK1, 182, 183, 235–245, 313
– NACK2, 182, 183, 238–242
– NACK-PQR pathway, 234, 239, 241–245
Nag (*N*-acetylglucosaminyl transferase), 218, 221
NAP1 (Nucleosome assembly protein 1), 67
Nautilocalyx, 143
NBS1, 112
NCD, 174, 179, 181, 183, 279, 281
NDD1, 15
NE (*see* Nuclear envelop)
NEB (*see* Nuclear envelope breakdown)
Necrotic lesion, 294
Neuroblast, 324
Nicotiana plumbaginifolia, 20
Nitrogen-fixing nodule, 80, 90, 93
NMCP1 (Nuclear matrix constituent protein 1), 215, 217, 218
Nod/Kid (Kinesin-10), 151
Non-crossover (NCO) product, 113
NPK1 (Nuclear- and phragmoplast-localized protein kinase 1), 233–245
NPSN (Novel plant SNARE)
– NPSN1, 295, 296
– NPSN11, 294–296, 298
NQK1 (MAPK kinase), 234, 237, 240, 241, 243–245
NRK1 (MAP kinase), 132, 183, 234, 237, 240–245
NSPN11, 262
NT-1 cell, 277
NtKIS1a, 82
NtMAP65, 183, 234, 243–245
NtMEK1, 240, 244
NtMybA1, 25
NtMybA2, 25
NtMybB, 25
Nucleotide-binding motif, 172
Nucleus, 35, 36, 38–41, 43–45, 50, 82, 84, 103, 110, 112, 113, 142–145, 150, 155, 156, 195, 208–210, 214–218, 220, 222, 233, 235–237, 256, 279, 281, 327, 329, 330, 332–334, 349, 370, 373

– Nuclear antigen factor, 26
– Nuclear cytoplasmic domain, 279, 281
– Nuclear division, 5, 103–106, 214, 277, 281, 335, 370, 386
– Nuclear envelope (NE), 44, 45, 109–111, 131, 144, 145, 154, 158, 169, 172–176, 180, 185, 207–210, 212–223, 235, 256, 279, 281, 329, 331, 332
– Nuclear envelope breakdown (NEB), 44, 45, 145, 154, 158, 169, 172, 174, 175, 180, 213, 219–221
– Nuclear import, 78, 209
– Nuclear localization signal, 185, 235, 236, 372
– Nuclear membrane, 19, 75, 207, 209, 216, 217, 219–221, 311
– Nuclear migration, 35, 36, 38, 40, 142, 143, 220
– Nuclear pore, 110, 111, 207, 209, 210, 214, 220–223
– Nuclear protein, 24, 370, 373
– Nuclear rotation, 333
– Nucleocytoplasmic transport, 67, 209, 214, 215
– Nucleolus, 110, 111, 216, 218, 256, 279, 281
– Nucleoplasm, 207
Nug-type GTPases, 209
NuMA (Nuclear mitotic apparatus protein), 153, 156
Nup153 (Nuclear pore complex protein 153), 223

Octamer (OCT) motif, 23, 24
One cell spacing rule, 347
Open mitosis, 207, 208, 210–213, 219, 220, 222
ORC2, 66, 368
Organelle, 125, 126, 134, 143, 150, 156, 157, 170, 195–199, 201–204, 208, 216, 251–253, 255, 290–292, 294, 297–299, 303, 314, 315
Organism, 3–8, 10, 13, 14, 16, 20, 33, 76, 81, 92, 94, 109, 157, 169, 171, 210–212, 215, 238, 243, 245, 246, 289, 299, 323, 324, 378–380
Organogenesis, 377, 378, 380, 386
Ortholog, 25, 78, 80, 81, 85, 107, 108, 132, 306
Oryza sativa, 171
Oryzalin, 21, 203

Os02g02820 (Rice AMS homolog), 370, 371, 373
Oscillating process, 14, 15
Oscillating mechanism, 13, 14, 16, 19
Osmium tetroxide, 253
Ostreococcus tauri, 10, 27, 384
Ovule, 66, 328, 336, 363, 365, 366, 368

p43^{Ntf6}, 313
Pachytene, 103, 108, 109, 112, 114, 115, 117
PAIR2, 107, 108
Pairing, 93,103–109, 113–116
PAKRP1, 182
PAM1 (PLURAL ABNORMALITIES OF MEIOSIS 1), 109–111, 216
Parenchyma cells, 364
Particle bombardment, 372
PARTING DANCERS (PTD), 114
PATL1 (PATELLIN 1), 311
Pavement cell, 239, 241, 343–345, 349, 352, 353, 356, 357
Pbs2, 240
PCNA (PROLIFERATING CELL NUCLEAR ANTIGEN), 26, 385
Pectin, 270, 271, 276
Pedicel, 347, 349, 351
Peptide, 171, 174, 175, 217, 240, 239
Perinuclear microtubules (MTs), 144, 154, 155
Perinuclear region, 128, 326
Period2, 5
Period4, 5
Peroxisome, 197, 201–204, 306
PH (Pleckstrin homology) domain, 303, 305, 306
Phaeophyceae, 327
Phagocytic event, 208
Phagocytosis, 209
Phalloidin, 46, 50, 132
Phaseolus coccineous, 76
PHD-finger motif, 372
Phenotype, 42, 48, 62–64, 67, 87–89, 106, 107, 111, 113, 115, 131, 152, 181, 241, 243, 245, 267, 294, 311, 347–353, 355, 365, 366, 368, 372, 381–386
Phosphatidylinositol (PI), 200
Phosphorylation, 6, 13–20, 23–25, 28, 33, 43, 45, 48, 60, 61, 63–65, 67, 78–81, 107, 132, 148, 171, 178, 183, 185, 220, 233–235, 238, 240, 241, 243–245, 313, 365, 381, 383
– Autophosphorylation, 365
– CDK-dependent phosphorylation, 16–20, 23–25, 148, 171, 178, 185
– Dephosphorylation, 6, 20, 33, 43, 48, 81, 381
– Hyperphosphorylation, 24, 64, 65, 86
– Phosphorylation site, 148, 171, 178, 185
– Protein phosphorylation, 13–17
– Tyrosine dephosphorylation, 381
Photoperiod, 3, 5
Photopolarization, 329, 330
Photoreceptor, 8, 330
Photosynthesis, 3, 5, 343
Photosynthetic endosymbiosis, 211
Photosynthetic organism, 3, 4, 6, 211, 327
Phragmoplast, 33–36, 41–52, 125, 127–133, 147, 149, 157, 158, 169, 170, 175, 177, 179–185, 197, 199–203, 212–214, 218, 219, 222, 233, 235, 236, 238, 240–246, 251–253, 255–265, 268, 271–273, 276, 277, 279–282, 298, 299, 303–306, 308, 309, 312–314, 334
– Mini-phragmoplast, 258, 279–281
– Phragmoplast array, 169, 258
– Phragmoplast guidance, 41, 44, 47–51, 158
– Phragmoplast-like structure, 131
– Phragmoplast microtubule (MT), 51, 130, 169, 170, 177, 179–185, 238, 242, 243, 245, 246, 251, 253, 255, 257, 258, 261, 263, 271, 272, 279, 303, 304, 313
– Phragmoplast midzone, 235, 244, 264
– Phragmoplast structure, 130, 303
Phragmoplastin, 262, 268, 298, 299, 303–306, 308, 309, 312–314
Phragmosome, 35, 36, 38–41, 49, 52, 143, 156, 197, 201–203
PhrIP1 (Phragmoplastin-interacting protein 1), 314
PHS1 (POOR HOMOLOGOUS SYNAPSIS1), 115, 116
Phycoplast, 213
Phylogenetic analysis, 171, 323, 336, 380
Phytohormone, 62, 75, 89, 90
Pinchase, 305–307
Pinus, 196
PITAIRE, 381
Planar fenestrated sheet, 256, 261, 275, 276, 279
Plant cell division, 34, 38–40, 125, 127, 134, 169, 171, 185, 235, 251

Subject Index

Plasma membrane (PM), 33, 35, 36, 46, 47, 180, 198, 200, 208, 221, 256, 259, 261, 270, 272, 275, 276, 279, 281, 289–299, 305–308, 310, 311, 314, 330, 333, 350, 366
- Acropetal, 296
- New, 270, 276, 308
- Parental, 180, 198, 261, 279, 281
- Recycling, 275, 307

Plasmodesmata, 256, 275, 277, 281, 311, 314, 363
Plastid, 66, 157, 195, 201–203, 211
PLETHORA (PLT), 26
Ploidy level, 381–383, 386
Ploidy-dependen, 80, 93
Pluripotent, 379
Plus end tracking protein (+TIP), 149, 150, 152
Polar auxin transport, 40
Polar caps, 143, 144, 146, 173
Polar organizers, 157, 158
Polar strands, 143
Polarity, 33, 35, 38–40, 52, 143, 151, 154–158, 179, 181, 324–328, 333, 335, 343, 345, 351
- Bipolar spindles, 36, 44, 45, 131, 151, 153–158, 178, 181, 196, 220, 333
- Cell polarity, 35, 38, 39, 143, 325, 328, 335, 343
- Microtubule polarity, 179, 181
- Multipolar spindle, 43, 45, 154
- Of cell growth, 40
- Spindle polarity, 151, 154, 155, 157, 158, 333

Polarized auxin flow, 40
Polarized cytokinesis, 36, 49, 50
Polarized exocytosis, 267
Polarized secretion, 265, 297
Poleward flux, 147, 153
Pollen, 42, 61, 63, 66, 131, 132, 241, 267, 312, 323, 331, 361–365, 367–371, 373
- Pollen cytokinesis, 267
- Pollen development, 312, 323, 361, 364, 368, 371, 373
- Pollen mitosis, 42, 131, 371
- Pollen mother cell (PMC), 362–370
- Pollen mutant, 61, 66, 241, 267
- Pollen tube, 132, 267, 331
- Pollen wall formation, 371

Poly (A)+ RNA, 335
Polycomb complex, 91, 384
Polycomb group gene, 91
Polymerase/depolymerase, 21, 26, 61, 66, 67, 127, 151, 183, 385
Polymerization/depolymerization, 42, 44, 45, 111, 127, 130, 131, 133, 145–147, 149, 170, 177, 179–183, 242, 243, 263, 304, 307, 330, 331
Polyploid cell, 75, 93
Polyploidization, 75, 116
Polyploidy, 82, 89
Polysaccharide, 200, 237, 266, 270–272, 276, 311, 331
Polytenic chromosome, 75, 81
Post-embryonic process, 344, 377, 379
Post-Golgi secretion, 199, 200, 297
Post-meiotic cytokinesis, 267
Post-transcriptional regulation, 14, 23, 361, 370
Post-translation regulation, 16, 28, 67, 187
PP2A (Protein phosphotase 2A), 43
PR (proline rich) domain, 304–306
Prasinophyceae, 212, 213
PRC1 (PROTEIN REGULATING CYTOKINESIS 1), 130, 148, 243–246
Pre-meiotic cell, 105
Pre-meiotic S-phase, 103, 106
Pre-meiotic interface, 105, 107, 111, 370
Premitotic nucleus, 38, 142, 329
Preprophase, 33–36, 39, 41–43, 48, 125, 127, 128, 143, 173, 197, 212, 221, 244, 277, 313, 315, 326, 327
Preprophase band (PPB), 33–36, 38–52, 125, 127, 129–132, 142–144, 149–152, 154–158, 173, 175, 180, 185, 186, 196, 198, 212, 221, 244, 255, 256, 277, 313, 315, 326, 327
Pre-replication complex (pre-RC), 66, 77, 78, 81, 385, 386
Pre-synaptic membrane t-SNARE complex, 295
Prevacuolar compartment, 199, 200
Primary arrest, 6
Primary cell wall, 127, 276
Primary motor, 177
Primary parietal cell (PPC), 362–364, 367, 368
Primary plasmodesmata, 275, 277
Primary sporogenous cell (PSC), 362–364
Programmed cell death, 86, 93
PROLIFERA (PRL), 386

Proliferating cell nuclear antigen (PCNA), 26
Proliferative division, 345
Prometaphase, 33, 127, 142, 145, 146, 148, 152, 154, 157, 175, 176, 198, 237, 258
Promoter, 4, 8, 9, 13, 15, 16, 23–26, 28, 64, 84, 135, 298, 312, 353, 354, 366, 369, 384
Pronucleus, 330, 332
Prophase, 34–36, 42–44, 48, 50, 103–106, 108, 111, 115, 117, 135, 143–146, 152–158, 169, 173–175, 196, 212, 215, 221, 235, 255, 256, 365, 367, 368
– Prophase I, 103–106, 108, 111, 115, 117, 365, 367, 368
– Prophase nucleus, 35, 44, 50, 143, 145, 156, 174, 175
– Prophase spindle, 143–145, 152, 154–158, 173
Propyzamide, 21, 127, 203, 237, 245
Protease, 117, 348, 370
Proteasome, 14–16, 18, 63, 64, 66, 129, 131, 386
Protein degradation, 13, 14, 18, 80, 89
Protein kinase, 16–18, 129, 131, 132, 233–237, 295, 313, 349, 365
Protein phosphatase, 16, 52
Protein phosphorylation, 13–17
Protein synthesis, 21, 92
Protein–protein interaction, 233, 238, 372
Proteoglycan, 270
Proteolysis, 14–16, 18–20, 23, 60, 63, 66, 76, 77, 88, 216, 378, 380
Proteome, 27
Proteomic study, 150, 299, 315
Proteomic technology, 148, 315
Prothalli cell, 323
Protist, 208, 214, 334
Protodermal cell, 344, 345
Protoplast, 65, 203, 296
PRZ1 (PROPORZ 1), 381
PSTAIRE motif, 380
Pteridophyte, 196
Pulling force, 156, 158, 295, 333
Pyramimonas, 212, 213

Q-SNARE, 310

R2R3 Myb protein, 348, 352
Rab GTPase, 200, 262, 267, 292, 294, 297, 314

RAD21, 108
RAD50, 112, 113, 115
RAD51, 110, 112, 113, 115, 116
Raf MapKKK 233, 238
RanBP2/Nup358, 214, 223
RanGAP 153, 210, 214, 215, 217, 218, 223, 311, 314
Ran GTPase, 78, 153, 209, 210, 214, 217, 222, 223, 314
– Ran cycle, 209, 214, 217, 222, 223
– RanGDP, 222
– RanGTP, 78, 153, 222, 223
Ras GTPAse, 209, 238
Rb-E2F pathway, 59, 60, 64–66, 68
RBR (Rb-related) protein, 65, 66, 79, 80, 86, 87, 91, 378, 383
– RBR/E2F, 80, 86
– RBR1, 79, 80, 86, 87, 385
– RBR2, 86
– RBR3, 86
– RBR-binding motif, 382
RCC1 GTP-exchange factor, 153, 222
REANT (Reconstructor and analyzer of 3-D structure), 134
REC8, 106, 108, 110, 117
RecA recombinase, 112, 113, 115
Recombinase, 106, 112
Recombination, 103–106, 108–117
Reductional division (meiosis I), 103
RepA (Replication initiation factor), 65, 79, 87, 384
Repetitive DNA sequence, 109, 115
Replication, 6, 13, 17, 18, 21, 22, 27, 59, 62–68, 76–82, 84–86, 89–91, 108, 354, 377, 379, 381–386
– Chromatin replication, 59, 66, 68
– DNA replication, 6, 17, 18, 21, 22, 59, 64–67, 77, 78, 80, 84, 85, 108, 354, 377, 385
– Endoreplication, 62, 63, 66, 89, 379, 381–386
– Genome replication, 77, 354, 379
– Incomplete replication 6, 13, 82
– Multiple replication origin, 66
– Replication origins, 66, 67, 76, 77, 79
Retinoblastoma protein (Rb), 9, 24, 59–61, 63–68, 76, 79, 383, 384
Retrograde traffic, 291
REVEILLE family of proteins, 8
Reverse genetics, 14, 61, 292, 356

Subject Index

Rhizoid, 323, 326–336
Rhodopsin, 330
Rho GTPase, 262, 267, 312–314
Ribonucleotide reductase, 22, 67
Ribosome, 92, 252, 255, 259, 261, 264
RLK (receptor-like kinase), 349–351, 355, 356, 365, 366, 373
- LRR (Leucine-rich repeat)-RLK, 349–351, 365, 366, 373
RNA, 24, 61, 65, 92, 93, 130, 238, 307, 314, 331, 335, 355, 366, 370, 385
- miRNA (microRNA), 361, 370, 382
- mRNA, 238, 314, 331, 355, 366, 370, 385
- RNA-binding protein, 314
- RNAi (RNA interference), 65, 66, 79, 113, 117, 130, 218, 222
RNR (Ribonucleotide-diphosphate reductase), 385
RNR2, 66, 67
ROOT HAIRLESS (RHL)
- RHL1, 386
- RHL2, 386
Root meristem, 20, 296, 328
Rop1 GTPase, 262, 312, 313
Rosette leaf, 89, 270, 346, 385
RPN1a, 386
R-SNARE, 293, 295, 296
RTH1, 266
R-VAMP72, 295, 296

Saccharomyces cerevisiae, 77, 78, 82, 109, 112, 130, 215
SBF (SCB binding factor), 15–18
SBF/MBF complex, 15
SC (Synaptonemal complex), 103, 104, 107, 111, 116, 117
SCC (Sister-chromatid cohesion), 106–108
SCD1 (STOMATAL CYTOKINESIS-DEFECTIVE 1), 314, 348, 354
SCF (Skp1-Cullin/CDC53-Fbox protein), 63, 79
- SCF-dependent ubiquitin ligase, 18
Schizosaccharomyces pombe, 78, 215
Scramblase, 298
SDD1 (STOMATAL DENSITY AND DISTRIBUTION 1), 346–350, 352, 354, 356
Search-and-capture pathway, 153–155, 157, 158
Sec1, 294, 295, 297, 309, 310, 348, 354

Sec-14, 311
Secretory pathway, 289–291, 299
Secretory vescicle, 199, 331
Seedling, 76, 239, 241, 293, 294, 296, 297
Self-assembly (SA) domain, 304
Self-organizational pathway, 153, 154, 156–158
Semi-quantitative RT-PCR, 370
Semi-sychronized culture, 47
Septum, 362, 363
Ser10, 107
SERCA-type ATPase, 217
SERK1 (SOMATIC EMBRYOGENESIS RECEPTOR-LIKE KINASE 1), 361, 365, 366, 368, 373
SERK2, 361, 365, 366, 368, 373
SET domain protein, 66
SH3 domain protein, 305
shibire locus (Dynamin), 305
Shoot meristem, 4, 20, 198, 199, 201, 350
SIAMESE (SIM), 82, 87–89, 382
Signaling cascade, 106, 233, 235, 313
Signaling pathway, 235, 313, 357, 364, 366, 368, 373
Signaling protein, 311, 313
Silique, 369, 371
Silvetia compressa, 326
SIMK (SALT-INDUCED MAP KINASE 1), 234
Single stranded DNA, 112
Single-end invasion (SEI), 112, 113, 116
Sinorhizobium meliloti, 80, 90
Sinusoidal kinetics, 22
Sister chromatid, 17, 19, 75, 81, 106–108, 113, 142, 145, 169, 172, 177, 257
Sister chromatid cohesion (SCC), 106–108
Skp1, 63, 107, 216
SM (Sec1p/Munc18) protein, 292, 294–297
SMC (Structural maintenance of chromosomes), 108
- SMC1, 108
- SMC3, 108
SMC (Subsidiary mother cell), 41
SNAP25, 293, 295, 310
SNAP33, 262, 294, 295, 298
SNARE, 265, 267, 292–294, 296, 297, 299, 310, 311
SOLO DANCERS (SDS), 106
Somatic cell, 116, 172, 180, 261, 268, 277, 323, 325, 327, 335, 363

Somatic embryo, 63
Somatic-type cell plate, 251, 252, 269, 277
Somatic-type cytokinesis, 183, 251, 255, 258, 261, 269, 270, 273, 275, 277, 281, 282
Source-sink relationship, 20
Soybean, 144, 176, 268, 303–305
Space-filling model, 307
Spc98/Spc97, 215
Sperm, 61, 82, 85, 222, 327–330, 332, 335, 361
S phase, 6, 15, 17–19, 21–26, 61, 62, 64–68, 77–79, 81, 82, 84–86, 88, 103, 106, 108, 127, 133, 354, 377, 379, 381, 383, 386
– G1/S transition, 6, 7, 9, 15, 17, 18, 23, 59–68, 76–78, 82, 88, 127, 378, 382, 383, 385
– S/G2 transition, 6, 7, 78, 142
– S-phase-specific gene, 15, 24, 62, 67, 79, 383
– S phase CDK, 18, 81, 84, 86
– S phase cyclin-CDK complex, 18
Spindle, 6, 13, 21, 33, 34, 36, 42–45, 49, 50, 52, 75, 117, 125, 127–132, 134, 141–158, 169, 170, 172–175, 177–179, 181, 183, 196, 199, 200, 202, 203, 207–210, 213–217, 219–223, 235, 242–245, 255, 257, 277, 279, 326, 329, 331, 333, 334
– Anaphase spindle, 130, 145, 147, 243, 279
– Bipolar spindles, 36, 44, 45, 131, 151, 153–158, 178, 181, 196, 220, 333
– Central spindle, 132, 242, 243, 245
– Metaphase spindle, 145, 146, 154, 157, 199, 200, 203, 329, 333
– Mitotic spindle, 21, 34, 52, 75, 125, 127–132, 134, 141–143, 148–153, 155, 157, 196, 207, 214, 223, 245, 255, 331
– Mitotic spindle inhibitor, 21
– Multipolar spindle, 43, 45, 154
– Prophase spindle, 143–145, 152, 154–158, 173
– Spindle assembly 43, 44, 141, 156, 158, 169, 172, 207, 208, 213, 214, 223, 235
– Spindle axis, 144, 257
– Spindle defect, 6, 243
– Spindle equator, 221
– Spindle matrix, 146
– Spindle microtubule (MT), 117, 132, 146, 147, 151, 152, 169, 170, 172–174, 196, 203, 219, 220, 255, 257, 334, 279

– Spindle midzone, 147, 152, 154, 178, 181, 183
– Spindle pole, 36, 44, 45, 129, 144, 145, 152–154, 156–158, 169, 170, 172–174, 177, 178, 196, 214–217, 219, 277, 333
– Telophase spindle, 213
SPIRAL (SPR1), 150
Spirogyra, 221, 222
SPO11, 111–113, 115
SPOROCYTELESS/NOZZLE (SPL/NZZ), 361, 363
Sporogenous cells, 361–364, 366–368, 373
Sporophytic cell, 336, 361, 362, 363, 364
SPTAIRE, 381
SSK1, 238, 240
SSK2, 238, 240
Stamen, 46, 51, 61, 175, 186, 203, 362, 364
– Stamen hair cell, 46, 51, 175, 186, 203
Starch, 83–85, 91, 93, 369
START, 6, 15, 17, 59
STD/TES/NACK2, 182, 183, 240
STE11, 238, 240, 246
Stem cell, 24, 61, 65, 343, 344, 352, 378–380, 383, 386
Sterol methyltransferase, 89
Stomata, 237, 239, 323, 343–347, 350–355
– Stomatal complex, 132, 344, 345, 351
– Stomatal development, 343–346, 348–350, 354–356, 381
– Stomatal differentiation, 343, 345, 347, 353, 354, 356, 357
– Stomatal lineage, 344, 346, 350, 351, 354, 355, 357, 381
– Stomatal patterning, 343, 344, 346–349, 352, 354–356
Stonium, 362
Stramenopile, 327
Streptophyta, 4
Structural maintenance of chromosomes (SMC), 108
STUD (STD), 239, 241
Subcellular localization, 23, 28, 43, 210, 238, 244, 303, 304
Subsidiary mother cells (SMC), 41
Subtelomeric heterochromatin repeat, 116
Subtilicin-like protease, 348
Subunit, 14, 15, 17–19, 43, 48, 61, 67, 89, 91, 108, 147, 152, 209, 265–267, 312, 313, 354, 386
Sucrose 21, 22, 312

Subject Index

SUGOSHINI (SGO1), 117
Suspension-cultured cell, 27, 44, 62, 128, 144, 148, 150, 155, 179, 181, 196
Suspensor, 75, 76, 328
SuSy (Sucrose synthase), 312
SUV39H1 (H3K9-methyltransferase), 66
Swi5p, 16
SWITCH (SWI), 105
- SWI1, 105, 106
- SWI4, 15
SWP (STRUWEELPETER), 384
SYN1, 108
Synapsis, 103–109, 113, 114, 116–118
Synaptobrevin, 293, 295
Synaptonemal complex (SC), 103, 104
Synchronization, 13, 21, 27, 44, 127, 327
Syncytial-type cell plate, 251, 252, 259, 264, 267, 269, 277, 279, 281, 282
Syncytial-type cytokinesis, 251, 258, 277, 279, 281
Synechococcus, 4, 5
Synergid cell, 75
Syntaxin, 26, 134, 262, 293, 295, 308–310, 354
Syntenic region, 109
SYP31, 311
SYP7, 295, 296, 298

TAB1, 238
TAK1, 238
TAM (TARDY ASYNCHRONOUS MEIOSIS), 63, 106
TAN1 (TANGLED 1), 48, 150
Tapetum, 75, 361–373
TAPETUM DETERMINANT 1 (TPD1), 361, 366, 368, 373
Taxol, 146, 150, 155, 243
TCP 26, 382
T-DNA, 182, 311, 353, 370
Telocentric chromosome, 110
Telomere, 108–112, 115, 213, 215, 216
Telomere bouquet, 108, 109, 111, 115, 216
Telomeric region, 107, 116
Telophase, 36, 103, 128, 133, 135, 147, 152, 197, 199, 201, 212, 213, 218, 219, 222, 236, 238, 241, 242, 244, 252, 256, 277, 290, 313, 314, 327, 329, 333–335, 365
- Telophase I, 103
- Telophase II, 365
Tetrad, 241, 282, 362, 369, 370, 372

TETRASPORE (TES), 182, 183, 239–241
Thallus, 326, 327, 329, 331, 333–336
Thymidine, 79
Tilia americana, 146
TIP (Tonoplast intrinsic protein), 134
+TIP (Plus end trafficking protein), 149, 150, 152
TKRP125, 131, 179, 181, 313
TMBP200, 149
TMM (TOO MANY MOUTHS), 346–352, 354, 356
Tobacco, 19–22, 25, 27, 51, 62, 65, 66, 79, 92, 126–128, 130–135, 148, 149, 152, 179, 181–183, 185, 196, 198, 202, 203, 216, 218, 221, 222, 233–235, 237–243, 246, 259, 277, 304, 305, 312, 313, 315, 381, 383, 384
Tobacco BY-2 (Bright Yellow 2) cell, 19–21, 25, 42, 44–47, 51, 62, 126–134, 147, 148, 150, 152, 179, 198–202, 218, 235–239, 241, 243–245, 259, 304, 305, 312, 315
TOC1 (TIME OF CAB EXPRESSION 1), 8–10
TOG1 (TUMOR OVEREXPRESSED GENE 2), 131
Tomogram, 252, 253, 255, 258, 272
TONNEAU (TON1), 155
Tonoplast, 134, 201, 281
TONSOKU (TSK), 386
Topoisomerase, 111, 386
Totipotent state, 379
TPD1 (TAPETUM DETERMINANT 1), 361, 366, 368, 373
TPX2, 153
Tracheary element, 129
Tradescantia, 46, 51, 170, 175, 186
Transcription, 8–10, 15, 16, 26, 27, 59–61, 63–65, 67, 79, 82, 85, 92, 235, 363
- Transcript, 14, 19, 21–23, 25, 27, 28, 336, 365, 366, 370, 372, 383
- Transcriptional activator, 64, 93, 353, 382
- Transcriptional regulation, 6, 13, 14, 16, 20, 22, 25, 26, 28, 353, 372, 380
- Transcriptional repressor, 9, 64, 89
- Transcription Factor, 8, 13–17, 21–26, 28, 64, 67, 79, 352, 353, 356, 361, 363, 370, 371, 373, 383, 384
Trans-Golgi network (TGN), 259, 291
Transition point, 6

Transmembrane, 217, 312, 331
Transmission electron microscopy (TEM), 112, 361
Transverse filament (TF), 117
trans-SNARE, 310
Trans-zeatin, 19
Triacylglycerol (TAG), 371
Trichome 75, 76, 82, 86–89, 91, 92, 314, 353, 382
Trophoblasts, 81
TRYPTICON (TRY), 89
t-SNARE (target membrane-associated SNARE), 265, 295, 310
Tubular network, 256, 259, 261, 264, 267, 273, 275, 276, 279, 280–282
Tubular structure, 134, 146, 219, 304–309, 311
Tubulase, 304–307
Tubulin, 129, 130, 145, 146, 173, 181, 182, 215, 216, 313
- Antibody, 51, 175, 178, 244, 304, 326, 332
- Complex, 130
- Fluorescent marker, 44, 46, 127, 128, 129, 147, 221
- Membrane associated, 111
- Polymerization, 42, 44, 130, 242
- Tubulin ring complex (TuRC), 129, 153, 215
Tubulo-vesicular network, 184, 256, 259, 261, 267, 271, 272, 275, 281, 299
TUNEL assay, 111
Turgor-driven valves, 343
TVM (Tubulovesicular membrane), 134, 135
TVN (Tubulovesicular network), 256, 309
Tyrosine, 20, 330, 381
Tyrosine dephosphorylation, 381

Ubiquitin, 16–18, 20, 60, 63, 77, 80, 88, 89, 131, 216
Ubiquitination, 19
Ubiquitin-dependent proteolysis, 16–18, 20
Ubiquitin-proteasome pathway, 131
UDP-glucose (UDPG), 299, 312
UDP-glucose transferase (UGT1), 299, 312
UDT1 (UNDEVELOPED TAPETUM 1), 361, 370, 371, 373
UGT1 312, 313, 385
ULTRAPETALA (ULT), 26
Uso1, 264, 265

Ustilago maydis, 220
UV, 3, 5, 263, 312, 343
- UV damage, 3
- UV filter, 312

Volvox carteri, 323, 324, 325
Vacuole, 125, 126, 134, 135, 197, 200, 201, 203, 218, 270, 276, 280, 281, 303, 304, 306, 309, 310
- Vacuolar membrane (VM), 125, 126, 133, 134, 201, 202
- Vacuolated cell, 34, 35, 36, 41, 50, 92, 143, 198, 362, 364
- Vacuolation, 92
VAMPS, 310
Vertebrate, 48, 62, 130
Vesicle, 46, 51, 125, 126, 147, 150, 169, 170, 179, 180, 182, 184–186, 197–200, 208, 209, 214, 216, 219, 222, 223, 237, 242, 252, 255–262, 264–267, 269–273, 275–277, 279, 289–295, 297–299, 303–311, 313, 331, 354
- Cell plate vesicle, 298, 306, 308
- Clathrin-coated vesicle, 259, 262, 267, 269, 273, 275, 291, 292, 306
- Exocytic vesicle, 310
- Golgi-derived vesicle, 125, 179, 184, 185, 197, 198, 242, 252, 256, 262, 265, 266, 270, 273, 279, 292, 303, 308
- Golgi vesicle, 199, 259, 291, 304, 309
- Secretory vesicle, 199, 331
- Vesicle budding, 46, 265, 269
- Vesicle delivery, 180, 182, 271, 272
- Vesicle docking, 297, 310
- Vesicle fusion, 184, 197, 214, 223, 237, 260, 262, 265, 267, 269–271, 275–277, 279, 293, 297, 303, 309–311, 354
- Vesicle-like structure, 185
- Vesicle tethering, 262, 265–267, 269, 275, 297
- Vesicle trafficking, 185, 186, 258, 289–294, 298–300, 306, 313, 354
- Vesicle transport, 46, 169, 170, 180, 185, 186, 208, 246, 264
Vinculin-like protein, 314
VLCFA (Very-long-chain fatty acid), 355
Volvox carteri, 323
VPS1, 303
v-SNARE (Vesicle-associated SNARE), 310
VTN (Vesiculo-tubular network), 308

Subject Index

Wax, 356
Wax biosynthesis gene, 356
WD40-repeat protein, 80, 354
WDV (wheat dwarf virus), 79, 87
WEE1, 6, 7, 9, 85
Wide tubular network, 259, 264, 267, 279–282
Wide-type, 67, 87, 88, 91, 110, 148, 239, 346, 350, 351, 365–370, 372
WLIM1 (LIM domain protein 1), 314
WPP domain protein, 218
WRKY transcription factor, 370

XCTK2 (Kinesin-14), 152, 153
Xenopus, 78, 131, 149, 152–154, 183, 219, 222
XKCM1, 149
XMAP215 (*Xenopus* MAP215), 131, 149
XRCC2, 113
XRCC3, 113, 115

YDA (YODA), 347–351, 353, 355
Yeast, 6, 14–17, 22, 23, 28, 43, 48, 59, 60, 63, 65, 66, 76, 78, 82, 85, 93, 106, 107, 109, 112–116, 129, 131, 148, 149, 210, 214–216, 223, 233–235, 238, 240, 241, 243, 265, 267, 294, 297, 303, 309–311, 313, 333–335, 380
– Budding yeast, 6, 16, 17, 22, 113, 333, 335
– Fission yeast, 6, 14, 17, 82, 85, 109, 334, 335
Yeast two-hybrid, 43, 48, 65, 238, 240, 241, 294
YFP (Yellow fluorescent protein), 48, 128, 200

Zea mays, 155
Zeatin, 19
Zinc-finger, 26
Zinc finger motifs, 26, 314
ZmRpd3I, 65, 66
Zygnematales, 213
Zygote, 297, 323–333, 335, 336, 349
zygotene, 103, 109–112, 114–116
ZYP1, 117

Printing: Krips bv, Meppel, The Netherlands
Binding: Stürtz, Würzburg, Germany